Statistics for Biology and Health

Series Editors:
M. Gail, K. Krickeberg, J. Samet, A. Tsiatis, W. Wong

Warren Ewens   Gregory Grant

Statisical Methods in Bioinformatics:
An Introduction

Second Edition

With 30 Figures

 Springer

Warren J. Ewens
Department of Biology
University of Pennsylvania
Philadelphia, PA 19104
USA
wewens@sas.upenn.edu

Gregory Grant
Penn Center for Bioinformatics
Computational Biology and Informatics
    Laboratory
University of Pennsylvania
Philadelphia, PA 19104
USA
ggrant@grant.org

Series Editors:

M. Gail
National Cancer Institute
Rockville, MD 20892
USA

K. Krickeberg
Le Chatelet
F-63270 Manglieu
France

J. Samet
Department of Epidemiology
School of Public Health
Johns Hopkins University
615 Wolfe Street
Baltimore, MD 21205-2103
USA

A. Tsiatis
Department of Statistics
North Carolina State University
Raleigh, NC 27695
USA

W. Wong
Sequoia Hall
Department of Statistics
Stanford University
390 Serra Mall
Stanford, CA 94305-4065
USA

ISBN   978-1-4419-2302-8   e-ISBN   978-0-387-26648-0

A C.I.P. Catalogue record for this book is available
from the Library of Congress.

9 8 7 6 5 4 3 2 1

springeronline.com

For Kathy and Elisabetta

Preface

We take *bioinformatics* to mean the emerging field of science growing from the application of mathematics, statistics, and information technology, including computers and the theory surrounding them, to the study and analysis of very large biological, and particularly genetic, data sets. The field has been fuelled by the increase in DNA data generation leading to the massive data sets already generated, and yet to be generated, in particular the data from the human and other genome projects.

Bioinformatics does not aim to lay down fundamental mathematical laws that govern biological systems parallel to those laid down in physics. Such laws, if they exist, are a long way from being determined for biological systems. Instead, at this stage the main utility of mathematics in the field is in the creation of tools that investigators can use to analyze data. For example, biologists need tools for finding genes in genomic DNA, and for estimating differences in how genes are expressed in different tissues. Such tools involve statistical modeling of biological systems, and it is our belief that there is a need for a book that introduces probability, statistics, and stochastic processes in the context of bioinformatics. We hope to fill that need here.

The material in this text assumes little or no background in biology. The basic notions of biology that one needs in order to understand the material are outlined in Appendix A. Some further details that are necessary to understand particular applications are given in the context of those applications. The necessary background in mathematics is introductory courses in calculus and linear algebra. In order to be clear about notation and terminology, as well as to organize several results that are needed in the text, a review of basic notions in mathematics is given in Appendix B. No computer science knowledge is assumed and no programming is necessary to understand the material.

Why are probability and statistics so important in bioinformatics? Bioinformatics involves the analysis of biological data. Many chance mechanisms are involved in the creation of these data, most importantly the many random processes inherent in biological evolution and the randomness inherent in any sampling process. Stochastic process theory involves the description of the evolution of random processes occurring over time or space. Biological evolution over eons has provided the outcome of one of the most complex stochastic processes imaginable, and one that requires complex stochastic process theory for its description and analysis.

Our aim is to give an introductory account of some of the probability theory, statistics, and stochastic process theory appropriate to computational biology and bioinformatics. This is not a "how-to" book, of which there are several in the literature, but it aims to fill a gap in the literature on the statistical and probabilistic aspects of bioinformatics. The first three chapters in this book contain standard introductory material suitable for any statistics course. Our main aim in these chapters is to establish notation and to provide material needed in later chapters, so as to make the book more-or-less self-contained. We have chosen to illustrate principles by simple examples, not necessarily having a biological relevance, since a simple statistical principle can often be obscured if it is placed in a complicated biological context. We are well aware of the possible shortcomings, particularly those involving "relevance," of this approach.

Despite the introductory aim of the introductory chapters, we have departed in them somewhat from well-trodden paths, focussing on material of interest in bioinformatics, for example the theory of the maximum of several random variables, moment-generating functions, geometric random variables and their various generalizations, together with information theory and entropy. We have provided, in Appendix B, some standard mathematical results that are needed as background for these and other concepts discussed in this book.

This text is by no means comprehensive. There are several books that cover some of the topics we consider at a more advanced level. The reader should approach this text in part as an introduction and a means of assessing his/her interest in and ability to pursue this field further. Thus we have not tried to cover a comprehensive list of topics, nor to cover those topics discussed in complete detail. No book can ever fulfill the task of providing a complete introduction to this subject, since bioinformatics is evolving too quickly. To learn this subject as it evolves one must ultimately turn to the literature of published articles in journals. We hope that this book will provide a first stepping stone leading the reader in that direction.

We also wish to appeal to trained statisticians and to give them an introduction to bioinformatics, since their contributions will be vital to the analysis of the biological data already at hand and, more important, to developing analyses for new forms of data that will arrive in the future. Such readers should be able to proceed to the later chapters of the book directly.

The statistical procedures currently used in this subject are often ad hoc, with different methods being used in different parts of the subject. For this reason we have attempted a more formal presentation of statistical methods, so as to give an account that does not encounter the problems of an ad hoc approach. We have also tried to provide as many threads running through the book as possible in order to overcome this problem and to integrate the material. One such thread is provided by aspects of the material on stochastic processes. BLAST (Basic Local Alignment Search Tool) is one of the most frequently used algorithms in applied statistics, one BLAST search being made every few seconds on average by bioinformatics researchers around the world. However, the stochastic process theory behind the statistical calculations used in this algorithm is not widely understood. We approach this theory by starting with random walks, and through these to sequential analysis theory and to Markov chains, and ultimately to BLAST. This sequence also leads to the theory of hidden Markov models and to evolutionary analyses.

We have chosen this thread for three reasons. The first is that BLAST theory is intrinsically important. The second, as just mentioned, is that this provides a coherent thread to the often unconnected aspects of stochastic process theory used in various areas of bioinformatics. The final reason is that, with the human genome data and the genomes of other important species complete, at least in first draft, we wish to emphasize procedures that lead to the analysis of these data. The analysis of these data will require new and currently unpredictable statistical analyses, and in particular, the theory for the most recent and sophisticated versions of BLAST, and for its further developments, will require new advanced theory.

So far as more practical matters are concerned, we are well aware of the need for precision in presenting any mathematically based topic. However, we are also aware of the perils of a too mathematically precise approach to probability, perhaps through measure theory, , in an applied field. Our approach has tended to be less rather than more pedantic, and detailed qualifications that interrupt the flow and might annoy the reader have been omitted. As one example, we assume throughout that all random variables we consider have finite moments of all orders. This assumption enables us to avoid many minor qualifications to the analysis we present.

So far as statistical theory is concerned, the focus in this book is on *discrete* as opposed to *continuous* random variables, since (especially with DNA and protein sequences) discrete random variables are more relevant to bioinformatics. However, some aspects of the theory of discrete random variables are difficult, with no limiting distribution theory available for the maximum of these random variables. In this case progress is made by using theory from continuous random variables to provide bounds and approximations. Thus continuous random variables are also discussed in some detail in the early chapters.

The focus in this book is, as stated above, on probability, statistics, and stochastic processes. We do, however, discuss aspects of the important algorithmic side of bioinformatics, especially when relevant to these probabilistic topics. In particular, the dynamic programming algorithm is introduced because of its use in various probability applications, especially in hidden Markov models. Several books are already available that are devoted to algorithmic aspects of the subject.

In a broad interpretation of the word "bioinformatics" there are several areas of the application of statistics to bioinformatics that we do not develop. Thus we do not cover aspects of the statistical theory in genetics associated with disease finding and linkage analysis. This subject deserves an entire book on its own. Nor do we discuss the increasingly important applications of bioinformatics in the stochastic theory of evolutionary population genetics. Again, each of these topics deserves a complete treatment of its own.

This book is based on lectures given to students in the two-semester course in bioinformatics and computational biology at the University of Pennsylvania given each year during the period 1995–2003. We are most indebted to Elisabetta Manduchi, from PCBI/CBIL, who helped at every stage in revising the material. We are also grateful to the late Christian Overton for guidance, inspiration, and friendship. We thank all other members of PCBI/CBIL who supported us in this task and patiently answered many questions, in particular Brian Brunk, Jonathan Schug, Chris Stoeckert, Jonathan Crabtree, Angel Pizarro, Deborah Pinney, Shannon McWeeney, Joan Mazzarelli, and Eugene Buehler. Bob Smythe pointed out an error in Section 5.4 of the first edition of this book which has been corrected in this edition. We thank Warren Gish for his help on BLAST, and for letting us reproduce his BLAST printout examples. We thank Chris Burge, Roger Day, Sandrine Dudoit, Terry Speed, Matt Werner, Alessandra Gallinari, Sam Sokolovski, Helen Murphy, Etienne Pardoux, Peter Petraitis, Stéphane Robin, Mike Morley, Ethan Fingerman, Aaron Shaver, Mireille Régnier, and Sue Wilson for their help. Finally, we thank students in the computational biology courses we have taught for their comments on the material, which we have often incorporated into this book. Any errors or omissions are, of course, our own responsibility. An archive of errata will be maintained at http://www.textbook-errata.org.

Warren J. Ewens
Gregory R. Grant
Philadelphia, Pennsylvania
February, 2001

Contents

1
Probability Theory (i): One Random Variable

1.1 Introduction

We start by reminding the reader that the aims of this chapter, and the two following, are described in the Preface.

The DNA in an organism consists of very long sequences from an alphabet of four letters (nucleotides), a, g, c, and t (for adenine, guanine, cytosine, and thymine, respectively).[1] These sequences are copied from generation to generation, and undergo changes within any population over the course of many generations, as random mutations arise and become fixed in the population. Therefore, two rather different sequences may derive from a common ancestor sequence. Suppose we have two small DNA sequences such as those in (1.1) below, perhaps from two different species, where the arrows indicate paired nucleotides that are the same in both sequences.

$$
\begin{array}{l}
\downarrow \quad \downarrow \qquad \downarrow \quad \downarrow\downarrow \quad \downarrow\downarrow \qquad\qquad \downarrow\downarrow\downarrow \\
g\ g\ a\ g\ a\ c\ t\ g\ t\ a\ g\ a\ c\ a\ g\ c\ t\ a\ a\ t\ g\ c\ t\ a\ t\ a \qquad (1.1) \\
g\ a\ a\ c\ g\ c\ c\ c\ t\ a\ g\ c\ c\ a\ c\ g\ a\ g\ c\ c\ c\ t\ t\ a\ t\ c
\end{array}
$$

We wish to gauge whether the two sequences show significantly more similarity than we would expect from two arbitrary segments of DNA from the two species, in order to obtain evidence as to whether they derive from

---

[1]The reader unfamiliar with the basic notions of biology should refer to Appendix A for the necessary background.

a remote common ancestor. This kind of calculation is very common in bioinformatics.

If the sequences were each generated at random, with the four letters a, g, c, and t having equal probabilities of occurring at any position, then the two sequences should tend to agree at about one quarter of the positions. The two sequences above agree at 11 out of 26 positions. How unlikely is this outcome if the sequences were generated at random? We cannot answer this question until we understand some properties of random sequences. It will be shown that under the assumptions of equal probabilities for a, g, c, and t at any site, and independence of all nucleotides involved, that the probability that there will be 11 or more matches in a sequence comparison of length 26 is approximately 0.04. (The concept of independence is discussed more fully in Section 2.2.)Therefore, our observation of 11 matches might give some evidence that something other than chance is at work. We may not however say in practice that this procedure provides strong evidence until we check that sequences arising in practice behave as we have assumed, that is with equal probabilities for a, g, c, and t at any site, and independence of all nucleotides involved.

The calculation in the previous paragraph is a statistical operation. That is, we have observed some data, in this case the number of matches between two sequences of length 26, and on the basis of some probability calculation and using some hypothetical value for some *parameter* (that the unknown probability of a match is $1/4$), we made a statement about our level of belief in this value of that parameter.

As a further point, the calculation made above concerned the probability of obtaining the observed number of matches or more. This was done since the true value of the parameter should be higher than the hypothetical value $1/4$ if indeed the two sequences are related.

Whatever statement is made on the basis of the observed value of some random variable depends on some probability calculation. No valid statistical operation can be carried out without first making the probability calculation appropriate to that operation. Thus a study of probability theory is essential for an understanding of statistics. Probability theory is important also on its own account, quite apart from its underpinning of statistics, particularly in bioinformatics. Thus because of the intrinsic importance of probability theory, and because of its relevance to statistics, this chapter provides an introduction to the probability theory relating to a single random variable. In the next chapter we extend this theory to the case of many random variables.

Statistics concerns the *optimal* methods of analyzing data generated from some chance mechanism. A significant component of this optimality requirement is the appropriate choice of what is to be computed from the data in order to carry out the statistical analysis. Should we focus, as above, on the total number of matches (in this case, 11) between the two sequences? Should we focus on the size of the longest observed run of matches (here 3, occurring in positions 9–11 and also positions 23–25)? We shall see later that the most frequently used statistical method for assessing the similarity of two sequences uses neither of these quantities. The question of what should be computed from complex data in order to make a statistical inference leads to sometimes difficult statistical and probabilistic calculations. We address these matters in Chapters 8 and 9.

The probability calculations that led to our conclusion were based on assumptions about the nature of unrelated sequences (equal probabilities and independence). The accuracy of such conclusions depends on the accuracy of the assumptions made. Methods to test for the accuracy of assumptions lie in the realm of statistics, and will be discussed at length later. The necessity of having to make simplifying assumptions, even when they do not hold, brings up one of the most important issues in the application of statistics to bioinformatics. This issue is discussed in Section 4.10.

1.2 Discrete Random Variables, Definitions

1.2.1 Probability Distributions and Parameters

In line with the comments made in the Preface, we give relatively informal definitions in this book for random variables, probability distributions, and parameters, rather than the formal definitions found in many statistics textbooks.

To motivate the discussion we refer to the sequence matching example of Section 1.1, where the two sequences agree at 11 of the 26 positions considered. In effect we asked the question: "If two sequences of length 26 are laid down at random, with each nucleotide type arising in any position with probability 1/4, how surprised would we be if we were to observe 26 or more matches?" This question was addressed by calculating the probability of obtaining 11 or more matches, assuming that indeed the two sequences were laid down at random, with a sufficiently small probability suggesting some form of non-randomness. Probability calculations of this type are intrinsic to testing hypotheses in any area of the biological sciences, and in particular in bioinformatics.

Probability calculations refer to random variables and to some conceptual experiment to be carried out in the future. In the above example this experiment is the planned generation of two sequences and the random variable is the number of matches that will be obtained. The probability distribution of this random variable is needed to answer the "How surprised are we?" question raised above. This then leads us to a formal definition of random variables and their probability distributions.

A *discrete random variable* is a numerical quantity that, in some experiment that involves some degree of randomness, takes one value from some discrete set of possible values. For example, the experiment might be the rolling of two six-sided dice, and the random variable the sum of the two numbers showing on the dice. In this case the possible values of the random variable are $2, 3, \ldots, 12$. In practice the possible values of a discrete random variable often consist of some subset of the integers $\{\ldots, -2, -1, 0, 1, 2, \ldots\}$, but the theory given below allows a general discrete set of possible values. In some cases there are infinitely many possible values for a random variable; for example, the random number of tosses of a coin until the first head appears can take any value $1, 2, 3, \ldots$.

By convention, random variables are written as uppercase symbols, often X, Y, and Z, while the observed values of a random variable are written in lowercase, for example x, y, and z. Thus Y might be the conceptual (random) number of matches between two randomly chosen DNA sequences of length 26 before they are actually obtained, and assuming the observed comparison given in (1.1) is a comparison of randomly chosen sequences, the observed value y of this random variable is 11.

We discuss both discrete and continuous random variables and will also discuss the relation between them. In order to clarify the distinction between the two, we frequently use the notation Y for a discrete random variable and X for a continuous random variable. If some discussion applies for both discrete and continuous random variables, we shall use the notation X for both.

The *probability distribution* of a discrete random variable Y is the set of values that this random variable can take, together with their associated probabilities. Probabilities are numbers between zero and one inclusive that always add to one when summed over all possible values of the random variable. An example is given in (1.2).

The probability distribution is often presented in the form of a table, as in (1.2), with a listing of the possible values that the random variable can take together with the probabilities of each value. This would be the probability distribution if, for example, if we toss a fair coin twice, and Y is the total number of heads that turn up.

Possible values of the random variable Y	0	1	2
Associated probabilities	.25	.50	.25

$$(1.2)$$

We show how these probabilities are calculated in the next section. In practice, the probabilities associated with the possible values of the random variable of interest are often unknown. For example, if a coin is biased, and the probability of a head on each toss is unknown to us, then the probabilities for 0, 1, or 2 heads when this coin is tossed twice are unknown.

We call the set of values that a discrete random variable can take with positive probability the *range* of that random variable. In the example above, the range of Y is the set $\{0, 1, 2\}$.

There are two other frequently used methods of presenting a probability distribution. The first is by using a "chart," or "diagram," in which the possible values of the random variable are indicated on the horizontal axis, and their corresponding probabilities by the heights of vertical lines, or bars, above each respective possible value (see Figure 1.1). When the possible values of the random variable are integers there is an appealing geometric interpretation. In this case the rectangles all have width one, and so the probability of a set of values is equal to the total area of the rectangles above them. This representation bears a strong analogy to that for the continuous case, to be discussed in Section 1.8.

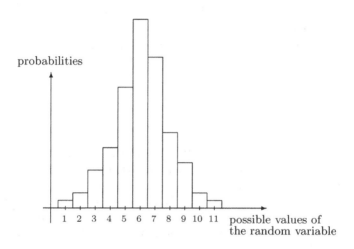

Figure 1.1.

However, a third method of presentation is more appropriate in theoretical work, namely through a mathematical function. In this functional approach we denote the probability that discrete random variable Y takes the value y by $P_Y(y)$. For example, $P_Y(4)$ is the probability that the random variable Y takes the value 4. The suffix "Y" indicates that the probability relates to the random variable Y. This is necessary since we often discuss probabilities associated with several random variables simultaneously. In the functional approach $P_Y(y)$ is written in the form

$$P_Y(y) = g(y), \quad y = y_1, y_2, y_3, \ldots, \tag{1.3}$$

where $g(y)$ is some specific mathematical function of y. Apart from specifying the mathematical function $g(y)$, it is also necessary to indicate the *range* of the random variable, that is, the set of values that it can take. In (1.3) above, $y_1, y_2, y_3, \ldots$ are the possible values of Y. The range of Y should not be confused with the range of the function $g(y)$. The range of Y is in fact the *domain* of $g(y)$.

As an example, suppose that the possible values of the random variable Y are 1, 2, and 3, and that

$$P_Y(y) = y^2/14, \quad y = 1, 2, 3.$$

Here $g(y) = y^2/14$. Of course, in this simple case we could rewrite the probabilities explicitly in tabular form:

Possible values of the random variable Y	1	2	3
Associated probabilities	1/14	4/14	9/14

(1.4)

However, the functional form is often more convenient, and further, for random variables with an infinite range, listing all possible values and their probabilities in tabular form is impossible.

As another example, suppose that Y can take the possible values 1, 2, and 3, and that now

$$P_Y(y) = \frac{\theta^{2y}}{\theta^2 + \theta^4 + \theta^6}, \quad y = 1, 2, 3,$$

where θ is some fixed nonzero real number. Whatever the value of θ, $P_Y(y) > 0$ for $y = 1, 2, 3$, and $P_Y(1) + P_Y(2) + P_Y(3) = 1$. Therefore, Y is a well-defined random variable. This is true even though the value of θ might be unknown to us. In such cases we refer to θ as a *parameter:* a parameter is some constant, usually unknown, involved in a probability distribution. The important thing to note is that Y is a well-defined random variable even though we may not know any of the actual probabilities of its possible values.

Another important function is the *cumulative distribution function* $F_Y(y)$ of the discrete random variable Y, often called simply the *distribution function.* This is the probability that the random variable Y takes a value y or less, so that

$$F_Y(y) = \sum_{y' \leq y} P_Y(y'). \tag{1.5}$$

An example of the functions $P_Y(y)$ and $F_Y(y)$ is given in Figure 1.2. Notice that the rightmost rectangle of the distribution function has height one. We reserve the notation $P_Y(y)$ throughout this book to denote the probability that the discrete random variable Y takes the value y, and the notation $F_Y(y)$ as defined in (1.5) to denote the cumulative distribution function of this discrete random variable.

1.2.2 Independence

The concept of independence is central in probability and statistics. We formally define independence of discrete random variables in Section 2.2. It is, however, convenient to give an informal definition here.

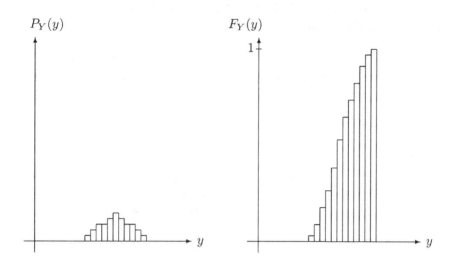

Figure 1.2. An example of a probability distribution (left) and its cumulative distribution function (right).

For the moment we take the word "independent" to have an intuitive meaning: Discrete random variables are independent if knowing the value of one of them does not affect in any way the probabilities associated with the possible values of any of the other random variables.

A common example of independent random variables is given by the outcomes of different rolls of a die. If a die is rolled any number of times, the number turning up on any one roll is assumed to be independent of the number turning up on any other roll.

As another example, suppose that two fair dice are to be rolled. Let one random variable S be the the sum of the two numbers showing on the dice, and the other random variable D be the difference of the two numbers. D can be any integer from -5 to 5, but if it is known that the observed value s of S equals 3, then it must be that the observed value d of D is ± 1, each with probability $1/2$. Thus knowing a value of S affects the probability distribution of D, and so they are not independent.

There is a similar concept of independence of "events," which we discuss in Section 1.12.5.

1.3 Six Important Discrete Probability Distributions

In this section we describe the six discrete probability distributions that occur most frequently in bioinformatics and computational biology. They are all presented in the functional form (1.3).

1.3.1 One Bernoulli Trial

A *Bernoulli trial* is a single trial with two possible outcomes, often called "success" and "failure." The probability of success is denoted by p and the probability of failure $1 - p$ is sometimes denoted by q.

A Bernoulli trial is so simple that it seems unnecessary to introduce a random variable associated with it. Nevertheless, it is useful to do so, and indeed there are two random variables that we shall find it useful to associate with a Bernoulli trial.

The first of these, the Bernoulli random variable, is the number of successes Y obtained on this trial. Clearly, $Y = 0$ with probability $1 - p$ and $Y = 1$ with probability p. The probability distribution of Y can then be written in the mathematical form (1.3) as

$$P_Y(y) = p^y(1 - p)^{1-y}, \quad y = 0, 1. \tag{1.6}$$

This random variable is natural when, for example, counting the number of matches between two aligned sequences. Sometimes it is more natural to consider a random variable whose possible values are ± 1. For example, a gambler might win one dollar (with probability p) or lose one dollar (with probability $1 - p$) depending on the result of a coin flip. We think of losing one dollar as winning minus one dollar. If the number of dollars the gambler wins (± 1) on any flip is denoted by S,

$$P_S(s) = p^{(1+s)/2}(1 - p)^{(1-s)/2}, \quad s = -1, +1. \tag{1.7}$$

1.3.2 The Binomial Distribution

The binomial random variable is the number of successes in a fixed number n of independent Bernoulli trials with the same probability of success for each trial. The number of heads in some fixed number of tosses of a coin is an example of a binomial random variable.

More precisely, the *binomial* distribution arises if all four of the following requirements hold. First, each trial must result in one of two possible outcomes, often called (as with a Bernoulli trial) "success" and "failure." Second, the various trials must be independent. Third, the probability of success must be the same on all trials. Finally, the number n of trials must be fixed in advance, not determined by the outcomes of the trials as they occur. The probability of success on any trial is denoted, as in the Bernoulli case, by p. We call p the parameter, and n the index, of this distribution.

The random variable of interest is the total number of successes in the n trials, denoted by Y. The probability distribution of Y is given by the formula

$$P_Y(y) = \binom{n}{y} p^y (1-p)^{n-y}, \quad y = 0, 1, 2, \ldots, n. \tag{1.8}$$

Appendix B.4 provides the definition of $\binom{n}{y}$. We derive this mathematical form for the binomial distribution when discussing the combinatorial term $\binom{n}{y}$ in Appendix B.6.

Figure 1.3 shows the probability distribution for two different binomials, the first for $n = 10$ and $p = \frac{1}{2}$, the second for $n = 20$ and $p = \frac{1}{4}$. The first is symmetric, while the second is not. The probabilities in the second distribution are nonzero up to and including $y = 20$, but the probabilities for values of y exceeding 12 are too small to be visible on the graph.

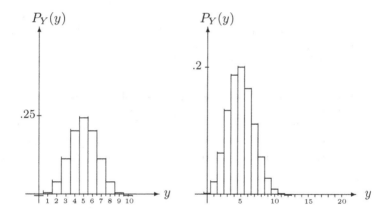

Figure 1.3. Two binomial distributions, the first with $n = 10$ and $p = \frac{1}{2}$, the second with $n = 20$ and $p = \frac{1}{4}$.

When using a binomial distribution one must be careful that the four defining conditions above all hold. This comment is relevant to the comparison of the two DNA sequences given in (1.1). Our assumptions leading to the probability of observing 11 or more matches were based on the assumption that the number of matches between randomly chosen sequences follows a binomial distribution. Suppose that a "success" is the event that the two nucleotides in corresponding positions in the two sequences match. It is not necessarily true that the probability of success is the same at all sites. Nor is it necessarily true that independence holds: It is a result of population genetics theory, for example, that the nucleotide frequencies at very close sites tend to evolve in a dependent fashion, leading to a potential non-independence of observing a success at close sites. Thus two of the central requirements for a binomial distribution might not hold for this sequence comparison. Depending on our purposes, however, it might still be desirable to make these assumptions as an approximation. When we construct probabilistic models, we almost invariably must make some simplifying assumptions about the processes leading to the data. This issue is discussed further in Section 4.10.

There are several further comments to make concerning Bernoulli trials and the binomial distribution. First, the single Bernoulli trial distribution (1.6) is the special case ($n = 1$) of the binomial distribution. Second, the quantity p in (1.8) is often an unknown parameter. Third, there is no simple formula for the cumulative distribution function $F_Y(y)$ (defined in (1.5)) for the binomial distribution. Probabilities associated with this function are often approximated by the "normal approximation to the binomial," discussed in Section 1.10.3, or are calculated numerically by summation, using equations (1.5) and (1.8). Current computing power makes direct calculation possible even for quite large values of n. Some perhaps unexpected aspects of the computation of this distribution function are given in Appendix C. Fourth, there is no unique "binomial distribution," but rather a family of distributions indexed by n and p. It is nevertheless standard to refer to *the* binomial distribution. A parallel comment applies for all distributions that we consider.

The final point illustrates a concept that will arise again in Chapter 4 and it is instructive to illustrate it with an example. A fair 400-sided die is rolled once. The probability that the number 1 turns up on the die is $p = 1/400$. Intuition might lead one to assume that the probability of seeing a 1 is approximately doubled if the die is rolled twice, is tripled if the die is rolled three times, and so on. The binomial distribution shows, more precisely, that the probability that the number 1 turns up at least once in two rolls is $2p - p^2$, not $2p$. With three rolls the probability is $3p - 3p^2 + p^3$, not $3p$. The intuition, however, is correct. When p is small, the probability of rolling a 1 at least once in n rolls is $np + o(np)$ (as $np \to 0$), and so it is very nearly np as long as np is small. (The "o" notation is defined in Appendix B.8.) The graph in Figure 1.4 shows the probability that the number 1 turns up at least once in n rolls. The graph is nearly linear in n up to about $n = 15$, and only after that does it significantly deviate from the straight line. This kind of approximation will be important when we discuss Poisson processes in Chapter 4.

1.3.3   *The Hypergeometric Distribution*

We introduce the hypergeometric distribution in an abstract context and then give a biological example where it can be used.

Suppose that an urn contains N objects, of which n are red and $N - n$ are white. Of these, m objects are taken out of the urn at random, in particular

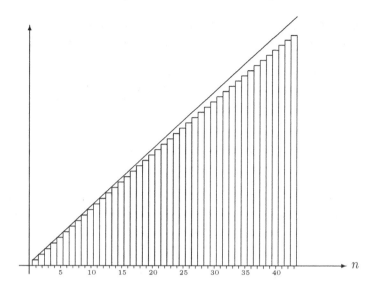

Figure 1.4. The probability of at least one success in n binomial trials, with probability $p = 1/400$ for success on each trial, as a function of n, is approximately linear when n is small.

without reference to the color of any object, and *without replacement*. The number of red objects taken out is a random variable Y, with probability distribution given by the formula

$$P_Y(y) = \frac{\binom{n}{y}\binom{N-n}{m-y}}{\binom{N}{m}}, \quad y = A, A+1, \ldots, B. \tag{1.9}$$

Here $A = \max(0, n + m - N)$, $B = \min(n, m)$. The upper bound B arises because the number of red objects taken out cannot exceed either the number of red objects in the urn or the number of objects taken out of the urn. The lower bound arises because the number of red objects drawn must be at least $n + m - N$ if the number of objects taken out of the urn exceeds the number of white objects initially in the urn.

The probability distribution in (1.9) is the *hypergeometric distribution*. For for values of y in the range $(A, A+1, \ldots, B)$, it may be derived by observing that the probability that y red objects are drawn out *in some specified order* is

$$\frac{n(n-1)\cdots(n-y+1)(N-n)(N-n-1)\cdots(N-n-m+y+1)}{N(N-1)\cdots(N-m+1)},$$
$$\tag{1.10}$$

which may be re-written as

$$\frac{n!(N-n)!(N-m)!}{(n-y)!(N-n-m+y)!N!}. \tag{1.11}$$

The hypergeometric probability in (1.9) is obtained upon multiplication of this quantity by the number $\binom{m}{y}$ of different orders in which the y red objects can be taken out of the urn.

Although a fixed number (m) of trials is conducted in this experiment and there are two possible outcomes on each trial ("red" or "white"), the number of red objects taken from the urn does *not* have the binomial distribution, because the outcomes of the various trials are not independent. If, for example, an unusually large number of red objects had been taken from the urn in the earlier trials, the probability of a red object on a later trial is less than it would have been if an unusually large number of white objects had been taken from the urn in the earlier trials. If each object were replaced in the urn immediately after it was drawn the requirements for a binomial distribution would hold and the number of red objects drawn out would have a binomial distribution.

Here is a more biologically relevant example. Suppose that N laboratory mice, n of which are males and $N - n$ females, are irradiated. We wish to test whether a certain mutation is more likely to arise in male mice than in females. After the radiation it is found that, in all, there are m new mutant mice in the joint sample of males and females, and thus $N - m$ mice which are non-mutant. Conditional on the event that the total number of mutant mice is m, the number Y of mutant males has the hypergeometric distribution (1.9) if there is no association between gender and the propensity to be a mutant.

This example differs from the "urn" example in that, while in the urn case the number m of objects to be taken out of the urn is known in advance, the eventual number m of mutant mice is not known in advance of the radiation experiment. Despite this, conditioning on the event that m mice in total are mutant implies that the hypergeometric distribution for Y still applies in the radiation example, under the assumption that there is no association between gender and the propensity to be a mutant – see Problem 1.4.

A further example, in which n and m are *both* unknown before the experiment, is the following. Suppose that a square box with sides of length 1 is divided into four rectangular compartments, with side lengths as shown in Figure 1.5. Suppose that N objects are thrown, independently and at random, into the box, so that the probability that any object is thrown into any compartment is equal to the area of that compartment: for example, the probability that any object is thrown into the upper right-hand compartment is $a(1 - b)$. Conditional on the event that n objects in total are thrown into the two top compartments and m objects in total are thrown into the two left-hand compartments, the number of objects thrown into the top left-hand compartment has the hypergeometric distribution (1.9) – see Problem 2.6. Thus the hypergeometric distribution arises in all three examples, once the appropriate conditioning is made in the second and third

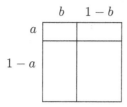

Figure 1.5. The "hypergeometric" box

examples. This fact is central to *Fisher's exact test*, discussed in more detail in Section 3.5.

There are two important concept of independence involved in the second and third examples. The first concept might be called individual-to-individual independence. In the second example, this is that, unconditional on the total number of mutant mice, the event that any one mouse is a mutant is independent of the event that any other mouse is a mutant. In the third example, this is that the compartment into which any object is thrown is independent of the compartment into which any other object is thrown.

The second concept of independence might be called category-to-category independence. In the mouse example, this is the assumption that a mouse's gender is independent of its propensity to be a mutant, or, equivalently, that there is no association between a mouse's gender and its mutant status. In the third example, this is that with the configuration of compartments shown in Figure 1.5, whether an object is thrown into one of the top two or one of the bottom two compartments is independent of whether it is thrown into one of the two left-hand or one of the two right-hand compartments. In other words, if we think of the box as divided into two rows and two columns, there is no association between which row any object is thrown into and which column it is thrown into.

It is this category-to-category assumption that is tested in Fisher's exact test. For example, in the mouse case, the assumption that there is no association between a mouse's gender and mutant status is the (null) hypothesis tested in this test.

1.3.4   The Uniform Distribution

Perhaps the simplest discrete probability distribution is the *uniform* distribution. In the case of most interest to us, a random variable Y has the *uniform distribution* if the possible values of Y are $a, a + 1, \ldots, a + b - 1$, for two positive integer constants a and b, with $b > 1$, and the probability that Y takes any specified one of these b possible values is b^{-1}. That is,

$$P_Y(y) = b^{-1}, \quad y = a, a + 1, \ldots, a + b - 1. \tag{1.12}$$

Properties of a random variable having this uniform distribution are discussed later.

A more general uniform distribution arises when the possible values of Y are $a, a+c, a+2c, \ldots, a+c(b-1)$, for any constant a, any constant $c > 0$, and any integer $b > 1$, so that

$$P_Y(y) = b^{-1}, \quad y = a, a+c, a+2c \ldots, a+c(b-1). \tag{1.13}$$

An important case arises when $a = 0$, $c(b-1) = 1$, so that

$$P_Y(y) = \frac{c}{c+1}, \quad y = 0, c, 2c, \ldots, 1. \tag{1.14}$$

1.3.5   The Geometric Distribution

The *geometric distribution* arises in a situation similar to that of the binomial. Suppose that a sequence of independent Bernoulli trials is conducted, each trial having probability p of success. The random variable of interest is the number Y of trials before but not including the first failure. The possible values of Y are $0, 1, 2, \ldots$. As will be shown in Section 1.12.5, the probability that several independent events all occur is the product of the probabilities of the individual events. Since if $Y = y$ there must have been y successes followed by one failure, it follows that

$$P_Y(y) = (1-p)p^y, \quad y = 0, 1, 2, \ldots. \tag{1.15}$$

The geometric distribution with $p = .7$ is shown in Figure 1.6. The cumulative distribution function $F_Y(y)$ can be calculated from (1.15) as

$$F_Y(y) = \text{Prob}(Y \leq y) = 1 - p^{y+1}, \quad y = 0, 1, 2, \ldots. \tag{1.16}$$

From this,

$$\text{Prob}(Y \geq y) = p^y, \quad y = 0, 1, 2, \ldots. \tag{1.17}$$

We also call Y the length of a "success run." It is often of interest to consider the length of a success run immediately following a failure. The length of such a run also has the distribution (1.15). Note that a possible value of Y is 0, and this value will arise if another failure occurs immediately after the initial failure considered.

One example of the use of success runs occurs in the comparison of the two sequences in (1.1), where the length of the longest success run in this comparison is three. We might wish to use geometric distribution probabilities to assess whether this is a significantly long run. "Significance" in this sense will be defined formally in Section 3.4, and aspects of this test will be discussed in Section 6.3.

The Shifted Geometric Distribution

In some contexts it will be more appropriate to consider the number of trials until the first failure, but now counting the trial on which this failure

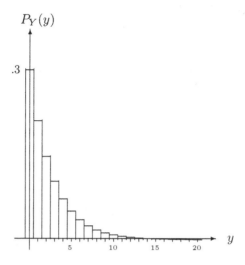

Figure 1.6. The geometric distribution with $p = .7$.

occurs. If Y is the number of trials defined in this way, equation (1.15) shows immediately that

$$P_Y(y) = (1-p)p^{y-1}, \quad y = 1, 2, 3, \ldots. \tag{1.18}$$

It is perhaps more customary to refer to this distribution as the geometric distribution, rather than to use this term for the distribution (1.15). However we use (1.15) more often than we use (1.18) in the remainder of this book, so we use the name "shifted geometric" for the distribution (1.18).

Geometric-Like Random Variables

Suppose that Y is a random variable with range $0, 1, 2, \ldots$ and that, as $y \to \infty$,

$$1 - F_Y(y-1) = \mathrm{Prob}(Y \geq y) \sim Cp^y, \tag{1.19}$$

for some fixed constant C, $0 < C < 1$. In this case we say that the random variable has the *geometric-like* distribution. The symbol "$\sim$" in (1.19) is the asymptotic symbol defined in Appendix B.8. From equation (1.17), the geometric distribution behaves as in (1.19) if $C = 1$ and the asymptotic relation is replaced by an equality.

Geometric-like distributions are not typically part of a standard treatment of elementary probability theory. However, we introduce them here, since they are central to BLAST theory.

1.3.6   *The Negative Binomial and the Generalized Geometric Distributions*

Suppose that a sequence of independent Bernoulli trials is conducted, each with a probability p of success. The binomial distribution arises when the number of trials n is fixed in advance, and the random variable is the number of successes in these n trials. In some applications the role of the number of trials and the number of successes is reversed, in that the number of successes is fixed in advance (at the value m), and the random variable N is the number of trials up to and including this mth success. The random variable N is then said to have the *negative binomial* distribution. The probability distribution of N is found as follows.

The probability $P_N(n)$ that $N = n$ is the probability that the first $n - 1$ trials result in exactly $m - 1$ successes and $n - m$ failures (in some order) and that trial n results in success. The former probability is found from (1.8) to be

$$\binom{n-1}{m-1} p^{m-1}(1-p)^{(n-1)-(m-1)}, \quad n = m, m+1, m+2, \ldots,$$

while the latter probability is simply p. Thus

$$P_N(n) = \binom{n-1}{m-1} p^m (1-p)^{n-m}, \quad n = m, m+1, m+2, \ldots. \quad (1.20)$$

An example for $p = .75$ and $m = 10$ is shown in Figure 1.7.

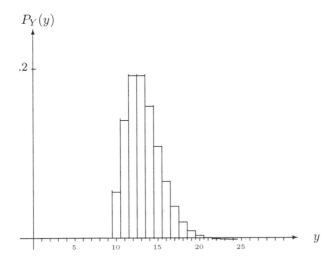

Figure 1.7. The negative binomial distribution with $p = .75$ and $m = 10$.

We shall later be interested in the case where we fix the number of failures in advance (at the value $k+1$), and consider the random number Y of trials preceding but not including this $(k+1)$th failure. An argument similar to that giving (1.20) shows that

$$P_Y(y) = \binom{y}{k} p^{y-k}(1-p)^{k+1}, \quad y = k, k+1, k+2, \ldots. \qquad (1.21)$$

Although (1.21) also describes a negative binomial distribution, it is also the generalization of the geometric distribution equation (1.15) to the case of general k, reducing to that distribution when $k = 0$. For this reason we use a non-standard terminology and refer to the distribution given in equation (1.21) as the *generalized geometric* distribution, with parameters p and k. (This distribution should not be confused with the geometric-like distribution defined in Section 1.3.5).

The cumulative distribution functions for these distributions do not admit a simple mathematical form. We discuss aspects of these cumulative distribution functions in Section 6.3 and Appendix C.

1.3.7 The Poisson Distribution

A random variable Y has a *Poisson* distribution (with parameter $\lambda > 0$) if

$$P_Y(y) = \frac{e^{-\lambda}\lambda^y}{y!}, \quad y = 0, 1, 2, \ldots. \qquad (1.22)$$

An example with $\lambda = 5$ is given in Figure 1.8.

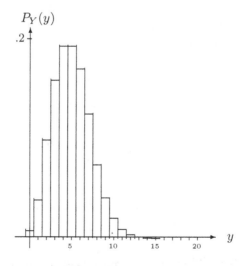

Figure 1.8. The Poisson distribution with $\lambda = 5$.

The fact that (1.22) does indeed provide a probability distribution follows from the basic identity

$$\sum_{n=0}^{\infty} \frac{\lambda^n}{n!} = e^{\lambda}$$

which holds for all λ (see Appendix B.12), together with the fact that when $\lambda > 0$, each term on the right-hand side of (1.22) is positive.

The Poisson distribution arises mainly in two contexts. The meaning of the word "random" is an elusive one, but the Poisson distribution arises first when events occur in some sense "randomly" in time or space. We formalize this concept of "time/space" randomness mathematically in Section 4.1 and show how the Poisson distribution arises through this concept of randomness.

The Poisson distribution also arises as a limiting form of the binomial distribution. If the number of trials n in a binomial distribution is large, the probability of success p on each trial is small, and the product $np = \lambda$ is moderate, then the binomial probability of y successes is very close to the probability that a Poisson random variable with parameter λ takes the value y. A more precise statement, with a formal mathematical derivation, is given in Section 4.2. This limiting property makes the Poisson distribution particularly useful, since the condition "n large, p small, np moderate" arises often in applications of the binomial distribution in bioinformatics, and the Poisson distribution is often easier to work with – for example the Poisson has one parameter while the binomial has two.

In the binomial case events can be thought of as occurring at n discrete time points. In contrast, in the Poisson case, the occurrences are in continuous time (e.g. clicks on a geiger counter). Beyond being an approximation to the binomial, the Poisson distribution also arises in its own right in certain biological processes; an example is described in Section 4.1.

1.3.8 Approximations

While the probability distributions described above arise frequently in applications, there are also many cases where the random variable of interest does not have one of these distributions. Sometimes the distribution of the random variable is complicated and might not be easily calculated. In such cases this distribution is often approximated by one of the distributions described above. If this is done it is important to be able to place an upper bound on the error involved in the approximation. There are several ways in which this bound can be defined. The method used most frequently in bioinformatics is to calculate the *total variation distance* between the true distribution and the approximating distribution. The total variation distance $d_{TV}(P_1, P_2)$ between two probability distributions $P_1(Y)$ and $P_2(Y)$ taking positive values for non-negative integer values of Y only is defined as follows. Let A be a set of non-negative integers and let $R_i(A)$ be the probability that Y takes a value in A when Y has probability distribution $P_i(Y)$. Then $d_{\text{TV}}(P_1, P_2)$ is the maximum value of $|R_1(A) - R_2(A)|$, the maximum being taken over all possible sets A. Thus $d_{\text{TV}}(P_1, P_2)$ is the maximum error one can make when calculating the probability of any set of values of Y if one distribution is assumed when the other is appropriate. An equivalent formula is given by

$$d_{\text{TV}}(P_1, P_2) = \frac{1}{2} \sum_y |P_1(y) - P_2(y)|, \tag{1.23}$$

the sum being taken over all non-negative integers. In many cases arising in practice an upper bound for $d_{\text{TV}}(P_1, P_2)$ is available for this sum by using the Chen-Stein method (Chen, (1975)); see Arratia et al. (1989) for a user-friendly account of this method. We return to this topic in Section 5.7.2.

1.4 The Mean of a Discrete Random Variable

The mean of a random variable is often confused with the concept of an average, and it is important to keep the distinction between the two concepts clear. The mean of a discrete random variable Y is defined as

$$\sum_y y P_Y(y), \tag{1.24}$$

the summation being over all possible values that the random variable Y can take (i.e., over its range). The expressions "the mean of the random variable Y" and "the mean of the probability distribution of the random variable Y" are equivalent and are used interchangeably, depending on which is more natural in the context. As an example, the mean of a random variable having the binomial distribution (1.8) is

$$\sum_{y=0}^{n} y \binom{n}{y} p^y (1 - p)^{n-y}, \tag{1.25}$$

and this can be shown (see Problem 1.2) to be np.

If the range of the random variable is infinite, the mean might not exist (that is, the sum in (1.24) might diverge). In all cases we consider, the mean exists (i.e., is some finite number), and since qualifications to the theory are needed when the mean might be infinite, we assume throughout, without further discussion, that all means of interest are finite.

There are several remarks to make regarding the mean of a discrete random variable.

(i) The notation μ_Y is often used for a mean of Y. When the random variable is clear from the context, as will usually be the case, we suppress the subscript Y. An alternative name for the mean of a random variable is the "expected value" of that random variable, and this leads to a second frequently used notation, namely $E(Y)$, for the expected value, or mean, of the random variable Y.

(ii) The word "average" is *not* an alternative for the word "mean," and has a quite different interpretation from that of "mean." Their relation and distinction is discussed further in Section 2.10.1.

(iii) In many practical situations the mean μ of a discrete random variable Y is unknown to us, because we do not know the numerical values of the probabilities $P_Y(y)$. That is to say, μ is a parameter. Of all the estimation and hypothesis testing procedures discussed in detail in Chapters 3, 8, and 9, estimation of, and testing hypotheses about, a mean are among the most important.

(iv) The mean is not necessarily a realizable value of a discrete random variable. For example, the distribution in (1.4) has mean

$$1 \times \frac{1}{14} + 2 \times \frac{4}{14} + 3 \times \frac{9}{14} = \frac{18}{7},$$

yet 18/7 is not a realizable value of the random variable.

(v) The concept of the mean of a discrete random variable Y can be generalized to the mean, or expected value, of any function $g(Y)$ of Y. The function $g(Y)$ is itself a random variable, and the expected value of $g(Y)$, denoted by $E(g(Y))$, is defined (using (1.3)) by

$$E(g(Y)) = \sum_z z P_{g(Y)}(z), \qquad (1.26)$$

where $P_{g(Y)}(z)$ is the probability that the random variable $g(Y)$ takes the value z and the sum is over all possible values z of $g(Y)$. This is also equal to

$$E(g(Y)) = \sum_y g(y) P_Y(y), \qquad (1.27)$$

where the sum is over all values y of Y, as can be seen from the equality

$$P_{g(Y)}(z) = \sum_{\substack{y \text{ such} \\ \text{that } g(y)=z}} P_Y(y).$$

Equation (1.27) is more convenient than (1.26) to use in practice, since it does not require us to find the probability distribution of $g(Y)$.

An important case of equation (1.27) is

$$E(Y^2) = \sum_y y^2 P_Y(y). \qquad (1.28)$$

This quantity is sometimes called the second moment of Y about the origin, with the mean being called the first moment about the origin. In general the second moment is different from the square of the first moment, that is $E(Y^2) \neq (E(Y))^2$.

A linearity property of the mean follows from equation (1.27): If Y is a random variable with mean μ, and if α and β are constants, then the random variable $\alpha + \beta Y$ has mean

$$E(\alpha+\beta Y) = \sum (\alpha+\beta y) P_Y(y) = \alpha \sum P_Y(y) + \beta \sum y P_Y(y) = \alpha+\beta\mu.$$

(vi) If the graph of a probability distribution function is symmetric about some vertical line $y = a$, the mean is equal to a.

Various problems at the end of this chapter ask for calculation of the means of the distributions described above. A summary of these means is given in Table 1.1 on page 23.

1.5 The Variance of a Discrete Random Variable

A quantity of importance equal to that of the mean of a random variable is its *variance*. The variance (denoted by σ_Y^2) of the discrete random variable Y is defined by

$$\sigma_Y^2 = \sum_y (y - \mu)^2 P_Y(y), \qquad (1.29)$$

the summation is taken over all possible values of Y. When the random variable is clear from the context, the subscript on σ^2 will be suppressed. The expressions "the variance of the random variable Y" and "the variance of the probability distribution of the random variable Y" are equivalent and are used interchangeably. The variance, like the mean, is often unknown to us.

The variance is a measure of the dispersion of the probability distribution of the random variable around its mean (see Figure 1.9).

There are several points to note concerning the variance of a discrete random variable.

(i) The *standard deviation* of a probability distribution is defined as the positive square root of the variance of that distribution, and (naturally enough) is denoted by σ. This is often more useful than the variance since is measured in the same units as the random variable having that probability distribution, and thus admits a more ready and often more convenient measure than the variance. For example, if the random variable is the weight of a randomly chosen individual, the standard deviation is measured in kilograms, whereas the variance has the almost incomprehensible unit of a kilogram squared.

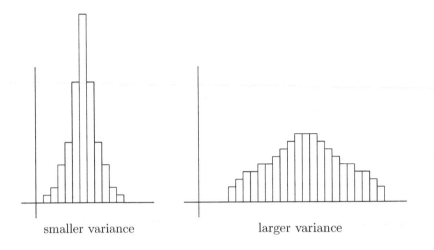

smaller variance   larger variance

Figure 1.9.

A quantity sometimes used for random variables taking positive values only is the *coefficient of variation*, defined as the standard deviation of the random variable divided by its mean.

(ii) If Y is a random variable with variance σ^2, and if α and β are constants, then the variance of the random variable $\alpha + \beta Y$ is $\beta^2 \sigma^2$ (see Problem 1.5).

(iii) An equivalent and often more convenient formula for the variance is

$$\sigma^2 = \left(\sum_y y^2 P_Y(y) \right) - \mu^2 = E(Y^2) - (E(Y))^2 , \qquad (1.30)$$

from which it follows that the second moment of Y about the origin, defined in (1.28), can be expressed as

$$E(Y^2) = \sigma^2 + \mu^2 \qquad (1.31)$$

(see Problem 1.6).

(iv) A further formula, particularly useful for random variables that take non-negative integer values, is

$$\sigma^2 = \left(\sum_y y(y-1) P_Y(y) \right) + \mu - \mu^2 \qquad (1.32)$$

(see Problem 1.6).

(v) Table 1.1 shows that the mean and the variance of a Poisson random variable are identical. In practice, equality or near equality of the mean and the variance of a random variable often suggest that its distribution is approximately Poisson.

Various problems at the end of this chapter ask for the calculation of the variances of the distributions described above, and Problem 1.6 asks for verification of equations (1.30) and (1.32).

As with the mean, the variance of any random variable we consider in this book is finite, so in order to avoid qualifications we assume from now on that every variance of interest to us is finite.

The means, variances, and probability distribution functions of the distributions discussed above are listed in Table 1.1.

distribution	mean	variance	probability distribution
Bernoulli	p	$p(1-p)$	$p^y(1-p)^{1-y}, \quad y = 0, 1$
Binomial	np	$np(1-p)$	$\binom{n}{y}p^y(1-p)^{n-y}, \quad y = 0, 1, \ldots, n$
Hypergeom.	mn/N	$\frac{mn(N-m)(N-n)}{N^2(N-1)}$	$\binom{n}{y}\binom{N-n}{m-y}/\binom{N}{m}, \quad y = A, A+1, \ldots, B.$
Uniform	$a+\frac{b-1}{2}$	$\frac{b^2-1}{12}$	$b^{-1}, \quad y = a, a+1, \ldots, a+b-1$
Geometric	$p/(1-p)$	$p/(1-p)^2$	$(1-p)p^y, \quad y = 0, 1, 2, \ldots$
Neg. Binom.	m/p	$m(1-p)/p^2$	$\binom{n-1}{m-1}p^m(1-p)^{n-m}, \quad n \geq m$
Poisson	λ	λ	$e^{-\lambda}\lambda^y/y!, \quad y = 0, 1, 2, \ldots$

Table 1.1. Means, variances, and probability distribution functions of discrete random variables. The interpretations of A and B are given in the text.

1.6   General Moments of a Probability Distribution

The mean and variance are special cases of *moments* of a discrete probability distribution, and in this section we explore these moments in more detail.

If r is any positive integer, then $E(Y^r)$ is called the *r*th *moment of the probability distribution about zero*, and is usually denoted by μ'_r. From (1.27),

$$E(Y^r) = \sum_y y^r P_Y(y). \qquad (1.33)$$

The mean is thus the first moment of the distribution about zero. Throughout this book we assume that for any random variable considered and for any r, the rth moment exists (i.e., the sum in (1.33) converges). This assumption holds for all of the random variables we consider. Note that $E(Y^r)$

is not the same as $(E(Y))^r$, and the numerical values of the two are equal only in certain special cases.

Of almost equal importance for discrete integer-valued random variables is the rth *factorial moment* of a probability distribution. This is denoted by $\mu'_{[r]}$ and is defined by

$$\mu'_{[r]} = E\left(Y(Y-1)\cdots(Y-r+1)\right)$$
$$= \sum_y \left(y(y-1)\cdots(y-r+1)\right) P_Y(y). \quad (1.34)$$

As an example, suppose that the random variable Y has the binomial distribution (1.8). Then the rth factorial moment is

$$E\left(Y(Y-1)\cdots(Y-r+1)\right)$$
$$= \sum_{y=0}^{n} y(y-1)\cdots(y-r+1)\binom{n}{y}p^y(1-p)^{n-y}, \quad (1.35)$$

which reduces (see Problem 1.7) to

$$n(n-1)\cdots(n-r+1)p^r. \quad (1.36)$$

The rth moment *about the mean* of the probability distribution, denoted by μ_r, is defined by

$$\mu_r = \sum_y (y-\mu)^r P_Y(y). \quad (1.37)$$

The first such moment, μ_1, is identically zero, and is thus not of interest. The second, μ_2, is the variance σ^2 of the random variable. The third, μ_3, is related to the *skewness* of the probability distribution. The skewness γ_1 of a distribution is a scale-free quantity, defined as

$$\gamma_1 = \mu_3/\sigma^3. \quad (1.38)$$

Distributions with positive skewness have long tails to the right, and distributions with negative skewness have long tails to the left.

1.7   The Probability-Generating Function

An important quantity for many of the discrete random variables we shall encounter is the probability-generating function, or pgf for short. The pgf for a discrete random variable Y is a function of a real variable t, given by

$$\mathbb{P}_Y(t) = E(t^Y) = \sum_y P_Y(y)\, t^y. \quad (1.39)$$

We discuss below some standard uses of a pgf; other applications of the concept are found later in the book.

The pgf of the random variable Y will be denoted by $p_Y(t)$, or some-times $q_Y(t)$, and when Y is clear from the context the subscript Y will be suppressed. The summation is taken over all possible values of Y. This sum always converges for $t = 1$ (to 1). If the sum only converges for $t = 1$, then $p(t)$ is not useful. For cases of interest $p(t)$ exists in some interval $(1 - a, 1 + a)$ of values of t surrounding 1, where $a > 0$. The variable t is sometimes called a "dummy" variable, since it has no meaning or interpre-tation of its own. Instead, it is a mathematical tool that can be used, in conjunction with Taylor series properties such as those given in Appendix B.12, to arrive efficiently at various desired results.

When the possible values of a random variable Y consist of finitely many non-negative integers then $p_Y(t)$ is a polynomial in t. If the possible values of a random variable consist of positive and negative integers, as for exam-ple the random variable S in Section 1.3.1, then then the sum above is not a Taylor series but is a Laurent series. For background on such series, see Appendix B.12 through B.14.

The original purpose for the pgf is to generate probabilities, as the name suggests. If the pgf of a random variable can be found easily, as is often the case, then the coefficient of t^r in the pgf is the probability that the random variable takes the value r.

Another use of the pgf is to derive *moments* of a probability distribution. This is done as follows. The derivative (when it exists) of the pgf $p_Y(t)$ of a discrete random variable with respect to t, taken at the value $t = 1$, is the mean of that random variable. That is,

$$\mu_Y = \left(\frac{d}{dt} p_Y(t) \right)_{t=1}. \tag{1.40}$$

The variance of a random variable with mean μ and whose probability distribution has pgf $p_Y(t)$ is found from

$$\sigma_Y^2 = \left(\frac{d^2}{dt^2} p_Y(t) \right)_{t=1} + \mu_Y - \mu_Y^2. \tag{1.41}$$

The verification of these identities is left as an exercise (see Problem 1.18).

As a straightforward example, the pgf of the Bernoulli random variable (1.6) is

$$1 - p + pt. \tag{1.42}$$

Similarly, the pgf of the random variable defined in (1.7) is

$$(1 - p)t^{-1} + pt. \tag{1.43}$$

As a less straightforward example, the pgf of a random variable having the binomial distribution (1.8) is

$$p(t) = \sum_{y=0}^{n} \binom{n}{y} p^y (1 - p)^{n-y} t^y, \tag{1.44}$$

and this can be shown (see Problem 1.17) to be

$$\mathbb{p}(t) = (1 - p + pt)^n. \tag{1.45}$$

In most cases of interest to us, a pgf is either a polynomial in t or an infinite Taylor series in t. In many cases such an infinite series converges for sufficiently small values of t. We therefore end with a useful theorem about pgfs whose proof is discussed in Appendix B.14, and which applies to all random variables of interest to us.

Theorem. Suppose that two discrete random variables Y_1 and Y_2 have pgfs $\mathbb{p}_1(t)$ and $\mathbb{p}_2(t)$, respectively, and that both pgfs converge in some open interval I containing 1. If $\mathbb{p}_1(t) = \mathbb{p}_2(t)$ for all t in I, then Y_1 and Y_2 have identical distributions.

This theorem implies immediately that if the pgf of a random variable converges in an open interval containing 1, then the pgf completely determines the distribution of the random variable.

1.8   Continuous Random Variables

Some random variables, by their nature, are discrete, such as the number of successes in n Bernoulli trials. Other random variables, by contrast, are continuous. Random variables such as the height or the blood pressure of a person chosen at random, or the time taken until an event occurs, are all of this type. We denote a continuous random variable by X, and the observed value of the random variable by x. Continuous random variables usually take any value in some continuous range of values, for example $-\infty < X < \infty$ or $L \leq X < H$.

Probabilities for continuous random variables are not allocated to specific values, but rather are allocated to intervals of values. Each continuous random variable X has an associated *density function* $f_X(x)$, and the probability that the random variable takes a value in some given interval (a, b) is obtained by integrating the density function over that interval. Specifically,

$$\text{Prob}(a < X < b) = \int_a^b f_X(x)\, dx. \tag{1.46}$$

See Figure 1.10. The set of values of x for which $f(x)$ is positive is called the *support* of the continuous random variable.

The definition (1.46) implies that the probability that a continuous random variable takes any specific nominated value is zero. It thus implies the further equalities

$$\text{Prob}(a < X < b) = \text{Prob}(a < X \leq b)$$
$$= \text{Prob}(a \leq X < b) = \text{Prob}(a \leq X \leq b). \tag{1.47}$$

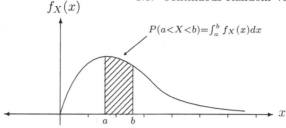

Figure 1.10. $P(a < X < b) = \int_a^b f_X(x)dx.$

The concept of a continuous random variable should not be confused with the concept of a continuous density function. Any positive valued function, continuous or not, which integrates to one over its domain, is a density function for a continuous random variable. By elementary calculus, when $f_X(\cdot)$ is continuous on $[x, x+h]$, (1.46) gives

$$\lim_{h \to 0^+} \frac{\text{Prob}(x < X < x+h)}{h} = f_X(x). \tag{1.48}$$

We will use this identity several times in what follows, as well as the approximation that follows from it, namely that when $f_X(\cdot)$ is continuous on $[x, x+h]$, and h is small,

$$\text{Prob}(x < X < x+h) \cong f_X(x)h. \tag{1.49}$$

(The symbol "$\cong$" is used here and throughout to indicate approximate equality of fixed quantities.)

As a particular case of equation (1.46), if the range of the continuous random variable X is either $L \leq X \leq H$, $L \leq X < H$, $L < X \leq H$, or $L < X < H$, then

$$\int_L^H f_X(x)\,dx = 1. \tag{1.50}$$

This L and H notation is used throughout this and the next section to mean the limits of the range of a continuous random variable.

An important function associated with a continuous random variable X is its cumulative distribution function $F_X(x)$, which is the probability that the random variable takes a value x or less. The cumulative distribution function is therefore given by

$$F_X(x) = \int_L^x f_X(u)du. \tag{1.51}$$

This definition implies that $0 \leq F_X(x) \leq 1$, that $F_X(x)$ is non-decreasing in x, and the fundamental theorem of calculus shows that when f is continuous

$$f_X(x) = \frac{d}{dx}F_X(x). \tag{1.52}$$

We reserve the notation $f_X(x)$ and $F_X(x)$ throughout this book to denote, respectively, the density function and the cumulative distribution function of the continuous random variable X.

1.9   The Mean, Variance, and Median of a Continuous Random Variable

1.9.1   Definitions

The mean μ_X and variance σ_X^2 of a continuous random variable having range (L, H) and density function $f_X(x)$ are defined by

$$\mu_X = \int_L^H x f_X(x) dx \qquad (1.53)$$

and

$$\sigma_X^2 = \int_L^H (x - \mu_X)^2 f_X(x) dx. \qquad (1.54)$$

We will suppress the subscripts when the identity of the random variable is clear. These definitions are the natural analogues of the corresponding definitions for a discrete random variable, and the remarks about the mean and the variance of a continuous random variable are very similar to those of a discrete random variable that were given on pages 20 through 23. In particular remarks (i), (ii), (iii), and (vi) for the mean of a discrete random variable remain unchanged for a continuous random variable (except for the replacement of "probability distribution function" with "density function" where necessary), and property (v) requires only replacing a summation by an integration. That is, the mean value $E(g(X))$ of the function $g(X)$ of the continuous random variable X is given by

$$E(g(X)) = \int_L^H g(x) f_X(x) dx, \qquad (1.55)$$

where $f_X(x)$ is the density function of X. One important case, parallel to the discrete random variable formulae (1.28) and (1.31), is that

$$E(X^2) = \int_L^H x^2 f_X(x) dx = \sigma^2 + \mu^2. \qquad (1.56)$$

All of the variance properties given for a discrete random variable apply also for a continuous random variable.

Higher moments for a continuous random variable are defined in a manner parallel to those for a discrete random variable discussed in Section 1.6. For example, the r-th moment μ_r about the mean of a continuous random variable X having mean μ and density function $f_X(x)$ is defined by

$$\mu_r = \int_L^H (x - \mu)^r f_X(x)dx, \tag{1.57}$$

which is the natural parallel of the discrete random variable definition (1.37).

For continuous random variables whose range is $(0, H)$ (for some positive H, possibly $+\infty$), an alternative formula for the mean (see Problem 1.14) is

$$\mu = \int_0^H (1 - F_X(x))\ dx. \tag{1.58}$$

For most continuous random variables arising in practice there is a unique *median*, denoted here by M, having the property that

$$\text{Prob}(X < M) = \text{Prob}(X > M) = \frac{1}{2}. \tag{1.59}$$

The mean and median coincide for random variables with symmetric density functions, both being at the point of symmetry, but the two are usually different for asymmetric density functions. Examples of mean, variance, and median calculations are given below.

1.9.2   Chebyshev's Inequality

The concepts of the mean and the variance of a random variable, discrete or continuous, allow us to prove an inequality that is of great use.

Let X be a random variable, discrete or continuous, having mean μ and variance σ^2. Then Chebyshev's inequality states that for any positive constant d,

$$\text{Prob}(|X - \mu| \geq d) \leq \frac{\sigma^2}{d^2}. \tag{1.60}$$

The proof of this statement is straightforward, and is given here for the continuous random variable case where the range of the random variable is $(-\infty, +\infty)$. The proof for any other range, and for a discrete random variable, is essentially identical. From the definition of σ^2,

$$\sigma^2 = \int_{-\infty}^{+\infty} (x - \mu)^2 f_X(x)\, dx$$

$$\geq \int_{-\infty}^{\mu - d} (x - \mu)^2 f_X(x)\, dx + \int_{\mu+d}^{+\infty} (x - \mu)^2 f_X(x)\, dx$$

$$\geq d^2 \int_{-\infty}^{\mu-d} f_X(x)\, dx + d^2 \int_{\mu+d}^{+\infty} f_X(x)\, dx$$

$$= d^2\, \text{Prob}(|X - \mu| \geq d).$$

The result (1.60) follows immediately.

1.10   Five Important Continuous Distributions

1.10.1   The Uniform Distribution

A continuous random variable X has the *uniform* distribution if, for some constants a and b with $a < b$, its range is one of the intervals $I = [a, b]$, (a, b), $(a, b]$, or $[a, b)$ (this interval notation is discussed in Appendix B.1) and its density function $f_X(x)$ is

$$f_X(x) = \frac{1}{b-a}, \quad \text{for } x \text{ in } I. \tag{1.61}$$

This density function is graphed in Figure 1.11. The mean and variance

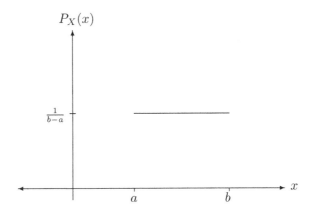

Figure 1.11. The density function of the uniform distribution $f_X(x) = \frac{1}{b-a}$.

are given by

$$\mu = \frac{a+b}{2}, \quad \sigma^2 = \frac{(b-a)^2}{12}. \tag{1.62}$$

The uniform distribution we consider most frequently is the particular case of (1.61) for which $a = 0$, $b = 1$. For this uniform distribution,

$$f_X(x) = 1, \quad F_X(x) = x, \quad 0 \le x \le 1. \tag{1.63}$$

From (1.62) the mean and variance of this distribution are, respectively, $1/2$ and $1/12$.

We use the term "uniform distribution" to describe any continuous uniform distribution and any discrete uniform distribution. The context will make clear whether the discrete or the continuous uniform distribution is intended.

1.10.2   The Normal Distribution

The most important continuous random variable is one having the *normal*, or *Gaussian*, distribution. The (continuous) random variable X has a normal distribution if it has range $(-\infty, \infty)$ and density function

$$f_X(x) = \frac{1}{\sqrt{2\pi}\beta}e^{-\frac{(x-\alpha)^2}{2\beta^2}},$$

where α and β are parameters of the distribution. The value of α is unrestricted, but it is required that $\beta > 0$.

It may be checked, either by using equation (1.53) or from symmetry, that the mean of this distribution is α. Symmetry also implies that the median of this distribution is α. It is also possible, although less straightforward, to then use equation (1.54) to show that the variance of the distribution is β^2. Thus the distribution is usually written in the more convenient form

$$f_X(x) = \frac{1}{\sqrt{2\pi}\sigma}e^{-\frac{(x-\mu)^2}{2\sigma^2}}, \tag{1.64}$$

so that the mean μ and the variance σ^2 are built into the mathematical form of the density function. Usually the normal distribution is presented directly in the form (1.64), but we have preferred the above approach, which derives the mean and variance from (1.53) and (1.54). A common notation is to refer to a random variable having the distribution (1.64) as an $N(\mu, \sigma^2)$ random variable.

A particularly important normal distribution is the one for which $\mu = 0$ and $\sigma^2 = 1$, whose density function is shown in Figure 1.12. This is sometimes called the *standard normal* distribution. Suppose that a random variable X is $N(\mu, \sigma^2)$. Then the "standardized" random variable Z, defined by $Z = (X - \mu)/\sigma$, is $N(0, 1)$. If x is the observed value of X, the value $z = (x - \mu)/\sigma$ is often called a z-score.

One of the original uses of this standardization is the following. If X is a random variable having a normal distribution with mean 6 and variance 16, we cannot find $\text{Prob}(5 < X < 8)$ by integrating the function in (1.64) in closed form. However, for the standardized variable Z, accurate approximations of such integrals are widely available in tables. These tables may be used, in conjunction with the standardization procedure, to find probabilities for random variable having any normal distribution. Thus in the above example, we can rewrite $\text{Prob}(5 < X < 8)$ in terms of the standard normal Z by

$$\text{Prob}(5 < X < 8) = \text{Prob}\left(\frac{5-6}{\sqrt{16}} < \frac{X-6}{\sqrt{16}} < \frac{8-6}{\sqrt{16}}\right)$$
$$= \text{Prob}(-0.25 < Z < 0.5), \tag{1.65}$$

and this probability is found from tables as 0.2902.

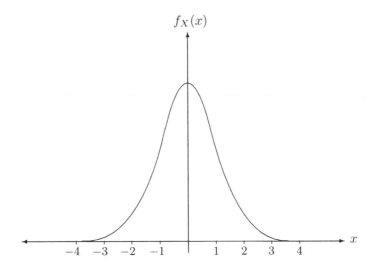

Figure 1.12. The density function for the standard normal distribution with $\mu = 0$, $\sigma = 1$.

Computers have all but replaced tables; however, z-scores are still useful for many reasons. For example, if we want to compare the significance of two values drawn from two different normal distributions, instead of calculating probabilities, we can simply compare their z-scores. The standardization can also used for deriving results theoretically, for example to show that the probability that an observation from *any* normal random variable is within two standard deviations from the mean is 0.9545, or approximately .95. This *two standard deviation rule* is useful as a quick approximation for probabilities concerning random variables whose probability distribution is in some sense close to a normal distribution (although it can also be misleading (see Problem 1.23)).

1.10.3 The Normal Approximation to a Discrete Distribution

One of the many uses of the normal distribution is to provide approximations for probabilities for certain discrete random variables. The first we consider is the normal approximation to the binomial. If the number of trials n in the binomial distribution (1.8) is very large, the normal distribution with mean np and variance $np(1-p)$ provides a very good approximation. Figure 1.13 shows the approximation of the binomial with $p = \frac{1}{4}$ and $n = 20$ by the normal with $\mu = 5$ and $\sigma^2 = \frac{15}{4}$.

The normal approximation to the binomial is a consequence of the central limit theorem, which is discussed in Section 2.10.1. How large n must be for any desired degree of accuracy depends on the value of p. For $p = \frac{1}{2}$

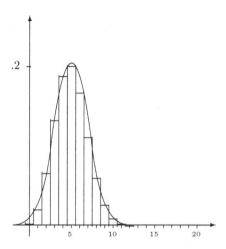

Figure 1.13. The normal approximation to the binomial random variable with $p = \frac{1}{4}$ and $n = 20$.

(the only case where the binomial probability distribution is symmetric) the approximation is good even for n as small as 20, as is shown in an example below. The closer p is to zero or one, the further the binomial distribution is from being symmetric, and the larger n must be for the normal distribution to fit the binomial distribution well.

If Y is binomial with parameters p and n, and X is normal with $\mu = np$ and $\sigma^2 = np(1-p)$, then for integers a and b, with $a < b$, the approximation is

$$\text{Prob}(a \leq Y \leq b) \cong \text{Prob}\left(a - \frac{1}{2} \leq X \leq b + \frac{1}{2}\right).$$

The reason for the term $\frac{1}{2}$ in this approximation can be seen from the graph in Figure 1.13. The rectangle whose area equals $\text{Prob}(Y = a)$ lies above the interval $(a - \frac{1}{2}, a + \frac{1}{2})$. Without the term $\frac{1}{2}$ in the approximation we would underestimate by $(\text{Prob}(Y = a) + \text{Prob}(Y = b))/2$. The term $\frac{1}{2}$ is known as the "continuity correction."

As an example, the probability that a binomial random variable with $p = \frac{1}{2}$ and $n = 20$ takes a value of 15 or more can be evaluated exactly from (1.8) to be 0.0207. The probability that a normal random variable with mean 10 and variance 5 (the appropriate mean and variance for this binomial distribution) exceeds the value 14.5 is 0.0222. This is a far more accurate approximation than that found without the continuity correction, which is 0.0127.

As a second example, the probability that a binomial random variable with $p = \frac{1}{4}$ and $n = 20$ takes a value of 9 or more is 0.0409. The normal approximation is 0.0353, not as accurate as the approximation in the case $p = \frac{1}{2}$ for the same value of n. Indeed as p decreases, n must increase in order to achieve a good approximation.

A second application is to use the normal distribution to approximate the Poisson distribution. As can be seen from the example of a Poisson distribution in Figure 1.8, the Poisson has a normal-like shape when $\lambda = 5$. It follows from the central limit theorem, discussed in Section 2.10.1, that when λ is large, the Poisson is closely approximated by a normal with mean λ and variance λ.

1.10.4   The Exponential Distribution

A third probability distribution arising often in computational biology is the *exponential* distribution. A (continuous) random variable having this distribution has range $[0, +\infty)$ and density function

$$f_X(x) = \lambda e^{-\lambda x}, \quad x \geq 0. \tag{1.66}$$

Here the single positive parameter λ characterizes the distribution. The graph of this density function is shown in Figure 1.14.

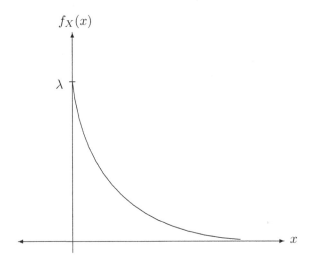

Figure 1.14. The density function of the exponential distribution, $f_X(x) = \lambda e^{-\lambda x}$.

The cumulative distribution function $F_X(x)$ is found by integration as

$$F_X(x) = 1 - e^{-\lambda x}, \quad x \geq 0. \tag{1.67}$$

The mean and variance of the exponential distribution are, respectively, $1/\lambda$ and $1/\lambda^2$ (see Problem 1.21). The median M, found by solving the equation $F_X(x) = 1/2$, is given by

$$M = \frac{\log 2}{\lambda}. \tag{1.68}$$

This value is just over two-thirds of the mean, implying (as the graph of the density function shows) that the exponential distribution is skewed to the right.

The Relation of the Exponential with the Geometric Distribution

The exponential distribution is the continuous analogue of the geometric distribution, as can be seen by comparing Figure 1.14 to Figure 1.6 (page 15). Therefore, the exponential may be used as a continuous approximation to the geometric. This is discussed further in Section 2.11.2.

A further relationship between the exponential and geometric distributions is the following. Suppose that the random variable X has the exponential distribution (1.66) and that $Y = \lfloor X \rfloor$ is the integer part of X (i.e., the greatest integer less than or equal to X). Then $\text{Prob}(Y = y) = \text{Prob}(y \le X < y + 1)$. By integration of (1.66), this probability is

$$\text{Prob}(Y = y) = (1 - e^{-\lambda}) e^{-\lambda y}, \quad y = 0, 1, 2, \ldots. \tag{1.69}$$

With the identification $p = e^{-\lambda}$, this is in the form of the geometric probability distribution (1.15). It follows from equation (1.16) that

$$\text{Prob}(Y \le y - 1) = 1 - e^{-\lambda y}, \quad y = 1, 2, \ldots. \tag{1.70}$$

This relation of the exponential distribution with the geometric distribution is useful in computational biology and bioinformatics, so that we shall sometimes refer to the geometric distribution as being given in the notation of equation (1.69) rather than in the notation of equation (1.15). In the notation of (1.69), the mean and variance of the geometric distribution are, respectively,

$$\mu = \frac{1}{e^\lambda - 1}, \quad \sigma^2 = \frac{e^\lambda}{(e^\lambda - 1)^2}. \tag{1.71}$$

The density function $f_D(d)$ of the *fractional part* D of a random variable X having the exponential distribution, defined by $D = X - \lfloor X \rfloor = X - Y$, can be shown to be

$$f_D(d) = \frac{\lambda e^{-\lambda d}}{1 - e^{-\lambda}}, \quad 0 \le d < 1. \tag{1.72}$$

This density function does not depend on the value of X. The mean and variance of this distribution are found from (1.53) and (1.54) to be, respectively,

$$\mu = \frac{1}{\lambda} - \frac{1}{e^{\lambda} - 1}, \quad \sigma^2 = \frac{1}{\lambda^2} - \frac{1}{(e^{\lambda} - 1)} - \frac{1}{(e^{\lambda} - 1)^2}. \tag{1.73}$$

More Comments on Geometric-Like Random Variables

Random variables having a geometric-like distribution were defined by the asymptotic relation (1.19). It is useful to express this relation in the notation developed in this section, that is, with p replaced by $e^{-\lambda}$. With this notation, a discrete random variable Y has a geometric-like distribution if, as $y \to \infty$,

$$\mathrm{Prob}(Y \ge y) \sim Ce^{-\lambda y}, \tag{1.74}$$

for some fixed positive constant $C < 1$. This equation, and geometric-like random variables, will be central to the theory of generalized random walks and of BLAST, developed respectively in Chapters 7 and 10.

1.10.5 The Gamma Distribution

The exponential distribution is a special case of the *gamma* distribution. The density function for the gamma distribution is

$$f_X(x) = \frac{\lambda^k x^{k-1} e^{-\lambda x}}{\Gamma(k)}, \quad x > 0. \tag{1.75}$$

Here λ and k are arbitrary positive parameters and $\Gamma(k)$ is the *gamma function* (see Appendix B.17). The value of k need not be an integer, but if it is, then $\Gamma(k) = (k-1)!$. The density function can take several different shapes, depending on the value of the parameter k (see Figure 1.15).

The mean μ and variance σ^2 of the gamma distribution are given by

$$\mu = \frac{k}{\lambda}, \quad \sigma^2 = \frac{k}{\lambda^2}. \tag{1.76}$$

The exponential distribution is the special case of the gamma distribution where $k = 1$. Another important special case is where $\lambda = \frac{1}{2}$ and $k = \frac{1}{2}\nu$, with ν a positive integer, so that

$$f_X(x) = \frac{1}{2^{\nu/2} \Gamma(\frac{1}{2}\nu)} x^{\frac{1}{2}\nu - 1} e^{-\frac{1}{2}x}, \quad x > 0. \tag{1.77}$$

This is called the *chi-square* distribution with ν degrees of freedom, and is important for statistical hypothesis testing. It can be shown (see Section 1.11) that if Z is a standard normal random variable, then Z^2 is a random variable having the chi-square distribution with one degree of freedom. The exponential distribution with $\lambda = \frac{1}{2}$ is the chi-square distribution with two degrees of freedom.

When k is a positive integer, the cumulative distribution function $F_X(x)$ of the gamma distribution may be found by repeated integration by parts,

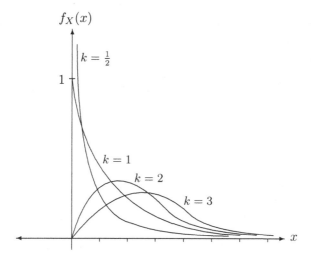

Figure 1.15. Examples of the density function of a gamma distribution with $\lambda = 1$ and $k = \frac{1}{2}, 1, 2, 3$.

and is given by the sum of k Poisson distribution terms. This result is developed in Section 4.3. When k is not an integer, no simple expression exists for the cumulative distribution function of the gamma distribution.

1.10.6 The Beta Distribution

A continuous random variable X has a beta distribution with positive parameters α and β if its density function is

$$f_X(x) = \frac{\Gamma(\alpha + \beta)}{\Gamma(\alpha)\Gamma(\beta)} x^{\alpha-1}(1 - x)^{\beta-1}, \quad 0 < x < 1. \qquad (1.78)$$

Here $\Gamma(\cdot)$ is the gamma function, discussed above in connection with the gamma distribution. The mean μ and the variance σ^2 of this distribution are

$$\mu = \frac{\alpha}{\alpha + \beta}, \quad \sigma^2 = \frac{\alpha\beta}{(\alpha + \beta)^2(\alpha + \beta + 1)}. \qquad (1.79)$$

The uniform distribution (1.63) is the special case of this distribution for the case $\alpha = \beta = 1$. Beta distributions, unlike Gaussian distributions, have finite range, and their shape can be varied widely by adjusting the two parameters, α and β.

1.11   The Moment-Generating Function

An important quantity for any random variable, discrete or continuous, or equivalently for its probability distribution, is its *moment-generating function*, or mgf for short. The moment-generating function is a function of a real variable θ and, for a discrete random variable Y, will be denoted by $m_Y(\theta)$, or $m(\theta)$ when the random variable is clear from the context. In this case

$$m_Y(\theta) = E(e^{\theta Y}) = \sum_y e^{\theta y} P_Y(y), \tag{1.80}$$

the sum being taken over all possible values of Y. Clearly, $m_Y(0) = 1$, and for all random variables that we consider, $m_Y(\theta)$ exists (i.e., the sum or integral defining it converges) for θ in some interval $(-a, a)$ about 0, $a > 0$. In this definition θ is a "dummy" variable, as was the quantity t in the pgf. For discrete random variables, the change of variables $t = e^\theta$ changes the probability-generating function to the moment-generating function.

The mgf of a continuous random variable X is the natural analogue of (1.80), namely

$$m(\theta) = E(e^{\theta X}) = \int_L^H e^{\theta x} f_X(x) dx, \tag{1.81}$$

where L and H are the boundaries of the range of X, as in Section 1.8.

The original purpose for the mgf was to generate moments, as the name suggests. Any moment of a probability distribution can be found by appropriate differentiation of the mgf $m(\theta)$ with respect to θ, with the derivative being evaluated at $\theta = 0$. That is, if a random variable has a distribution whose mgf is $m(\theta)$, the mean μ of that random variable is given by

$$\mu = \left(\frac{d\,m(\theta)}{d\theta} \right)_{\theta=0}, \tag{1.82}$$

and the variance is given by

$$\sigma^2 = \left(\frac{d^2\,m(\theta)}{d\theta^2} \right)_{\theta=0} - \mu^2. \tag{1.83}$$

An analogue of the uniqueness property of pgfs discussed on page 26 holds for mgfs: If two mgfs agree in an open interval containing zero, the corresponding probability distributions are identical. In some cases an mgf is defined, that is exists, only for certain values of θ, while in other cases it exists for all θ. In the former case we restrict θ to those values for which the mgf is defined.

Example 1. The mgf of the exponential distribution (1.66) is

$$\int_0^{+\infty} e^{\theta x} \lambda e^{-\lambda x} dx = \frac{\lambda}{\lambda - \theta}. \tag{1.84}$$

The integral on the left-hand side converges only when $\theta < \lambda$, so that θ is assumed to be restricted to these values.

Equations (1.82) and (1.84) jointly show that the mean of the exponential distribution (1.66) is

$$\mu = \left(\frac{\lambda}{(\lambda - \theta)^2} \right)_{\theta=0} = \frac{1}{\lambda},$$

and equations (1.83) and (1.84) jointly show that the variance is

$$\sigma^2 = 2 \left(\frac{\lambda}{(\lambda - \theta)^3} \right)_{\theta=0} - \mu^2 = \frac{1}{\lambda^2},$$

confirming the results of Problem 1.21, where these moments are found by integration.

Example 2. The mgf of the normal distribution (1.64) is

$$\int_{-\infty}^{+\infty} \frac{1}{\sqrt{2\pi}\sigma} e^{\theta x - \frac{(x-\mu)^2}{2\sigma^2}} \, dx.$$

Completing the square in the exponent, we find that this reduces to

$$e^{\mu\theta + \frac{1}{2}\sigma^2\theta^2}. \tag{1.85}$$

In this case θ can take any value in $(-\infty, \infty)$. Equations (1.82) and (1.85) jointly show that the mean of the normal distribution is

$$\left((\mu + \sigma^2\theta)e^{\mu\theta + \frac{1}{2}\sigma^2\theta^2} \right)_{\theta=0} = \mu,$$

as claimed in the discussion surrounding (1.85). A further differentiation and application of (1.83) confirms that the variance of the normal distribution (1.85) is σ^2.

Example 3. The mgf of the chi-square distribution (1.77) is

$$\int_0^{+\infty} e^{\theta x} \frac{1}{2^{\nu/2}\Gamma(\frac{1}{2}\nu)} x^{\frac{1}{2}\nu-1} e^{-\frac{1}{2}x} \, dx. \tag{1.86}$$

Taking $\theta < 1/2$, the change of variable $y = x(1 - 2\theta)$ transforms this into

$$(1 - 2\theta)^{-\nu/2} \int_0^{+\infty} \frac{1}{2^{\nu/2}\Gamma(\frac{1}{2}\nu)} y^{\frac{1}{2}\nu-1} e^{-\frac{1}{2}y} \, dy, \tag{1.87}$$

and since the integral in this expression is 1, being the integral of the original chi-square distribution, the required mgf is

$$(1 - 2\theta)^{-\nu/2}. \tag{1.88}$$

Application of equations (1.82) and (1.83) show that the mean of the chi-square distribution is ν, the number of degrees of freedom associated with the distribution, and that the variance is 2ν.

The moment-generating function is useful when it can be evaluated as a comparatively simple function, as in the above examples. This is not always possible; for the beta distribution (1.78), for example, the moment-generating function does not in general take a simple form.

It is often more useful to use the logarithm of the mgf than the mgf itself. If X is a random variable whose distribution has mgf $m(\theta)$, then the mean μ and the variance σ^2 of X are found from

$$\mu = \left(\frac{d \log m(\theta)}{d\theta}\right)_{\theta=0}, \quad \sigma^2 = \left(\frac{d^2 \log m(\theta)}{d\theta^2}\right)_{\theta=0}. \tag{1.89}$$

We conclude this section with two important properties of mgfs. First, let X be any continuous random variable with density function $f_X(x)$, $L < x < H$, and let $g(X)$ be any function of X. Then $g(X)$ is itself a random variable, and from (1.55) the moment-generating function of $g(X)$ can be written in terms of the density function for X as

$$m_{g(X)}(\theta) = \int_L^H e^{\theta g(x)} f_X(x)\, dx. \tag{1.90}$$

A similar definition holds for a discrete random variable, namely

$$m_{g(Y)}(\theta) = \sum_y e^{\theta g(y)} P_Y(y), \tag{1.91}$$

the summation being over all possible values of the random variable Y.

Example. If z has a normal distribution with mean 0, variance 1, the mgf of Z^2 is

$$m_{Z^2}(\theta) = E(e^{\theta Z^2}) = \int_{-\infty}^{+\infty} \frac{1}{\sqrt{2\pi}} e^{\theta x^2 - x^2/2}\, dx, \quad \theta < 1/2. \tag{1.92}$$

The change of variable $y = x(1 - 2\theta)^{1/2}$ shows that this is $(1 - 2\theta)^{-1/2}$. Comparison of this with (1.88), together with the uniqueness theorem discussed after (1.8), shows that Z^2 has a chi-square distribution with one degree of freedom. □

The second property of mgfs is important for BLAST theory, and we state it as a theorem.

Theorem 1.1. Let Y be a discrete random variable with mgf $m(\theta)$. Suppose that Y can take at least one negative value (say $-a$) with positive probability $P_Y(-a)$ and at least one positive value (say b) with positive probability $P_Y(b)$, and that the mean of Y is non-zero. Then there exists a unique non-zero value θ^* of θ such that $m(\theta^*) = 1$.

Proof. We prove the theorem for the case where the mgf is defined for all θ in $(-\infty, \infty)$. The general case is proved in a similar way.

Since all terms in the sum (1.80) defining $m(\theta)$ are positive,

$$m(\theta) > P_Y(-a)e^{-\theta a} \quad \text{and} \quad m(\theta) > P_Y(b)e^{\theta b}.$$

Thus $m(\theta) \to +\infty$ as $\theta \to -\infty$ and also as $\theta \to +\infty$. Further,

$$\frac{d^2 m(\theta)}{d\theta^2} = \sum_y y^2 P_Y(y)e^{\theta y}, \qquad (1.93)$$

and since this is positive for all θ, the curve of $m(\theta)$, as a function of θ, is convex. The definition (1.80) of an mgf shows that $m(0) = 1$. By (1.82), the mean of Y is the slope of $m(\theta)$ at $\theta = 0$, and this is non-zero by assumption. The above shows that if the mean of Y is negative, then the slope of the graph of $m(\theta)$ at $\theta = 0$ is negative, and together with the other properties of $m(\theta)$ determined above, its graph must be approximately as shown in the left-hand graph in Figure 1.16. Consequently, there is a unique positive value θ^* of θ such that $m(\theta^*) = 1$.

If the mean of Y is positive, then the graph of $m(\theta)$ is approximately as shown in the right-hand graph in Figure 1.16, and so in this case also there is a unique negative value θ^* of θ such that $m(\theta^*) = 1$.

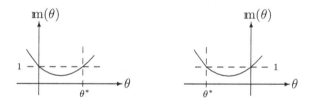

Figure 1.16.

The same conclusions also hold for continuous random variables, making the appropriate changes to the statement of the theorem.

1.12 Events

1.12.1 What Are Events?

So far, the development of probability theory has been in terms of random variables, that is, quantities that by definition take one or another numerical value. In many contexts, however, it is more natural, or more convenient, to consider probabilities relating to events rather than to random variables. For example, if we examine the nucleotides in two aligned DNA sequences, it might be more natural to say that the event A_j occurs if the two sequences have the same nucleotide at position j, rather than to define a random variable I_j, taking the value $+1$ if the two sequences have the same nucleotide at this position and taking the value 0 if they do not. This section provides some elementary probability theory relating to events.

Our definition of an event is deliberately casual, since in bioinformatics a detailed sample-space-based definition is not necessary. We simply think of an event as something that either will or will not occur when some experiment is performed. For example, if we plan to roll a die, A_1 might be the event that the number turning up will be even and A_2 might be the event that the number turning up will be a 4, 5, or 6. The experiment in this case is the actual rolling of the die, and none, one, or both of these two events will have occurred after the experiment is carried out.

A *certain event* is an event that must happen, such as rolling a number less than seven on one regular die. A null or empty event is one that cannot happen, such as rolling a number greater than seven on a regular die. The empty event for a random experiment will denoted $\emptyset$.

In different contexts an event might be described in quite different ways, but in fact is the same event. For example, in one context S_1 might be defined as the event that the number turning up on a die is less than 7, and in another context S_2 might be defined as the event that the number is divisible by 1. S_1 and S_2 are just different ways of describing the certain event.

1.12.2 *Complements, Unions, and Intersections*

Before considering the probability theory associated with events we consider some aspects of events themselves.

Associated with any event A is its complementary event "not A," usually denoted by $\bar{A}$. The event $\bar{A}$ is the event that A does not occur. For example, if A is the event "an even number turns up on the roll of a die," then $\bar{A}$ is the event "an odd number turns up."

If A_1 and A_2 are two events, the *union* of A_1 and A_2 is the event that either A_1 or A_2 (or both) occurs. The union of the events A_1 and A_2 is denoted by $A_1 \cup A_2$. For any event A, the event $A \cup \bar{A}$ is the certain event. More generally, in some cases at least one of A_1 or A_2 *must* occur, so that $A_1 \cup A_2$ is the certain event. In such a case we say that A_1 and A_2 are *exhaustive*.

The *intersection* of two events A_1 and A_2 is the event that both occur. The intersection of the events A_1 and A_2 is denoted by $A_1 \cap A_2$ or, more frequently, simply by $A_1 A_2$.

In some cases the events A_1 and A_2 cannot occur together, that is, they are *mutually exclusive*. In this case the intersection of A_1 and A_2 is empty. For any event A, $A \cap \bar{A} = \emptyset$.

These definitions extend in a natural way to any number of events. Thus the union of the events $A_1, A_2, \ldots, A_n$, denoted by $A_1 \cup A_2 \cup \cdots \cup A_n$, is the event that at least one of the events $A_1, A_2, \ldots, A_n$ occurs. The intersection of the events $A_1, A_2, \ldots, A_n$, denoted by $A_1 A_2 \cdots A_n$, is the event that all of the events $A_1, A_2, \ldots, A_n$ occur.

More complicated events such as $(A_1 A_2) \cup A_3$ may be defined. These lead to various useful identities, of which we list four here:

(1) $\overline{A_1 \cup A_2} = \bar{A}_1 \cap \bar{A}_2,$
(2) $\overline{A_1 \cap A_2} = \bar{A}_1 \cup \bar{A}_2,$
(3) $A_1 \cap (A_2 \cup A_3) = (A_1 \cap (A_2)) \cup (A_1 \cap (A_3)),$
(4) $A_1 \cup (A_2 \cap A_3) = (A_1 \cup (A_2)) \cap (A_1 \cup (A_3)).$

The first two of these are known as DeMorgan's laws. A proof of (1) is that $\overline{A_1 \cup A_2}$ occurs $\Leftrightarrow A_1 \cup A_2$ does not occur $\Leftrightarrow A_1$ does not occur *and* A_2 does not occur $\Leftrightarrow \overline{A_1}$ occurs *and* $\overline{A_2}$ occurs. $\Leftrightarrow \bar{A}_1 \cap \bar{A}_2$ occurs.

1.12.3 Probabilities of Events

Probability refers to the assignment of a real number $P(A)$ to each event A. These numbers must satisfy the following "axioms of probability":

(1) $P(A) \geq 0$ for any event A.
(2) For any certain event S, $P(S) = 1$.
(3) If $A_1 \cap A_2 = \emptyset$, then

$$P(A_1 \cup A_2) = P(A_1) + P(A_2).$$

It follows from the axioms that for any event A, $0 \leq P(A) \leq 1$ and $P(\bar{A}) = 1 - P(A)$.

Often one wishes to find the probability that at least one of the events $A_1, A_2, \ldots, A_n$ occurs. If the events are mutually exclusive for all $i \neq j$, then it follows by mathematical induction from the above axioms that this probability is

$$P(A_1 \cup A_2 \cup \cdots \cup A_n) = \sum_i P(A_i). \tag{1.94}$$

A more complicated formula arises when some of the events $A_1, A_2, \ldots, A_n$ are not mutually exclusive. The simplest case concerns two events, A_1 and A_2. The sum $P(A_1) + P(A_2)$ counts the probability of the intersection $A_1 A_2$ twice, whereas it should be counted only once (see Figure 1.17). This argument leads to

$$P(A_1 \cup A_2) = P(A_1) + P(A_2) - P(A_1 A_2). \tag{1.95}$$

This formula can be generalized to find the probability of the union of an arbitrary number of events in terms of probabilities of intersections. This

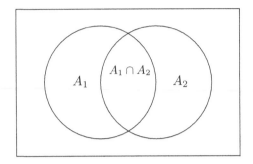

Figure 1.17. Diagram showing $P(A_1 \cup A_2) = P(A_1) + P(A_2) - P(A_1 \cap A_2)$.

is discussed at length in Feller (1957), and we reproduce his results here without proof.

Define $S_1, S_2, \ldots, S_n$ by the formulae

$$S_1 = \sum_i P(A_i), \quad S_2 = \sum_{i<j} P(A_i A_j), \quad S_3 = \sum_{i<j<k} P(A_i A_j A_k), \quad \ldots \tag{1.96}$$

Then the probability P_1 of the union $A_1 \cup A_2 \cup \cdots \cup A_n$ is given by

$$P_1 = S_1 - S_2 + S_3 - \cdots + (-1)^{n-1} S_n. \tag{1.97}$$

This implies that the probability that none of the events $A_1, A_2, \ldots, A_n$ occurs is

$$1 - S_1 + S_2 - S_3 - \cdots + (-1)^n S_n. \tag{1.98}$$

One simple consequence of equation (1.97) is that, whether the events are independent or not,

$$P_1 \leq S_1. \tag{1.99}$$

This is seen from Figure 1.17 for the case of two events and is easily shown to hold for any number of events.

1.12.4   Conditional Probabilities

Suppose that a fair die is rolled once. If it is known that the number turning up is less than or equal to three, elementary arguments show that the probability that it is odd is $\frac{2}{3}$. We write this

$$P(\text{number is odd} \mid \text{number is less than or equal to 3}) = \frac{2}{3}. \tag{1.100}$$

This is an example of a *conditional probability* calculation.

More precisely, suppose that A_1 and A_2 are two events and that $P(A_2) \neq 0$. Then the *conditional* probability that the event A_1 occurs, given that the event A_2 occurs, denoted by $P(A_1 \,|\, A_2)$, is defined by

$$P(A_1 \,|\, A_2) = \frac{P(A_1 A_2)}{P(A_2)}. \qquad (1.101)$$

This is a crucial formula. The intuition behind the formula can be seen from Figure 1.17, page 44: If it is given that the event A_2 occurs, the only part of the diagram of relevance is the right-hand circle, corresponding to the event A_2. Of the probability associated with this circle, the proportion $P(A_1 A_2)/P(A_2)$ corresponds to the event that A_1 occurs. This proportion is the right-hand side in (1.101).

Conditional probability calculations are often counterintuitive (some examples are given in the Problems), and it is essential that the definition (1.101) be used when calculating them.

The calculation in (1.100) can be arrived at from (1.101) from the fact that if A_1 is the event that the number is odd and A_2 is the event that the number is less than or equal to three, then $A_1 A_2$ is the event that the number is 1 or 3. The probability of $A_1 A_2$ is then $\frac{1}{3}$ and that of A_2 is $\frac{1}{2}$. Equation (1.100) follows from this.

In using (1.101), care must be exercised in assessing what the intersection event $A_1 A_2$ means. For example, suppose that a die (fair or unfair) is rolled twice. Let A_1 be the event that the number turning up on the first roll is y_1 and the number turning up on the second roll is y_2, and let A_2 be the event that the sum of the two numbers is y_3. Then the event $A_1 A_2$ is empty unless $y_1 + y_2 = y_3$, so that when $y_1 + y_2 \neq y_3$, $P(A_1 \,|\, A_2) = 0$. When $y_1 + y_2 = y_3$, the event A_1 implies the event A_2, so that $A_1 A_2 = A_1$. In this case $P(A_1 \,|\, A_2) = P(A_1)/P(A_2)$.

It should be noted that $P(A_1 \,|\, A_2)$ does not necessarily equal $P(A_2 \,|\, A_1)$. For example, if a fair die is rolled once and A_1 is the event that the number turning up is not a one, and A_2 is the event that the number is odd, then $P(A_1 \,|\, A_2) = \frac{2}{3}$, whereas $P(A_2 \,|\, A_1) = \frac{2}{5}$.

Equation (1.101) has various important consequences. It shows immediately, for example, that

$$P(A_1 A_2) = P(A_2)P(A_1 \,|\, A_2), \qquad (1.102)$$

and in this form its truth is almost self-evident: The probability that both A_1 and A_2 occur is the probability that A_2 occurs multiplied by the probability that A_1 occurs *given* that A_2 occurs. Further, the left-hand side in equation (1.102) is symmetric in A_1 and A_2, so that the right-hand side may be replaced by $P(A_1)P(A_2 \,|\, A_1)$. This implies that

$$P(A_1)P(A_2 \,|\, A_1) = P(A_2)P(A_1 \,|\, A_2), \qquad (1.103)$$

or

$$P(A_2 \,|\, A_1) = \frac{P(A_2)P(A_1 \,|\, A_2)}{P(A_1)}. \qquad (1.104)$$

This is known as Bayes' formula, which allows for the writing of $P(A_2 \mid A_1)$ in terms of $P(A_1 \mid A_2)$. In this form the formula is used in Bayesian statistics; see Section 3.9.

A second implication of equation (1.101) is the following. Suppose that $B_1, B_2, \ldots, B_k$ is a set of mutually exclusive and exhaustive events, so that exactly one of these events must occur. Then for any event A,

$$P(A) = \sum_{j=1}^{k} P(AB_j).$$

Equation (1.101) then implies

$$P(A) = \sum_{j=1}^{k} P(B_j)P(A \mid B_j), \tag{1.105}$$

which is used frequently in this book as it is often the most convenient way to calculate $P(A)$.

All the above formulae for events have direct analogues for probabilities of random variables. In fact, the latter probabilities may be derived immediately from the former. These are discussed at length in the next chapter.

1.12.5 Independence of Events

An event A_1 is *independent* of another event A_2 if

$$P(A_1) = P(A_1 \mid A_2). \tag{1.106}$$

The natural interpretation of this requirement is that knowledge of the occurrence of A_2 does not affect the probability of the occurrence of A_1. The definition of $P(A_1 \mid A_2)$ in (1.101) shows that A_1 is independent of A_2 if and only if

$$P(A_1 A_2) = P(A_1) \cdot P(A_2). \tag{1.107}$$

This equation provides a useful way to check for independence. Note that it follows from this that if A_1 is independent of A_2, then A_2 is independent of A_1, so that we simply say that A_1 and A_2 are independent.

As an example, if a fair die is to be rolled, the events A_1 (an even number turns up) and A_2 (4, 5, or 6 turns up) are not independent. Each of these events separately has probability $\frac{1}{2}$, and the probability of their intersection (the event that either a 4 or a 6 will turn up) is $\frac{1}{3}$. Thus equation (1.107) does not hold, and the events are not independent. On the other hand, if A_3 is the event that a 3, 4, 5, or 6 turns up, then A_1 and A_3 are independent. These statements all assume that the die is fair. If the die is unfair, they will usually no longer hold.

More generally, the events $A_1, A_2, \ldots, A_n$ are defined to be independent if and only if the requirements

$$P(A_i A_j) = P(A_i)P(A_j) \text{ for all distinct } i, j, \qquad (1.108)$$
$$P(A_i A_j A_k) = P(A_i)P(A_j)P(A_k) \text{ for all distinct } i, j, k, \qquad (1.109)$$
$$P(A_i A_j A_k A_l) = P(A_i)P(A_j)P(A_k)P(A_l) \text{ for all distinct } i, j, k, l,$$

$$\cdots$$

$$P(A_1 A_2 A_3 \cdots A_n) = P(A_1)P(A_2)P(A_3) \cdots P(A_n), \qquad (1.110)$$

all hold.

As in the case of two random variables, this definition is equivalent to saying that the probability of any collection of the events is not affected by knowledge of any collection of the remaining events.

It is not sufficient for independence of the events $A_1, A_2, \ldots, A_n$ that the "pairwise" conditions (1.108) hold: All of the conditions (1.108)–(1.110) must be satisfied for the events to be independent. It is possible that the pairwise conditions (1.108) hold but that the "triple-wise" conditions (1.109) do not hold. The following example demonstrates this.

Example. Suppose that two fair dice are thrown, and that the events A_1, A_2, and A_3 are, respectively, "number on first die is odd," "number on second die is odd," "sum of the two numbers is odd." Clearly, both A_1 and A_2 have probability $\frac{1}{2}$ and it is not hard to show that A_3 also has probability $\frac{1}{2}$. Each pairwise intersection event has probability $\frac{1}{4}$; the event $A_1 A_3$, for example, that the first number is odd and the sum of the two numbers is odd, is the event "first number is odd, second number is even," and this has probability $\frac{1}{4}$. The probability of the triple-wise event $A_1 A_2 A_3$ is zero, since this event cannot occur. However, $P(A_1)P(A_2)P(A_3) = \frac{1}{8}$, so that condition (1.109) is not met, and the three events are not independent.

If the events A_1 and A_2 are independent, then so are their respective complementary events $\bar{A}_1$ and $\bar{A}_2$. This follows using the laws of probability, DeMorgan's laws (see page 43), and the independence of A_1 and A_2:

$$P(\bar{A}_1 \bar{A}_2) = 1 - P(A_1 \cup A_2)$$
$$= 1 - P(A_1) - P(A_2) + P(A_1 A_2) = 1 - P(A_1) - P(A_2) + P(A_1)P(A_2)$$
$$= \{1 - P(A_1)\}\{1 - P(A_2)\} = P(\bar{A}_1)P(\bar{A}_2).$$

This argument can be extended to any number of events. The probability that none of the independent events $A_1, A_2 \ldots A_n$ occurs is

$$\prod_{j=1}^{n} \{1 - P(A_j)\}. \qquad (1.111)$$

1.13 The Memoryless Property of the Geometric and the Exponential Distributions

Suppose that y_1 and y_2 are two positive integers and that a random variable Y having the geometric distribution (1.15) takes a value y_1 or more. What is the probability that it takes a value $y_1 + y_2$ or more? The intersection of the events $Y \geq y_1$ and $Y \geq y_1 + y_2$ is the event that $Y \geq y_1 + y_2$. In this case the conditional probability formula (1.101) becomes

$$\frac{\text{Prob}(Y \geq y_1 + y_2)}{\text{Prob}(Y \geq y_1)}, \tag{1.112}$$

and from equation (1.16), this is $p^{y_1+y_2}/p^{y_1} = p^{y_2}$. But this is just the unconditional probability that $Y \geq y_2$. Therefore,

$$\text{Prob}(Y \geq y_1 + y_2 \,|\, Y \geq y_1) = \text{Prob}(Y \geq y_2).$$

This is called the *memoryless* property of the geometric distribution. The memoryless concept is discussed below in connection with the exponential distribution.

The exponential distribution, like the geometric distribution, also has the memoryless property. This is shown as follows. If X is a random variable having the exponential distribution (1.54) and if $x_2 > 0$, then from (1.67),

$$\text{Prob}(X \geq x_2) = e^{-\lambda x_2}. \tag{1.113}$$

From the conditional probability formula (1.101), if also $x_1 > 0$, then

$$\text{Prob}(X \geq x_1 + x_2 \,|\, X \geq x_1) = \frac{e^{-\lambda(x_1+x_2)}}{e^{-\lambda x_1}} = e^{-\lambda x_2}. \tag{1.114}$$

Thus

$$\text{Prob}(X \geq x_1 + x_2 \,|\, X \geq x_1) = \text{Prob}(X \geq x_2). \tag{1.115}$$

An application of this property follows.

Cellular proteins and RNA molecules are regulated in many ways, involving the degradation and generation of molecules in a cell. Some molecules may eventually break down with age, while others are actively degraded. Molecules that are not actively degraded can be equally likely to degrade spontaneously at any given time. This phenomenon is, in effect, the memoryless property. The (random) lifetime of such a molecule is modeled by an exponential distribution.

A further property of the exponential distribution is relevant to this behavior. Suppose that h is small. The probability that a random variable having the exponential distribution (1.66) takes a value in the interval $(x, x + h)$, given that its value exceeds x, is, from (1.101) and (1.67),

$$\frac{\int_x^{x+h} \lambda e^{-\lambda u} \, du}{e^{-\lambda x}} = 1 - e^{-\lambda h} = \lambda h + o(h), \tag{1.116}$$

where the second equality follows from (B.21) and the $o(h)$ notation is discussed in Appendix B.8. In the case of cellular molecules, this implies that having lived to age at least x, the probability that a molecule degrades in the time interval $(x, x + h)$, where h is small, is approximately proportional to the length h of the interval. This example will be pursued further in Sections 2.11.1 and 4.1.

1.14   Entropy and Related Concepts

1.14.1   Entropy

Suppose that Y is a discrete random variable with probability distribution $P_Y(y)$. The *entropy* $H(\mathbf{P}_Y)$ of this probability distribution is defined by

$$H(\mathbf{P}_Y) = -\sum_y P_Y(y) \log P_Y(y), \tag{1.117}$$

the sum (as with all sums in this section) being taken over all values of y in the range of Y. Since this quantity depends only on the probabilities of the various values of Y, and not on the actual values themselves, it can be thought of, as the notation indicates, as being a function of the probability distribution $\mathbf{P}_Y = \{P_Y(y)\}$ rather than the random variable Y.

In some areas of computational biology the base of the logarithm in this definition is taken to be 2. The reason for this, in terms of "bits" of information, is discussed in Appendix B.10. Although we will use the base 2 in the definition (1.117) later in this book, for the moment we use natural logarithms (and the notation "log" for these).

The entropy of a probability distribution is a measure of how close to uniform that distribution is, and thus, in a sense, of the unpredictability of any observed value of a random variable having that distribution. If there are s possible values for the random variable, the entropy is maximized when $P_Y(y) = s^{-1}$. In this case it takes the value $\log s$, and the value to be assumed by Y is in a sense maximally unpredictable. At the other extreme, if only one value of Y is possible, the entropy of the distribution is zero, and the value to be assumed by the random variable Y is then completely predictable.

Despite the fact that both the entropy and the variance of a probability distribution measure in some sense the uncertainty of the value of a random variable having that distribution, the entropy has an interpretation different from that of the variance. The entropy is defined only by the probabilities of the possible values of the random variable and not by the values themselves. On the other hand, the variance depends on these values. Thus if Y_1 takes the values 1 and 2 each with probability 0.5, and Y_2 takes the values 1 and 100 each with probability 0.5, the distributions of Y_1 and Y_2 have equal entropies but quite different variances.

1.14.2   Relative Entropy

Several procedures in bioinformatics use the *relative entropy* of two different probability distributions $P_{Y_0}(y)$ and $P_{Y_1}(y)$. It is convenient to denote
the first of these probability distributions by $\boldsymbol{P}_0$ and the second by $\boldsymbol{P}_1$.
We assume that these are different distributions but have the same range,
that is, that $P_{Y_0}(y) > 0$ if and only if $P_{Y_1}(y) > 0$. There are two relative
entropies associated with these two distributions, namely $H(\boldsymbol{P}_0||\boldsymbol{P}_1)$ and
$H(\boldsymbol{P}_1||\boldsymbol{P}_0)$, defined respectively by

$$H(\boldsymbol{P}_0||\boldsymbol{P}_1) = \sum_y P_{Y_0}(y) \log \frac{P_{Y_0}(y)}{P_{Y_1}(y)}, \qquad (1.118)$$

$$H(\boldsymbol{P}_1||\boldsymbol{P}_0) = \sum_y P_{Y_1}(y) \log \frac{P_{Y_1}(y)}{P_{Y_0}(y)}, \qquad (1.119)$$

the summation in both cases being over all values in the (common) range
of the two probability distributions. A proof that both $H(\boldsymbol{P}_0||\boldsymbol{P}_1)$ and
$H(\boldsymbol{P}_1||\boldsymbol{P}_0)$ are positive can be based on the fact that for positive x,
$-\log x \geq 1 - x$, with equality holding only when $x = 1$. Then, for example,

$$H(\boldsymbol{P}_0||\boldsymbol{P}_1) = \sum_y P_{Y_0}(y) \left(-\log \frac{P_{Y_1}(y)}{P_{Y_0}(y)} \right)$$

$$\geq \sum_y P_{Y_0}(y) \left(1 - \frac{P_{Y_1}(y)}{P_{Y_0}(y)} \right)$$

$$= \sum_y P_{Y_0}(y) - \sum_y P_{Y_1}(y) = 0, \qquad (1.120)$$

with equality holding only when $P_{Y_0}(y) = P_{Y_1}(y)$.

The relative entropy of two probability distributions measures in some
sense the dissimilarity between them. However, since $H(\boldsymbol{P}_0||\boldsymbol{P}_1)$ is not
equal to $H(\boldsymbol{P}_1||\boldsymbol{P}_0)$, the sometimes-used practice of calling one or the
other of these quantities the distance between the two distributions is
not appropriate. On the other hand, the quantity $J(\boldsymbol{P}_0||\boldsymbol{P}_1)$, defined by
$J(\boldsymbol{P}_0||\boldsymbol{P}_1) = H(\boldsymbol{P}_0||\boldsymbol{P}_1) + H(\boldsymbol{P}_1||\boldsymbol{P}_0)$, does satisfy this requirement. In the
information theory literature, $J(\boldsymbol{P}_0||\boldsymbol{P}_1)$ is called the *divergence* between
the two distributions (see Kullback (1978)).

1.14.3   Scores and Support

We define the *support* $S_{0,1}(y)$ given by an observed value y of Y in favor
of $\boldsymbol{P}_0$ over $\boldsymbol{P}_1$ by

$$S_{0,1}(y) = \log \left(P_{Y_0}(y)/P_{Y_1}(y) \right), \qquad (1.121)$$

and similarly we define $S_{1,0}(y)$, given by

$$S_{1,0}(y) = \log\left(P_{Y_1}(y)/P_{Y_0}(y)\right), \tag{1.122}$$

as the support given by the observed value y of Y in favor of $\boldsymbol{P}_1$ over $\boldsymbol{P}_0$. The term "support" derives from the fact that if $S_{0,1}(y) > 0$, the observed value y of Y is more likely under $\boldsymbol{P}_0$ than it is under $\boldsymbol{P}_1$. If it happens, for some specific value of y of Y, that $P_{Y_0}(y) = P_{Y_1}(y)$, then both measures of support are zero, and this accords with the commonsense view that if Y is equally likely to take the value y under the two distributions, then observing the value y of Y affords no support for one distribution over the other. The definition of support also has the reasonable property that for any observed value y, $S_{0,1}(y) = -S_{1,0}(y)$. A general description of the concept of support is given in Edwards (1992).

If the true distribution of Y is $\boldsymbol{P}_0$, then $S_{0,1}(Y)$ is a random variable whose mean $I_{0,1}$ is found from equation (1.26) to be

$$I_{0,1} = \sum_y P_{Y_0}(y) S_{0,1}(y), \tag{1.123}$$

which is identical to $H(\boldsymbol{P}_0 \| \boldsymbol{P}_1)$. Correspondingly, if the true distribution is $\boldsymbol{P}_1$, then the mean support for $\boldsymbol{P}_1$ over $\boldsymbol{P}_0$ is

$$I_{1,0} = \sum_y P_{Y_1}(y) S_{1,0}(y), \tag{1.124}$$

and this is identical to $H(\boldsymbol{P}_1 \| \boldsymbol{P}_0)$.

One single observation is not enough for us to assess which distribution we believe is more likely to be correct. In equation (2.78) we define the concept of the accumulated support afforded by a number of observations. In Section 9.9.1 we discuss the question of how much accumulated support is needed to reasonably decide between two values of a binomial parameter, and in Section 10.6 we shall discuss accumulated support in the context of a BLAST procedure. We later call $S_{1,0}(y)$ a "score" statistic, and in Section 10.2.4 we consider the problem of the optimal choice of score statistics for the substitution matrices used in BLAST.

1.15 Transformations

Suppose that Y_1 is a discrete random variable and that $g(t)$ is a function of a real variable which is one-to-one on the range of Y_1. Then if $Y_2 = g(Y_1)$,

$$\text{Prob}(Y_2 = y_2) = \text{Prob}(Y_1 = y_1).$$

This equation makes it easy to find the probability distribution of Y_2 from that of Y_1. In the case where the transformation is not one-to-one, a simple addition of probabilities yields the distribution of Y_2. For example, if Y_1

can take the values -1, 0, and $+1$, each with probability $\frac{1}{3}$, and if $Y_2 = Y_1^2$, then the possible values of Y_2 are 0 and 1, and Y_2 takes these values with respective probabilities $\frac{1}{3}$ and $\frac{2}{3}$.

The corresponding calculation for a continuous random variable is less straightforward. Suppose that X_1 is a continuous random variable and that $X_2 = g(X_1)$ is a one-to-one function of X_1. In most cases of interest this means that $g(x)$ is either a monotonically increasing, or a monotonically decreasing, function of x, and we assume that one of these two cases holds. Let $f_1(x)$ be the density function of X_1 and let $f_2(x)$ be the density function of X_2, which we wish to find from that of X_1. Let $F_1(x)$ and $F_2(x)$ be the two respective cumulative distribution functions. (Here for typographic reasons we depart from the notation previously adopted for density functions and cumulative distribution functions.)

Suppose first that $g(x)$ is a monotonically increasing function of x and that $X_2 = g(X_1)$. Then

$$F_2(x) = \text{Prob}(X_2 < x) = \text{Prob}(X_1 < g^{-1}(x)) = F_1(g^{-1}(x)), \qquad (1.125)$$

where $g^{-1}(x)$ is the inverse function of $g(x)$. Differentiating both sides with respect to x and using the chain rule, we get

$$f_2(x) = \left(f_1(y) \div \frac{dg(y)}{dy} \right)_{y=g^{-1}(x)}. \qquad (1.126)$$

When X_2 is a monotonically decreasing function of X_1, similar arguments show that

$$f_2(x) = -\left(f_1(y) \div \frac{dg(y)}{dy} \right)_{y=g^{-1}(x)}. \qquad (1.127)$$

Both cases are covered by the equation

$$f_2(x) = \left| \left(f_1(y) \div \frac{dg(y)}{dy} \right) \right|_{y=g^{-1}(x)}. \qquad (1.128)$$

As an example, suppose that X_1 has density function $f_1(x) = 2x$, $0 < x < 1$, and that $X_2 = g(X_1) = X_1^3$. Then $dg(y)/dy = 3y^2$ and equation (1.128) shows that

$$f_2(x) = \left| \frac{2y}{3y^2} \right|_{y=\sqrt[3]{x}} = \frac{2}{3}x^{-1/3}, \quad 0 < x < 1. \qquad (1.129)$$

The form of this density function is quite different from that of X_1.

Theorem 1.2. A particularly important function of X_1 is $X_2 = F(X_1)$, where $F(X_1)$ is the distribution function of X_1. Here application of (1.128) shows that

$$f_2(x_2) = 1, \quad 0 < x_2 < 1. \qquad (1.130)$$

That is, X_2 has the uniform distribution in $(0, 1)$, whatever the distribution of X_1 might be. This is an important observation and is used several times in this book.

1.16 Empirical Methods

We often encounter random variables with distributions that cannot be calculated mathematically, or at least whose calculation appears difficult. In other cases the mean and/or the variance can be calculated but not the entire probability distribution: An example is given in Section 2.9.2. Often we cannot even calculate these quantities. One consequence of currently available computational power, however, is that, with a good model, we can simulate an experiment many times and obtain data from each simulated replication. We can then approximate probability distributions and their means and variances using these simulation data. The "plug-in" method discussed in Section 8.6 also uses this approach.

We illustrate the simulation approach with an example relating to the problem of gene mapping. Suppose we are given a DNA sequence consisting of L consecutive nucleotides. For convenience we say that this sequence is of "length" L. A number of possibly overlapping segments of this sequence are observed, the first segment consisting of i_1 consecutive nucleotides, the second consisting of i_2 consecutive nucleotides, and so on. We will say that these segments are of respective lengths $i_1, i_2, \ldots$. Figure 1.18 shows the four possible positions of a single segment of length $i_j = 3$ when $L = 6$. In general there are $L - i_j + 1$ possible positions for a segment of length i_j.

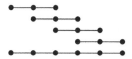

Figure 1.18. The 4 possible positions of a segment of length 3 in a sequence of length 6.

The direct-IBD mapping procedure of Cheung et al. (1998) produced data which can be represented as a collection of such segments. Cheung et al. aimed at testing whether the various segments in the data showed a significant tendency to cover the same positions. This was done by finding the maximum number M of segments that cover the same position. A sufficiently large value of M then supports the hypothesis that significant overlap of the segments does occur.

The elements of the theory used for testing hypotheses of this type will be described in Section 3.4. For our purposes now it is sufficient to say that the hypothesis testing procedure requires the calculation of the probability distribution of M when the segments are placed at random on the DNA sequence of length L. (By "random" we mean, for example, that the segment of length i_j occurs in each of the $L - i_j + 1$ possible positions with equal probability.) From this it is required to find the "random placement" probability that M is equal to or exceeds the observed value m of M.

When the number n of segments is large, it is difficult if not impossible to find this probability by mathematical methods. Thus Cheung et al. used approximate and conservation calculations to find an upper bound for this probability.

The approximations used were sufficient for the example studied; however as a general procedure the approximation is less efficient than is desirable. Another approach is to use simulation. If n and L are sufficiently small, it is possible to enumerate all possible configurations of the n segments by computer and from this obtain the exact distribution by counting. When n and L are large this approach is not feasible. An alternative and relatively simple approach is to replicate the experiment of randomly placing the n segments on a sequence of DNA of length L many times, and in this way obtain an empirical distribution of M, under the randomness assumption. The larger the number of replications, the closer the empirical distribution of M will be to the true one.

This is known as a Monte Carlo approach. There are simple computer programs for generating the uniform random variables necessary to do this simulation (see, for example, Press et al. (1992)).

For $n = 7$, $L = 100$, and $(i_1, \ldots, i_7) = (6, 5, 6, 5, 5, 8, 4)$ a simulation involving 10,000 replications gave the empirical distribution shown in Figure 1.19.

Suppose that, in a data set with $L = 100$ and 7 segments, with respective lengths i_1, ..., i_7 equal to 6,5,6,5,5,8,4, the observed value m of M is 4. From the empirical distribution, the estimated "randomness" probability that $M \geq 4$ is quite small, being .0163. We might then be unwilling to believe that the segments in the data were placed randomly, preferring the hypothesis that significant evidence of overlap occurs for some reason.

An empirical approach such as this was taken in Grant et al. (1999) to refine the calculations in Cheung et al. (1998), which led to a probability estimate that was an order of magnitude more precise than the Cheung et al. value.

In the previous example, all that was needed was to generate random numbers from a uniform distribution. Often it is necessary, to perform the appropriate simulations, to generate numbers from other distributions. The result at the end of the previous section allows us to transform uniformly distributed random variables into random variables following any continuous distribution whose cumulative distribution function is available. Suppose that we wish to generate random values of a continuous random variable X_1 having cumulative distribution function $F_1(x)$. Theorem 1.2

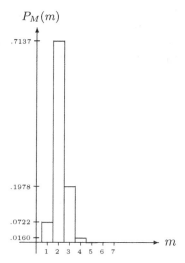

Figure 1.19. Empirical probabilities for M when $n = 7$ and $L = 100$. Estimated probabilities are: $P_M(1) = .0722$, $P_M(2) = .7137$, $P_M(3) = .1978$, $P_M(4) = .0160$, $P_M(5) = .0003$. The values $M = 6$ and 7 were not observed in the simulation.

shows that if we generate a uniformly distributed random variable X_2, and then define X_1 by $X_1 = F_1^{-1}(X_2)$, where F_1^{-1} is the inverse function of F_1, then X_1 is a random variable having the required density function. For example, (1.67) shows that the cumulative distribution function of a random variable having the exponential distribution (1.66) is $F_1(x) = 1 - e^{-\lambda x}$, $x \geq 0$. Thus if X_2 has a uniform distribution in $(0, 1)$, and if X_1 is defined by $X_1 = -\frac{1}{\lambda} \log(1 - X_2)$ (so that $X_2 = 1 - e^{-\lambda X_1} = F_1(X_1)$), then X_1 is a random variable having the exponential distribution (1.66). We can thus easily simulate exponentially distributed data.

A useful application of empirical methods in computational biology and bioinformatics concerns the distribution of test statistics used in statistical inference. The DNA segment example above provides an illustration of such an application. Further examples of this approach are given later in this book, in particular in connection with the complex theory of BLAST and of inference concerning phylogenetic trees. In some cases these examples demonstrate, using empirical methods, that various approximations commonly used for the distributions of test statistics are not as accurate as would be desired.

One might ask why, if empirical methods are so convenient, we should go to the trouble to calculate theoretical properties of distributions. One reason is that an exact theoretical distribution is preferred to an approximate empirical distribution is because the mathematical form of the distribution is often informative. Second, a drawback of empirical methods is that they can be slow and might take an unacceptably long time to run. Sometimes estimates have to be calculated in a matter of microseconds. When empirical methods are not adequate for this task, theoretical methods, even at the possible cost of some accuracy, become necessary.

Problems

1.1. If Y is a random variable having the distribution (1.2), show $E(Y^2) \neq (E(Y))^2$.

1.2. Use equation (1.36) to prove that the mean of a random variable having the binomial distribution is np, and together with equation (1.32) show that the variance of this random variable is $np(1 - p)$.

1.3. Use definition (1.24) and equation (B.18) to show that the mean of a random variable having the geometric distribution (1.15) is p/q, where $q = 1 - p$. From this, show immediately that the mean of the distribution (1.18) is q^{-1}. Also show that the mean of a random variable having the Poisson distribution (1.22) is λ.

1.4. This problem refers to the radiation example in Section 1.3.3. Suppose that the probability that a mouse of either gender is a mutant is p. The number of mutant male mice has a binomial distribution with parameter p and index n and the number of mutant female mice has a binomial distribution with parameter p and index $N - n$. The total number number of mutant mice has a binomial distribution with parameter p and index N.

Let A_1 be the event that y male mice are mutants and A_2 be the event that, in all, m mice are mutants. Use the conditional probability formula (1.101) to derive the hypergeometric distribution in Section 1.3.3 for the number of mutant male mice, given the total number m of mutant mice. *Hint:* The event A_1A_2 is identical to the event that y males and $m - y$ females are mutants. Calculate the probability of the event A_1A_2 using this fact.

1.5. Derive property (ii) of the variance on page 22 from the variance definition (1.29).

1.6. Show that equations (1.30) and (1.32) follow from the variance definition (1.29).

1.7. Prove (1.36) as follows: First, evaluate the expression (1.35) for the case $r \leq n$. Next, show that

$$y(y-1)\cdots(y-r+1)\binom{n}{y} = n(n-1)\cdots(n-r+1)\binom{n-r}{y-r}.$$

From this, equation (1.35) reduces to

$$E\big(Y(Y-1)\cdots(Y-r+1)\big)$$

$$= n(n-1)\cdots(n-r+1)p^r \sum_{y=r}^{n} \binom{n-r}{y-r}p^{y-r}(1-p)^{n-y}. \quad (1.131)$$

Finally, show that the sum in this expression equals 1.

1.8. Let Y be a random variable having a Poisson distribution with parameter λ. Show that the probability that the observed values y of Y is an even number exceeds $1/2$. *Hint:* use the exponential series in equation (B.20) for $x = \lambda$ and also for $x = -\lambda$. What is the asymptotic value of this probability as $\lambda \to \infty$?

1.9. Let Y be a random variable having a Poisson distribution with parameter λ. If λ is small (say less than 0.05), show that the probability that $Y \geq 1$ is approximately λ. *Hint:* Let A be the event $Y = 0$, so that the probability required is $P(\bar{A})$. Find $P(A)$ and then use the formula $P(\bar{A}) = 1 - P(A)$. Also use the approximation (B.21).

1.10. Suppose that the random variable Y has the Poisson distribution (1.22). Find the mean of the function $Y(Y-1)$. Then generalize your result to find the rth factorial moment of this distribution, that is,

$$E\big(Y(Y-1)\cdots(Y-r+1)\big).$$

Hint: Use an approach analogous to that leading to equation (1.36).

1.11. Use formula (1.32) and the results of Problem 1.10 for the case $r = 2$ to find the variance of a random variable having the Poisson distribution (1.22).

1.12. Suppose that calls arrive at a telephone exchange according to a Poisson process with parameter λ. If the lengths of the various calls are iid random variables, each with cumulative distribution function $F_X(x)$, find the probability distribution of the number of calls in progress at any time.

1.13. Suppose that a sequence of independent Bernoulli trials is conducted, with the same probability of success on each trial. The event that $r-1$ or fewer successes occur in the first n trials implies the event that the number of trials needed to achieve r successes is $n+1$ or more. Use this result to find a relation between the cumulative distribution functions of the binomial and the generalized geometric distributions.

1.14. Prove equation (1.58). *Hint:* Write the right-hand side as

$$\int_0^H 1 \cdot (1 - F_X(x)) \, dx;$$

then use integration by parts.

1.15. Is Chebyshev's inequality (1.60) useful when $d = \sigma$? When X has a normal distribution, mean μ, variance σ^2, compare the exact probability that $|X - \mu| \geq d$ with the values found in Chebyshev's inequality for the cases $d = 2\sigma$, $d = 3\sigma$.

1.16. Use equation (B.19) and further manipulation to show that the variance of a random variable having the geometric distribution (1.15) is $p/q + (p/q)^2 = p/q^2$, where $q = 1 - p$. Argue, without using mathematics, that the same variance applies for a random variable having the probability distribution (1.18).

1.17. Derive the expression (1.45) for the pgf of the binomial distribution. Also, show that the pgfs of the geometric, the shifted geometric and the Poisson distribution are, respectively, $(1-p)/(1-pt)$, $(1-p)t/(1-pt)$ and $e^{\lambda(t-1)}$.

1.18. Use definition (1.24), and equations (1.32) and (1.39) to prove equations (1.40) and (1.41).

1.19. Use the results of Problems 1.17 and 1.18 to rederive the results of Problems 1.2 and 1.3.

1.20. Use the pgfs found in Problem 1.17, together with (1.40) and (1.41), to find the variance of a random variable having (i) the Poisson distribution, (ii) the geometric distribution, and check that your results agree with those found respectively in Problem 1.11 and Problem 1.16.

1.21. Use equations (1.53) and (1.54) to prove that the mean of the exponential distribution (1.66) is $1/\lambda$ and that the variance is $1/\lambda^2$. (*Hint:* Use equations (B.42) and (B.43).) Confirm your calculation for the mean by using equations (1.67) and (1.58).

1.22. Show that the mean and variance of a uniform random variable are as given in (1.62).

1.23. Show that the probability that a random variable having the exponential distribution (1.66) is within two standard deviations of its mean is close to 0.95, (so that the *two standard deviation rule* of Section 1.10.2, applying for a normal random variable, also applies for an exponential random variable). Despite this fact, the normal distribution and the exponential distribution have quite different shapes. The probability that a normal random variable having mean μ and variance σ^2 is less than $\mu - \sigma$ is about 0.1587. What is the corresponding value for an exponential random variable?

1.24. Find the skewness of the exponential distribution (1.66), thus showing that it is a positive constant independent of λ.

1.25. Find the mean and variance given in (1.73) by using integration methods and (1.72).

1.26. Show that the mean and variance of a random variable having the chi-square distribution (1.77) with ν degrees of freedom are, respectively, ν and 2ν.

1.27. Find the mgfs of the binomial, the geometric, the Poisson, and the uniform distributions.

1.28. Use definitions (1.24) and (1.80) to prove (1.82), (1.83), and (1.89).

1.29. Suppose a discrete random variable Y takes the values -2 with probability 0.2, -1 with probability 0.3, and $+1$ with probability 0.5. Show that the mean of this random variable is negative. Confirm the result of Theorem 1.1, page 40, by showing that there is a unique positive solution θ^* for θ of the equation $m(\theta) = 1$, where $m(\theta)$ is the mgf of Y.

1.30. Suppose that X is a random variable having a normal distribution with mean μ, variance σ^2. Use the mgf of X to find, in one line, the mean of the function e^X.

1.31. A fair die is to be rolled once. Define A_1 to be the event that an even number turns up and A_2 the event that a 4, 5, or 6 turns up. Define the event $A_1 \cup A_2$ and find its probability using a direct argument. Now check that the answer that you found agrees with that found by applying equation (1.95).

1.32. Two fair coins are tossed and you are told that at least one coin came up heads. What is the probability that both coins came up heads? Note: The two "instinctive" answers $\frac{1}{2}$ and $\frac{1}{4}$ are both incorrect: it is necessary to use the conditional probability formula (1.101).

1.33. Suppose that the random variable Y has the geometric-like distribution (1.19). Assuming for the moment that the probability given in (1.19) is exact rather than asymptotic, calculate $\mathrm{Prob}(Y \geq y_1)$ for any $y_1 \geq 1$ and also calculate $\mathrm{Prob}(Y \geq y_1 + 1)$. Then use equation (1.101) to find the conditional probability that $Y \geq y_1 + 1$, given that $Y \geq y_1$. Compare your result with that following (1.112).

   Noting now that the probability in (1.19) is asymptotic rather than exact, what asymptotic statement can you make?

1.34. Suppose an unfair die has the property that the probability that the number i turns up is $\frac{i}{21}$. Are the events "an even number turns up" and "a number 4 or less turns up" independent? What is the corresponding result for a fair die?

1.35. The classic example of the use of equations (1.97)–(1.98) arises through the procedure of taking a random permutation of the integers $(1, 2, \ldots, n)$. In any such random permutation, any given integer i will match its original position with probability n^{-1}. More generally, if A_i is the event that the integer i does indeed match its original position, then

$$P(A_i) = n^{-1},$$
$$P(A_i A_j) = \big(n(n-1)\big)^{-1},$$
$$P(A_i A_j A_k) = \big(n(n-1)(n-2)\big)^{-1},$$

and so on. Use equation (1.98) to show that the probability that no number in the random permutation is in its original position is

$$1 - 1 + \frac{1}{2!} - \frac{1}{3!} + \cdots + (-1)^n \frac{1}{n!}. \tag{1.132}$$

Find the limit of this probability as $n \to +\infty$.

1.36. Suppose that the various triple-wise intersection probabilities associated with three events A_1, A_2 and A_3 are generated from the formulae

$$P(A_1 A_2 A_3) = pqr + d,$$
$$P(A_1 A_2 \bar{A}_3) = pq(1 - r) - d,$$
$$P(A_1 \bar{A}_2 A_3) = p(1 - q)r - d,$$
$$P(A_1 \bar{A}_2 \bar{A}_3) = p(1 - q)(1 - r) + d,$$
$$P(\bar{A}_1 A_2 A_3) = (1 - p)qr - d,$$
$$P(\bar{A}_1 A_2 \bar{A}_3) = (1 - p)q(1 - r) + d,$$
$$P(\bar{A}_1 \bar{A}_2 A_3) = (1 - p)(1 - q)r + d,$$
$$P(\bar{A}_1 \bar{A}_2 \bar{A}_3) = (1 - p)(1 - q)(1 - r) - d,$$

where $0 < p, q, r < 1$ and d is chosen small enough so that all probabilities are positive. Show that the requirement (1.108) holds for such a probability scheme, but that the only case for which the events A_1, A_2, and A_3 are independent is when $d = 0$.

1.37. Let $P_{Y_0}(y)$ and $P_{Y_1}(y)$ be two probability distributions defined on the same range R. It can be shown that $-\sum_{y \in R} P_{Y_0}(y) \log P_{Y_1}(y)$ is maximized (with respect to choice of $P_{Y_1}(y)$) by the choice $P_{Y_1}(y) = P_{Y_0}(y)$. Use this fact to show that the relative entropy (1.118) is non-negative. *Hint:* Write $\log(P_{Y_0}(y)/P_{Y_1}(y))$ as $\log P_{Y_0}(y) - \log P_{Y_1}(y)$.

1.38. Suppose that X_1 has the exponential distribution (1.66). Show that $X_2 = \log X_1$ has a variance that does not dependent on λ.

2
Probability Theory (ii): Many Random Variables

2.1 Introduction

In almost every application of statistical methods we consider the analysis of many observations. For example we might measure the level of proteins in a cell for many proteins simultaneously, or we might measure the level of one protein many times in many different cells, in order to carry out some statistical hypothesis testing or estimation procedure. In BLAST we compare a "query" sequence with a large number of database sequences, leading to a large number of match scores. In order to understand the theory for testing hypotheses and estimating parameters in bioinformatics applications, we must therefore consider the theory of many random variables.

Perhaps the most basic issue concerning the theory of many random variables is that of independence. The concept of independence was referred to informally in Sections 1.1 and 1.2.2, and independence of events was defined in Section 1.12.5. This concept will be made precise below, but informally, several random variables are independent if knowledge of the values of some of them does not affect the probability of the values of the others. For example, the numbers appearing on different rolls of a die are independent. Thus if we know that a die is fair, the fact that ten rolls have produced ten ones does not change the fact that the probability of obtaining a one on the next roll is 1/6. (On the other hand, if we do not know that the die is fair, and observe ten ones on ten rolls, we might reasonably infer that the die is biased. This however is a different matter from doubting independence of the outcomes of the various rolls.)

Similarly, measurements of the level of proteins in cells from two unrelated individuals are generally assumed to be independent. However, measurements of the level of several proteins simultaneously in the same cell may not be independent. For example, two proteins might always have approximately the same level in any given cell, so knowledge of one of the proteins in a cell changes the probability of the level of the unknown one in the same cell.

In the die example, the random variables Y_i, $i = 1, \ldots, 6$, giving the respective numbers of times that the number i appears in a fixed number of rolls of a six-sided die, comprise a dependent set of six random variables, since, for example, knowing the value of any five of them determines the sixth.

Consider n random variables $X_1, X_2, \ldots, X_n$ associated with an experiment. These can be either continuous or discrete. The *joint range R* is the set of all vectors $(x_1, x_2, \ldots, x_n)$ in n-dimensional space that the random vector $(X_1, X_2, \ldots, X_n)$ can take. It is important to note that if x_1 is in the range of X_1 and x_2 is in the range of X_2, it is not necessarily the case that (x_1, x_2) is in the range of (X_1, X_2). For example, suppose we observe two cells until they both die. Let X_1 be the time it takes until the first of the two cells dies, and let X_2 be the time it takes until both cells die. The range of both X_1 and X_2 is $(0, \infty)$, but it is impossible that $X_1 > X_2$.

We shall first discuss how probabilities are allocated for many discrete random variables. We continue to use the notation established in Chapter 1, with Y referring to a discrete random variable. We also write the random variables $Y_1, Y_2, \ldots, Y_n$ in vector form as $\boldsymbol{Y} = (Y_1, Y_2, \ldots, Y_n)$. Throughout this chapter we shall reserve the boldface notation for vectors of random variables or their possible values.

We associate with the vector $\boldsymbol{Y}$ of discrete random variables the *joint probability distribution* $P_{\boldsymbol{Y}}(\boldsymbol{y})$, defined by

$$P_{\boldsymbol{Y}}(\boldsymbol{y}) = \mathrm{Prob}(Y_1 = y_1, Y_2 = y_2, \ldots, Y_n = y_n), \qquad (2.1)$$

for any vector $\boldsymbol{y} = (y_1, y_2, \ldots, y_n)$ in the joint range of $Y_1, Y_2, \ldots, Y_n$.

In the continuous case, probabilities are allocated to regions of the joint range of the random variables $X_1, X_2, \ldots, X_n$, and not to individual points. This generalizes the corresponding procedure for a single continuous random variable. That is, we associate with the continuous random variables $X_1, X_2, \ldots, X_n$ with joint range R a *joint density function* $f_{\boldsymbol{X}}(x_1, x_2, \ldots, x_n)$, defined on R, such that

$$\mathrm{Prob}((X_1, X_2, \ldots, X_n) \text{ is in } Q)$$

$$= \int \cdots \int_Q f_{\boldsymbol{X}}(x_1, x_2, \ldots, x_n) dx_1 dx_2 \cdots dx_n, \qquad (2.2)$$

for any region $Q \subseteq R$ for which the integral in (2.2) exists (see Figure 2.1).

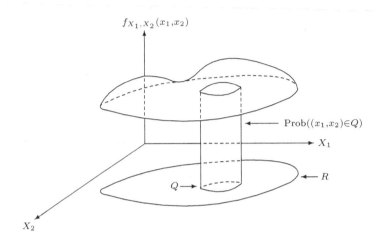

Figure 2.1. The probability that $(x_1, x_2) \in Q$ is the integral over Q of the joint density function f_{X_1, X_2} of X_1 and X_2.

2.2 The Independent Case

Consider n discrete random variables Y_1, $Y_2, \ldots$, Y_n. Each Y_i has its own probability distribution $P_{Y_i}(y_i) = \text{Prob}(Y_i = y_i)$, with range R_i. The random variables $Y_1, Y_2, \ldots, Y_n$ are said to be *independent* if their joint range R is the Euclidean product of the respective ranges R_i, and if

$$P_{\boldsymbol{Y}}(\boldsymbol{y}) = \prod_{i=1}^{n} P_{Y_i}(y_i) \tag{2.3}$$

for all $\boldsymbol{y} = (y_1, y_2, \ldots, y_n)$ in R. The equivalence of (2.3) with the conceptual interpretation of independence described on page 62 is made precise in Section 2.6 below.

The n continuous random variables $X_1, X_2, \ldots, X_n$ with individual respective density functions $f_{X_i}(x_i), (i = 1, 2, \ldots, n)$ and ranges R_i are independent if their joint range R equals the Euclidean product of the respective ranges R_i and

$$f_{\boldsymbol{X}}(x_1, x_2, \ldots, x_n) = \prod_{i=1}^{n} f_{X_i}(x_i) \tag{2.4}$$

for all $(x_1, \ldots, x_n)$ in the joint range of $(X_1, \ldots, X_n)$. The conceptual meaning of independence of continuous random variables is the same as that for discrete random variables.

It often occurs that the independent random variables $X_1, X_2, \ldots, X_n$, discrete or continuous, all have the same probability distribution. If this is the case, we say that they are *iid* (independently and identically distributed). As an example, in the case of rolling a die n times, we normally assume that the numbers appearing on rolls 1, 2, $\ldots$, n are iid.

The condition for independence of several random variables might appear to be less stringent than those given in (1.108)–(1.110) for the independence of events $A_1, A_2, \ldots, A_n$, but in fact the conditions for independence of events and that for independence of random variables are identical. This is so for discrete random variables since, if A_i is the event that $Y_i = y_i$, the independence requirement (2.3) implies that all the conditions (1.108)–(1.110) are satisfied. Verification of this is left as an exercise (Problem 2.1). A parallel remark holds for continuous random variables, where the event A_i is of the form $a_i \leq X_i \leq b_i$.

Example 1. If $Y_1, Y_2, \ldots, Y_n$ are independent random variables each having the Poisson distribution (1.22), with parameters $\lambda_1, \ldots, \lambda_n$, respectively, then (2.3) implies that

$$P_{\boldsymbol{Y}}(\boldsymbol{y}) = \frac{e^{-\sum \lambda_i} \prod \lambda_i^{y_i}}{\prod (y_i!)}, \tag{2.5}$$

the summation and both products being from 1 to n. If $Y_1, Y_2, \ldots, Y_n$ are iid, each having a Poisson distribution with parameter λ, then

$$P_{\boldsymbol{Y}}(\boldsymbol{y}) = \frac{e^{-n\lambda} \lambda^{\sum y_i}}{\prod (y_i!)}. \tag{2.6}$$

Example 2. If the probability of success on each trial in a sequence of n independent Bernoulli trials is p, and if $Y_i = 1$ when trial i results in success and $Y_i = 0$ when trial i results in failure, then $Y_1, Y_2, \ldots, Y_n$ are iid, and by (2.3)

$$P_{\boldsymbol{Y}}(\boldsymbol{y}) = p^{\sum y_i} (1-p)^{n-\sum y_i} = p^y (1-p)^{n-y}, \tag{2.7}$$

where $y = \sum y_i$ is the total number of successes. The probability (2.7) differs from the binomial probability (1.8), since (1.8) includes the combinatorial term $\binom{n}{y}$. The binomial distribution gives the probability of y successes and $n - y$ failures regardless of their order, while (2.7) gives the probability of y successes and $n - y$ failures in a nominated order. As shown in Appendix B.4, the binomial term $\binom{n}{y}$ is the number of orderings of y successes and $n - y$ failures, and this observation explains the difference between the two probabilities.

Example 3. If $X_1, X_2, \ldots, X_n$ are iid, each having a normal distribution with mean μ and variance σ_i^2 – in which case we say that $X_1, X_2, \ldots, X_n$ are NID(μ, σ^2) – then by (2.4) their joint density function is

$$f_{\boldsymbol{X}}(x_1, x_2, \ldots, x_n) = \prod_{i=1}^{n} \frac{1}{\sqrt{2\pi}\sigma} e^{-\frac{(x_i - \mu)^2}{2\sigma^2}} = \left(\frac{1}{\sqrt{2\pi}\sigma}\right)^n e^{-\frac{\sum(x_i - \mu)^2}{2\sigma^2}},$$

$$-\infty < x_i < +\infty. \quad (2.8)$$

Example 4. The result corresponding to Example 3 for the joint distribution of n iid random variables having the exponential distribution (1.66) is

$$f_{\boldsymbol{X}}(x_1, x_2, \ldots, x_n) = \lambda^n e^{-\lambda \sum x_i}, \quad 0 \le x_i < +\infty. \quad (2.9)$$

2.3   Generating Functions

Probability-generating functions and moment-generating functions are well defined for both independent and dependent random variables. However, we shall use them almost entirely for independent random variables, so we discuss them here in connection with the theory of independent random variables.

2.3.1   *Properties of Probability-Generating Functions*

Let $Y_1, Y_2, \ldots, Y_n$ be independent discrete random variables, having probability distributions with respective pgfs $\mathbb{p}_1(t), \mathbb{p}_2(t), \ldots, \mathbb{p}_n(t)$. Then the sum S_n, defined by

$$S_n = Y_1 + Y_2 + \cdots + Y_n, \quad (2.10)$$

is itself a random variable, and the pgf of its probability distribution is the product

$$\mathbb{p}_{S_n}(t) = \mathbb{p}_1(t) \cdot \mathbb{p}_2(t) \cdots \mathbb{p}_n(t). \quad (2.11)$$

The most direct proof of this statement uses the induction method described in Appendix B.18, and we outline it here. In the case $n = 2$, the probability that $S_2 = j$ is $\sum_i \mathrm{Prob}(Y_1 = i) \cdot \mathrm{Prob}(Y_2 = j - i)$. This sum is the coefficient of t^j in $\mathbb{p}_1(t) \cdot \mathbb{p}_2(t)$, so that (2.11) is true for the case $n = 2$. Suppose now that (2.11) is true for $n = m$ for some $m \ge 2$, so that

$$\mathbb{p}_{S_m}(t) = \mathbb{p}_1(t) \cdot \mathbb{p}_2(t) \cdots \mathbb{p}_m(t). \quad (2.12)$$

We write $Y_1 + \cdots + Y_{m+1}$ as $Y' + Y_{m+1}$, where $Y' = Y_1 + \cdots + Y_m$. By the induction hypothesis, the pgf of Y' is as given in (2.12). The case $n = 2$ then shows that the pgf of $Y_1 + \cdots + Y_{m+1}$, that is of $Y' + Y_{m+1}$, is $(p_1(t) \cdot p_2(t) \cdots p_m(t)) \cdot p_{m+1}(t)$. This is of the same form as the right-hand side of (2.12), but with m replaced by $m + 1$. By induction, (2.11) is therefore true for all n.

An important particular case of (2.11) occurs when the random variables Y_i have the same probability distribution. If the common pgf of the random variables Y_i is $p(t)$, then

$$p_{S_n}(t) = \text{pgf of } S_n = \big(p(t)\big)^n. \qquad (2.13)$$

Suppose we recognize the pgf found in (2.11) as the pgf of a known random variable Y. The *uniqueness* theorem for pgfs given in Section 1.7 shows that $Y_1 + Y_2 + \cdots + Y_n$ has the same distribution as Y. Use of this theorem is often the most straightforward way of finding the distribution of the sum of several random variables.

Example 1. If $Y_1, Y_2, \ldots, Y_n$ are independent Poisson random variables, with respective parameters $\lambda_1, \lambda_2, \ldots, \lambda_n$, then equation (2.11), together with one of the calculations in Problem 1.17, shows that the pgf of $\sum_i Y_i$ is $e^{\sum_i \lambda_i (t-1)}$. The uniqueness property shows immediately that $\sum_i Y_i$ has a Poisson distribution with parameter $\sum_i \lambda_i$.

This result can be understood in an intuitive way by an example. Properties of the spontaneous degradation of cellular molecules were discussed in Section 1.13. As will be shown in Chapter 4, if a cell holds the number of a certain type of spontaneously degrading molecule constant, by synthesizing them at the same rate they degrade, then the number of molecules to degrade in any given time interval is a random variable having a Poisson distribution. Now consider two non-overlapping time intervals. We would expect that the distribution of the total number of molecules to degrade in the two time intervals of length A and B is identical to the distribution of the number of molecules to degrade in a single time interval whose length is $A + B$. The above result shows that this is indeed the case.

Example 2. The pgf (1.45) of the binomial distribution can be derived from equations (1.42) and (2.13) as follows. The number of successes in n trials is a sum, namely the sum of the number of successes on trial 1 (either 0 or 1) plus the number of successes on trial 2, (again 0 or 1), plus the number of successes on trial 3, and so on. From (1.42), the pgf of the number of successes on any one trial is $1 - p + pt$. Thus from (2.13), the pgf of the total number of successes in n trials is $(1 - p + pt)^n$, and this is (1.45).

By expanding (1.45) in powers of t and considering the coefficient of t^y in this expansion, we arrive at (1.8) for the probability of y successes in n trials. This calculation gives us an alternative way of arriving at the binomial distribution formula (1.8).

2.3.2   Properties of Moment-Generating Functions

Moment-generating functions (mgfs) have properties parallel to those for probability-generating functions discussed in Section 2.3.1. These properties are identical for both discrete random variables and continuous random variables, and we develop them using continuous random variable notation.

If $X_1, X_2, \ldots, X_n$ are independent random variables, discrete or continuous, whose distributions have respective mgfs $\mathrm{m}_1(\theta), \ldots, \mathrm{m}_n(\theta)$, then their sum $S_n = X_1 + X_2 + \cdots + X_n$ is a random variable whose mgf is

$$\mathrm{m}_{S_n}(\theta) = \mathrm{m}_1(\theta) \cdot \mathrm{m}_2(\theta) \cdots \mathrm{m}_n(\theta). \tag{2.14}$$

In the discrete case the proof of this is similar to the proof of (2.12). In the continuous case the equation is still valid but the proof is less straightforward, involving manipulations of integrals. In particular, if $X_1, X_2, \ldots, X_n$ are iid random variables whose common probability distribution has mgf $\mathrm{m}(\theta)$, then

$$\mathrm{m}_{S_n}(\theta) = (\mathrm{m}(\theta))^n. \tag{2.15}$$

Equations (1.82), (1.83), and (2.15) are useful for finding properties of sums of iid random variables. For example, suppose that $X_1, X_2, \ldots, X_n$ are iid random variables, each having mean μ and variance σ^2. From (2.15), the derivative of the mgf of their sum S_n is

$$n \left(\mathrm{m}(\theta)\right)^{n-1} \frac{d\mathrm{m}(\theta)}{d\theta}. \tag{2.16}$$

Carrying out a further differentiation and then putting $\theta = 0$, we find from (1.82) and (1.83) that

$$\text{mean of } S_n = n\mu \tag{2.17}$$

and

$$\text{variance of } S_n = n\sigma^2. \tag{2.18}$$

When $X_1, X_2, \ldots, X_n$ are independent random variables with possibly different distributions, with respective means $\mu_1, \mu_2, \ldots, \mu_n$ and variances $\sigma_1^2, \sigma_2^2, \ldots, \sigma_n^2$, equations (1.82), (1.83), and (2.14) show that

$$\text{mean of } S_n = \mu_1 + \mu_2 + \cdots + \mu_n \tag{2.19}$$

and

$$\text{variance of } S_n = \sigma_1^2 + \sigma_2^2 + \cdots + \sigma_n^2. \tag{2.20}$$

More generally, if $S_n = a_1 X_1 + a_2 X_2 + \cdots + a_n X_n$, then

$$\text{mean of } S_n = a_1\mu_1 + a_2\mu_2 + \cdots + a_n\mu_n, \tag{2.21}$$

$$\text{variance of } S_n = a_1^2\sigma_1^2 + a_2^2\sigma_2^2 + \cdots + a_n^2\sigma_n^2. \tag{2.22}$$

As will be shown in Section 2.7, (2.21) also holds when $X_1, X_2, \ldots, X_n$ are dependent, but other than in exceptional cases, (2.22) does not.

As stated in Section 1.11, moment-generating functions for continuous random variables have a uniqueness property similar to that of discrete random variables. As in the discrete case, often the best way to find the distribution of some random variable is to find its mgf and recognize it as the mgf of a known distribution. This is often done in conjunction with equation (2.14) or (2.15).

Example 1. The mgf of the gamma distribution (1.75) is

$$\int_0^{+\infty} \frac{\lambda^k\, x^{k-1} e^{-\lambda x}\, e^{\theta x}}{\Gamma(k)}\, dx = \left(\frac{\lambda}{\lambda - \theta}\right)^k. \tag{2.23}$$

The fact that this is the kth power of the mgf of the exponential distribution implies that the sum of k independent random variables, each having the exponential distribution (1.66), is a random variable having the gamma distribution (1.75).

Example 2. A result closely related to that of Example 1, deriving from the mgf (1.88) of the chi-square distribution

$$(1 - 2\theta)^{-\nu/2},$$

is that the sum of independent chi-square random variables is itself a chi-square random variable with degrees of freedom equal to the sum of the degrees of freedom of the various chi-squares in the sum. The results of the examples in Section 1.11 then show that if $Z_1, Z_2, \ldots, Z_n$ are independent normal random variables, each having mean 0 and variance 1, (that is, they are NID(0,1)), then $Z_1^2 + Z_2^2 + \cdots + Z_n^2$ has a chi-square distribution with n degrees of freedom. An extension of this result is that if $X_1, X_2, \ldots, X_n$ are NID(μ, σ^2), and if $\bar{X} = (X_1 + \cdots + X_n)/n$ is their average, then it can be shown that $\sum (X_i - \bar{X})^2 / \sigma^2$ has a chi-square distribution with $n - 1$ degrees of freedom. Results such as this are central to the theory behind the ANOVA analyses discussed in Section 9.3.3.

Example 3. In Chapter 7 we shall need to consider "random walks" that consist of a sequence of random steps; each step goes up (a value of $+1$) with probability p and down (a value of -1) with probability $q = 1 - p$. Thus the value of a random step is similar to the number of successes in a Bernoulli trial, except that its possible values are -1 and $+1$ rather than 0 and 1. The mgf of a random step S is

$$\mathrm{m}(\theta) = qe^{-\theta} + pe^{\theta}. \tag{2.24}$$

We can check (2.15) by considering the position, or displacement, of this random walk after two steps have been taken. The two-step random variable is the sum of two one-step random variables. The total displacement is either -2 (with probability q^2), 0 (with probability $2pq$), or $+2$ (with probability p^2). Thus the mgf of this "two-step" displacement is

$$q^2 e^{-2\theta} + 2pq + p^2 e^{2\theta},$$

and this is indeed the square of the expression given in (2.24). This result can be extended to the case of three steps (Problem 2.15) and any higher number of steps.

2.4   The Dependent Case

The theory for dependent random variables is naturally more complex than that for independent random variables. While the joint distribution of independent random variables can be described simply by multiplying the respective distributions, as in (2.3) and (2.4), obtaining formulas for the joint distribution of dependent random variables is often difficult, even when the distributions of the individual random variables are known.

In this section we define first some quantities which measure certain kinds of dependence. We then give some important examples of dependent random variables whose joint distributions are known.

2.4.1   Covariance and Correlation

Two important concepts concerning the relationship between two random variables are their covariance and their correlation. We introduce these concepts first for continuous random variables. Let X_1 and X_2 be continuous random variables with means μ_1 and μ_2 and variances σ_1^2 and σ_2^2, respectively. The *covariance* σ_{X_1,X_2} of X_1 and X_2 is defined by

$$\sigma_{X_1,X_2} = \iint_R (x_1 - \mu_1)(x_2 - \mu_2) f_{\mathbf{X}}(x_1, x_2) dx_1 dx_2, \qquad (2.25)$$

the integration being taken over the joint range R of X_1 and X_2. The *correlation* ρ_{X_1,X_2} between X_1 and X_2 is then defined by

$$\rho_{X_1,X_2} = \frac{\sigma_{X_1,X_2}}{\sigma_1 \sigma_2}. \qquad (2.26)$$

The covariance between two discrete random variables Y_1 and Y_2 is defined as

$$\sigma_{Y_1,Y_2} = \sum_{y_1,y_2} (y_1 - \mu_1)(y_2 - \mu_2) P_{\mathbf{Y}}(y_1, y_2), \qquad (2.27)$$

where μ_i is the mean of Y_i $(i = 1, 2)$ and the summation is taken over all (y_1, y_2) in the joint range R of Y_1 and Y_2.

The correlation ρ_{Y_1,Y_2} between Y_1 and Y_2 is then defined by

$$\rho_{Y_1,Y_2} = \frac{\sigma_{Y_1,Y_2}}{\sigma_1 \sigma_2}. \tag{2.28}$$

The numerical value of any correlation is always in the interval $[-1,+1]$ (see Problem 2.13). The covariance and correlation between independent random variables is always zero.

On the other hand, the correlation ρ_{X_1,X_2} can be zero, even if X_1 and X_2 are dependent. For example, if X_1 has a uniform distribution in $[-1,+1]$ and $X_2 = X_1^2$, the correlation between X_1 and X_2 is zero. This arises because the relation between X_1 and X_2 is quadratic and has no linear component. For a further example, see Problem 2.5.

In both the discrete and the continuous case correlation is a measure of the degree of *linear* association between the random variables involved. In other words, if X_1 and X_2 are dependent in such a way that X_2 tends to be close to $aX_1 + b$ for some constants $a > 0$, b, then the random variables will have positive correlation, and if $a < 0$ they will tend to have negative correlation. In extreme case when the linear relationship is an actual equality $X_2 = aX_1 + b$, then the correlation between X_1 and X_2 is $+1$ if $a > 0$ and -1 if $a < 0$. An example is given in the next section.

In some contexts it is necessary to find the covariance of a random variable with itself. It is natural in the discrete case to make the definition

$$P_{Y,Y}(y,y) = P_Y(y), \tag{2.29}$$

since both sides of this equation give the probability that the random variable Y takes the value y. Use of this identity in (2.27) shows that the covariance of Y with itself is its variance. A similar conclusion holds for a continuous random variable.

2.4.2 The Multinomial Distribution

An important example of a joint discrete probability distribution where the individual random variables are dependent is the multinomial distribution. This is a direct generalization of the binomial distribution to the case where there are n possible outcomes on each of m independent trials, with $n \geq 3$. It is assumed that the probability of outcome i $(i = 1, 2, \ldots, m)$ is the same for all trials, and this probability is denoted by p_i.

The random variables of interest are the numbers $Y_1, Y_2, \ldots Y_n$, where Y_i as the number of times that outcome i occurs in the m trials $(i = 1, 2, \ldots, n)$. Each Y_i considered on its own has a binomial distribution with mean mp_i and variance $mp_i(1 - p_i)$. The probability that $Y_i = y_i$, $i = 1, 2, \ldots, n$, is given by the *multinomial distribution formula*

$$P_{\boldsymbol{Y}}(\boldsymbol{y}) = \frac{m!}{\prod_i(y_i!)} \prod_i p_i^{y_i}, \tag{2.30}$$

both products being over $i = 1, 2, \ldots, n$. This formula can be derived in a manner generalizing the derivation of the binomial distribution.

A simple example is provided by the outcomes of m rolls of a possibly unfair die, where the probability that the number i turns up on any roll of the die is p_i.. Here $n = 6$ and Y_i is the number of times that the number i turns up in the m rolls.

In the multinomial distribution the random variables $Y_1, Y_2, \ldots, Y_n$ are dependent, since necessarily $\sum_i Y_i = m$. The most extreme case of dependence occurs for the case $n = 2$, where the value of Y_2 is completely determined by Y_1 as $Y_2 = m - Y_1$. Here it is clear that (2.3) does not hold: for example, $\text{Prob}(Y_1 = 1) > 0$ and $\text{Prob}(Y_2 = m) > 0$, but $\text{Prob}(Y_1 = 1, Y_2 = m) = 0$.

Application of (2.27) shows that the covariance between Y_i and Y_j $(i \neq j)$ in the multinomial distribution (2.30) is

$$\sigma_{Y_i Y_j} = -m p_i p_j. \tag{2.31}$$

From this and (2.28), the correlation between Y_i and Y_j $(i \neq j)$ is

$$\rho_{Y_i, Y_j} = -\frac{\sqrt{p_i p_j}}{\sqrt{(1 - p_i)(1 - p_j)}}. \tag{2.32}$$

In the extreme case when $n = 2$, this correlation is $-\sqrt{p_1 p_2}/\sqrt{p_2 p_1} = -1$, the largest negative values that a correlation can take. This occurs because $y_1 = m - y_2$ is a linear function of y_2 with negative slope.

2.4.3   The Multivariate Normal Distribution

The variances and covariances of n random variables $X_1, X_2, \ldots, X_k$ can be gathered into a symmetric $k \times k$ *variance–covariance matrix* Σ, whose ith diagonal element is the variance σ_i^2 of X_i and whose (i, j) element $(i \neq j)$ is the covariance between X_i and X_j. The continuous random variables $X_1, X_2, \ldots, X_k$ have the *multivariate normal distribution* with mean vector $\boldsymbol{\mu} = (\mu_1, \ldots, \mu_k)$ and variance–covariance matrix Σ if their joint density function is

$$f_{\boldsymbol{X}}(\boldsymbol{x}) = \frac{1}{(2\pi)^{k/2}|\Sigma|^{1/2}} e^{-\frac{1}{2}(\boldsymbol{x} - \boldsymbol{\mu})' \Sigma^{-1} (\boldsymbol{x} - \boldsymbol{\mu})}, \quad -\infty < x_i < +\infty. \tag{2.33}$$

Here $|\Sigma|$ is the determinant of the matrix Σ and μ_i is the mean of X_i. Note that the exponent is the product of a row vector, a matrix, and a column vector, and is therefore a scalar.

The multivariate normal distribution arises often in practice for the same reason that the (univariate) normal distribution arises, since a random vector $\boldsymbol{X} = (X_1, \ldots, X_k)$ often has, either exactly or approximately, the multivariate normal distribution.

2.5 Marginal Distributions

Suppose Y_1 and Y_2 are two discrete random variables. Their joint distribution assigns probabilities to pairs (y_1, y_2) in the joint range of Y_1 and Y_2. The probability that $Y_1 = y_1$ can be recovered from this joint distribution from the formula

$$\text{Prob}(Y_1 = y_1) = \sum_{y_2} \text{Prob}(Y_1 = y_1, Y_2 = y_2),$$

where the sum is over all y_2 in the range of Y_2. This yields the probability that $Y_1 = y_1$ as the sum of the probabilities of all possible pairs (y_1, y_2). This procedure gives the *marginal* probability distribution of Y_1, which is the probability distribution of Y_1 ignoring Y_2. It suggests how marginal distributions can be calculated in general, and we now discuss this in more detail.

The marginal distribution of the subset $(Y_1, Y_2, \ldots, Y_i)$ of n discrete random variables $(Y_1, Y_2, \ldots, Y_n)$ is the distribution of $Y_1, Y_2, \ldots, Y_i$ ignoring $Y_{i+1}, Y_{i+2}, \ldots, Y_n$. If the joint distribution of $Y_1, Y_2, \ldots, Y_n$ is given, this marginal distribution is found in the discrete case as above by summation, and in the continuous case by integration. Thus in the discrete case

$$\text{Prob}(Y_1 = y_1, \ldots, Y_i = y_i) = \sum_{y_{i+1}, \ldots, y_n} \text{Prob}(Y_1 = y_1, \ldots, Y_n = y_n),$$

(2.34)

the summation being over all possible values $y_{i+1}, \ldots, y_n$ such that $y_1, \ldots, y_n$ is in the joint range of $Y_1, \ldots, Y_n$.

In the above example we have found the marginal distribution of $Y_1, Y_2, \ldots, Y_i$ for notational convenience only. A formula parallel to (2.34) applies for the marginal distribution of any subset of $Y_1, Y_2, \ldots, Y_n$.

Example 1. As an example of a marginal distribution, suppose that $Y_1, Y_2, \ldots, Y_n$ have the multinomial distribution (2.30). In a later application we shall want the marginal distribution of $Y_1, Y_2, \ldots, Y_i$, for $i < n$. Equation (2.34) can be used to give

$$\text{Prob}(Y_1 = y_1, \ldots, Y_i = y_i)$$
$$= \frac{m!}{y_1! \cdots y_i! m!} p_1^{y_1} \cdots p_i^{y_i} (1 - p_1 - \cdots - p_i)^k, \quad (2.35)$$

where $k = m - y_1 - \cdots - y_i$ (see Problem 2.10).

The analogue of (2.34) in the continuous case is the following. Define $\boldsymbol{X}_i$ by $\boldsymbol{X}_i = (x_1, \ldots, x_i)$. If $X_1, \ldots, X_n$ have joint density function $f_{\boldsymbol{X}_n}(x_1, \ldots, x_n)$, the marginal density function $f_{\boldsymbol{X}_i}(x_1, \ldots, x_i)$ of $X_1, \ldots, X_i$ is given by

$$f_{\mathbf{X}_i}(x_1,\ldots,x_i) = \int \cdots \int_R f_{\mathbf{X}}(x_1, x_2, \ldots, x_n) dx_{i+1} \cdots dx_n, \qquad (2.36)$$

the domain R of integration being over all possible values $x_{i+1}, x_{i+2}, \ldots, x_n$ such that $(x_1, \ldots, x_n)$ is in the joint range of $X_1, \ldots, X_n$. Care must be taken in finding the correct domain of integration, since this can depend on the values of $x_1, x_2, \ldots, x_i$.

For independent random variables, problems of the appropriate domain of integration do not arise: The range of any one random variable is in this case independent of the values of all other random variables.

In practice the procedure for independent random variables of going from the joint density function to the marginal density functions is often reversed: Given that the random variables are independent, then the joint probability distribution or density function is obtained, from (2.3) or (2.4), as the product of the individual density functions.

Example 2. Suppose that the joint density function of X_1 and X_2 is

$$f_{(X_1, X_2)}(x_1, x_2) = \frac{1}{2}, \qquad (2.37)$$

with the range of X_1 and X_2 being the square region bounded by the lines $X_2 = X_1$, $X_2 = -X_1$, $X_1 + X_2 = 2$ and $X_1 - X_2 = 2$ (see Figure 2.2).

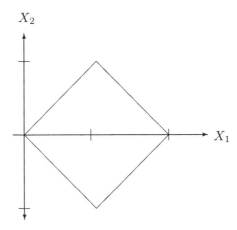

Figure 2.2. The region bounded by the lines $X_2 = X_1$, $X_2 = -X_1$, $X_1 + X_2 = 2$ and $X_1 - X_2 = 2$.

The marginal density function of X_1 is found by integrating out X_2 in the joint density function (2.37). When this is done, different terminals in the integration are needed for the two ranges $0 \leq X_1 \leq 1$ and $1 \leq X_1 \leq 2$.

Specifically, the density function $f_{X_1}(x)$ of X_1 is

$$f_{X_1}(x_1) = \begin{cases} \int_{-x_1}^{+x_1} \frac{1}{2}\, dx_2 = x_1, & 0 \le X_1 \le 1 \\ \int_{x_1-2}^{2-x_1} \frac{1}{2}\, dx_2 = 2 - x_1, & 1 \le X_1 \le 2. \end{cases}$$

Similarly,

$$f_{X_2}(x_2) = \begin{cases} 1 + x_2, & -1 \le x_2 \le 0 \\ 1 - x_2, & 0 \le x_2 \le 1. \end{cases}$$

Since the joint range is X_1 and X_2 is not a rectangle with sides parallel to the axes, X_1 and X_2 cannot be independent.

Example 3. Suppose that $X_1, X_2, \ldots, X_n$ have the multivariate normal distribution (2.33). Then any subset of $X_1, X_2, \ldots, X_n$ has a multivariate normal distribution, with means found from the appropriate elements in the mean vector $\boldsymbol{\mu}$ in (2.33) and variances and covariances taken from the appropriate elements in the variance–covariance matrix Σ. For example, the marginal distribution of $X_1, X_2, \ldots, X_j$, for any $j < n$, is multivariate normal with mean vector given by the first j elements in $\boldsymbol{\mu}$ and variance–covariance matrix given by the $j \times j$ upper left-hand sub-matrix of Σ.

2.6   Conditional Distributions

The conditional probability formula for events (1.101) has a direct analogue in conditional distributions of random variables, and the conditional probability formulae given in (2.38), (2.39), and (2.43) below all derive from (1.101).

The intuition behind the concept of conditional probabilities of events was discussed in Section 1.12.4. For random variables there are many kinds of conditional distributions that arise. In the simplest case, if Y_1 and Y_2 are two discrete random variables, then the conditional probability that $Y_2 = y_2$, given that $Y_1 = y_1$, denoted $\mathrm{Prob}(Y_2 = y_2 \,|\, Y_1 = y_1)$, is given by

$$\mathrm{Prob}(Y_2 = y_2 \,|\, Y_1 = y_1) = \frac{\mathrm{Prob}(Y_2 = y_2, Y_1 = y_1)}{\mathrm{Prob}(Y_1 = y_1)}, \qquad (2.38)$$

where it is assumed that $\mathrm{Prob}(Y_1 = y_1) > 0$.

More generally, for discrete random variables $Y_1, Y_2, \ldots, Y_n$, the *conditional probability distribution* of $Y_{i+1}, \ldots, Y_n$ given $Y_1 = y_1, \ldots, Y_i = y_i$ is

$$\mathrm{Prob}(Y_{i+1} = y_{i+1}, \ldots, Y_n = y_n \,|\, Y_1 = y_1, \ldots, Y_i = y_i)$$
$$= \frac{\mathrm{Prob}(Y_1 = y_1, \ldots, Y_n = y_n)}{\mathrm{Prob}(Y_1 = y_1, \ldots, Y_i = y_i)}, \qquad (2.39)$$

where the denominator of the right-hand side is again assumed to be positive. In the above example we have found the conditional probability of $Y_{i+1}, Y_{i+1}, \ldots, Y_n$ for notational convenience only, a formula parallel to (2.39) applies for any subset of $Y_1, Y_2, \ldots, Y_n$.

The formula corresponding to (2.39) in the continuous case is that the *conditional density function* of $\boldsymbol{X} = (X_{i+1}, \ldots, X_n)$, given $\boldsymbol{X}_i = (X_1, \ldots, X_i)$, is

$$f_{\boldsymbol{X}|\boldsymbol{X}_i} = \frac{f_{\boldsymbol{X}}(x_1, \ldots, x_n)}{f_{\boldsymbol{X}_i}(x_1, \ldots, x_i)}, \tag{2.40}$$

where the denominator is assumed to be positive.

When the random variables $Y_1, Y_2, \ldots, Y_n$ are independent, equation (2.3) shows that

$$\mathrm{Prob}(Y_{i+1} = y_{i+1}, \ldots, Y_n = y_n \mid Y_1 = y_1, \ldots, Y_i = y_i)$$
$$= \frac{\prod_{j=1}^n \mathrm{Prob}(Y_j = y_j)}{\prod_{j=1}^i \mathrm{Prob}(Y_j = y_j)} = \mathrm{Prob}(Y_{i+1} = y_{i+1}, \ldots, Y_n = y_n). \tag{2.41}$$

More generally, the conditional probability distribution of any subset of independent random variables, given the remaining random variables, is identical to the marginal distribution of that subset. An analogous statement applies for the continuous case.

Conversely, if (2.41) holds, we say $Y_{i+1}, Y_{i+2}, \ldots, Y_n$ is independent of Y_1, $Y_2, \ldots, Y_i$. In this case knowing the values of $Y_1, Y_2, \ldots, Y_i$ does not change the probabilities of the values of the random variables $Y_{i+1}, Y_{i+2}, \ldots, Y_n$. This is perhaps the most intuitive interpretation of the word "independent," though this is a weaker kind of independence than that captured in (2.3) and (2.4). Definition (2.3) can be cast in terms of (2.41) however. Indeed if (2.41) holds *for all* subsets of the $Y_1, Y_2, \ldots, Y_n$, given the remaining ones, then definition (2.3) is satisfied. A similar statement holds for the continuous case.

More interesting cases of equation (2.39) and (2.40) arise when the random variables are dependent.

Example 1. Suppose that the random variables $Y_1, Y_2, \ldots, Y_n$ have the multinomial distribution (2.30) and we wish to find the conditional probability of $(Y_{i+1}, Y_{i+2}, \ldots, Y_n)$, given the $Y_1 = y_1, Y_2 = y_2, \ldots, Y_i = y_i$. Application of the probability formula (2.30), the marginal probability formula (2.35), and the conditional probability formula (2.39) shows that

$$\mathrm{Prob}(Y_{i+1} = y_{i+1}, \ldots, Y_n = y_n \mid Y_1 = y_1, \ldots, Y_i = y_i)$$
$$= \frac{k!}{\prod_{j=i+1}^n (y_j!)} \prod_{j=i+1}^n q_j^{y_j}, \tag{2.42}$$

where $k = m - y_1 - y_2 - \cdots - y_i = y_{i+1} + y_{i+2} + \cdots + y_n$ and $q_j = p_j/(p_{i+1} + p_{i+2} + \cdots + p_n)$. This is itself a multinomial distribution, having suitably amended parameter values. $\qquad\square$

Another kind of conditional distribution involving random variables X_1, $\ldots$, X_n is the joint distribution of X_1, $\ldots$, X_n given the value of some function or functions of X_1, $\ldots$, X_n. The following is an example.

Example 2. A particularly important case arises for the conditional distribution of $Y_1, Y_2, \ldots, Y_n$ given their sum $S_n = \sum_j Y_j$. Even if the random variables $Y_1, Y_2, \ldots, Y_n$ are independent, the random variables $Y_1, Y_2, \ldots, Y_n$ and S_n are not, since S_n is determined by $Y_1, Y_2, \ldots, Y_n$. The required conditional distribution is found as follows. The event that $Y_1 = y_1, \ldots$, $Y_n = y_n$ and that their sum S_n is equal to s, where $s = y_1 + \cdots + y_n$, is the same as the event that $Y_1 = y_1, \ldots$, $Y_n = y_n$. Thus

$$\text{Prob}(Y_1 = y_1, \ldots, Y_n = y_n \mid S_n = y_1 + \cdots + y_n = s)$$
$$= \frac{\text{Prob}(Y_1 = y_1, \ldots, Y_n = y_n)}{\text{Prob}(S_n = s)}. \qquad (2.43)$$

As an example of the use of this formula, suppose that $Y_1, Y_2, \ldots, Y_n$ are independent Poisson random variables, having respective parameters $\lambda_1, \lambda_2, \ldots, \lambda_n$. Their sum has a Poisson distribution with parameter $\sum_i \lambda_i$ (Problem 2.2). In this case the right-hand side in equation (2.43) becomes

$$\frac{e^{-\sum \lambda_i} \prod \lambda_i^{y_i} / \prod(y_i!)}{e^{-\sum \lambda_i} (\sum \lambda_i)^{\sum y_i} / (\sum y_i)!}, \qquad (2.44)$$

all sums and products being over $i = 1, 2, \ldots, n$. This reduces to

$$\frac{(\sum y_i)!}{\prod(y_i!)} \prod_i \left(\frac{\lambda_i}{\sum_j \lambda_j} \right)^{y_i}, \qquad (2.45)$$

and with appropriate changes in notation, this becomes the multinomial probability distribution (2.30). In the particular case where $\lambda_1 = \lambda_2 = \cdots = \lambda_n = \lambda$, this conditional probability does not depend on λ.

The formula corresponding to (2.43) in the continuous case is that the conditional density function of $\boldsymbol{X} = X_1, \ldots, X_n$, given that their sum $S_n = X_1 + X_2 + \cdots + X_n$ takes the value s, is

$$f_{\boldsymbol{X} \mid S_n = s}(x_1, \ldots, x_n \mid s) = \frac{f_{\boldsymbol{X}}(x_1, \ldots, x_n)}{f_{S_n}(s)}, \qquad (2.46)$$

where the numerator on the right-hand side is the joint density function of $X_1, \ldots, X_n$ and the denominator is the density function of S_n.

As an example, if $X_1, \ldots, X_n$ are iid random variables having the exponential distribution (1.66), their joint density function (given in (2.9)) is

$$f_{\mathbf{X}}(x_1, x_2, \ldots, x_n) = \lambda^n e^{-\lambda \sum x_i}.$$

The sum S_n has a gamma distribution (see page 69), so its distribution is found from (1.75) by replacing k by n. That is, since n is an integer,

$$f_{S_n}(s) = \frac{\lambda^n s^{n-1} e^{-\lambda s}}{(n-1)!}.$$

The conditional density function of $X_1, \ldots, X_n$, given that $S_n = s$, is, from (2.46),

$$f_{\mathbf{X}|S_n=s}(x_1, \ldots, x_n \mid S_n = s) = \frac{(n-1)!}{s^{n-1}}, \qquad (2.47)$$

where the joint range of $X_1, \ldots, X_n$ given $S_n = s$ is the set of $(x_1, x_2, \ldots, x_n)$ such that $x_i > 0$, $\sum x_i = s$. This distribution is constant (as a function of $x_1, x_2, \ldots, x_n$) and does not depend on λ. □

Another important conditional density function arises when the range of a random variable is restricted in some way. Suppose, for example, that we wish to find the conditional density function of a continuous random variable X having density function $f_X(x)$, given that the value of X is in some interval (a, b) contained within the range (L, H) of X. Let $(x, x + h)$ be a small interval contained in (a, b). Then since the event $(x < X < x + h)$ and $(a < X < b)$ is identical to the event $(x < X < x + h)$, equation (1.101) gives

$$\mathrm{Prob}(x < X < x + h \mid a < X < b) = \frac{\mathrm{Prob}(x < X < x + h)}{\mathrm{Prob}(a < X < b)}.$$

Dividing both sides by h and letting $h \to 0$, we find (by (1.48)) that the conditional density function of X, given that $(a < X < b)$, is

$$f_{X|a<X<b}(x) = \frac{f_X(x)}{\int_a^b f_X(u)\, du}. \qquad (2.48)$$

A generalization of (2.48) to the case of many random variables $X_1, \ldots, X_n$ is that the conditional joint density function of $X_1, \ldots, X_n$, given that $(X_1, \ldots, X_n) \in Q$, for some region Q of the joint range of $X_1, \ldots, X_n$, is

$$f_{X|X \in Q}(x) = \frac{f_{X_1, \ldots, X_n}(x_1, \ldots, x_n)}{\int_Q f_{X_1, \ldots, X_n}(x_1, \ldots, x_n)\, dx_1 \cdots dx_n}. \qquad (2.49)$$

Example 3. The conditional density function of a continuous random variable X having the exponential distribution (1.66), given that $0 \le X < 1$, is

$$\frac{\lambda e^{-\lambda x}}{1 - e^{-\lambda}}, \quad 0 \le x < 1. \qquad (2.50)$$

This is the same as the density function of the "fractional part" D of X, given in (1.72). This is not a coincidence, and the result follows from, and can be found directly by using, the "memoryless" property of the exponential distribution (see Problem 2.11).

As a more general example, the conditional distribution of an exponential random variable X, given that $0 < X < L$, is

$$\frac{\lambda e^{-\lambda x}}{1 - e^{-\lambda L}}, \quad 0 \le x < L. \tag{2.51}$$

The mean of this distribution is

$$\int_0^L x \, \frac{\lambda e^{-\lambda x}}{1 - e^{-\lambda L}} \, dx = \frac{1}{\lambda} - \frac{L}{e^{\lambda L} - 1}. \tag{2.52}$$

We use this expression in Section 5.1 when considering shotgun sequencing.

Example 3. Suppose that $X_1, X_2, \ldots, X_n$ have the multivariate normal distribution (2.33). Then the conditional distribution of any subset of $X_1, X_2, \ldots, X_n$, for example $X_1, X_2, \ldots, X_i$, given the values $x_{i+1}, x_{i+2}, \ldots, x_n$ of $X_{i+1}, X_{i+2}, \ldots, X_n$, is a multivariate normal distribution. However the conditional means, variances and covariances of $X_1, X_2, \ldots, X_i$ in general depend on $x_{i+1}, x_{i+2}, \ldots, x_n$ as well as the elements in the mean vector $\boldsymbol{\mu}$ and the variance–covariance matrix Σ.

The case $n = 2$ provides a simple example. Given that $X_2 = x_2$, the mean of X_1 is

$$\mu_1 + \rho_{12}(x_2 - \mu_2)\frac{\sigma_1}{\sigma_2}$$

and the variance of X_1 is $\sigma_1^2(1 - \rho_{12}^2)$. These are respectively equal to the (marginal) mean and variance of X_1 if $\rho_{12} = 0$. Unless $x_2 = \mu_2$ the mean of the conditional distribution of X_1 differs from the mean of the marginal distribution of X_1 if $\rho_{12} \neq 0$, while the variance of the conditional distribution of X_1 always differs from the variance of the marginal distribution of X_1 if $\rho_{12} \neq 0$.

These results conform to common sense. For example, if X_1 and X_2 are positively correlated and if the observed value x_2 of X_2 exceeds its mean μ_2, the mean of X_1 exceeds its unconditional value μ_1. Whether the correlation between X_1 and X_2 is positive or negative, information about the value of X_2 provides some information about the likely values of X_1, thus decreasing the conditional variance of X_1 below its unconditional value.

2.7 Expected Values of Functions of Many Random Variables

Suppose that $\boldsymbol{X} = (X_1, X_2, \ldots, X_n)$ is a vector of continuous random variables, dependent or independent, with joint range R and joint density function $f_{\boldsymbol{X}}(x_1, x_2, \ldots, x_n)$. Let $g(X_1, X_2, \ldots, X_n)$ be some function of X_1, $X_2, \ldots, X_n$. Then $g(X_1, X_2, \ldots, X_n)$ is a random variable with range

$$Q = \{g(x_1, x_2, \ldots, x_n) \text{ such that } (x_1, x_2, \ldots, x_n) \in R\},$$

and density function $f_g(g)$ defined for g in Q. The expected (or mean) value of $g(X_1, X_2, \ldots, X_n)$ is defined as

$$E(g(X_1, X_2, \ldots, X_n)) = \int \cdots \int_Q g f_g(g) dg, \qquad (2.53)$$

and this equals

$$= \int \cdots \int_R g(x_1, x_2, \ldots, x_n) f_{\boldsymbol{X}}(x_1, x_2, \ldots, x_n) dx_1 dx_2 \cdots dx_n, \qquad (2.54)$$

by a similar argument to that leading to the analogous result (1.27) for functions of one discrete random variable. If $g(X_1, X_2, \ldots, X_n)$ is a function of $\boldsymbol{X}_i = (X_1, X_2, \ldots, X_i)$ only, written for convenience as $h(X_1, \ldots, X_i)$, its expected value can be found either from equation (2.54) or from the equation

$$E(h(X_1, X_2, \ldots, X_i))$$
$$= \int \cdots \int_{R'} h(x_1, x_2, \ldots, x_i) f_{\boldsymbol{X}_i}(x_1, x_2, \ldots, x_i) dx_1 dx_2 \cdots dx_i, \qquad (2.55)$$

where $f_{\boldsymbol{X}_i}(x_1, x_2, \ldots, x_i)$ is the marginal density function of $X_1, X_2, \ldots,$ X_i and R' is the joint range of $(X_1, X_2, \ldots, X_i)$. On some occasions one of these equations is easier to use, and on other occasions the other is easier.

The parallel result for discrete random variables is that if $\boldsymbol{Y} = (Y_1, Y_2, \ldots, Y_n)$ is a vector of discrete random variables, dependent or independent, with joint range R and joint probability distribution $P_{\boldsymbol{Y}}(y_1, y_2, \ldots, y_n)$, then

$$E(g(Y_1, Y_2, \ldots, Y_n)) = \sum_R g(y_1, y_2, \ldots, y_n) P_{\boldsymbol{Y}}(y_1, y_2, \ldots, y_n). \qquad (2.56)$$

If $g(Y_1, Y_2, \ldots, Y_n)$ is a function of $\boldsymbol{Y}_i = (Y_1, Y_2, \ldots, Y_i)$ only, say $h(Y_1, \ldots, Y_i)$, its expected value can be found either from equation (2.56) or from the equation

$$E(h(Y_1, Y_2, \ldots, Y_i)) = \sum_{R'} h(y_1, y_2, \ldots, y_i) P_{\boldsymbol{Y}_i}(y_1, y_2, \ldots, y_i), \qquad (2.57)$$

where $P_{\mathbf{Y}_i}(y_1, y_2, \ldots, y_i)$ is the marginal distribution of $Y_1, Y_2, \ldots, Y_i$ and R' is the joint range of $(Y_1, Y_2, \ldots, Y_i)$.

Independence of two or more random variables implies several simplifying properties. One useful property concerns expected values of products: If X_1 and X_2 are independent random variables, discrete or continuous, then

$$E(X_1 X_2) = E(X_1)E(X_2). \qquad (2.58)$$

The proof of this claim follows from (2.4) and (2.54) in the continuous case and (2.3) and (2.57) in the discrete case. The converse statement is not necessarily true: In some cases equation (2.58) holds for dependent random variables.

In Sections 10.3.3 and 12.2.3 we shall need the idea of a *conditional expectation*. For discrete random variables Y_1 and Y_2, the conditional expectation of Y_1 given Y_2 is the mean of the conditional distribution of Y_1 given Y_2. If the joint distribution of Y_1 and Y_2 is $P_{\mathbf{Y}}(y_1, y_2)$ and the marginal distribution of Y_2 is $P_{Y_2}(y_2)$, this mean is denoted by $E(Y_1 \mid Y_2 = y_2)$, and is calculated (see equation (2.38)) from

$$E(Y_1 \mid Y_2 = y_2) = \frac{\sum_{y_1} y_1 P_{\mathbf{Y}}(y_1, y_2)}{P_{Y_2}(y_2)}. \qquad (2.59)$$

The summation is taken over all possible values that Y_1 can take, given the value y_2 of Y_2; this might differ from one value of y_2 to another. This definition implies that the conditional expectation of Y_1 is in general a function of the conditioning value y_2.

If X_1 and X_2 are continuous random variables, and the conditional density function of X_1 given X_2 is $f_{X_1 \mid X_2}(x_1 \mid x_2)$, the conditional mean of X_1 given X_2 is

$$\int_R x_1 f_{X_1 \mid X_2}(x_1 \mid x_2) dx_1, \qquad (2.60)$$

the range of integration being the range of possible values of X_1 given the value x_2 of X_2.

Two calculation of the mean and variance of a linear combination of random variables is particularly important. Suppose $X_1, X_2, \ldots, X_n$ are random variables, discrete or continuous, with means $\mu_1, \mu_2, \ldots, \mu_n$, variances $\sigma_1^2, \sigma_2^2, \ldots, \sigma_n^2$, and covariances σ_{ij}. If $a_1, a_2, \ldots, a_n$ are given constants, then

$$E(a_1 X_1 + \cdots + a_n X_n) = \sum_{i=1}^{n} a_i \mu_i \qquad (2.61)$$

and

$$\mathrm{Var}(a_1 X_1 + \cdots + a_n X_n) = \sum_{i=1}^{n} a_i^2 \sigma_i^2 + \sum_{i \neq j} a_i a_j \sigma_{ij}. \qquad (2.62)$$

These results generalize equations (2.21) (2.22) since they allow for dependence between the X_i's. We outline the proof of (2.61) and (2.62) here. This outline is given for the continuous case; a parallel proof applies in the discrete case.

First, by equation (2.54), the mean of $a_1 X_1 + \cdots + a_n X_n$ is given by

$$E(a_1 X_1 + \cdots + a_n X_n)$$
$$= \int \cdots \int_R (a_1 x_1 + \cdots + a_n x_n) f_X(x_1, x_2, \ldots, x_n) dx_1 dx_2 \cdots dx_n, \quad (2.63)$$

which is equal to

$$\sum_{i=1}^n a_i \int_R x_i f_X(x_1, \ldots, x_n) dx_1 \cdots dx_n, \quad (2.64)$$

from standard integration formulas. From equation (2.54),

$$\int_R x_i f_X(x_1, \ldots, x_n) dx_1 \cdots dx_n = \mu_i, \quad (2.65)$$

so that

$$E(a_1 X_1 + \cdots + a_n X_n) = a_1 \mu_1 + \cdots + a_n \mu_n. \quad (2.66)$$

This mean is independent of the correlation between any two of the random variables.

The variance of $(a_1 X_1 + \cdots + a_n X_n)$ is given by

$$\mathrm{Var}(a_1 X_1 + \cdots + a_n X_n)$$
$$= \int \cdots \int_R (a_1(x_1 - \mu_1) + \cdots + a_n(x_n - \mu_n))^2 f_X(x_1, \ldots, x_n) dx_1 \cdots dx_n.$$
$$(2.67)$$

Equation (2.62) follows by expanding the squared term on the right-hand side and then using the definition of variance and covariance, found from (2.54).

As an important special case, if the random variables $X_1, X_2, \ldots, X_n$ are independent, then

$$\mathrm{Var}(a_1 X_1 + \cdots + a_n X_n) = \sum_{i=1}^n a_i^2 \sigma_i^2. \quad (2.68)$$

Essentially identical calculations, replacing integrations by summations, and with the same conclusions, apply for discrete random variables.

2.8 Asymptotic Distributions

Let $X_1, X_2, X_3, \ldots$ be an infinite sequence of random variables, discrete or continuous, with respective cumulative distribution functions $F_1(x), F_2(x), F_3(x), \ldots$. There are various concepts of convergence for such a sequence. We consider here the concept of *convergence in distribution*. Suppose that X is a random variable whose distribution function is $F_X(x)$. Then if

$$\lim_{n \to \infty} F_n(x) = F_X(x)$$

for all points of continuity of $F_X(x)$, then we say that the sequence $\{X_n\}$ converges in distribution to X. The random variables $X, X_1, X_2, X_3, \ldots$ do not need to have the same range for this definition to make sense, because the domain of a cumulative distribution function is always $(-\infty, \infty)$, even for discrete random variables. In fact the concept of convergence in distribution often arises when the random variables $X_1, X_2, X_3, \ldots$ are discrete and X is continuous.

All of the convergence results in this book, and in particular the central limit theorem described in Section 2.10.1, refer to convergence in distribution. If (as with the central limit theorem) the random variable X has a normal distribution, then we say that the sequence is *asymptotically normal*.

2.9 Indicator Random Variables

2.9.1 Definitions

There is a useful discrete random variable associated with an event A called the *indicator* function of that event, for which we shall normally use the special notation I, or, if needed, I_j for the event A_j. We use the notation I_A when it is necessary to indicate the dependence on A. This random variable is defined by

$$I_A = \begin{cases} 1 & \text{if the event } A \text{ occurs,} \\ 0 & \text{if the event } A \text{ does not occur.} \end{cases}$$

It follows from this definition that I_A is a random variable having the Bernoulli distribution (1.6). The mean of I_A is simply the probability $P(A)$ that the event A occurs.

Let $A_1, A_2, \ldots, A_n$ be events, $I_1, I_2, \ldots, I_n$ their respective indicator random variables, and p_i be the probability of the event A_i. Then

$$\sum_{i=1}^{n} I_i$$

is the total number of the events that occur. The mean of a sum of random variables is the sum of the means, whether or not the random variables are independent, as proven in Section 2.7. Therefore, the mean of the number of the events $A_1, A_2, \ldots, A_n$ that occur is

$$E\left(\sum_{i=1}^{n} I_i\right) = \sum_{i=1}^{n} E(I_i) = p_1 + p_2 + \cdots + p_n. \qquad (2.69)$$

If $p_1 = p_2 = \cdots = p_n = p$, then $\sum I_i$ can be thought of as a generalization of a binomial random variable to the case of possibly dependent trials. In this case

$$E\left(\sum_{i=1}^{n} I_i\right) = np, \qquad (2.70)$$

the same as the binomial distribution mean found in Problem 1.2. In the derivation of the binomial distribution we required the trials to be independent, but the result just found shows that the mean np applies whether or not this is the case.

When the events $A_1, \ldots, A_n$ are independent, the variance of the number of the events $A_1, \ldots, A_n$ that occur is, from (2.20),

$$p_1(1 - p_1) + \cdots + p_n(1 - p_n). \qquad (2.71)$$

If $p_1 = p_2 = \cdots = p_n = p$, then the random variable $\sum_{i=1}^{n} I_i$ has a binomial distribution. From (2.71), the variance is $np(1 - p)$, which agrees with the binomial distribution variance found in Problem 1.2. Even when $p_1 = p_2 = \cdots = p_n = p$, the variance in the dependent case need not be, and usually is not, $np(1 - p)$.

2.9.2 Example: Sequencing EST Libraries

In this example we apply (2.69) to answer a question regarding the generation of a database of ESTs (expressed sequence tags).

Genes are expressed by a two-step process. First they are transcribed from the DNA into molecules called messenger RNA (mRNA), and then the mRNAs are translated into proteins, which are sequences made up of the 20 amino acids. (For further details, see Appendix A.) Any gene is said to give rise to one or more[1] *species* of mRNA specific to that gene. An EST is a sequence on the order of a hundred or more contiguous bases of an mRNA. A typical cell has hundreds of thousands of mRNAs, representing thousands of different genes. Some genes are expressed at a high level in the cell, and thus there are many duplicate copies in the cell of the mRNA

---

[1] Some genes give rise to more than one mRNA, due to alternative splicing of introns and exons.

corresponding to such genes. Other genes are expressed at a low level and thus there are relatively few mRNAs from those genes in the cell. Any gene represented by L mRNAs will be said to be at abundance level L.

A database of ESTs is generated by repeated random sampling from the collection of mRNAs in the cell. Because of the way in which this is carried out, the sampling can be taken to be with replacement. The more abundant mRNA species will be selected more often, and the rarer a transcript is in the cell the less likely it is to be sampled at all. We would like have an idea of how the number of samples affects the proportion of the rarer transcripts that will be found.

To illustrate (2.69), consider the following simplified example. Suppose there are 10,000 different species of mRNA in the cell, falling into four abundance classes $L = 5, 50, 200$, and 1000. Suppose there are 4000 species at abundance level 5, 3250 species at level 50, 2500 species at level 200, and 250 at level 1000. This distribution of species into abundance levels is given in Table 2.1. Suppose a database of S ESTs is to be created, so

copies per cell (abundance level L)	number of different mRNA species	number of mRNAs per abundance level
5	4000	20,000
50	3250	162,500
200	2500	500,000
1000	250	250,000
total	10,000	N=932,500

Table 2.1.

that a random sample of S ESTs is taken from the pool of all mRNAs in the cell. Define $J_L(S)$ to be the number of different species of abundance level L mRNA in the database. We are interested in the expected value $E(J_L(S))$, since from this expected value we can assess how large S should be in order that the mean percentage of any specified abundance class of mRNAs in the database reaches some specified value. For example, one might be interested in determining how many ESTs must be generated in order to expect to have seen at least 50% of the different mRNA species that are expressed at the abundance level L.

The calculation of $E(J_L(S))$ using (1.24) is not easy, since obtaining the values $P_Y(y)$ in this case is difficult. However, the theory of indicator random variables leads to an immediate calculation for this mean. Let a be an mRNA species and define

$$I_a = \begin{cases} 1, & \text{if } a \text{ has been seen in the } S \text{ samples,} \\ 0, & \text{if } a \text{ has not been seen in the } S \text{ samples.} \end{cases}$$

Then the number of different species from abundance class L in the database of size S is $\sum_a I_a$, where the sum is taken over all mRNA species a in abundance class L. The I_a's are not independent, but they have the same mean for all a in abundance class L, and thus $\sum_a I_a$ is a generalization of a binomial random variable to the case of dependent trials, as discussed at the end of Section 2.9.1. Let p_L be the common mean of the I_a's in abundance class L. If there are n_L species of mRNA in abundance class L, then by (2.70), the mean is $n_L p_L$. The probability p_L is most easily calculated as $1 - r_L$, where r_L is the probability that the specified species is not in the database. Then $r_L = (1 - L/N)^S$, where N is the total number of mRNAs in the cell. Thus the mean number of species of abundance level L seen in the sample is

$$ n_L \left(1 - \left(1 - \frac{L}{N} \right)^S \right) . $$

The total number of mRNAs in the data in the above table is 932,500. Thus, for example, the mean number of species of abundance level 50 expected in a sample of 10,000 is

$$ 3250 \left(1 - \left(1 - \frac{50}{932{,}500} \right)^{10{,}000} \right) = 1348.75 . $$

This represents 41.5% of the different species in this abundance class. Table 2.2 gives percentage values for each abundance class for various values of S. It is interesting to note that even with $S = 50{,}000$, the mean percentage of species seen in the lowest abundance class is only 23.52%.

abundance level	Size of EST database					
	1,000	5,000	10,000	50,000	250,000	1,000,000
5	0.53	2.65	5.22	23.52	73.83	99.53
50	5.22	23.52	41.5	93.15	100	100
200	19.31	65.78	88.29	100	100	100
1000	65.8	99.53	100	100	100	100

Table 2.2. Expected percent of mRNAs in each abundance class to be seen in an EST database of size S, given the distribution of Table 2.1.

As this example shows, it is difficult to obtain many of the lower abundance mRNAs by sampling in this way. If the goal is simply to find as many different species of mRNAs as possible, then one can sequence EST libraries from many different cell types, since the low-abundance mRNAs in one cell type might be expressed in moderate or high abundance in another cell type. However, some genes are always expressed at a low level, regardless of cell type. And often the goal is to know which genes are expressed in the specific cell type in question. In such cases there are wet bench procedures (called subtraction and normalization) to remove or lower the levels of the higher abundance classes. However, with these methods, any measure of relative abundance is lost.

2.10 Derived Random Variables (i): Sums, Averages, and Minima

We have discussed the joint distribution of n independent discrete random variables and the joint density function of n independent continuous random variables. These random variables can be used to define various *derived* random variables, for example their sum, their average, their minimum, and their maximum. The joint distribution of the n random variables determines, implicitly, the distribution of any such derived random variable. Many properties of derived random variables can be found easily; for example, equations (2.19) and (2.20) show that the mean and the variance of the sum of n independent random variables can be written down from the means and variances of the individual random variables.

Many statistical operations are carried out using derived random variables. To carry out these operations it is often necessary to know the probability distribution of these random variables. In the discrete case, the calculations are usually straightforward in principle, although in practice they sometimes involve complicated summations involving combinatorial terms. The calculations in the continuous case involve transformation techniques and are less straightforward. Examples of transformation methods will be given in Section 2.13.

2.10.1 Sums and Averages

Suppose that $X_1, X_2, \ldots, X_n$ are iid random variables, discrete or continuous, each having a probability distribution with mean μ and variance σ^2. Perhaps the two most important random variables derived from X_1, $X_2, \ldots, X_n$ are their sum S_n, defined by

$$S_n = X_1 + X_2 + \cdots + X_n, \tag{2.72}$$

and their average $\bar{X}$, defined by

$$\bar{X} = \frac{X_1 + X_2 + \cdots + X_n}{n}. \tag{2.73}$$

It is important to emphasize that both of these quantities are *random variables*. In particular, the average $\bar{X}$ is a random variable, and must be distinguished carefully, as noted in Section 1.4, from a mean μ, which is a parameter.

The mean and variance of S_n are given in (2.17) and (2.18) as $n\mu$ and $n\sigma^2$, respectively. The corresponding formulae for the average are

$$\text{mean of } \bar{X} = \mu, \quad \text{variance of } \bar{X} = \frac{\sigma^2}{n}. \tag{2.74}$$

These formulae can be generalized in two ways. First, the same principle as is indicated in equations (2.74) applies to any well-behaved function of X. Thus if $g(X)$ is a function of X with finite mean $E(g(X)) = \mu_g$ and variance σ_g^2, then the mean and variance of the average $\bar{g}$ of $g(X_1)$, $g(X_2), \ldots, g(X_n)$ are

$$\text{mean of } \bar{g} = \mu_g, \quad \text{variance of } \bar{g} = \frac{\sigma_g^2}{n}. \tag{2.75}$$

The second generalization is the following. Suppose that $\boldsymbol{X}_1, \boldsymbol{X}_2, \ldots, \boldsymbol{X}_n$ are n iid vectors each of k random variables. The random variables within each vector can be discrete or continuous, independent or dependent. Write $\boldsymbol{X}_i = (X_{i1}, X_{i2}, \ldots, X_{ik})$ and suppose that $g(\boldsymbol{X}_i)$ is some function of X_{i1}, $X_{i2}, \ldots, X_{ik}$ with mean and variance μ_g and σ_g^2, respectively. We can define the average $\bar{g}$ from these vectors by

$$\bar{g} = \frac{g(\boldsymbol{X}_1) + g(\boldsymbol{X}_2) + \cdots + g(\boldsymbol{X}_n)}{n}. \tag{2.76}$$

Then the mean and variance of $\bar{g}$ are as given in (2.75).

It is important to emphasize that equations (2.74) and (2.75) are derived under the iid assumption. The statements about the mean remain the same in the case of dependent random variables. Generalizations of the variance statements to the dependent case can be found from (2.62).

Sums and averages have many important properties, of which we mention several important ones here.

The normal distribution case. When $X_1, X_2, \ldots, X_n$ are NID(μ, σ^2), both $\bar{X}$ and S_n also have a normal distribution, with respective means and variances given in (2.74), (2.17), and (2.18). The following remarkable result which is the simplest and most important version of the central limit theorem, shows that $\bar{X}$ and S_n are nearly normal for large n, regardless of the distribution of the iid random variables $X_1, X_2, \ldots, X_n$ as long as they have finite mean and variance.

The central limit theorem. Assume that $X_1, X_2, \ldots, X_n$ are iid, each with finite mean μ and finite variance σ^2. Then as $n \to \infty$, the standardized random variable $(S_n - n\mu)/(\sqrt{n}\sigma)$, which is identical to the standardized random variable $(\bar{X} - \mu)\sqrt{n}/\sigma$, converges in distribution (as defined in Section 2.8) to a random variable having the standard normal distribution $N(0,1)$.

This result holds regardless of the (common) distribution of $X_1, X_2, \ldots$. The breadth of application of the central limit theorem is one reason for the importance of the normal distribution, since many random variables of interest to us are either sums or averages. The proof of this theorem is beyond the scope of this book.

Particularly important cases of S_n and $\bar{X}$ arise when each X_i is a Bernoulli random variable with parameter p (see Section 1.3.1), and thus with mean p and variance $p(1-p)$. Here S_n is the total number of successes, and $\bar{X}$ is the *proportion* of successes, in n trials. Equation (2.74) shows that

$$\text{mean of } \bar{X} = p, \quad \text{variance of } \bar{X} = \frac{p(1-p)}{n}, \qquad (2.77)$$

and the central limit theorem shows that the distribution of $\bar{X}$ is approximately normal, with this mean and variance.

Accumulated support. In equations (1.121) and (1.122) we discussed the support that the observed value of one single random variable Y gave for one probability distribution over another. In practice we normally use the observed values of many iid random variables before assessing which of two distributions the random variable of interest comes from. Suppose then that $Y_1, Y_2, \ldots, Y_n$ are iid random variables, and denote their observed values by $y_1, y_2, \ldots, y_n$. Then the total support given by these observations for distribution P_1 over P_0 is defined as the sum of the supports given by each individual observation; specifically,

$$\text{accumulated support for } P_1 \text{ over } P_0 = \sum_{j=1}^{n} S_{1,0}(y_j). \qquad (2.78)$$

This accumulated support is the sum of the observed values of n iid random variables, and several of its properties follow from the properties of sums. In particular, the mean accumulated support given by n iid observations is n times the mean support given by one observation.

The reason for the additive definition of support will become clearer when we develop the concepts of support and accumulated support in the context of likelihood ratio tests (Chapter 3), substitution matrices (Chapter 6), sequential analysis (Chapter 9), and finally BLAST theory (Chapter 10).

Means, averages, and Chebyshev's inequality. Chebyshev's inequality is often not useful in finding information about the probability behavior of $\bar{X}$, the central limit theorem being more useful for this purpose. For example, if $X_1, \ldots, X_{100}$ each have a distribution with mean μ and variance 25, Chebyshev's inequality states that $\text{Prob}(|\bar{X} - \mu)| \geq 1) \leq 0.25$, while the central limit theorem gives the sharper statement that $\text{Prob}(|\bar{X} - \mu)| \geq 1) \cong 0.0456$.

Random n. It is assumed above that the number n of random variables making up the sum S_n is fixed. In some cases, however, this number is also a random variable, which we denote by N. We assume that N is independent of X_1, $X_2, \ldots$, X_n. In such cases the derivation of the distribution of the sum, which we denote by S, is best found through pgfs. For the case of independent discrete random variables the argument is as follows.

Suppose that $\text{Prob}(N = n) = P_n$, so that the pgf $\mathbb{p}(t)$ of N is $\mathbb{p}(t) = \sum_n P_n t^n$. We assume that the iid random variables X_1, $X_2, \ldots$ in the sum S each have the probability distribution of a random variable X, whose pgf is $\mathbb{q}(t)$. Then from (1.105),

$$\text{Prob}(S = s) = \sum_n P_n \cdot \text{Prob}(S = s \,|\, N = n). \qquad (2.79)$$

Equation (2.13) shows the probability that $S = s$ given that $N = n$ is the coefficient of t^s in $[\mathbb{q}(t)]^n$. Then from (2.79),

$$\text{Prob}(S = s) = \text{ coefficient of } t^s \text{ in } \sum_n P_n \cdot (\mathbb{q}(t))^n$$

$$= \text{ coefficient of } t^s \text{ in } \mathbb{p}(\mathbb{q}(t)). \qquad (2.80)$$

Thus the pgf of S is $\mathbb{p}(\mathbb{q}(t))$, and from this we can find in principle (and often easily in practice) the complete probability distribution of S.

The mean and variance of S can be found using (1.40) and (1.41), together with the chain rule of differentiation. The mean of S is

$$E(S) = E(N)E(X), \qquad (2.81)$$

as might be expected, and the variance is found by using (1.41), replacing $\mathbb{p}(t)$ by $\mathbb{p}(\mathbb{q}(t))$ (see Problem 2.14).

An example of a case for which (2.81) holds even when N is *not* independent of X_1, X_2, $\ldots$ is given in equation (7.23).

2.10.2 The Minimum of n Random Variables

Let X be a continuous random variable and suppose that $X_1, X_2, \ldots, X_n$ are iid random variables each with the same distribution as X. We denote their minimum by $X_{\min}$. The minimum of $X_1, X_2, \ldots, X_n$ does not have the original distribution of the individual X_i's, and its density function is found as follows. To say that $X_{\min}$ is greater than or equal to some number x is equivalent to saying that all of the values $X_1, X_2, \ldots, X_n$ are greater than or equal to x. Using the independence of $X_1, X_2, \ldots, X_n$ we get

$$\text{Prob}(X_{\min} \geq x) = (\text{Prob}(X \geq x))^n . \qquad (2.82)$$

If we denote the density function and the cumulative distribution function of X by $f_X(x)$ and $F_X(x)$, respectively, and the distribution function of $X_{\min}$ by $F_{\min}(x)$, then equation (2.82) can be written

$$1 - F_{\min}(x) = (1 - F_X(x))^n . \qquad (2.83)$$

The density function $f_{\min}(x)$ of $X_{\min}$ is found by differentiation (see equation (1.52)) to be

$$f_{\min}(x) = n \, f_X(x) \, (1 - F_X(x))^{n-1}. \tag{2.84}$$

Therefore, in the iid case the situation is straightforward. In Section 5.5 we analyze a case concerning dependent random variables where the calculations are more complicated. We now consider some examples of the minimum of iid random variables.

Example 1. As an example of an application of (2.82), suppose that $X_1, X_2, \ldots, X_n$ are independent random variables each having the exponential distribution (1.66). Then (1.113) and (2.82) show that

$$\mathrm{Prob}(X_{\min} \geq x) = e^{-n\lambda x}, \quad x \geq 0, \tag{2.85}$$

so that

$$F_{\min}(x) = 1 - e^{-n\lambda x}, \quad x \geq 0.$$

Differentiation of both sides of this equation shows that the density function of $X_{\min}$ is

$$f_{\min}(x) = n\lambda e^{-n\lambda x}, \quad x \geq 0. \tag{2.86}$$

Thus $X_{\min}$ itself has an exponential distribution, but with parameter $n\lambda$ rather than λ. The result of Problem 1.21 then shows that

$$\text{mean of } X_{\min} = \frac{1}{n\lambda}, \quad \text{variance of } X_{\min} = \frac{1}{(n\lambda)^2}. \tag{2.87}$$

The possibility of modeling the lifetime of spontaneously degrading proteins by an exponential distribution was discussed in Section 1.13, page 48. If there are n such molecules in a cell, equation (2.86) shows that the time until the first such molecule degrades also has an exponential distribution.

Example 2. *The uniform distribution.* As a second example of the application of (2.82), and with a change in notation to one more convenient in a later application, suppose that $X_1, X_2, \ldots, X_n$ are independent random variables each having the uniform distribution with range $[0, L]$. Then from (2.84) the density function of $X_{\min}$ is

$$f_{\min}(x) = \frac{n(L - x)^{n-1}}{L^n}, \quad 0 \leq x \leq L. \tag{2.88}$$

Example 3. *The normal distribution.* Since the cumulative distribution function of the normal distribution cannot be written in terms of elementary functions, a closed form for the density function of $X_{\min}$ is not available. However, the mean and standard deviation of $X_{\min}$ can be approximated accurately by numerical methods and are available in tables for selected values of n.

2.11   Derived Random Variables (ii): The Maximum of n Random Variables

2.11.1   Distributional Properties: Continuous Random Variables

Several important tests in bioinformatics are carried out by examining the observed value of the *maximum* of several random variables. For this reason we consider this maximum in far greater detail than the corresponding minimum.

Suppose that $X_1, X_2, \ldots, X_n$ are iid continuous random variables and that the maximum of these random variables is denoted by $X_{\max}$. The density function of $X_{\max}$ is, as with $X_{\min}$, different from that of the individual X_i's, and is found by an argument analogous to that used in finding the density function of $X_{\min}$.

To say that $X_{\max}$ is less than or equal to some number x is equivalent to saying that all of the values $X_1, X_2, \ldots, X_n$ are less than or equal to x. This implies that, if X is a random variable having the same distribution as each X_i, then, by independence,

$$\text{Prob}(X_{\max} \leq x) = \big(\text{Prob}(X \leq x)\big)^n. \qquad (2.89)$$

In terms of cumulative distribution functions, if X has density function $f_X(x)$ and cumulative distribution function $F_X(x)$, then

$$F_{\max}(x) = \big(F_X(x)\big)^n. \qquad (2.90)$$

Equivalently,

$$\text{Prob}(X_{\max} \geq x) = 1 - \big(F_X(x)\big)^n. \qquad (2.91)$$

The density function $f_{\max}(x)$ of $X_{\max}$ is, from (2.90),

$$f_{\max}(x) = n f_X(x) \big(F_X(x)\big)^{n-1}. \qquad (2.92)$$

Example 1. The uniform distribution. As an example of the application of (2.92), consider the maximum of n independent random variables, each having the uniform distribution with range $[0, L]$. The density function of $X_{\max}$ is

$$f_{\max}(x) = \frac{n\,x^{n-1}}{L^n}, \quad 0 \leq x \leq L. \qquad (2.93)$$

Elementary integration shows that the mean and variance of this distribution are, respectively,

$$\frac{nL}{(n+1)}, \quad \frac{nL^2}{(n+1)^2(n+2)}. \qquad (2.94)$$

Example 2. The exponential distribution. The density function of the maximum of n independent random variables, each having the exponential distribution (1.66), is, from (1.66), (1.67), and (2.92),

$$f_{max}(x) = n\lambda e^{-\lambda x}(1 - e^{-\lambda x})^{n-1}, \quad 0 \le x < +\infty. \tag{2.95}$$

The cumulative distribution function $F_{max}(x)$ of X_{max} is

$$F_{max}(x) = (1 - e^{-\lambda x})^n. \tag{2.96}$$

It is not easy, for the exponential distribution, to find the mean and variance of X_{max} using (1.53), (1.54), and (2.95).[1] However, a simple indirect way of finding this mean and variance, relying on the memoryless property of the exponential distribution and the mean and the variance of the *minimum* of n exponential random variables, is illustrated by the following example.

As in Section 1.13, we assume that the lifetimes of certain cellular proteins until degradation have an exponential distribution. Assuming this, and starting at time 0, we follow the fate of a cohort of n proteins. The mean time until at least one protein will degrade is the mean of the minimum of n exponential random variables, and from (2.87) this mean is $1/(n\lambda)$. By similar arguments the variance of this time is $1/(n\lambda)^2$.

The memoryless property of the exponential distribution implies that the mean value of the further time until the next protein degrades is independent of the time taken until the first one degraded. However, there are now only $n - 1$ proteins involved, so that the mean of the time until the next degradation is, by the same argument as that just used, $1/((n-1)\lambda)$, and the variance of this time is $1/\big((n-1)\lambda\big)^2$. Continuation of this argument shows that the time until the final protein degrades has mean

$$\frac{1}{\lambda} + \frac{1}{2\lambda} + \cdots + \frac{1}{n\lambda} \tag{2.97}$$

and, by independence, the variance is

$$\frac{1}{\lambda^2} + \frac{1}{(2\lambda)^2} + \cdots + \frac{1}{(n\lambda)^2}. \tag{2.98}$$

These are, respectively, the mean and variance of X_{max}.

It is curious that as $n \to \infty$ the expression for the mean of X_{max} diverges to $+\infty$, whereas the expression for the variance converges (see equations (B.7)–(B.9)). For large n, (B.7) and (B.9) show that

$$\text{mean of } X_{max} \approx (\gamma + \log n)/\lambda, \quad \text{variance of } X_{max} \approx \pi^2/6\lambda^2, \tag{2.99}$$

where γ is Euler's constant $0.577216\ldots$. (The "$\approx$" notation is defined in Appendix B.8.) Thus the mean of X_{max} grows (very slowly) with n at the

---

[1] At least compared to the straightforward calculation for X_{min}. This shows that the theory for maximum and minimum must be handled separately.

approximate rate $(\log n)/\lambda$. The variance becomes essentially constant, remaining within 1% of $\pi^2/6\lambda^2$ for all $n \geq 100$. The fact that the variance of $X_{\max}$ does not diverge to $+\infty$ as $n \to \infty$ is important, and we shall return to this point later.

2.11.2 Distributional Properties: Discrete Random Variables

We now consider properties of maxima of iid discrete integer-valued random variables. Our approach is based on equation (2.89), which applies also in the discrete case with the change of notation from X to Y. In other words, if $Y_{\max}$ is the maximum of n discrete iid random variables $Y_1, Y_2, \ldots, Y_n$ with common cumulative distribution function $F_Y(y)$, then

$$\text{Prob}(Y_{\max} \leq y) = \big(F_Y(y)\big)^n, \tag{2.100}$$

so that

$$\text{Prob}(Y_{\max} \geq y) = 1 - \big(F_Y(y-1)\big)^n \tag{2.101}$$

and

$$\text{Prob}(Y_{\max} = y) = \big(F_Y(y)\big)^n - \big(F_Y(y-1)\big)^n. \tag{2.102}$$

The right-hand side in this equation involves two functions, each raised to the power n. It follows from this that when n is large, sudden changes in the distribution function of $Y_{\max}$ can occur from one value of y to the next when $F_Y(y)$ is close to 1. For example, if $F_Y(20) = 0.99$ and $F_Y(19) = 0.98$ and $n = 100$, the right-hand side in (2.102) is about 0.36, so that the probability that $Y_{\max} = 20$ is 0.36. Similar substantial probabilities attach to values of $Y_{\max}$ close to 20 and very small probabilities attach to other values of $Y_{\max}$. The mean of $Y_{\max}$ would in this case be close to 20 and this implies that, when n is large, the probability distribution of $Y_{\max}$ is tightly concentrated around its mean.

From equation (2.102), the mean $\mu_{\max}$ and the variance and $\sigma_{\max}^2$ of $Y_{\max}$ are given, respectively, by

$$\mu_{\max} = \sum_y y\big((F_Y(y))^n - (F_Y(y-1))^n\big) \tag{2.103}$$

and

$$\sigma_{\max}^2 = \sum_y y^2\big((F_Y(y))^n - (F_Y(y-1))^n\big) - \mu_{\max}^2. \tag{2.104}$$

Since several statistical procedures in bioinformatics use the maximum of n discrete random variables, an evaluation of the probabilities (2.102), (2.103), and (2.104) is often required. Since n is often large and $F_Y(y-1)$ and $F_Y(y)$ are raised to the nth power, small errors in estimating $F_Y(y-1)$ and $F_Y(y)$ will compound into large errors in (2.102), (2.103), and (2.104), so very precise estimates of $F_Y(y-1)$ and $F_Y(y)$ are needed. We return to this issue in Section 3.7.

The Geometric Distribution

We now discuss properties of the maximum of n iid geometrically distributed random variables in some detail. We will use this theory in Section 6.3 when we study the significance of alignments. It also serves as an introduction to the theory of the maximum of iid geometric-like random variables to be used in BLAST theory. (Geometric-like random variables are defined on page 15.)

Suppose that $Y_1, Y_2, \ldots, Y_n$ are independent random variables each having the geometric distribution (1.15). Then equations (1.16) and (2.100) show that when y is an integer,

$$\text{Prob}(Y_{\max} \leq y) = (1 - p^{y+1})^n, \tag{2.105}$$

and from this it follows that

$$\text{Prob}(Y_{\max} \geq y) = 1 - (1 - p^y)^n \tag{2.106}$$

and

$$\text{Prob}(Y_{\max} = y) = (1 - p^{y+1})^n - (1 - p^y)^n. \tag{2.107}$$

Thus the mean $\mu_{\max}$ of $Y_{\max}$ is given by

$$\mu_{\max} = \sum_{y=0}^{\infty} y \left((1 - p^{y+1})^n - (1 - p^y)^n \right), \tag{2.108}$$

and the variance $\sigma^2_{\max}$ of $Y_{\max}$ is given by

$$\sigma^2_{\max} = \sum_{y=0}^{\infty} y^2 \left((1 - p^{y+1})^n - (1 - p^y)^n \right) - \mu^2_{\max}. \tag{2.109}$$

The relation between the geometric distribution and the exponential distribution often makes it convenient to write the geometric distribution in the reparametrized form (1.69), that is, with p replaced by $e^{-\lambda}$. In this notation,

$$\text{Prob}(Y_{\max} \leq y) = (1 - e^{-\lambda(y+1)})^n, \tag{2.110}$$

$$\text{Prob}(Y_{\max} \geq y) = 1 - (1 - e^{-\lambda y})^n, \tag{2.111}$$

$$\text{Prob}(Y_{\max} = y) = (1 - e^{-\lambda(y+1)})^n - (1 - e^{-\lambda y})^n, \tag{2.112}$$

$$\mu_{\max} = \sum_{y=0}^{\infty} y \left((1 - e^{-\lambda(y+1)})^n - (1 - e^{-\lambda y})^n \right), \tag{2.113}$$

$$\sigma^2_{\max} = \sum_{y=0}^{\infty} y^2 \left((1 - e^{-\lambda(y+1)})^n - (1 - e^{-\lambda y})^n \right) - \mu^2_{\max}. \tag{2.114}$$

We use the two sets of notation interchangeably, since for some purposes one notation is preferred and for other purposes the other is preferred.

The calculation of μ_{max} and σ^2_{max} as given in equations (2.108) and (2.109), and similar calculations for other discrete random variables, often requires significant computing effort for large values of n, and an approach using continuous distribution approximations to discrete distributions was typically used to find approximate properties of quantities like Y_{max}. Current computing power implies that direct computation, using only properties of the discrete random variable itself, is now often possible. Nevertheless, continuous approximations are still useful and are valuable for theoretical considerations, and we begin our discussion of them by considering an approximation of the distribution of the maximum of geometric random variables by using the maximum of exponential random variables.

This approximation is based on the fact (discussed on page 35) that the integer part of a random variable having the exponential distribution (1.66) has the geometric distribution (1.69). The corresponding result for maxima is that the integer part of the maximum of n exponential random variables has the distribution of the maximum of n random variables having the geometric distribution. This is seen as follows.

If X_{max} is the largest of n random variables having the exponential distribution, and $\lfloor X_{max} \rfloor$ is the integer part of X_{max}, then

$$\text{Prob}(\lfloor X_{max} \rfloor = y) = \text{Prob}(y \leq X_{max} < y + 1). \tag{2.115}$$

From equation (2.96) this is

$$(1 - e^{-\lambda(y+1)})^n - (1 - e^{-\lambda y})^n, \tag{2.116}$$

which is identical to the required expression (2.112).

When $n > 1$ the density function of D, the fractional part X_{max} - $\lfloor X_{max} \rfloor$ of X_{max}, is not identical to (1.72) and does not have mean and variance given by (1.73). Instead, even for comparatively small n, the density function of D can be shown to be very close to the uniform distribution (1.63), and this approximation becomes increasingly accurate as n increases. Thus to a close approximation D has a mean of $\frac{1}{2}$ and variance $\frac{1}{12}$ when n is large. Various properties of $Y_{max} = \lfloor X_{max} \rfloor$ can then be obtained by writing $Y_{max} = X_{max} - D$. For example, equations (2.66) and (2.97) imply that to a close approximation,

$$E(Y_{max}) = E(X_{max}) - E(D) \approx \frac{1}{\lambda} + \frac{1}{2\lambda} + \cdots + \frac{1}{n\lambda} - \frac{1}{2}. \tag{2.117}$$

Thus from equation (B.7),

$$E(Y_{max}) \approx \frac{\gamma + \log n}{\lambda} - \frac{1}{2}. \tag{2.118}$$

Further, it can be shown that the covariance between D and X_{max} approaches zero as $n \to +\infty$. Thus equations (2.68) and (2.98), together with the variance given immediately below (1.63), imply that to a close approximation,

$$\text{Var}(Y_{\max}) \approx \text{Var}(X_{\max}) + \text{Var}(D)$$

$$\approx \frac{1}{\lambda^2} + \frac{1}{(2\lambda)^2} + \cdots + \frac{1}{(n\lambda)^2} + \frac{1}{12}, \qquad (2.119)$$

so that for large n, from equation (B.9),

$$\text{Var}(Y_{\max}) \approx \frac{\pi^2}{6\lambda^2} + \frac{1}{12}. \qquad (2.120)$$

This expression shows that the variance of $Y_{\max}$ shares the property of the variance of the maximum $X_{\max}$ of n exponential random variables in that it does not diverge to $+\infty$ as $n \to \infty$.

The accuracy of these approximations can be assessed from the values given in Table 2.3, which displays the respective approximations (2.117) and (2.118) for the mean and the respective approximations (2.119) and (2.120) for the variance of $Y_{\max}$.

These values show the accuracy of these various approximations increasing as n increases. In Section 6.3 we use the approximation (2.118) for the mean and (2.120) for the variance when $n = 75{,}000$, and for this value of n these approximations are extremely accurate.

2.11.3 An Asymptotic Formula for the Distribution of $X_{\max}$

Let $X_1, X_2, \dots, X_n$ be iid random variables, each having the exponential distribution (1.66) and define $X_{\max}$ as the maximum of $X_1, X_2, \dots, X_n$. The approximate mean and variance of $X_{\max}$ are given in (2.99).

Our first aim is to define a "centered" random variable corresponding to $X_{\max}$. The leading term in the mean of $X_{\max}$ is $(\log n)/\lambda$ and the standard deviation of $X_{\max}$ is proportional to $1/\lambda$. The centered random variable U that we then construct is of the form

$$U = \frac{X_{\max} - \frac{\log n}{\lambda}}{\frac{1}{\lambda}} = \lambda X_{\max} - \log n. \qquad (2.121)$$

The random variable U is *not* standardized in the sense of the standardized quantity Z described in Section 1.10.2, and its mean is not 0 and its variance is not 1. In fact, (2.99), shows that the mean of U is approached γ (Euler's constant), and the variance of U approaches $\pi^2/6$, as $n \to \infty$.

From equation (2.96),

$$\text{Prob}(U \le u) = \text{Prob}(\lambda X_{\max} - \ln n \le u) \qquad (2.122)$$

$$= \text{Prob}(X_{\max} \le (u + \ln n)/\lambda) \qquad (2.123)$$

$$= (1 - e^{-u - \ln n})^n \qquad (2.124)$$

$$= \left(1 - e^{-u}/n\right)^n, \qquad (2.125)$$

$n = 5$

	Mean			Variance		
	exact	approximations		exact	approximations	
p	(2.108)	(2.117)	(2.118)	(2.109)	(2.119)	(2.120)
$1/4$	1.140	1.147	1.077	0.867	0.845	0.939
$1/3$	1.576	1.578	1.490	1.307	1.296	1.446
e^{-1}	1.782	1.783	1.687	1.554	1.547	1.728
$1/2$	2.794	2.794	2.655	3.133	3.130	3.507

$n = 10$

	Mean			Variance		
	exact	approximations		exact	approximations	
p	(2.108)	(2.117)	(2.118)	(2.109)	(2.119)	(2.120)
$1/4$	1.616	1.612	1.577	0.881	0.890	0.939
$1/3$	2.167	2.166	2.121	1.363	1.367	1.446
e^{-1}	2.429	2.429	2.380	1.631	1.633	1.728
$1/2$	3.726	3.726	3.655	3.309	3.309	3.507

$n = 20$

	Mean			Variance		
	exact	approximations		exact	approximations	
p	(2.108)	(2.117)	(2.118)	(2.109)	(2.119)	(2.120)
$1/4$	2.093	2.095	2.077	0.919	0.914	0.939
$1/3$	2.775	2.775	2.752	1.408	1.406	1.446
e^{-1}	3.098	3.098	3.073	1.680	1.680	1.728
$1/2$	4.690	4.690	4.655	3.406	3.406	3.507

Table 2.3. Exact values and two approximations for the mean (2.117), (2.118) and variance (2.119), (2.120) of the maximum ($Y_{\max}$) of $n = 5$, 10, and 20 geometric random variables, for selected values of $p = e^{-\lambda}$. Relevant equation numbers are noted.

and as $n \to \infty$ we obtain the limiting result

$$\text{Prob}\,(U \le u) \;=\; e^{-e^{-u}} \tag{2.126}$$

or equivalently

$$\text{Prob}\,(U \ge u) \;=\; 1 - e^{-e^{-u}} \tag{2.127}$$

As $u \to -\infty$ the right-hand side in (2.126) approaches 0 and as $u \to +\infty$ the right-hand side in (2.126) approaches 1. Thus the right-hand side in (2.126) is a cumulative distribution function of a random variable with range $(-\infty, +\infty)$ and we consider now a random variable having this distribution function. Adopting the notation of the finite n case, we denote this random variable by U.

The density function of U, found by differentiating the right-hand side in (2.126), is

$$f_U(u) = e^{-u-e^{-u}}, \quad -\infty < u < +\infty, \quad (2.128)$$

A random variable with this density function has mean γ (Euler's constant) and variance $\pi^2/6$ (see Problem 2.17). These values agree with the result of the discussion below (2.121).

The density function (2.128) is an important one in probability theory and statistics: When $X_{\max}$ is the maximum of n iid continuous random variables having any distribution with finite moments of all orders, support of the form $(A, +\infty)$ for some finite value A, and when the asymptotic relations

$$\text{mean of } X_{\max} \sim a \log n, \quad \text{variance of } X_{\max} \sim b, \quad (2.129)$$

hold as $n \to \infty$, where a and b are finite constants, the asymptotic $(n \to \infty)$ distribution of a centered random variable U derived from $X_{\max}$, having mean γ and variance $\pi^2/6$, is the same as (2.128).

This implies that when any such random variable $X_{\max}$ and any number x are both close to $\mu_{\max}$,

$$\text{Prob}(X_{\max} \le x) \sim e^{-e^{-\left(\pi(x-\mu_{\max})/(\sigma_{\max}\sqrt{6})+\gamma\right)}}, \quad (2.130)$$

or equivalently

$$\text{Prob}(X_{\max} \ge x) \sim 1 - e^{-e^{-\left(\pi(x-\mu_{\max})/(\sigma_{\max}\sqrt{6})+\gamma\right)}}. \quad (2.131)$$

These calculations are used in Sections (3.7.2) and (5.5), and they also form the basis of the hypothesis testing procedure in BLAST, as we will see in Chapter 10.

2.12 Order Statistics

2.12.1 Definition

The random variables $X_{\min}$ and $X_{\max}$ are examples of *order statistics* of continuous random variables. In this section we discuss these order statistics in more detail.

Suppose that X is a continuous random variable and that $X_1, \ldots, X_n$ are iid random variables each with the same distribution as X, with density function $f_X(x)$ and cumulative distribution function $F_X(x)$. Let $X_{(1)}$ be the smallest of the X_i's, $X_{(2)}$ the second smallest, and so on up to $X_{(n)}$, the largest. We call these the *order statistics*. The order statistic $X_{(1)}$ is identical to $X_{\min}$, and $X_{(n)}$ is identical to $X_{\max}$. Because the probability that two independent continuous random variables both take the same value is zero, these order statistics are distinct.

The density function of $X_{(i)}$ is found as follows. We let h be small and ignore events whose probability is $o(h)$. Then the event that $u < X_{(i)} < u + h$, is the event that $i - 1$ of the random variables are less than u, that one of the random variables is between u and $u + h$, and that the remaining random variables exceed $u + h$. This is a multinomial event with n trials and $k = 3$ outcomes on each trial. If the approximation

$$\text{Prob}(y < X < y + h) \cong f_X(y)h$$

is used when h is small (see (1.49)), the expression (2.30) shows that the probability of the event $u < X_{(i)} < u + h$ is

$$\frac{n!}{(i-1)!(n-i)!}\left(F_X(u)\right)^{i-1} f_X(u)\, h\, \left(1 - F_X(u + h)\right)^{n-i}. \qquad (2.132)$$

Thus by (1.48)

$$f_{X_{(i)}}(x_{(i)}) = \frac{n!}{(i-1)!(n-i)!}\left(F_X(x_{(i)})\right)^{i-1} f_X(x_{(i)})\left(1 - F_X(x_{(i)})\right)^{n-i}. \qquad (2.133)$$

Here $x_{(i)}$ denotes an observed value of $X_{(i)}$. The expressions (2.84) and (2.92) are particular cases of this density function.

Suppose next that $i < j$ and that h_1 and h_2 are small. Then arguing as above, and ignoring terms of order $o(h_1)$ and $o(h_2)$, the probability of the joint event that $u < X_{(i)} < u + h_1$ and that $v < X_{(j)} < v + h_2$ is the probability of the event that $i - 1$ of the random variables are less than u, that one random variable is between u and $u + h_1$, that $j - i - 1$ of the random variables lie between $u + h_1$ and v, that one random variable is between v and $v + h_2$, and that $n - j$ random variables exceed $v + h_2$. From this argument the joint density function $f_{X_{(i)},X_{(j)}}(x_{(i)}, x_{(j)})$ of $X_{(i)}$ and $X_{(j)}$ is

$$\frac{n!}{(i-1)!(j-i-1)!(n-j)!}\left(F_X(x_{(i)})\right)^{i-1} f_X(x_{(i)})$$
$$\times \left(F_X(x_{(j)}) - F_X(x_{(i)})\right)^{j-i-1} f_X(x_{(j)})\left(1 - F_X(x_{(j)})\right)^{n-j}. \qquad (2.134)$$

An important particular case is that where $i = 1$, $j = n$. Here (2.134) reduces to the simpler form

$$f_{X_{(1)},X_{(n)}}(x_{(1)}, x_{(n)})$$
$$= n(n-1)f_X(x_{(1)})\left(F_X(x_{(n)}) - F_X(x_{(1)})\right)^{n-2} f_X(x_{(n)}). \qquad (2.135)$$

One can continue in this way, finding the joint density function of any number of order statistics. Eventually, the joint density function

$$f_{X_{(1)},X_{(2)},\ldots,X_{(n)}}(x_{(1)}, x_{(2)}, \ldots, x_{(n)})$$

of all n order statistics is found as

$$f_{X_{(1)}, X_{(2)}, \ldots, X_{(n)}}(x_{(1)}, x_{(2)}, \ldots, x_{(n)}) = n! \prod_{i=1}^{n} f_X(x_{(i)}). \qquad (2.136)$$

(see Problem 2.21.)

The meaning of this equation is seen more clearly when it is used to find the conditional distribution of the original observations $X_1, X_2, \ldots, X_n$, given the observed values $x_{(1)}, x_{(2)}, \ldots, x_{(n)}$ of the order statistics. Using the conditional probability formula (2.39),

$$\text{Prob}(X_1 = x_{(a_1)}, \ldots, X_n = x_{(a_n)} \mid X_{(1)} = x_{(1)}, \ldots, X_{(n)} = x_{(n)}),$$

where $a_1 \neq a_2 \cdots \neq a_n$ is some permutation of the numbers $1, 2, \ldots, n$ is, from (2.136), $1/n!$. This simply states that if the order statistics are given, all of the $n!$ allocations of the actual observations to the order statistics' values are equally likely, each having probability $1/n!$. This observation is the basis of *permutation tests*, discussed in more detail in Chapter 3.

Order statistics for discrete random variables have much more complicated distributions than those for continuous random variables because of the possibility that two or more of the random variables take the same value. Fortunately we do not need to consider them here.

2.12.2 Example: The Uniform Distribution

As an example of order statistics, suppose that $X_1, X_2, \ldots, X_n$ are iid, each having the uniform distribution with range $(0, L)$. Then (2.133) shows that the density function of $X_{(i)}$ is

$$f_{X_{(i)}}(x_{(i)}) = \frac{n!}{(i-1)!(n-i)!} x_{(i)}^{i-1} (L - x_{(i)})^{n-i} L^{-n}. \qquad (2.137)$$

In the case $L = 1$, this is a beta distribution with parameters $\alpha = i$, $\beta = n - i + 1$. From equation (1.79), the mean and variance of $X_{(i)}$ are

$$\text{mean of } X_{(i)} = \frac{i}{n+1}, \quad \text{variance of } X_{(i)} = \frac{i(n-i+1)}{(n+1)^2(n+2)}. \qquad (2.138)$$

For general L,

$$\text{mean of } X_{(i)} = \frac{Li}{n+1}, \quad \text{variance of } X_{(i)} = \frac{L^2 i(n-i+1)}{(n+1)^2(n+2)}. \qquad (2.139)$$

The particular case $i = 1$ is important in various statistical procedures. Equation (2.137) shows that

$$f_{X_{(1)}}(x_{(1)}) = n(1 - x_{(1)})^{n-1}. \qquad (2.140)$$

This follows also from equation (2.84). If some probability α is given, the value $K(n, \alpha)$ such that $\text{Prob}(X_{(1)} \leq K(n, \alpha)) = \alpha$ is found by integration as

$$K(n, \alpha) = 1 - \sqrt[n]{1 - \alpha}. \qquad (2.141)$$

This expression will also appear in Section 3.11, and we return to it in Section 13.3, where it forms the basis of an approach to hypothesis testing in microarray analysis.

It is also important to find the joint density function of two order statistics from the uniform distribution. Equation (2.134) shows that the joint density function of $X_{(i)}$ and $X_{(j)}$ is

$$f_{X_{(i)}, X_{(j)}}(x_{(i)}, x_{(j)})$$
$$= \frac{n!}{(i-1)!\,(j-i-1)!(n-j)!} x_{(i)}^{i-1} (x_{(j)} - x_{(i)})^{j-i-1} (L - x_{(j)})^{n-j} L^{-n}. \qquad (2.142)$$

In the particular case $i = 1$, $j = n$, this is

$$f_{X_{(1)}, X_{(n)}}(x_{(1)}, x_{(n)}) = n(n-1)(x_{(n)} - x_{(1)})^{n-2} L^{-n}. \qquad (2.143)$$

Finally, equation (2.136) shows that the joint density function of all n order statistics is given by

$$f_{X_{(1)}, X_{(2)}, \ldots, X_{(n)}}(x_{(1)}, x_{(2)}, \ldots, x_{(n)}) = n! L^{-n}. \qquad (2.144)$$

This density function, which is a particular case of (2.136), is constant over the joint range of $X_{(1)}, X_{(2)}, \ldots, X_{(n)}$.

2.12.3   The Sample Median

The sample median $\hat{M}$ is an important random variable, and in this section we define it and briefly consider some of its properties.

Consider n iid continuous random variables $X_1, X_2, \ldots, X_n$, or equivalently their order statistics $X_{(1)}, X_{(2)}, \ldots, X_{(n)}$. When n is odd, so that we can write $n = 2m + 1$, the sample median $\hat{M}$ is defined as $X_{(m+1)}$: Half of the observed sample values are less than $\hat{m} = x_{(m+1)}$, and half exceed this value. When n is even, so that we can write $n = 2m$, the convention is to define the sample median $\hat{M}$ as $(X_{(m)} + X_{(m+1)})/2$, so that the observed value $\hat{m}$ of the sample median is the average $(x_{(m)} + x_{(m+1)})/2$ of the two central observations.

The sample median $\hat{M}$ is a random variable, and when n is odd, its probability distribution is identical to that of $X_{(m+1)}$, found from equation (2.132) with $n = 2m + 1$, $i = m + 1$. From this distribution the mean, the variance, and other properties of the sample median can in principle be found. In practice this might involve difficult problems of integration. These problems can be even greater when n is even. An example of a direct way of finding the mean and variance of $X_{(m+1)}$ when the iid random variables $X_1, X_2, \ldots, X_n$ have an exponential distribution is given in Problem 2.23.

When the random variables are discrete it becomes possible that the observed values of several random variables are equal. Because of this the theory for the distribution of the sample median for discrete random variables is more complex than it is for continuous random variables, and therefore we do not consider the discrete case here.

2.13 Transformations

Suppose that $X_1, X_2, \ldots, X_n$ are continuous random variables and let $V_1 = V_1(X_1, X_2, \ldots, X_n)$, $V_2 = V_2(X_1, X_2, \ldots, X_n), \ldots, V_n = V_n(X_1, X_2, \ldots, X_n)$ be functions of $X_1, X_2, \ldots, X_n$. These functions define a mapping from $(x_1, x_2, \ldots, x_n)$ to $(V_1(x_1, x_2, \ldots, x_n), V_2(x_1, x_2, \ldots, x_n), \ldots, V_n(x_1, x_2, \ldots, x_n))$, from a subset of n-dimensional space to some other subset of n-dimensional space. Suppose the mapping is one-to-one and differentiable with differentiable inverse. Then the two Jacobians defined below exist and are always non-zero.

We are interested in finding the joint density function of $V_1, V_2, \ldots, V_n$. If the joint density function of $X_1, X_2, \ldots, X_n$ is $f_{\boldsymbol{X}}(x_1, x_2, \ldots, x_n)$, arguments extending those given in Section 1.15 to the n-dimensional case show that the joint density function of $V_1, V_2, \ldots, V_n$ is given by

$$f_{\boldsymbol{V}}(v_1, v_2, \ldots, v_n) = f_{\boldsymbol{X}}(x_1, x_2, \ldots, x_n)|J^{-1}|, \tag{2.145}$$

where J is the Jacobian

$$J = \begin{vmatrix} \frac{\partial v_1}{\partial x_1} & \frac{\partial v_1}{\partial x_2} & \cdots & \frac{\partial v_1}{\partial x_n} \\ \frac{\partial v_2}{\partial x_1} & \frac{\partial v_2}{\partial x_2} & \cdots & \frac{\partial v_2}{\partial x_n} \\ \vdots & \vdots & \ddots & \vdots \\ \frac{\partial v_n}{\partial x_1} & \frac{\partial v_n}{\partial x_2} & \cdots & \frac{\partial v_n}{\partial x_n} \end{vmatrix}$$

and the right-hand side in (2.145) is computed as a function of $v_1, v_2, \ldots, v_n$. An equivalent formula is

$$f_{\boldsymbol{V}}(v_1, v_2, \ldots, v_n) = f_{\boldsymbol{X}}(x_1, x_2, \ldots, x_n)|J^*|, \tag{2.146}$$

where J^* is the Jacobian

$$J^* = \begin{vmatrix} \frac{\partial x_1}{\partial v_1} & \frac{\partial x_1}{\partial v_2} & \cdots & \frac{\partial x_1}{\partial v_n} \\ \frac{\partial x_2}{\partial v_1} & \frac{\partial x_2}{\partial v_2} & \cdots & \frac{\partial x_2}{\partial v_n} \\ \vdots & \vdots & \ddots & \vdots \\ \frac{\partial x_n}{\partial v_1} & \frac{\partial x_n}{\partial v_2} & \cdots & \frac{\partial x_n}{\partial v_n} \end{vmatrix}$$

with the right-hand side in (2.146) again being expressed as a function of $v_1, v_2, \ldots, v_n$. Sometimes (2.145) is the easier formula to use, sometimes (2.146).

The transformation from the joint density function of $X_1, \ldots, X_n$ to that of $V_1, \ldots, V_n$ is often used to find the density function of a *single* random variable. Suppose that we wish to find the density function of V_1 only. We first choose $n - 1$ "dummy" variables $V_2, \ldots, V_n$, of no direct interest to us. We then use transformation techniques to find the joint density function of $V_1, \ldots, V_n$. Having found this, we integrate out $V_2, \ldots, V_n$ to find the desired (marginal) density function of V_1, as described in equation (2.36). With a sufficiently careful choice of the dummy variables, this seemingly roundabout procedure is often the most efficient way of finding this density function. As noted in the discussion below equation (2.36), care must be taken in finding the correct domain of integration for $V_2, \ldots, V_n$, since this can depend on the value of V_1.

We give three examples of the use of transformations below. The first two underlie the much of the theory involved with ANOVA (see Section 9.3.3), while the third is relevant to BLAST.

Example 1. Suppose that X_1 has a chi-square distribution with ν_1 degrees of freedom, that X_2 has a chi-square distribution with ν_2 degrees of freedom and that X_1 and X_2 are independent. Our aim is to show that $X_1 + X_2$ has a chi-square distribution with $\nu_1 + \nu_2$ degrees of freedom. To do this we put $U = X_1 + X_2$ and $V = X_2$, find the joint density function of U and V by the transformation approach, then integrate out V to find the marginal density function of U. (Note: A far simpler approach than using transformation theory is to use moment-generating functions, as discussed in Examples 1 and 2 of Section 2.3.2. Here our aim is to illustrate the transformation technique.)

Since X_1 and X_2 are independent, their joint density function is, from (1.77),

$$CX_1^{\frac{1}{2}\nu_1 - 1} e^{-\frac{1}{2}X_1} X_2^{\frac{1}{2}\nu_2 - 1} e^{-\frac{1}{2}X_2}, \quad X_1, X_2 > 0, \qquad (2.147)$$

where

$$C = \frac{1}{2^{(\nu_1 + \nu_2)/2} \, \Gamma(\frac{1}{2}\nu_1) \Gamma(\frac{1}{2}\nu_2)}.$$

The absolute value of the Jacobian of the transformation from (X_1, X_2) to (U, V) is 1, so that the joint density function of U and V is given by (2.147) with X_1 replaced by $u - v$ and X_2 replaced by v, that is by

$$C(u - v)^{\frac{1}{2}\nu_1 - 1} v^{\frac{1}{2}\nu_2 - 1} e^{-\frac{1}{2}u}, \quad 0 < v < u. \qquad (2.148)$$

The domain of U and V ia as indicated in (2.148) because X_2 is positive, so that V cannot exceed U. The (marginal) density function of U is found by integrating the joint density function (2.148) with respect to v over the range $0 < v < u$, to obtain

$$Ce^{-\frac{1}{2}u} \int_0^u (u-v)^{\frac{1}{2}\nu_1-1} v^{\frac{1}{2}\nu_2-1} \, dv. \tag{2.149}$$

The change of variable $v = ut$ shows that this is equal to

$$Cu^{\frac{1}{2}(\nu_1+\nu_2)-1} e^{-\frac{1}{2}u} \int_0^1 t^{\frac{1}{2}\nu_2-1}(1-t)^{\frac{1}{2}\nu_1-1} \, dt, \tag{2.150}$$

and the form of the beta distribution (1.78) shows that the value of the integral in (2.150) is

$$\frac{\Gamma(\frac{1}{2}\nu_1)\Gamma(\frac{1}{2}\nu_2)}{\Gamma(\frac{1}{2}(\nu_1+\nu_2))}.$$

This leads to the density function

$$\frac{1}{2^{(\nu_1+\nu_2)}\Gamma(\frac{1}{2}(\nu_1+\nu_2))} u^{\frac{1}{2}(\nu_1+\nu_2)-1} e^{-\frac{1}{2}u} \tag{2.151}$$

for $U = X_1 + X_2$, showing (see (1.77)) that $X_1 + X_2$ has a chi-square distribution with $\nu_1 + \nu_2$ degrees of freedom.

This result can be generalized immediately to show that the sum of any number of independent chi-square random variables is itself a chi-square random variable, with degrees of freedom equal to the sum of the degrees of freedom of the constituent chi-square random variables.

Example 2. A second example involving the chi-square distribution is the following. Suppose as in Example 1 that X_1 has a chi-square distribution with ν_1 degrees of freedom, that X_2 has a chi-square distribution with ν_2 degrees of freedom and that X_1 and X_2 are independent. Define F by $F = \frac{X_1}{X_2} \times \frac{\nu_2}{\nu_1}$. By introducing the random variable $G = X_2$, and subsequently integrating with respect to G in the joint density function of F and G, it can be shown that the density function of F is

$$\left(\frac{\nu_2}{\nu_1}\right)^{\nu_2/2} \frac{\Gamma((\nu_1+\nu_2)/2)}{\Gamma(\nu_1/2)\Gamma(\nu_2/2)} \cdot \frac{F^{\nu_1/2-1}}{(\nu_2/\nu_1 + F)^{(\nu_1+\nu_2)/2}}. \tag{2.152}$$

This is the so-called F distribution with ν_1, ν_2 degrees of freedom. It is central to many statistical procedures, in particular the ANOVA procedures discussed in Section 9.5. Because of this, selected percentage points of this distribution, for combinations of values of ν_1 and ν_2 found in practice, are extensively tabulated. The derivation of (2.152) is left as an exercise in transformation theory (see Problem 2.28).

Example 3. An example of the use of a dummy variable in evaluating a density function by transformation methods, of direct relevance to BLAST, is given in Appendix D.

Problems

2.1 Let $\boldsymbol{Y} = (Y_1, Y_2, \ldots, Y_n)$ be a random vector such that

$$P_{\boldsymbol{Y}}(\boldsymbol{y}) = \prod_{i=1}^{n} P_{Y_i}(y_i) \qquad (2.153)$$

for all possible combinations of values of $\boldsymbol{y} = (y_1, y_2, \ldots, y_n)$. If A_i is the event that $Y_i = y_i$, then show (2.153) implies that all the conditions (1.108)–(1.110) are satisfied.

2.2. Suppose that $Y_1, Y_2, \ldots, Y_n$ are independent random variables, each having a Poisson distribution, the parameter of the distribution of Y_j being λ_j. Find the pgf of the distribution of their sum S_n, and thus show that the probability distribution of S_n is Poisson with parameter $\sum \lambda_j$.

2.3. Use (2.14) and (1.89) to prove (2.19) and (2.20), and use (2.15) and (1.89) to prove (2.17) and (2.18).

2.4. Suppose we are given a DNA sequence consisting of 10 consecutive nucleotides. Three segments of this sequence are to be chosen at random, one consisting of 3 consecutive nucleotides, a second consisting of 4 consecutive nucleotides, and the third consisting of 5 consecutive nucleotides. By "random" we mean the segment of 3 nucleotides can be in any of the 8 possible positions with equal probability, and similarly the segment of 4 nucleotides can be in any of the 7 possible positions with equal probability, and the segment of 5 in any of the 6 possible positions with equal probability. Let Y be the number of positions (out of 10) that are in all three segments. Then Y has observable values 0, 1, 2, 3. What is the expected value of Y, $E(Y)$? *Hint:* Use indicator random variables.

2.5. Show that although the random variables in Example 2 in Section 2.5 are dependent, the covariance between them is 0.

2.6. This problem refers to the example of throwing objects into a box as discussed in Section 1.3.3. The aim is to show that the probability $P(y|n, m)$ that y objects are thrown into the upper left-hand compartment, given that n objects in total are thrown into the top two compartments and m objects in total are thrown into the two left-hand compartments, is the hypergeometric probability (1.9). To do this, write $P(y|n, m)$ as

$$P(y|n, m) = \frac{P(y, n, m)}{P(n, m)}.$$

The numerator on the right-hand side is identical to the multinomial probability that there are y objects in the upper left-hand compartment,

$n - y$ in the upper right-hand compartment, $m - y$ in the lower left-hand compartment and $N - n - m + y$ in the lower right-hand compartment, namely

$$C(ab)^y \{a(1 - b)\}^{n-y} \{(1 - a)b\}^{m-y} \{(1 - a)(1 - b)\}^{N-n-m+y},$$

where

$$C = \frac{N!}{y!(n - y)!(m - y)!(N - n - m + y)!}.$$

Use this distribution, together with the fact that n and m are independent binomial random variables both with index N and with respective parameters a and b, to obtain the desired result.

2.7. Show that, if Y_1 is the number of successes in a binomial distribution with $n = 2$ trials, so that $Y_2 = 2 - Y_1$ is the number of failures, the correlation between Y_1^2 and Y_2 is not zero.

2.8. Use (2.48) to generalize the result given in (2.50) by showing that if X is a random variable having the exponential probability distribution (1.66), and if it is given that $a \leq X < a+1$, then $X - a$ has the distribution (2.50).

2.9. Use (2.15), together with the formula for the mgf of a normal random variable, mean μ, variance σ^2 (found in Problem 1.27) to find the mgf of the distribution of the sum of n independent random variables, each having this normal distribution. What conclusion do you draw about the distribution of this sum?

2.10. Prove equations (2.35) and (2.42).

2.11. Prove (2.50) directly using the memoryless property of the exponential distribution.

2.12. Use equation (2.51) to derive (2.52).

2.13. Suppose X_1 and X_2 are possibly dependent random variables, each with mean 0, variance 1 (so that $EX_i^2 = 1$ for $i = 1, 2$). Since $(X_1 - X_2)^2$ is never negative, $E\left((X_1 - X_2)^2\right) \geq 0$. By expanding the term $(X_1 - X_2)^2$, show that $E(X_1 X_2) \leq 1$. Apply a similar argument to $(X_1 + X_2)^2$ to show that $E(X_1 X_2) \geq -1$. It follows from these two facts that $|E(X_1 X_2)| \leq 1$.

Now suppose X_1 and X_2 are arbitrary random variables with respective means and standard deviations μ_i and σ_i, $i = 1, 2$. Apply the above conclusion to $X_i' = (X_i - \mu_i)/\sigma_i$ (for $i = 1, 2$) to show that $|\rho| \leq 1$, where ρ is the correlation between X_1 and X_2.

2.14. Use equation (2.80) to establish equation (2.81), and carry out one further differentiation to show that the variance of S is

$$E(N)\operatorname{Var}(X) + E(X)^2 \operatorname{Var}(N).$$

2.15. For the random walk taking a step up with probability p and a step down with probability $q = 1 - p$, find the possible values of the total displacement of the random walk after three steps, together with their probabilities. Thus find the mgf of this displacement, and check that it is the cube of the expression given in (2.24).

2.16. Prove equation (2.96) by appropriate integration.

2.17. Show that the mgf of the "extreme value" density (2.128) is $\Gamma(1 - \theta)$, and thus find the mean and variance of this distribution. *Hint:* In finding the mgf, make the change of variable $v = e^{-u}$, and be careful about using the appropriate terminals in the ensuing integration. To find the mean and variance, use equations (1.82) and (1.89) and the properties of the gamma function given in Appendix B.17.

2.18. Let $X_1, X_2, \ldots, X_n$ be iid random variables coming from a continuous probability distribution with median θ. Find the probability distribution of the number of these random variables that are less than θ.

2.19. Let $X_1, X_2, \ldots, X_n$ be independent random variables, each having the exponential distribution (1.66). Use (2.133) to find the density function of $X_{(2)}$. From this, find the mean and variance of $X_{(2)}$ and relate these to the calculations leading to equations (2.97) and (2.98).

2.20 The following problem is relevant to linkage analysis. Suppose that a parent of genetic type Mm has three children. Then the parent transmits the M gene to each child with probability $\frac{1}{2}$, and the genes that are transmitted to each of the three children are independent. Let $I_1 = 1$ if children 1 and 2 had the same gene transmitted (that is, both received M or both received m), and $I_1 = 0$ otherwise. Similarly, let $I_2 = 1$ if children 1 and 3 had the same gene transmitted, $I_2 = 0$ otherwise, and let $I_3 = 1$ if children 2 and 3 had the same gene transmitted, $I_3 = 0$ otherwise.

   (i) Show that while these three random variables are pairwise independent, they are not independent.

   (ii) Show that despite the conclusion of (i), the variance of $I_1 + I_2 + I_3$ is the sum of the variances of I_1, I_2, and I_3.

   (iii) Explain your result in terms of the various covariances between I_1, I_2, and I_3 and the pairwise independence of I_1, I_2, and I_3.

2.21. Prove the results given in (2.136).

2.22. Prove the results given in (2.139).

2.23. Let $X_1, X_2, \ldots, X_n$ be iid random variables, each having the exponential distribution (1.66). For $n = 2m + 1$ an odd integer, use the argument that led to equations (2.97) and (2.98) to show that the mean and variance of $X_{(m+1)}$ (the so-called sample median) are, respectively,

$$E(X_{(m+1)}) = \frac{1}{(2m + 1)\lambda} + \frac{1}{(2m)\lambda} + \cdots + \frac{1}{(m + 1)\lambda},$$

$$\mathrm{Var}(X_{(m+1)}) = \frac{1}{(2m + 1)^2\lambda^2} + \frac{1}{(2m)^2\lambda^2} + \cdots + \frac{1}{(m + 1)^2\lambda^2}.$$

2.24. *Continuation.* Use the approximation (B.7) to show that as $n \to +\infty$, the mean of $X_{(m+1)}$ approaches the exponential distribution median $\frac{\log 2}{\lambda}$ given in equation (1.68).

2.25. *Continuation.* Use the asymptotic results

$$1 + \frac{1}{2} + \cdots + \frac{1}{n} = \log n + \gamma + \frac{1}{2n} + o\left(\frac{1}{n}\right)$$

and

$$\log\left(\frac{2m + 1}{m}\right) = \log 2 + \frac{1}{2m} + o\left(\frac{1}{m}\right)$$

to obtain a more precise result than that in Problem 2.24, namely

$$\text{mean of } X_{(m+1)} = \frac{\log 2}{\lambda} + \frac{1}{(4m + 2)\lambda} + o\left(\frac{1}{m}\right).$$

2.26. *Continuation.* When $n = 2m$ is even, the sample median is defined as $(X_{(m)} + X_{(m+1)})/2$. Use the asymptotic results given in Problem 2.25 to show that when n is even,

$$\text{mean of sample median} = \frac{\log 2}{\lambda} + \frac{1}{2n\lambda} + o\left(\frac{1}{n}\right).$$

2.27. *Continuation.* Use the approximation

$$\frac{1}{a^2} + \frac{1}{(a + 1)^2} + \cdots + \frac{1}{b^2} \cong \frac{1}{a - \frac{1}{2}} - \frac{1}{b + \frac{1}{2}}$$

to show that the variance of the sample median $X_{(m+1)}$ is approximately $\frac{1}{n\lambda^2}$ when n is large and odd.

2.28. Carry out the calculations leading to the F distribution given in (2.152).

Hint: If X_i $(i = 1, 2)$ has a chi-squared distribution with ν_i degrees of freedom, and X_1 and X_2 are independent, then from (1.77) the joint density function of X_1 and X_2 is

$$f_{X_1, X_2}(x_1, x_2) = C x_1^{\frac{1}{2}\nu_1 - 1} x_2^{\frac{1}{2}\nu_2 - 1} e^{-\frac{1}{2}(x_1 + x_2)},$$

where

$$C = \left(2^{(\nu_1 + \nu_2)/2} \Gamma(\tfrac{1}{2}\nu_1) \Gamma(\tfrac{1}{2}\nu_2) \right)^{-1}.$$

Now use transformation methods to find the joint density function of $F = X_1\nu_2 / X_2\nu_1$ and $G = X_2$. The Jacobian J of the transformation is $\nu_2/(X_2\nu_1)$. The joint density function of F and G is then found to be

$$C \left(\frac{\nu_1}{\nu_2} \right)^{\nu_1/2} F^{\nu_1/2 - 1} G^{(\nu_1 + \nu_2)/2 - 1} \exp\left(-\frac{1}{2} G(1 + F \frac{\nu_1}{\nu_2}) \right), \quad F, G > 0.$$

Then integrate out G to find the density function of F.

3
Statistics (i): An Introduction to Statistical Inference

3.1   Introduction

Statistics is the method by which we analyze data in whose generation chance has played some part. In practice, it consists of two main areas, namely estimation and hypothesis testing; more specifically estimating parameters and testing hypotheses about parameters. We often associate a so-called *confidence interval* with an estimate of a parameter, as discussed in Section 3.3.1. A confidence interval is a range of values within which the true value of the parameter lies with some specified probability. For example, we may wish to estimate, on the basis of a comparatively small sample, the proportion of purines in the genome of some species, and to find a confidence interval for that proportion. We might also wish to test the hypothesis that the proportion of purines in two species is identical, again using data taken from two comparatively small samples.

Both estimation and hypothesis testing are used extensively in bioinformatics. In this chapter we give a brief introduction to estimation and hypothesis testing ideas: A more complete discussion of the underlying theory is given in Chapters 8 and 9.

A fundamental requirement in any statistical procedure is that the data to be analyzed derive from a random sample of the population of interest and to which the inferences eventually made relate. If this requirement is not satisfied, no statistical inferences made from the statistical analysis is justified. Here and throughout we assume that the random sampling requirement is satisfied.

3.2   Classical and Bayesian Methods

There are two main approaches to both estimation and hypothesis testing, each deriving from its own broad view of the way in which we should conduct statistical inference. These are the "classical," or "frequentist," approach on the one hand, and the "Bayesian" approach on the other. There is controversy among some statisticians as to the theoretical underpinnings of each of these approaches. Here we give only a brief sketch of their respective approaches and will not attempt to summarize the views of both camps in detail, especially since there is no unique Bayesian or classical position.

Two arguments that Bayesians often advance to support the Bayesian approach to statistical inference are as follows.

(i) The Bayesian approach asks the right question in a hypothesis testing procedure, namely, "What is the probability that this hypothesis is true, given the data?" rather than the classical approach, which asks a question like, "Assuming that this hypothesis is true, what is the probability of the observed data?"

(ii) Prior knowledge and reasonable prior concepts can be built into a Bayesian analysis. For example, suppose that a fair-looking coin is tossed three times and gives three heads. The classical approach estimate of the probability of a head is 1. A Bayesian might well claim that this estimate is unreasonable, and does not take into account the information that the coin seems reasonably symmetric.

Behind the Bayesian approach there is the broad feeling that a probability is a measure of belief in a proposition, rather than the frequentist interpretation that a probability of an event is in some sense the long-term frequency with which it occurs.

The main arguments for using a classical approach to statistical inference can perhaps best be stated by giving the classical approach counter-argument to the Bayesian positions outlined above.

An outline of these is as follows. Bayesian theory requires that prior assumptions must be made, for example the nature of the distribution of a parameter, but the form of this distribution is chosen for mathematical convenience rather than from any objective scientific basis. Further, the prior distribution that a Bayesian needs for his/her inferences about new phenomena cannot be known with certainty. These problems are not overcome by using so-called uninformative priors. Further, the very concept that a hypothesis can have a prior probability of being true is disputed by frequentists for some forms of hypotheses. The subjective nature of the approach, so that two different investigators might come to different con-

clusions from the same data because of their different prior beliefs, is also
of concern.

Bayesians have replies to these views, and the debate continues. We make
no comment on these positions here, noting that in some cases classical
methods can produce clearly unsatisfactory conclusions and in other cases
Bayesian can also (see Wasserman (2004), in particular pages 185–189).
Thus we favor the pragmatic approach adopted by many workers in bioinformatics. The focus in this book is, however, on classical hypothesis testing
methods, since these methods are currently more widely used in bioinformatics than are Bayesian methods. This is true in particular of BLAST and
in microarray analyses, two topics that we discuss at length. On the other
hand we describe in Section 6.6 a procedure that uses Bayesian concepts
and that might well not work if classical methods were employed.

3.3 Classical Estimation Methods

In much of the discussion in Chapters 1 and 2 the values of the various
parameters entering the probability distributions considered were taken as
being known. In practice these parameters are usually unknown, and must
be estimated from data. In this section we consider introductory aspects of
standard estimation procedures, and defer a more theoretical discussion to
Chapter 8. Much of the theory concerning estimation of parameters is the
same for both discrete and continuous random variables, so in this section
we use the notation X for both.

Let X be a random variable having a probability distribution $P_X(x; \theta)$
(for discrete random variables) or density function $f_X(x; \theta)$ (for continuous
random variables), depending (as the notation implies) on some unknown
parameter θ. How may we estimate θ from the observed value x of X?

The observed value x on its own will usually not be sufficient to provide
a good estimate. We must repeat the experiment that generated this value
an (ideally large) number of times to give n observations $x_1, x_2, \ldots, x_n$.
We think of these as the observed values of n iid random variables
$X_1, X_2, \ldots, X_n$, each X_i having a probability distribution $P_{X_i}(x; \theta)$ identical to $P_X(x; \theta)$ (for discrete random variables) or density function $f_{X_i}(x; \theta)$
identical to $f_X(x; \theta)$ (for continuous random variables). The iid assumption
is used throughout this section.

An *estimator* of the parameter θ is some function of the random
variables $X_1, X_2, \ldots, X_n$, and thus may be written $\hat{\theta}(X_1, X_2, \ldots, X_n)$, a
notation that emphasizes that this estimator is itself a random variable.
For convenience we generally use the shorthand notation $\hat{\theta}$. The quan-
tity $\hat{\theta}(x_1, x_2, \ldots, x_n)$, calculated from the observed values $x_1, x_2, \ldots, x_n$ of
$X_1, X_2, \ldots, X_n$, is called the *estimate* of θ.

Various desirable criteria have been proposed for an estimator to satisfy, and we now discuss some of these.

3.3.1 Unbiased Estimation

One desirable property of an estimator is that it be *unbiased*. An estimator $\hat{\theta}$ is said to be unbiased estimator of θ if its mean value $E(\hat{\theta})$ is equal to θ. In this section we consider unbiased estimation of the mean μ and the variance σ^2 of any probability distribution, estimation of the probability p of the Bernoulli distribution (1.6), and estimation of the parameters $\{p_i\}$ in the multinomial distribution (2.30). In doing so we use the generally accepted parameter notation for these examples, such as μ for a mean, rather than the generic notation θ.

Unbiasedness is not the only criterion for a "good" estimator. If an estimator $\hat{\theta}$ of θ is unbiased, we would also want the variance of $\hat{\theta}$ to be small, since if it is, the observed value of $\hat{\theta}$ calculated from from data should be close to θ. It would also be desirable if $\hat{\theta}$ has, either exactly or approximately, a normal distribution, since then well-known properties of this distribution can be used to provide properties of $\hat{\theta}$. Fortunately, several of the estimators we consider are unbiased, have a small variance, and have an approximately normal distribution.

It is natural to estimate the mean μ of a probability distribution by the average $\bar{X}$. Since the mean value of $\bar{X}$ is μ (from equation (2.74)), $\bar{X}$ is an unbiased estimator of μ. Since the variance of $\bar{X}$ decreases as the sample size n increases (see (2.74)), this variance is small when n is large, the observed average $\bar{x}$ is more and more likely to be close to μ as n increases. Finally, the central limit theorem of Section 2.10.1, page 88 shows that the distribution of $\bar{X}$ is approximately normal when n is large. If the random variables $X_1, X_2, \ldots, X_n$ have a normal distribution, the distribution of $\bar{X}$ is normal for any value of n.

The two standard deviation rule of Section 1.10.2, page 32, deriving from properties of the normal distribution, then shows that for large n,

$$\text{Prob}\left(\mu - \frac{2\sigma}{\sqrt{n}} < \bar{X} < \mu + \frac{2\sigma}{\sqrt{n}}\right) \cong 0.95. \tag{3.1}$$

3.3.2 Confidence Intervals

The inequalities (3.1) can be written in the equivalent form

$$\text{Prob}\left(\bar{X} - \frac{2\sigma}{\sqrt{n}} < \mu < \bar{X} + \frac{2\sigma}{\sqrt{n}}\right) \cong 0.95, \tag{3.2}$$

which provides an approximate 95% *confidence interval* for μ, in the sense that the probability that the random interval

$$\left(\bar{X} - \frac{2\sigma}{\sqrt{n}}, \bar{X} + \frac{2\sigma}{\sqrt{n}} \right) \qquad (3.3)$$

contains μ is approximately 95%. Given the observed values $x_1, x_2, \ldots, x_n$ of $X_1, X_2, \ldots, X_n$, the observed value of this interval is

$$\left(\bar{x} - \frac{2\sigma}{\sqrt{n}}, \bar{x} + \frac{2\sigma}{\sqrt{n}} \right). \qquad (3.4)$$

This interval is valuable in providing a measure of accuracy of the estimate $\bar{x}$ of μ. To be told that the estimate of a mean is 14.7 and that it is approximately 95% likely that the mean is between 14.3 and 15.1 is far more useful information than being told only that the estimate of a mean is 14.7.

Often the variance σ^2 is unknown, so that (3.4) is not immediately applicable. However, an unbiased estimator of the variance σ^2 of any distribution is provided by the estimator $\hat{\sigma}^2$, defined by

$$\hat{\sigma}^2 = \frac{\sum_{i=1}^{n} (X_i - \bar{X})^2}{n - 1}. \qquad (3.5)$$

The appearance of the term $n - 1$ in the denominator of this estimator is perhaps initially surprising. To check that $\hat{\sigma}^2$ is an unbiased estimator of σ^2, we write the numerator of the right-hand side expression in (3.5) as

$$\sum_{i=1}^{n} \left((X_i - \mu - (\bar{X} - \mu) \right)^2 = \sum_{i=1}^{n} (X_i - \mu)^2 - n(\bar{X} - \mu)^2.$$

The expected value of each term in the sum on the right-hand side in this expression is, by definition, σ^2, and since the variance of $\bar{X}$ is σ^2/n, the expected value of the final term on the right-hand side in this expression is σ^2. The expected value of the right-hand side is thus $(n - 1)\sigma^2$, and this leads to the desired result.

Corresponding to (3.5), the estimate s^2 of σ^2 found from observed data values $x_1, x_2, \ldots, x_n$ is

$$s^2 = \frac{\sum_{i=1}^{n} (x_i - \bar{x})^2}{n - 1}. \qquad (3.6)$$

Although $\hat{\sigma}^2$ is an unbiased estimator of σ^2, its variance depends on the value of the fourth moment about the mean of the probability distribution of X_i (see the definition (1.37) in the discrete case and (1.57) in the continuous case), and can in some cases be large unless the sample size n is itself very large.

From these results,

$$\frac{\hat{\sigma}^2}{n} = \frac{\sum_{i=1}^{n} (X_i - \bar{X})^2}{n(n - 1)}$$

is an unbiased estimator of the variance σ^2/n of $\hat{\mu} = \bar{X}$, and is estimated from the data by

$$\frac{\sum_{i=1}^{n}(x_i - \bar{x})^2}{n(n-1)}. \tag{3.7}$$

An approximate 95% confidence interval for μ is then

$$\left(\bar{X} - \frac{2S}{\sqrt{n}}, \bar{X} + \frac{2S}{\sqrt{n}}\right), \tag{3.8}$$

and given the data, the observed value of this interval is

$$\left(\bar{x} - \frac{2s}{\sqrt{n}}, \bar{x} + \frac{2s}{\sqrt{n}}\right). \tag{3.9}$$

Such an estimated confidence interval is useful, since it provides a measure of the accuracy of the estimate $\bar{x}$. On the other hand, the potentially large variance of $\hat{\sigma}^2$ implies that it should be used with caution.

The estimation of a binomial parameter p is usually carried out by using the theory of the binomial distribution directly and not by using the above general theory. Use of the theory above will lead to slightly different results, as we show below. Let Y have the binomial distribution parameter p and index n. Equation (2.77) shows that the mean value of Y/n is p and the variance of Y/n is $p(1-p)/n$. Thus

$$\hat{p} = Y/n \tag{3.10}$$

is an unbiased estimator of p. If y successes were obtained when the trials were carried out, the estimate of p is y/n and the generally–used estimate of the variance of $\hat{p}$ is

$$\frac{\hat{p}(1-\hat{p})}{n} = \frac{y(n-y)}{n^3}. \tag{3.11}$$

An approximate 95% confidence interval for p is

$$\left(\frac{y}{n} - 2\sqrt{\frac{y(n-y)}{n^3}}, \frac{y}{n} + 2\sqrt{\frac{y(n-y)}{n^3}}\right), \tag{3.12}$$

applicable when the distribution of $\hat{p}$ is approximately normal.

The estimator (3.11) is a biased estimator of $p(1-p)/n$ (see Problem 3.2). An unbiased estimate is found by replacing the denominator in (3.11) by $n^2(n-1)$, following the format of (3.5).

The estimation of the multinomial parameter p_i is carried out in a similar way.

3.3.3   Biased Estimators

The four estimators $\bar{X}$, S^2, $\hat{p}$, and $\hat{p}_i$ discussed above are unbiased estimators of the parameters μ, σ^2, p, and p_i, respectively. In some cases *biased estimators* of a parameter are of interest: $\hat{\theta}$ is a biased estimator of θ if the mean value $E(\hat{\theta})$ of $\hat{\theta}$ differs from θ, and its *bias* is defined as $E(\hat{\theta}) - \theta$.

When $\hat{\theta}$ is a biased estimator of θ its accuracy is usually assessed by its mean square error (MSE) rather than its variance. The MSE is defined by

$$\text{MSE}(\hat{\theta}) = E\left((\hat{\theta} - \theta)^2\right). \tag{3.13}$$

Thus the MSE of an unbiased estimator is its variance, and more generally it can be shown that

$$\text{MSE}(\hat{\theta}) = \text{Var}(\hat{\theta}) + \left(E(\hat{\theta}) - \theta\right)^2. \tag{3.14}$$

It often happens, when $\hat{\theta}$ is a function of n random variables, that the bias of $\hat{\theta}$ is proportional to n^{-1}, that is that

$$E(\hat{\theta}) = \theta + O(n^{-1}). \tag{3.15}$$

(See Appendix B.8 for the O notation.) In this case $\hat{\theta}$ is asymptotically ($n \to \infty$) unbiased, and it follows from equation (3.14) that the MSE and the variance of $\hat{\theta}$ differ by a term proportional to n^{-2}, so that when n is large the two are close.

A biased estimator can be of interest for two reasons. First, a parameter might not admit an unbiased estimate. For example, although there is an unbiased estimator of the parameter p in a binomial distribution, there is no unbiased estimator of p^{-1} (see Problem 3.3). A further example is discussed below. Second, a biased estimator of a parameter might be preferred to an unbiased estimator if its mean square error is smaller than the variance of the unbiased estimator. An example of this is given in Problem 8.6.

An important example of biased estimation is provided by the estimation of the correlation ρ_{12} (defined in (2.26)) between any two random variables X_1 and X_2 having a bivariate normal distribution (the case $k = 2$ of (2.33)). Given n pairs of random variables (X_{11}, X_{21}), $(X_{12}.X_{22})$, ..., (X_{1n}, X_{2n}) from this distribution, the standard estimator $\hat{\rho}_{12}$ of ρ_{12} is defined by

$$\hat{\rho}_{12} = \frac{\hat{\sigma}_{12}}{\hat{\sigma}_1 \hat{\sigma}_2}, \tag{3.16}$$

where $\hat{\sigma}_1^2$ and $\hat{\sigma}_2^2$ are defined as in (3.5) and $\hat{\sigma}_{12}$, the estimate of the covariance between the random variables X_1 and X_2, is defined by

$$\hat{\sigma}_{12} = \frac{\sum_{i=1}^n (X_{1i} - \bar{X}_1)(X_{2i} - \bar{X}_2)}{n - 1}. \tag{3.17}$$

The estimator (3.16) is a biased estimator of ρ_{12}, and no unbiased estimate of ρ_{12} exists. The bias is of order n^{-1} and is thus significant when n is small. Nevertheless, correlation estimates of the form (3.16) are used in bioinformatics in some cases when n is small, as discussed in Section 13.5, and this bias can then have important consequences.

3.4   Classical Hypothesis Testing

3.4.1   General Principles

Classical statistical hypothesis testing involves the test of a *null hypothesis* against an *alternative hypothesis*. The procedure consists of five steps, the first four of which are completed before the data to be used for the test are gathered, and relate to probabilistic calculations that set up the statistical inference process.

We illustrate these steps by using the two DNA sequences given in (1.1). We call this the "sequence-matching" example and will refer to it several times throughout the book.

Step 1

The first step in a hypothesis testing procedure is to declare the relevant null hypothesis H_0 and the relevant alternative hypothesis H_1. The choice of null and alternative hypotheses should be made before the data are seen. To decide on a hypothesis as a result of the data is to introduce a bias into the procedure, invalidating any conclusion that might be drawn from it. Our aim is eventually to accept or to reject the null hypothesis as the result of an objective statistical procedure, using data in our decision.

It is important to clarify the meaning of the expression "the null hypothesis is accepted." In the conservative approach to statistical hypothesis testing as outlined below, this expression means that there is no statistically significant evidence for rejecting the null hypothesis in favor of the alternative hypothesis. For reasons discussed below, the null hypothesis is often a particular case of the alternative hypothesis, and when it is, the alternative hypothesis must explain the data at least as well as the null hypothesis. Despite this, the null hypothesis might well be accepted, in the above sense, in that the alternative hypothesis might not explain the data significantly better than does the null hypothesis. A better expression for "accepting" is thus "not rejecting."

It is important to note the words "in favor of the alternative hypothesis" in the above. Suppose that the null hypothesis is that the probability of success p in a binomial distribution is $1/2$ and the alternative is that this parameter exceeds $1/2$. Suppose further that in 1,000 trials, only 348 successes are observed. The null hypothesis is accepted in favor of the alternative since the alternative hypothesis does not explain this result significantly better than does the null hypothesis – in fact it explains it less well than does the null hypothesis. Nevertheless, it would be unreasonable to believe that the null hypothesis is true: the data clearly suggest that $p < 1/2$. Thus accepting a null hypothesis in favor of some alternative does not necessarily imply in an absolute sense that the null hypothesis provides a reasonable explanation for the data observed.

We illustrate a choice of null and alternative hypotheses with a contrived example. Suppose that, in the sequence-matching case discussed above, the probability that a given nucleotide arises at a given site is 0.25. The null hypothesis might specify that the two sequences were generated at random with respect each other, which implies that the probability of a match between the two nucleotides at any site is 0.25. The alternative hypothesis might specify that the probability of a match at any site is some value larger than 0.25, for example 0.35, as might occur if the two sequences are related. Thus if the (unknown) probability of a match at any site is p, the null hypothesis claims that $p = 0.25$ and the alternative hypothesis claims that $p = 0.35$.

A hypothesis can be *simple* or *composite*. A simple hypothesis specifies the numerical values of all unknown parameters in the probability distribution of interest. In the above example, both null and alternative hypotheses are simple. A composite alternative does not specify all numerical values of all the unknown parameters. In the sequence-matching example, the alternative hypothesis "p exceeds 0.25" is composite. It is also *one-sided* ($p > 0.25$) as opposed to *two-sided* ($p \neq 0.25$).

In the above example, the alternative hypothesis $p > 0.25$ is a natural one. However, for technical reasons associated with the hypothesis testing theory developed in Chapter 9, it is often advantageous to make the null hypothesis a particular case of the alternative hypothesis, in which case we say it is nested within the alternative hypothesis. If this is done, then in the example of the previous paragraph, the one-sided alternative $p > 0.25$ would be replaced by $p \geq 0.25$ and the two-sided alternative $p \neq 0.25$ would be replaced by "p unspecified." In practice there is no change to the testing procedures if the null hypothesis is nested within the alternative hypothesis in this way, and we shall freely use both the nested notation such as $p \geq 0.25$ and the non-nested notation such as $p > 0.25$ interchangeably.

The sequence matching case illustrates the fact that tests of hypotheses usually involve the value of some unknown parameter (or parameters). We generically denote the parameter of interest by θ, although in some cases we use a more specific notation (such as μ for a mean). The nature of the alternative hypothesis is determined by the context of the test, in particular whether it is one-sided up (that is the unknown parameter θ exceeds some specified value θ_0), one-sided down ($\theta < \theta_0$), or two-sided ($\theta \neq \theta_0$). In many cases in bioinformatics the natural alternative is both composite and one-sided. The sequence-matching case is an example: Unless there is some reason to choose a specific alternative such as $p = 0.35$, it seems more reasonable to choose the composite alternative $p \geq 0.25$.

Step 2

Since the decision to accept or reject H_0 will be made on the basis of data derived from some random process, it is possible that an incorrect decision will be made, that is, to reject H_0 when it is true (a *Type I error*), or to accept H_0 when it is false (a *Type II error*). When testing a null hypothesis against an alternative it is not possible to ensure that the probabilities of making a Type I error and a Type II error are both arbitrarily small unless we are able to make the number of observations as large as we please. In practice we are seldom able to do this. This dilemma is resolved in practice by observing that there is often an asymmetry in the implications of making the two types of error. In the sequence-matching case, for example, there might be more concern about making the false positive claim of a similarity between the two sequences when there is no such similarity, and less concern about making the false negative conclusion that there is no similarity when there is. For this reason, a frequently adopted procedure is to focus on the Type I error, and to fix the numerical value α of this error at some acceptably low level (usually 1% or 5%), and not to attempt to control the numerical value of the Type II error. The choice of the values 1% and 5% is reasonable, but is also clearly arbitrary. The choice 1% is a more conservative one than the choice 5%. Step 2 of the hypothesis testing procedure consists in choosing the numerical value for the Type I error.

Step 3

The third step in the hypothesis testing procedure consists in determining a *test statistic*. This is the quantity calculated from the data whose numerical value leads to acceptance or rejection of the null hypothesis. In the sequence-matching example one possible test statistic is the total number Y of matches. This is a reasonable choice, but in more complicated cases the choice of a test statistic is not straightforward. The theory in Chapter 9 focuses on deriving test statistics that, for a given Type I error, minimize the probability of our making a Type II error, given the number of observations to be made. There is a substantial body of statistical theory associated with such an optimal choice of a test statistic, discussed in detail in Chapter 9.

Step 4

The next step in the procedure consists in determining those observed values of the test statistic that lead to rejection of H_0. This choice is made so as to ensure that the test has the numerical value for the Type I error chosen in Step 2. We illustrate this step with the sequence-matching example. Suppose that, in this example, the total number Y of matches is chosen as the test statistic. In both the case of a simple alternative hypothesis such as "$p = 0.35$" and in the case of the composite alternative hypothesis "$p \geq 0.25$," the null hypothesis $p = 0.25$ is rejected in favor of the alternative when the observed value y of Y is sufficiently large, that is, if y is greater than or equal to some *significance point* K. If for example the Type I error is chosen as 5%, K is found from the requirement

$$\text{Prob(null hypothesis is rejected when it is true)}$$
$$= \text{Prob}(Y \geq K \,|\, p = 0.25) = 0.05. \qquad (3.18)$$

In practice, when discrete random variables are involved, it may be impossible to arrive at a procedure having exactly the Type I error chosen. This difficulty arises here: It is impossible to find a value of K such that (3.18) is satisfied exactly. In practice, the choice of K is made by a conservative procedure: When $\alpha = .05$, $p = .25$ and $n = 100$, $\text{Prob}(Y \geq 32) = .069$ and $\text{Prob}(Y \geq 33) = .044$, and we use the conservative value 33 for K. This difficulty is to be taken as understood in all testing procedures when the test statistic is a discrete random variable.

For very long sequences, a normal approximation to the binomial might be employed. For example, if in the above example both sequences have length 1,000,000, and the Type I error is chosen as 5%, K might be determined in practice by the requirement

$$\text{Prob}\left(X \geq K - \frac{1}{2} \right) = 0.05, \qquad (3.19)$$

where X is a random variable having a normal distribution with mean $1{,}000{,}000(0.25) = 250{,}000$ and variance $1{,}000{,}000(0.25)(0.75) = 187{,}500$ and the continuity correction $\frac{1}{2}$ has been employed. The resulting value of K is 250,712.81; in practice the conservative value 250,713 would be used.

In the above example the null hypothesis is rejected if Y is sufficiently large. If the alternative hypothesis had specified a value of p that is less than 0.25, then the null hypothesis would be rejected for sufficiently small Y.

In many test procedures the null hypothesis does not specify the numerical values of all the parameters involved in the distribution of the random variables involved in the test procedure. In such a case problems can arise in the testing procedure since there might be no unique significance point (such as K above) having the property that the probability that the test statistic exceeds k is equal to the Type I error no matter what the values of the parameters not specified by the null hypothesis. This problem is illustrated in a practical case in Section 3.5.2.

Step 5

The final step in the testing procedure is to obtain the data, and to determine whether the observed value of the test statistic is equal to or more extreme than the significance point calculated in Step 4, and to reject the null hypothesis if it is. Otherwise the null hypothesis is accepted.

3.4.2   P-Values

A testing procedure equivalent to that just described involves the calculation of a so-called *P-value*, or *achieved significance level*. Here Step 4, the
calculation of the significance point such as K in the example described,
is not carried out. Instead, once the data are obtained, we calculate the
null hypothesis probability of obtaining the observed value, or one more
extreme, of the test statistic. This probability is called the *P*-value. If the
P-value is *less than or equal to* the chosen Type I error, the null hypothesis
is rejected. This procedure always leads to a conclusion identical to that
based on the significance point approach.

For example, the null hypothesis $(p = 0.25)$ probability of observing 11
matches *or more* in a sequence comparison of length 26 (as in (1.1)) is
found from the binomial distribution to be about 0.04. This is the *P*-value
associated with the observed number 11. If in the length 1,000 sequence-
matching case there are 278 matches, the *P*-value might be found, using
the normal approximation to the binomial distribution, as

$$\text{Prob}(X \geq 277.5),$$

where X has a normal distribution with mean 250 and variance 187.5. This
gives a *P*-value of 0.022. If the Type I error had been chosen to be 1%,
the null hypothesis would then not be rejected. This conclusion agrees with
that obtained using the significance point approach, since the significance
point K found above is 283, greater than the observed value 278.

The *P*-value calculation for a two-sided alternative hypothesis is a little
more complicated. Suppose for example that we wish to test whether a
coin is fair, and obtain 58 heads from 100 tosses. The *P*-value is then the
probability of obtaining 58 or more, or 42 or fewer, heads, since values 42
or fewer are more extreme, for a two-sided alternative, than the observed
value 58.

Before the experiment is conducted, the eventual *P*-value is a random
variable. If the test statistic is continuous and the null hypothesis is true,
the probability distribution of this *P*-value is the continuous uniform distribution (1.63) on $[0, 1]$. This follows from the fact that a *P*-value is the
probability that the test statistic is at least as extreme as its observed value.
In the case where the alternative hypothesis corresponds to small values
of the test statistic, the *P*-value is $F_X(x)$, where x is the observed value
of the test statistic and $F_X(x)$ is the cumulative distribution function of
the test statistic when the null hypothesis is true. Theorem 1.2 then shows
that the *P*-value has the uniform distribution (1.63), as claimed. A similar
argument holds when the alternative hypothesis corresponds to large values of the test statistic, where the *P*-value (for a continuous test statistic)
is $1 - F_X(x)$. In both cases the *P*-value satisfies the equation

$$\text{Prob}(P\text{-value} \leq x | H_0 \text{ true}) = x. \tag{3.20}$$

This implies that the P-value has the uniform distribution on $[0, 1]$ when the null hypothesis is true, as claimed.

The situation for a test statistic having a discrete distribution is not so straightforward. As a simple example, suppose that a coin is to be tossed n times in order to test the null hypothesis that it is fair against the alternative hypothesis that it is biased towards tails. The test statistic is the number of heads observed. There are $n + 1$ possible P-values, given by

$$\sum_{j=0}^{i} \binom{n}{j} \left(\frac{1}{2}\right)^{n}, \quad i = 0, 1, 2, \ldots, n, \tag{3.21}$$

the P-value $\sum_{j=0}^{i} \binom{n}{j}(\frac{1}{2})^n$ arising when the n tosses result in i heads. This P-value has null hypothesis probability $\binom{n}{i}(\frac{1}{2})^n$, which depends nontrivially on the value of i. Thus the P-value does not have a (discrete) uniform distribution. Further, although equation (3.20) holds for values of x of the form $x = i/n$, it does not hold for other values of x, for which case

$$\text{Prob}(P\text{-value} \leq x | H_0 \text{ true}) < x. \tag{3.22}$$

Thus for any value of x in $[0,1]$,

$$\text{Prob}(P\text{-value} \leq x | H_0 \text{ true}) \leq x. \tag{3.23}$$

This result can be shown to be true for any test statistic having a discrete probability distribution. We return to this conclusion in Section 3.12.

One feature of any P-value is that it can never take the value 0. This point is discussed again in the context of a discrete test statistic in Section 3.8.1.

3.4.3 Power Calculations

For the test of a simple null hypothesis against a simple alternative, with a fixed number of observations, the choice of the Type I error implicitly determines the numerical value β of the Type II error, or equivalently of the *power* of the test, defined as the probability $1 - \beta$ of rejecting the null hypothesis when the alternative is true.

When the alternative hypothesis is composite, the probability that the null hypothesis is rejected will normally depend on the actual value of the parameter (or parameters) concerned in the test. There is therefore no unique value for the Type II error, and no unique value for the power of the test, under the alternative hypothesis. In this case the principle adopted in Step 3, namely that of choosing a test statistic that maximizes the power of the test, becomes more difficult to apply than in the case when the alternative hypothesis is simple. This problem is discussed further in Chapter 9.

3.5 Hypothesis Testing: Examples

3.5.1 Example 1. Testing for a Mean: the One-Sample Case

A classic test in statistics concerns the unknown mean μ of a normal distribution. Suppose first that the variance σ^2 of this distribution is known. One case of this is the test of the null hypothesis $\mu = \mu_0$ against the one-sided alternative hypothesis $\mu > \mu_0$. If this test is carried out using the observed values of random variables $X_1, X_2, \ldots, X_n$ having the normal distribution in question, the statistical theory of Chapter 9 leads to the use of $\bar{X}$ as an optimal test statistic and the rejection of the null hypothesis if the observed value $\bar{x}$ of $\bar{X}$ is "too much larger" than μ_0. The random variable $\bar{X}$ has known variance σ^2/n and mean μ_0 if the null hypothesis is true. The standardization procedure described in Section 1.10.2 then shows that the random variable Z, defined by

$$Z = \frac{(\bar{X} - \mu_0)\sqrt{n}}{\sigma} \tag{3.24}$$

has the standard normal distribution when the null hypothesis is true. Since the probability that such a random variable exceeds 1.645 is 0.05, a desired Type I error 5% is achieved if the null hypothesis is rejected when

$$\frac{(\bar{x} - \mu_0)\sqrt{n}}{\sigma} \geq 1.645, \tag{3.25}$$

where $\bar{x}$ is the observed value of $\bar{X}$ once the data are obtained. Equivalently, the null hypothesis is rejected if

$$\bar{x} \geq \mu_0 + 1.645\sigma/\sqrt{n}. \tag{3.26}$$

If the alternative hypothesis had been $\mu \leq \mu_0$, the null hypothesis would be rejected if the observed value $\bar{x} \leq \mu_0 - 1.645\sigma/\sqrt{n}$. If the alternative hypothesis had been two-sided, so that no specification is made for the value of μ, the null hypothesis would be rejected if $|\bar{x} - \mu_0| \geq 1.96\sigma/\sqrt{n}$. This shows that the nature of the alternative hypothesis determines the values of the test statistic that lead to rejection of the null hypothesis. It will be shown in Section 3.7 that in some cases it can also determine the choice of the test statistic itself.

A more realistic situation arises when σ^2 is unknown, in which case a *one-sample t-test* is used. Here we estimate the unknown variance σ^2 by s^2, defined in (3.6), and use as test statistic the one-sample t statistic, defined by

$$t = \frac{(\bar{x} - \mu_0)\sqrt{n}}{s}. \tag{3.27}$$

Under the assumption that $X_1, X_2, \ldots, X_n$ are $\text{NID}(\mu, \sigma^2)$, the null hypothesis distribution of T, defined by

$$T = \frac{(\bar{X} - \mu_0)\sqrt{n}}{S}, \tag{3.28}$$

is well known (as the t distribution with $n - 1$ degrees of freedom). The density function of T is independent of μ_0 and σ^2, being

$$f_T(t) = \frac{\Gamma\left(\frac{n+1}{2}\right)}{\sqrt{n\pi}\,\Gamma\left(\frac{n}{2}\right)\left(1 + \frac{t^2}{n}\right)^{(n+1)/2}}, \quad -\infty < t < +\infty. \tag{3.29}$$

An outline of the derivation of this density function is given in Problem 3.7.

It is perhaps remarkable that this density function is independent of the value of σ^2. The value σ^2 is not specified under the null hypothesis, and this implies that significance points of t can be calculated no matter what the value of σ^2 might be. These significance points have been calculated from (3.29) for a variety of values of n and the chosen Type I error, and are widely available.

The t distribution (3.29) differs from standard normal distribution applying for the statistic Z, so that the significance points appropriate for Z are not appropriate for T. However the t distribution converges to the standard normal distribution as $n \to \infty$.

Since the null hypothesis distribution of T is independent of the values of μ_0 and σ^2, T is said to be a *pivotal quantity*. It is because of the pivotal nature of T that explicit significance points of the t distribution can be found, whatever the values of μ_0 and σ^2 might be.

3.5.2 Example 2. The Two-Sample t-Test

A protein coding gene is a segment of the DNA that codes for a particular protein (or proteins). In any given cell type at any given time, this protein may or may not be needed. Each cell will generate the proteins it needs, which will usually be some small subset of all possible proteins. If a protein is generated in a cell, we say that the gene coding for this protein is *expressed* in that cell type. Furthermore, any given protein can be expressed at many different levels. One cell type might need more copies of a particular protein than another cell type. When this happens we say that the gene is *differentially expressed* between the two cell types. There are several techniques for measuring the level of gene expression in a cell type. All of these methods are subject to both biological and experimental variability. Therefore, one cannot simply measure the level of expression once in each cell type to test for differential expression. Instead, one must repeat each experiment several times and perform a statistical test of the hypothesis that they are expressed at the same or different levels.

Suppose that the mean expression levels of a given gene in two cell types, for example normal and tumor (cancerous) cells, are to be compared. In statistical terms, this comparison can be framed as the test of the equality of two unknown means. For the moment we assume that the (unknown) variance of expression level in normal cells is identical to that in tumor cells. To test for equality of the two means, we plan to measure the expression levels of m cells of one type and compare these with the expression levels of n cells of another type. Suppose that, before the experiment, the measurements $X_{11}, X_{12}, \ldots, X_{1m}$ from the first cell type are thought of as m NID(μ_1, σ^2) random variables, and the measurements $X_{21}, X_{22}, \ldots, X_{2n}$ from the second cell type are thought of as n NID(μ_2, σ^2) random variables. The null hypothesis states that $\mu_1 = \mu_2 (= \mu$, unspecified). We assume for the moment that the alternative hypothesis leaves both μ_1 and μ_2 unspecified, so that our eventual test is two-sided.

The theory in Chapter 9 shows that, under the assumptions made, the optimal test statistic is T, defined now by

$$T = \frac{(\bar{X}_1 - \bar{X}_2)\sqrt{mn}}{S\sqrt{m+n}}, \tag{3.30}$$

with S defined by

$$S^2 = \frac{\sum\limits_{i=1}^{m}(X_{1i} - \bar{X}_1)^2 + \sum\limits_{i=1}^{n}(X_{2i} - \bar{X}_2)^2}{m+n-2}. \tag{3.31}$$

The form of this test statistic can been understood by observing that the variance of $\bar{X}_1 - \bar{X}_2$ is $\sigma^2/m + \sigma^2/n$. If we had known the variance σ^2, we could use as test statistic the quantity Z, defined by

$$Z = \frac{\bar{X}_1 - \bar{X}_2}{\sqrt{\sigma^2/m + \sigma^2/n}} = \frac{(\bar{X}_1 - \bar{X}_2)\sqrt{mn}}{\sigma\sqrt{m+n}}. \tag{3.32}$$

Since σ^2 is unknown, it is estimated by the pooled estimator S^2, using observations from both normal and tumor cells, and in general from the two groups being compared. This leads to the T statistic in (3.30).

The null hypothesis probability distribution of T is independent of both the value for the (common) mean unspecified under the null hypothesis and of the unknown variance σ^2. This implies that T (defined by (3.30)) is a pivotal quantity. The null hypothesis distribution of T is the t distribution (3.29) with $m + n - 2$ degrees of freedom, and this enables a convenient assessment of the significance of the observed value t of T, defined as

$$t = \frac{(\bar{x}_1 - \bar{x}_2)\sqrt{mn}}{s\sqrt{m+n}}, \tag{3.33}$$

with s defined by

$$s^2 = \frac{\sum_{i=1}^{m}(x_{1i} - \bar{x}_1)^2 + \sum_{i=1}^{n}(x_{2i} - \bar{x}_2)^2}{m + n - 2}. \tag{3.34}$$

For the two-sided test discussed above, significantly large positive or large negative values of t lead to the rejection of the null hypothesis. When the alternative hypothesis is $\mu_1 \geq \mu_2$, significantly large positive values of t lead to the rejection of the null hypothesis, and when the alternative hypothesis is $\mu_1 \leq \mu_2$, significantly large negative values of t lead to the rejection of the null hypothesis.

In reality, expression levels cannot generally be expected to have normal distributions, nor should the variances of the two types generally be expected to be equal. These two assumptions were made in the above t-test procedure, and the significance points of the t distribution are calculated assuming that both assumptions hold. Thus in practice it might not be appropriate to use the t-test to test for differential expression. In general, if the normal distribution assumption is unjustified we should use the non-parametric tests: these are discussed in Section 3.8.2 and in Chapter 13.

The optimality property of the two-sample t-test procedure described above derives from statistical theory – see Chapter 9. The theoretical development assumes that the variances of the random variables in the two groups considered are equal. When, as is often the case in practice, these two variances cannot reasonably be taken as being equal, the theoretical approach of Chapter 9 fails to lead to a testing procedure for which the test statistic has the same distribution for all parameter values not specified by the null hypothesis. That is, no pivotal quantity analogous to equal variance case T as defined in (3.30) exists. This implies that there is no well-defined null hypothesis probability distribution available analogous to that in (3.29) from which significance points can be obtained, whatever the unknown variances in the two groups might be. Because of this, approximate heuristic procedures are required.

One frequently used procedure is as follows. Under the null hypothesis, $\bar{X}_1$ and $\bar{X}_2$ have normal distributions with the same mean and respective variances σ_1^2/m and σ_2^2/n, so that the difference $\bar{X}_1 - \bar{X}_2$ has a normal distribution with mean zero and variance

$$\frac{\sigma_1^2}{m} + \frac{\sigma_2^2}{n}. \tag{3.35}$$

The variances σ_1^2 and σ_2^2 are unknown, but have estimators S_1^2 and S_2^2, where

$$S_1^2 = \frac{\sum_{i=1}^{m}(X_{i1} - \bar{X}_1)^2}{m-1}, \quad S_2^2 = \frac{\sum_{i=1}^{n}(X_{i2} - \bar{X}_2)^2}{n-1}. \tag{3.36}$$

One then computes the statistic T', defined by

$$T' = \frac{\bar{X}_1 - \bar{X}_2}{\sqrt{A + B}}, \qquad (3.37)$$

where

$$A = \frac{S_1^2}{m}, \quad B = \frac{S_2^2}{n}.$$

When the null hypothesis of equal means is true, T' has an approximate t distribution with degrees of freedom given by the largest integer less than or equal to ν (see Lehmann (1986)), where ν defined by

$$\nu = \frac{(A + B)^2}{\frac{A^2}{m-1} + \frac{B^2}{n-1}}.$$

When $m = n$, T' is identical to the T statistic (3.30). However, in this case the number of degrees of freedom appropriate for t' is not equal to the number $2(n - 1)$ applying when the two variances are assumed to be equal: The value of ν lies in the interval $[n - 1, 2(n - 1)]$, the actual value depending on the ratio of S_1^2/S_2^2.

Markowski and Markowski (1990) show for the case $m = n$ that even when the variances in the two groups differ, use of the "equal variance" t-test procedure leads to a very small error.

An important case of the two-sample t test arises if $n = m$ and the random variables X_{1i} and X_{2i} are logically paired, for example being expression levels of normal and tumor cells taken from the same person. In this "paired t-test" case the test is carried out by using the differences $D_i = X_{1i} - X_{2i}$ and basing the test entirely on these differences. This reduces the test to a one-sample t-test with test statistic T as defined in (3.28) and with X_i replaced by D_i and μ_0 set equal to 0. The test statistic is then

$$T = \frac{\bar{D}\sqrt{n}}{S_D}, \qquad (3.38)$$

where S_D^2 defined by the right-hand side in (3.5), with X_i replaced by D_i and $\bar{X}$ by $\bar{D}$.

The advantage of the pairing procedure is that the variance estimate S_D^2 measures only cell type to cell type variation, and eliminates person-to-person variation. If there is significant person-to-person variation, this provides a more powerful test of cell type to cell type variation. In this procedure we see the beginnings of the concept of the Analysis of Variance (ANOVA). In an ANOVA procedure the variation in a body of data is broken down into separate components, each measuring one source of variation, and the significance of one potential source of variation can be investigated free of any influence of other potential sources of variation. The ANOVA concept is developed at length in Section 9.5.

3.5.3 Example 3. Tests on Variances

In Section 3.5.2 we considered two tests, each comparing the means of two groups of random variables. These tests differ depending on whether or not one is prepared to assume that the variances of the random variables in the two groups are equal. This makes it important to describe a test for equality of variances.

We suppose that $X_{11}, X_{12}, \ldots, X_{1m}$ are $NID(\mu_1, \sigma_1^2)$ and $X_{21}, X_{22}, \ldots, X_{2n}$ are $NID(\mu_2, \sigma_2^2)$. We wish to test the null hypothesis $\sigma_1^2 = \sigma_2^2$. To do this we consider the ratio S_1^2/S_2^2 of the two variance estimators S_1^2 and S_2^2 defined in (3.36). Under the null hypothesis this ratio has the F distribution with $(m-1, n-1)$ degrees of freedom, developed in Section 2.13, whatever values the unknown means μ_1 and μ_2 take. If for example the alternative hypothesis were $\sigma_1^2 > \sigma_2^2$, significantly large values of the observed value of this ratio would lead to rejection of the null hypothesis. Significance points of F for Type I errors arising in practice are extensively tabulated, allowing a ready evaluation of whether the observed value of the ratio is indeed significantly large.

We will meet the F test in Section 9.5 in the context of ANOVA (the analysis of variance), where (perhaps unexpectedly) it is used as a test for the equality of several means, rather than as a test for the equality of two variances.

3.5.4 Example 4. Testing for the Parameters in a Multinomial Distribution

In this example we consider a test of the null hypothesis that prescribes specific values for the probabilities $\{p_i\}$ in the multinomial distribution (2.30). The alternative hypothesis considered here is composite and leaves these probabilities unspecified. This can be used, for example, to test for prescribed probabilities for the four nucleotides in a DNA sequence.

Let Y_i be the number of observations in category i. A test statistic often used for this testing procedure is X^2, defined by

$$X^2 = \sum_{i=1}^{k} \frac{(Y_i - np_i)^2}{np_i}. \tag{3.39}$$

Sufficiently large values of the observed value

$$\sum_{i=1}^{k} \frac{(y_i - np_i)^2}{np_i} \tag{3.40}$$

of X^2 lead to rejection of the null hypothesis. The quantity (3.40) may be thought of as a measure of the discrepancy between the observed values $\{y_i\}$ and the respective null hypothesis expected values $\{np_i\}$.

When the null hypothesis is true and n is large, X^2 has approximately the chi-square distribution (1.77) with $\nu = k - 1$ degrees of freedom. The proof of this claim, and the corresponding claim for the chi-square statistic associated with Table 3.1, is beyond the level of the material in this book. The statistic X^2 is frequently referred to as the "chi-square statistic," and this explains the notation X^2. Tables of the significance points of the chi-square distribution are widely available for all values of ν likely to arise in practice.

The choice of the test statistic (3.39) is not arrived at from the statistical theory to be discussed in Chapter 9. Instead, that theory and the discussion following (9.25) leads to the test statistic

$$2 \sum_i Y_i \log \frac{Y_i}{np_i}. \tag{3.41}$$

Despite the optimality theory associated with the statistic (3.41), common practice is to use the statistic (3.39) for the testing procedure. This occurs largely for historical reasons. When the null hypothesis is true and the sample size is large, the numerical values of the statistics (3.39) and (3.41) are usually quite close (see Problem 3.6), and this can be thought of as a justification for the use of (3.39).

3.5.5   Example 5. Association tests.

In this section we consider tests of association. Specifically, observations are categorized into one of an number of "row" categories and also into one of a number of "column" categories. The tests we consider may be thought of as tests of association of the the categorization by rows and the categorization by columns. We start with the case of two rows and two columns.

Two-by-Two Tables: Fisher's Exact Test

We illustrate the test of association for a two-by-two table with the "genders and mutations" example of Section 1.3.3. We wish to test the null hypothesis that there is no association between gender and propensity to be a mutant, the alternative hypothesis of interest being that males are more likely to be mutants than are females. As in Section 1.3.3, we suppose that n male mice and $N - n$ female mice are irradiated, and that in all a total of m mutant mice is observed and thus a total of $N - m$ non-mutants. These four totals are taken as given.

To illustrate the calculations, suppose that $n = 8$, that $N = 20$ and that $m = 9$. Of the males, $y = 6$ are mutants. (We use very small numbers to illustrate the computations.) These data may be arranged in the form of a two-by-two contingency table, as shown below.

	mutant	non-mutant	total
male	6	2	8
female	3	9	12
total	9	11	20

The example of throwing objects into a box given at the end of section 1.3.3 provides an appropriate paradigm for this experiment. The requirement of the independence of the throws in that example becomes the requirement that the event that any one mouse is a mutant is independent of the event that any other mouse is a mutant. The fact that the column into which an object is thrown is independent of the row into which it is thrown corresponds to the assumption that the null hypothesis in the radiation experiment is true. The P-value is the probability of observing a value 6 or larger in the upper left-hand cell in the table, assuming that the null hypothesis is true and that the four marginal totals 8, 12, 9, and 11 are given. The hypergeometric formula (1.9) shows that this probability is

$$\frac{\binom{8}{6}\binom{12}{3}}{\binom{20}{9}} + \frac{\binom{8}{7}\binom{12}{2}}{\binom{20}{9}} + \frac{\binom{8}{8}\binom{12}{1}}{\binom{20}{9}},$$

or about 0.039890. With a Type I error of 5%, we would reject the null hypothesis. This is an example of *Fisher's exact test*, and the P-value calculated is exact.

Until computer packages became available, this exact procedure was used only when the numbers in the two-by-two table such as that shown were comparatively small, since hand calculation of the hypergeometric probabilities for large numbers is prohibitive. Computer packages now allow calculations for quite large numbers, and for very large numbers many packages use the approximate chi-square procedure described in the next section, which is quite accurate when the numbers in the table are large.

Two-sided tests are carried out by the obvious extension of the procedure described above.

The hypergeometric distribution from which the P-value is calculated applies *only* if various criteria are met. In particular it is required that the individual readings which add up to the counts in the four cells in the table be independent. The same comment applies for tables of general size, considered below: unfortunately, this requirement is often overlooked in the application of association tests in bioinformatics.

A final observation is that the procedure described can only test for *association*, and any subsequent claim of a cause and effect relation must be made on grounds other than the association test itself.

		column					
		1	2	3	$\cdots$	c	Total
row	1	Y_{11}	Y_{12}	Y_{13}	$\cdots$	Y_{1c}	$y_{1\cdot}$
	2	Y_{21}	Y_{22}	Y_{23}	$\cdots$	Y_{2c}	$y_{2\cdot}$
	$\vdots$	$\vdots$	$\vdots$	$\vdots$	$\ddots$	$\vdots$	$\vdots$
	r	Y_{r1}	Y_{r2}	Y_{r3}	$\cdots$	Y_{rc}	$y_{r\cdot}$
	Total	$y_{\cdot 1}$	$y_{\cdot 2}$	$y_{\cdot 3}$	$\cdots$	$y_{\cdot c}$	y

Table 3.1. Two-way table data.

Tables of Arbitrary Size: the Chi-Square Test

A situation more general than that described in the previous section arises when categorization count data arise in the form of a two-way table with an arbitrary number r of rows and an arbitrary number c or columns, such as Table 3.1. The row and column totals are not of direct interest and can often be chosen in advance. They are thus written in lower case in Table 3.1. As with Fisher's exact test, the testing procedure described below is valid only if the y observations leading to the counts $\{Y_{jk}\}$ in the table are all independent of each other. This important fact is often overlooked in the application of chi-square procedures in bioinformatics.

Given the row and column totals, it can be shown that when the null hypothesis is true, Y_{jk} is a random variable with mean value E_{jk}, where $E_{jk} = y_{j\cdot}y_{\cdot k}/y$. [1] Thus when the null hypothesis is true, the observed value of Y_{jk} should be close to E_{jk}. This argument leads to the frequently used chi-square test statistic

$$\sum_{jk} \frac{(Y_{jk} - E_{jk})^2}{E_{jk}}, \tag{3.42}$$

which can be regarded as a measure of the difference of the Y_{jk} and E_{jk} values. When the null hypothesis is true and the independence requirement discussed above holds, the statistic (3.42) has an asymptotic chi-square distribution with $\nu = (r-1)(c-1)$ degrees of freedom.

Some examples of the application of this test in a genetic context are given in Sections 5.2, 5.3.4, and 6.1, and a theoretical discussion is given in Example 4 of Section 9.4.

The theory of Section 9.4 leads to the use of the test statistic

$$2 \sum_{jk} Y_{jk} \log \frac{Y_{jk}}{E_{jk}} \tag{3.43}$$

---

[1] The notation E arises for historical reasons, since the E_{jk} are usually described as "expected values," and should not be confused with the use of the notation E found elsewhere in this book to describe a random variable.

as the appropriate test statistic, rather than the statistic (3.42). As with the statistic (3.40), the statistic (3.42) is generally used for historical reasons only. The numerical values of the two statistics (3.42) and (3.43) are usually close when the null hypothesis is true and the sample size is large.

We conclude this section by remarking that the chi-square procedures described above are two-sided tests: the alternative hypothesis is always that there is an association of some kind between row and column modes of classification. When there are only two row and two column classifications a one-sided alternative can be tested, using Fisher's exact test.

3.6 Likelihood Ratios, Information, and Support

In this section we briefly discuss aspects of the choice of test statistic in a hypothesis testing procedure. It will be shown in Chapter 9 that in testing a simple null hypothesis against a simple alternative hypothesis, a reasonable optimality argument leads to the use of the ratio of the probability of the data under the alternative hypothesis to the probability of data under the null hypothesis as the test statistic. For obvious reasons, this is called the *likelihood ratio*.

As a simple example, we consider the sequence-matching example discussed in Section 3.4.1, and define the indicator variable Y_i by $Y_i = 1$ if the pair in position i match and $Y_i = 0$ if they do not, for $i = 1, 2, \ldots, n$. If the null hypothesis probability of a match is 0.25 and the alternative hypothesis probability is 0.35, the likelihood ratio is

$$\frac{(0.35)^{\sum Y_i}(0.65)^{(n-\sum Y_i)}}{(0.25)^{\sum Y_i}(0.75)^{(n-\sum Y_i)}}. \tag{3.44}$$

This simplifies to

$$\left(\frac{7}{5}\right)^{Y}\left(\frac{13}{15}\right)^{n-Y}, \tag{3.45}$$

where $Y = \sum Y_i$ is the (random) total number of matches. This is a monotonically increasing function of Y, the total number of matches, and thus use of the likelihood ratio as test statistic is equivalent to the use of Y as test statistic. This explains the choice of Y as test statistic in step 3 of the hypothesis testing procedure described in Section 3.4.1. More complicated examples are discussed in Chapter 9.

The logarithm of the expression in (3.45) is

$$Y \log\left(\frac{7}{5}\right) + (n - Y)\log\left(\frac{13}{15}\right). \tag{3.46}$$

The definition of accumulated support in equation (2.78) shows that this is the accumulated support, after n pairs have been observed, for the claim that the probability of a match is 0.35 against the claim that it is 0.25. If the true probability of a match is p, the expected value of the expression in (3.46) is

$$n \left(p \log \left(\frac{7}{5} \right) + (1 - p) \log \left(\frac{13}{15} \right) \right). \tag{3.47}$$

This is zero when $p = 0.2984$, approximately half-way between the values 0.25 and 0.35. For values of p close to 0.2984, many observations will be needed before a preference between the values 0.25 and 0.35 can be established.

We shall return to the concept of accumulated support in Section 9.9, when considering sequential tests of hypotheses.

3.7   Hypothesis Testing Using a Maximum as Test Statistic

The use of the maximum of several random variables as a test statistic arises in several areas in bioinformatics, and in particular in BLAST (Chapter 10), so in this section we introduce aspects of hypothesis testing theory using a maximum as a test statistic.

3.7.1   The Normal Distribution

Let $X_1, X_2, \ldots, X_n$ be independent normal random variables, each having variance 1. Suppose the null hypothesis states that the mean of each of these random variables is 0, while the alternative hypothesis claims that one of these random variables has mean μ greater than 0, the remaining variables having mean 0. It is not, however, stated by the alternative hypothesis which of the random variables has mean μ. An example where this situation arises in practice is discussed in Example 2 of Section 9.4.

This is a different alternative hypothesis from the one considered in Example 1 of Section 3.5. The statistical principles given in Chapter 9 that lead to $\bar{X}$ as optimal test statistic in that example lead in this case to $X_{\max}$, the maximum of $X_1, X_2, \ldots, X_n$, as optimal test statistic. If the desired Type I error is α and K is computed so that $\mathrm{Prob}(X_{\max} \geq K \mid \text{null hypothesis true}) = \alpha$, the null hypothesis is rejected if $X_{\max} \geq K$. Equivalently, if $x_{\max}$ is the observed value of $X_{\max}$, the null hypothesis is rejected if the P-value $\mathrm{Prob}(X_{\max} \geq x_{\max} \mid \text{null hypothesis true})$ is less than α.

The calculation of the value of K is found in principle from the null hypothesis probability distribution of the maximum $X_{\max}$ of n NID$(0,1)$ random variables. No simple exact form is available for this distribution.

Various approximations can be found in the literature, usually based either on simulation or on the asymptotic density function (2.128) for the maximum of n iid continuous random variables. However, a direct calculation of P-values is possible using normal distribution tables and equation (2.91). For example, suppose that $n = 100$ and that $x_{\max}$, the observed value of $X_{\max}$, is 3.6. The probability that an $N(0,1)$ random variable is less than 3.6 is approximately 0.9998409, and equation (2.91) then shows that the P-value associated with the observed value 3.6 of $X_{\max}$ is approximately $1 - (0.9998409)^{100} \cong 0.0158$, which is significant if the Type I error is 0.05. If $n = 500$ and $x_{\max} = 3.6$, the approximate P-value is $1 - (0.9998409)^{500} \cong 0.0765$, which is not significant if the Type I error is 0.05.

The calculations just described rely on our ability to calculate the cumulative distribution function of the random variable of interest. For random variables for which this is difficult, another approach might be employed. For a discrete random variable, an approximation using the total variation distance (1.23) introduced in Section 1.3.8 might be tried. This procedure, however, should be used with caution. Two probability distributions might have a small total variation distance between them and yet the probability distribution of the maximum of n independent random variables from one distribution might have a large total variation distance from the probability distribution of the maximum of n independent random variables from the other distribution. As an example, consider the two probability distributions in Table 3.2. The total variation distance between the two distributions shown in the table is less than 10^{-4}. Despite this, it is almost certain that the maximum of 10^{6} random variables taken from the first distribution is 1, whereas it is almost certain that the maximum of 10^{6} random variables taken from the second distribution is 2. As a result, the total variation distance between the probability distributions of the two maxima is then almost 1, as large as is possible. Thus care must be taken when using the total variation distance concept of individual random variables when a procedure is employed using the maximum of n of these random variables.

Possible values	0	1	2
Probabilities for distribution 1	$1 - 10^{-4} - 10^{-8}$	10^{-4}	10^{-8}
Probabilities for distribution 2	$1 - 10^{-4} - 10^{-8}$	10^{-8}	10^{-4}

Table 3.2.

3.7.2 P-values for the Maximum of Geometric Random Variables

Suppose that Y_{max} is the maximum of n iid random variables, each having the geometric distribution (1.69). The probability that $Y_{max} \geq y$, for any given value of y, is given in (2.111). The exponential limiting process in equation (B.3) in Appendix B shows that, to a close approximation, the P-value associated with an observed value y_{max} of Y_{max} is

$$P\text{-value} \approx 1 - e^{-n\,e^{-\lambda y_{max}}} \tag{3.48}$$

as $n \to \infty$. This asymptotic relation can be written in the form

$$P\text{-value} \approx 1 - e^{-e^{-\lambda(y_{max} - \log n/\lambda)}} \tag{3.49}$$

as $n \to \infty$. If we denote by μ_{max} and σ^2_{max} the approximate mean and variance of Y_{max}, given respectively in (2.118) and (2.120), we get

$$\frac{\log n}{\lambda} \approx \mu^* - \frac{\gamma}{\lambda}, \quad \lambda \approx \frac{\pi}{\sigma^* \sqrt{6}}$$

as $n \to \infty$, where

$$\mu^* = \mu_{max} + \frac{1}{2}, \quad (\sigma^*)^2 = \sigma^2_{max} - \frac{1}{12}. \tag{3.50}$$

The approximation (3.49) can then be written in the more cumbersome, but more easily generalized, form

$$P\text{-value} \approx 1 - e^{-e^{-\left(\pi(y-\mu^*)/(\sigma^*\sqrt{6})+\gamma\right)}}. \tag{3.51}$$

The similarity between the approximations (3.51) and (2.131) is clear. However, (2.131) applies for the maximum of a wide range of continuous random variables, and is not directly applicable for discrete random variables. This implies that it is important to assess the use of the approximation (3.51) to find the P-value associated with the observed value y of the maximum Y_{max} of n iid discrete random variables. This approximation is made in several areas of bioinformatics; a specific example will be discussed in Section 6.3 in connection with the comparison of two DNA sequences.

Suppose that μ_{max} and σ^2_{max} denote the mean and variance of Y_{max}, and μ^* and $(\sigma^*)^2$ are defined as in equation (3.50). In Table 3.3 we display the mean μ_{max} and the variance σ^2_{max} of the maximum of 75,000 independent random variables, each having the generalized geometric distribution (1.21). These are calculated from equations (2.103) and (2.104) and the theory in Appendix C. The calculations assume that $p = \frac{1}{4}$ and cover the cases $k = 0, 1, \ldots, 5$. The means and variances in Table 3.3 can now be used, in conjunction with (3.50) and (3.51), to obtain approximate P-values for discrete generalized geometric random variables, which can then be compared with *exact* P-values calculated from equation (2.101) and the theory of Appendix C. Both sets of values are displayed in Table 3.4.

k	0	1	2	3	4	5
$\mu_{\max}$	8.013	10.559	12.812	14.919	16.933	18.883
$\sigma^2_{\max}$	0.939	1.055	1.152	1.244	1.331	1.413

Table 3.3. The mean $\mu_{\max}$ and the variance $\sigma^2_{\max}$ of the maximum of 75,000 iid generalized geometric random variables, for various values of k. $p = \frac{1}{4}$.

	$k = 0$		$k = 1$		$k = 2$	
y	Exact	Approx	Exact	Approx	Exact	Approx
7	0.990	0.990				
8	0.682	0.682				
9	0.249	0.249	1.000	1.000		
10	0.069	0.069	0.891	0.892		
11	0.018	0.018	0.456	0.455	1.000	1.000
12	0.004	0.004	0.152	0.152	0.940	0.934
13	0.001	0.001	0.044	0.044	0.563	0.558
14			0.012	0.012	0.214	0.218
15			0.003	0.003	0.067	0.071
16					0.006	0.007

	$k = 3$		$k = 4$		$k = 5$	
y	Exact	Approx	Exact	Approx	Exact	Approx
14	0.950	0.952				
15	0.604	0.603				
16	0.247	0.245	0.943	0.937		
17	0.082	0.082	0.604	0.597		
18	0.025	0.026	0.256	0.258	0.923	0.918
19	0.008	0.008	0.089	0.094	0.597	0.572
20	0.002	0.002	0.029	0.032	0.249	0.251
21			0.009	0.010	0.089	0.093
22			0.003	0.003	0.030	0.033

Table 3.4. Exact and approximate P-values for the maximum of 75,000 iid generalized geometric random variables, for various values of k and y. Exact values are from equation (2.101) and the theory of Appendix C. Approximations are obtained using equations (3.50), (3.51), and the values in Table 3.3 with $p = \frac{1}{4}$.

The case $k = 0$ corresponds to the geometric distribution, for which the approximation using (3.51) is very accurate. The values in Table 3.4 show that the approximation using (3.51) is also good for all values of k up to 5, at least for the given range of values of y, which covers cases of practical interest. On the other hand, the values in Table 3.4 show that the accuracy of the approximation near the mean of $Y_{\max}$ decreases slightly as k increases.

These calculations suggest that, although no theory is available for the discrete random variable case, useful approximations can be obtained, at least for the generalized geometric distribution, from the approach described.

Apart from assessing the accuracy of the approximation (3.51), the calculations in Table 3.4 illustrate a point made in Section 2.11.2, that very sharp changes in the cumulative distribution function of the maximum of n iid random variables occur near the mean when n is large. For example, in the case $k = 5$, the probability that $Y_{\max} \geq 18$ is 0.923 and the the probability that $Y_{\max} \geq 21$ is 0.093. This implies that unless very accurate approximations are used for the mean and variance in (3.51), serious errors in P-value approximations can arise when the approach via (3.51) is used. We return to this matter in Section 6.3, where we apply the maximum of generalized geometric distributions to testing the significance of a DNA sequence alignment.

3.8   Nonparametric Alternatives to the One-Sample and Two-Sample t-Tests

The two-sample equal variance t-test described in Section 3.5.2 is an example of a distribution-dependent test. In the t-test we are concerned with random variables having the normal distribution. The aim is to test hypotheses about the mean of this distribution or the means of several normal distributions, using the observed value of some test statistic. The test statistic is often determined by theoretical and optimality arguments – in the case of the t-test, these are described in Chapter 9. The null hypothesis probability distribution of this test statistic is calculated, based on the assumed normal distribution of the random variables in the test, and this allows the calculation of P-value for the observed value of the test statistic. The null hypothesis is then rejected if this P-value is less than the chosen Type I error.

Problems do exist with this approach, as exemplified by the unequal variance case discussed in Section 3.5.2, but these are not the subject of the present discussion. For the moment we assume a clear-cut procedure as exemplified by the equal variance t-test.

Any P-value calculation relies on the assumption that the distributions assumed for the random variables involved in the test are correct. If this assumption is not justified, the use of the calculated P-value leads to an invalid testing procedure, in that the actual P-value will almost certainly not be equal to the one calculated. For example, if the observations come from an exponential distribution, the probability that the t statistic exceeds any specific value in t tables is not accurately provided by the values given in the table.

This difficulty leads to hypothesis testing methods that do not rely on any specific assumption about the form of the probability distribution of the random variables involved in the test. Such methods are called *non-parametric*, or (perhaps more accurately) *distribution-free*. We now describe two non-parametric procedures that are sometimes used as alternatives to the two-sample t-test and one non-parametric procedure that is sometimes used as an alternative to the one-sample t-test.

3.8.1   The Two-Sample Permutation Test

The two-sample permutation test is often preferred to the two-sample t-test of Section 3.5.2, since use of the t test strictly assumes that the data analyzed have a normal distribution, whereas use of the permutation test does not imply this assumption. When the data approximately have a normal distribution the t test will usually lead only to small errors, but in some bioinformatics applications one might be unwilling to make the normal distribution assumption. When the data do have a normal distribution, with the same variances in the two groups, use of the two-sample permutation procedure gives results close to those obtained using the equal variance two-sample t-test when the sample size is sufficiently large.

Second, it is not necessary to find the probability distribution of the test statistic used in this or any other permutation procedure, a calculation that can be quite difficult for parametric tests.

We now describe the permutation procedure. We assume two groups of random variables, $X_{11}, X_{12}, \ldots, X_{1m}$ in the first group and $X_{21}, X_{22}, \ldots, X_{2n}$ in the second. $X_{11}, X_{12}, \ldots, X_{1m}$ are assumed to be iid, as are $X_{21}, X_{22}, \ldots, X_{2n}$, with X_{1i} possibly having a distribution different from that of X_{2j}. It is assumed that X_{1i} is independent of the X_{2j} for all (i, j). The null hypothesis tested in the permutation procedure is that the common probability distribution of X_{1i} is identical to that of common distribution of the X_{2j}.

There are $Q = \binom{m+n}{m}$ possible ways in which the observed values x_{11}, $x_{12}, \ldots, x_{1m}, x_{21}, x_{22}, \ldots, x_{2n}$ these random variables can be placed into two groups, with m observed values in one group and the remaining n in the other. Each such rearrangement is called a *permutation*. When the null hypothesis is true, all Q of these have the same probability. (This is intuitively clear, and can be derived rigorously from the comments following equation (2.136).)

For each such permutation we calculate the value of some test statistic, the choice of which we discuss further below. For a Type I error α, the null hypothesis is rejected if the observed value of the statistic is among the $100\alpha\%$ most extreme permutation values of the test statistic.

There is no clear-cut guidance for the choice of test statistic. However, regardless of which statistic is used, the procedure outlined above will give the exact Type I error rate α desired, as shown below. The power of the test, however, will depend on the choice of statistic. The null hypothesis in a permutation procedure is that the complete probability distribution of X_{1i} is the same as that of X_{2j}. If the alternative hypothesis is that the distributions are identical except for a shift of one with respect to the other, either the t or the t' statistic should provide a powerful test procedure. If the alternative hypothesis had been that the means of the two distributions are the same but their variances differ, the F-statistic described in Section 3.5.3 should provide a powerful test.

The logic behind the permutation procedure is as follows. When the null hypothesis is true, all Q permutation values of the test statistic have the same probability, so that the actual observed value of the test statistic should rarely be extreme. More precisely, it will be among the $100\alpha\%$ most extreme values with probability α. When null hypothesis is not true, a random allocation of the $m + n$ observations into the two groups will tend to diminish differences between the two sets of "observations" in the two groups. This will tend to move the permutation values of the test statistic towards the value to be expected under the null hypothesis, and the un-permuted actual value of the test statistic now will tend to be an extreme one.

We now discuss the concept of a "permutation P-value" for the case of continuous random variables. We write the statistic that is calculated in each permutation as s, and suppose, to be concrete, that a one-sided test is carried out, with sufficiently large positive values of s leading to rejection of the null hypothesis. (The argument for a two-sided test is similar to that given here.)

One possible definition of the permutation P-value is $j/(Q-1)$, where j is the number of permutations for which the permutation value of s is greater than actual observed value of s. This permutation P-value takes one of the values $0, 1/(Q-1), 2/(Q-1), \ldots, 1$. Suppose that the order statistics of all Q values of s, including the value actually observed, are $s_{(1)}, s_{(2)}, \ldots, s_{(Q)}$. If the null hypothesis is true, the probability that the observed value of s is equal to $s_{(i)}$ is $1/Q$ for all $i, i = 1, 2, \ldots, Q$. The event that the permutation P-value as defined above is equal to $j/(Q-1)$ is identical to the event that the observed value of s is t_{Q-s}, so that it has probability $1/Q$. Thus when the null hypothesis is true, the permutation P-value as defined above has the discrete uniform distribution with range $\{0, 1/(Q-1), 2/(Q-1), \ldots, 1\}$. The discussion of the null hypothesis properties of P-values in Section 3.4.2 shows that this is one desirable feature of this definition of a P-value. This argument also verifies the claim made above, that rejection of the null hypothesis if the observed value of s is among the $100\alpha\%$ most extreme permutation values of the test statistic leads to a Type I error α, since when the null hypothesis is true the uniform distribution of the P-value implies that the probability that the observed value of s is among the $100\alpha\%$ most extreme permutation values is α.

While the P-value as defined above has a (discrete) uniform distribution when the null hypothesis is true, it has the undesirable feature that there is a probability $1/Q$ that that it takes the value 0, even if the null hypothesis is true. A conservative procedure avoiding this is to define the permutation P-value as j/Q, where j is the number of permutations, including the actual one defined by the data, for which the permutation value of s is greater than or equal to the actual observed value of s. This conservative procedure is generally used in practice. A further reason for using this procedure is that for a test statistic having a discrete distribution, a P-value of 0 is impossible (see Section 3.4.2).

Other definitions of the permutation P-value appear in the literature. For example, Wasserman (2004) defines the permutation P-value as $(j-1)/Q$. However, under this definition the permutation P-value does not have a discrete uniform distribution of the form (1.14), and further, it can take the value 0.

We turn now to computational matters. If the equal variance t statistic (3.30) had been chosen as the test statistic s, it is not necessary to compute the value of t for each permutation, and a simpler equivalent procedure requires only the calculation of the difference $d = \bar{x}_1 - \bar{x}_2$ for each permutation. We now demonstrate this for the two-sided test; the one-sided case is left as an exercise (see Problem 3.12).

In the two-sided case it is equivalent to use t^2 instead of t as the test statistic, since t^2 is a monotonic function of $|t|$. Standard algebra shows that the numerator of the joint variance estimator (3.31) can be written as

$$S^2 = \sum_{i=1}^{m} X_{1i}^2 + \sum_{i=1}^{n} X_{2i}^2 - (m+n)\bar{\bar{X}}^2 - \frac{mn}{m+n}D^2.$$

The sum of the first three terms on the right-hand side is invariant under permutation, and we write it as C. This means that the square of the observed value of the t-statistic (3.30) can be written as

$$t^2 = \frac{mn(m+n-2)^2}{m+n}\frac{d^2}{C - \frac{mn}{m+n}d^2}. \tag{3.52}$$

This is a monotonic increasing function of d^2, and the result follows.

A further computational convenience is that it is only necessary to calculate the average of the "observations" in the *first* group rather than the difference $\bar{x}_1 - \bar{x}_2$. This is because the sum of the observations in the two groups is invariant under permutation, so that the average of the "observations" in the second group, and hence $\bar{x}_1 - \bar{x}_2$, is determined by the average of the "observations" in the first group.

The permutation procedure clearly involves substantial computation unless both n and m are small, since the number of different permutations is extremely large even for relatively small values of n and m. Modern computing power makes this an increasingly unimportant problem. When

m and n are jointly so large that computation of all Q permutations is not feasible, close approximations to P-values and other quantities may be found from a random sample of a large number of permutations.

Difficulties arise with the permutation procedure using the t statistic when the numbers m and n of observations in the two groups are small, so that the number of possible permutations of the data is not large. In the case $m = n = 3$, for example, there are 20 possible permutation values of t, and for each positive value there is a corresponding negative value of the same magnitude. It is therefore impossible to achieve significance with Type I error of 5% for a two-sided test, since even if the observed value of the test statistic is the most extreme positive of the 20 possible permutation values, there will be a corresponding negative value of equal magnitude, and the P-value estimate is $2/20 = 0.10$. When $n = m = 4$ there are 70 possible permutations of the data and thus a P-value of $2/70 \cong 0.03$ would arise it the observed value of t is the most extreme one (positive or negative). If observed value of t is the next most extreme one (positive or negative), the P-value estimate is $4/70 \cong 0.06$, which is not significant with a Type I error of 5%. We call the problem discussed above a "granularity" problem.

We conclude with a brief discussion of the relative merits of using the t statistic (3.30) or the "unequal variance" statistic t' defined in (3.37). The statistic t' is used in cases where the variances in the two groups are thought to differ. However, as emphasized above, the permutation procedure null hypothesis is that the complete distributions of the random variables in the two groups are identical; in particular, the variances as well as the means are assumed equal under the null hypothesis. Thus use of t' as test statistic in the permutation procedure might seem illogical.

3.8.2 The Mann–Whitney Test

The *Mann–Whitney* test (sometimes called the "Wilcoxon two-sample test") is a frequently used non-parametric alternative to the two-sample t-test discussed in Section 3.5.2, and we assume here the same random variables as for that test.

In the Mann–Whitney test the observed values $x_{11}, x_{12}, \ldots, x_{1m}$ and $x_{21}, x_{22}, \ldots, x_{2n}$ of the $m + n$ random variables are listed in increasing order, and each observation is associated with its rank in this list. Thus each observation is associated with one of the numbers $1, 2, \ldots, m + n$. (If ties exist a slightly amended procedure is used.) Various (equivalent) forms of the test statistic for this test exist: here we take it to be the sum of the ranks of the observations from the first group. The sum of the ranks of all $m + n$ observations is $(m + n)(m + n + 1)/2$ (see Appendix B.18). The observations in the first group provide a fraction $m/(m + n)$ of all observations, and when the null hypothesis is true, the expected value of the sum of the ranks of the observations in the first group is the fraction $m/(m+n)$ of the sum $(m+n)(m+n+1)/2$ of all ranks. Thus this expected value is $m(m + n + 1)/2$. A more advanced calculation (see Problem 3.9) shows that the null hypothesis variance of the sum of the ranks for the first group in the sample is $mn(m + n + 1)/12$. The central limit theorem shows that, for large sample sizes, (in practice when m and n both exceed about 20), the distribution of this sum is very close to a normal distribution. From the null hypothesis mean variance of the sum of ranks, together with its observed value, a z-score can then be calculated to test the null hypothesis. For smaller sample sizes an exact P-value can be calculated for any observed value of the Mann–Whitney test statistic, as described below.

As with the permutation test, the null hypothesis of the Mann–Whitney test is that the probability distribution of any random variable in the first group is identical to that of any random variable in the second group. It is not enough for the validity of the procedure that the means of the two distributions be the same. Even if the means of the two distributions are the same, if other characteristics of the two distributions differ, the distribution of the test statistic does not have the mean and variance given above. This is shown theoretically by Babu and Padmanabhan (2002). One consequence of this is that the probability of rejecting the null hypothesis that the means of the two groups are equal, even when this hypothesis is true, might exceed the chosen Type I error. We return to this issue in Section 3.8.4.

The permutation test and the Mann–Whitney test might initially appear to provide two different non-parametric alternatives to the two-sample t-test. However the two procedures are more closely related than might at first appear. The null hypothesis mean $m(m + n + 1)/2$ and variance $mn(m + n + 1)/12$ given above are the mean and variance of the Mann–Whitney test statistic under all possible permutations of the data. Thus the Mann–Whitney procedure is simply the permutation procedure applied to the ranks of the observations rather than to the observations themselves. From this it follows that a Mann–Whitney P-value can be found as the proportion of times, under permutation, that a value of the Mann–Whitney test statistic is obtained equal to or more extreme than the observed value.

The fact that the Mann–Whitney test is in effect a permutation test implies that the theoretical analysis of Babu and Padmanabhan (2002) for the Mann–Whitney test applies also to permutation tests.

3.8.3   The Wilcoxon Signed-Rank Test

The Wilcoxon signed-rank procedure is a test for the value of the *median* M of a continuous random variable, defined in (1.59), with no specific assumption being made about the form of the probability distribution of the random variables involved in the test. For a symmetric distribution the mean and the median coincide, so for a symmetric probability distribution the test may also be thought of as a test for the value of the mean.

The null hypothesis tested in the Wilcoxon signed-rank procedure is that the median of the distribution of the observations is some specified value M_0. Given the observed values $x_1, x_2, \ldots, x_n$ of n iid random variables from the distribution being tested, we first calculate the differences $x_i - M_0$, $i = 1, 2, \ldots, n$. In general, some of these differences will be positive, some negative. From these differences we calculate the absolute values $|x_i - M_0|$, $(x_1, x_2, \ldots, x_n)$. These absolute values are then ranked from smallest to largest and then given respective ranks $1, 2, \ldots, n$, with the smallest absolute difference receiving rank 1 and the largest rank n. (As with the Mann–Whitney test, an amended procedure is available when ties exist.) The sum of the ranks of the originally positive differences is then calculated, and this is the (observed value of) the test statistic in the Wilcoxon signed-rank procedure. If the null hypothesis is true, the probability that any difference $X_i - M_0$ is positive is $1/2$. This implies that, since the sum of all the ranks is $n(n + 1)/2$, (see Section B.18), the null hypothesis mean value of this test statistic is $n(n + 1)/4$. A more complex argument (see Problem 3.8) shows that the null hypothesis variance of this test statistic is $n(n + 1)(2n + 1)/24$. Further, the central limit theorem implies that the test statistic, being a sum, has approximately a normal distribution when n is large. If the Type I error is 5%, the alternative hypothesis is that the median exceeds M_0 and n is large, the observed value of the sum of ranks would be judged to be significant if it exceeds $n(n + 1)/4 + 1.645\sqrt{n(n + 1)(2n + 1)/24}$.

The Wilcoxon signed-rank procedure, as with the Mann–Whitney procedure, is in effect also a permutation test. We can consider all 2^n possible assignments to the sign ($+$ or $-$) of each $x_j - M_0$, calculate the value of the test statistic for each of these 2^n assignments, and reject the null hypothesis if the observed value is a significantly extreme member of these 2^n values. The mean and the variance given above for the test statistic are simply the mean and variance relative to all these permutation values.

For small n it is desirable to carry out the Wilcoxon signed-rank procedure as a permutation procedure, and this procedure allows the calculation of an exact P-value.

The Wilcoxon signed-rank procedure can be used to compare two probability distributions. This is done, for example, in the Affymetrix microarray procedure, as discussed in Chapter 13. It is therefore important to discuss this case in some detail. Suppose that paired data (as for the paired t-test) arise, but because we are unwilling to assume that the observations have a normal distribution, we do not wish to use the paired t-test procedure and the statistic (3.38). In this case the difference $d_j = x_{1j} - x_{2j}$ is first calculated for each pair of observations (x_{1j}, x_{2j}), $j = 1, 2, \ldots, n$. Given the observed values $d_1, d_2, \ldots, d_n$ the absolute values $|d_1|, |d_2|, \ldots, |d_n|$ are then calculated. The test then proceeds as in the Wilcoxon procedure described above, with $x_i - M_0$ replaced by $|d_i|$ and M_0 replaced by 0.

The null hypothesis tested by this procedure is that the median of the probability distribution of the differences D_i is 0. This is in principle different from the null hypothesis that the means, or the medians, in the two groups being compared are equal. This matter is discussed further in Section 3.8.4, where the question of what is assumed in a non-parametric test is taken up .

3.8.4   What is Assumed in a Non-Parametric Test?

Non-parametric tests are used frequently in bioinformatics, particularly in the analysis of microarray data (see Chapter 13). However, in the microarray literature incorrect assertions are sometimes made about what hypothesis any given non-parametric procedure actually tests. This section is devoted to a discussion of this issue.

Although non-parametric tests are attractive because they do not rely on the assumption that the random variables involved have some specified distribution, for example a normal distribution, they nevertheless do make distributional assumptions. We now discuss these assumptions for the non-parametric tests described above.

The null hypothesis in the one-sample Wilcoxon signed-rank test is that the median of the distribution of the random variables involved is some specified value M_0. When this test is used as a non-parametric alternative to the two-sample paired t-test, as described in Section 3.8.3, the formal null hypothesis tested is that the median of the differences D_i of the paired observations in the test is 0. This hypothesis does apply if the null hypothesis of interest is that the *complete distribution* of any observation in the first group is equal to that of any observation in the second group. However it does not necessarily apply if the null hypothesis of interest is that the *mean* of any observation in the first group is equal to the mean of any observation in the second group. Thus this null hypothesis is not validly tested by the Wilcoxon signed-rank test using differences.

The null hypothesis in the Mann–Whitney test, and also in the permutation test, is that the probability distributions of the observations in the two groups are identical. Thus these two tests are valid only under this null hypothesis. In particular they are not valid tests of the null hypothesis that the means in the two groups are equal. This is despite an assumption to the contrary sometimes made in the microarray literature.

The reason why a permutation test is not a valid test of the equality of two means is that, even when the means of two distributions are equal, the true Type I error of the permutation tests when used as tests for equality of the means is not necessarily the value chosen. The same remark applies to the Mann–Whitney test. This matter has been investigated numerically by Modarres et al. (2004), using simulation. As one example of their results, if the variance in one group is 10 times larger than that in the other, then for $n = 6, m = 12$, the true Type I error for the Mann–Whitney test is typically either about half, or about twice, the chosen value, depending on whether the group with the larger variance has 6 or 12 observations. The same comment is true for the permutation test when the "equal variance" t statistic is used in the procedure. Even when the "unequal variance" t' statistic is used in the permutation test the same problem arises. On the other hand, this problem is far less severe when the sample sizes in the two groups are equal. These and other results confirm the theoretical predictions of Babu and Padmanabhan (2002). Further details, and numerical examples, are provided by Modarres et al. (2004).

3.9   The Bayesian Approach to Hypothesis Testing

Let H be some hypothesis and let $\text{Prob}(H)$ be the probability that H is true, before any data are seen. This is called the *prior* probability of H, and some of the controversies referred to in Section 3.2 refer to the meaningfulness of such a probability or to its numerical value if meaningful. We proceed here assuming that this probability is meaningful and that its numerical value is specified.

We introduce the Bayesian argument with an example. Suppose that a bag contains ten coins, three of which are fair, the remaining seven having probability 0.6 of giving heads when flipped. A coin is taken at random from the bag and flipped five times. All five flips give heads. The *prior probability* $\text{Prob}(H)$ that the coin is fair is 0.3, and the *posterior* probability $\text{Prob}(H \mid D)$ that it is fair, given the data D that all five flips gave heads, is given by

$$\text{Prob}(H \mid D) = \frac{0.3(0.5)^5}{0.3(0.5)^5 + 0.7(0.6)^5} = 0.147. \tag{3.53}$$

The prior probability 0.3 has been decreased as a result of the flips of the coin.

The above calculation is a direct one and is not specifically Bayesian. The Bayesian approach arises when the prior probabilities are not obvious from the context, but are arrived at perhaps from experience or simply from a prior measure of belief. We illustrate the details below, where we generalize the above argument to the case when the aim is to amend the prior probability of some one of a finite number of *hypotheses* in the light of observed data.

The Case of a Finite Number of Hypotheses

Suppose that we have $h + 1$ different hypotheses, $H_0, H_1, \ldots, H_h$, and allocate respective *prior probabilities* $\pi_0, \pi_1, \ldots, \pi_h, \sum_j \pi_j = 1$ to these. These prior probabilities have been chosen perhaps by experience, perhaps from some prior measure of belief. From these prior probabilities and some observed data D, we wish to find the *the posterior probabilities* of these hypotheses.

Using the logic leading to the calculation in (3.53), the posterior probability of H_i is

$$\text{Prob}(H_i \mid D) = \frac{\text{Prob}(H_i)\,\text{Prob}(D \mid H_i)}{\text{Prob}(D)}. \tag{3.54}$$

By (1.105), the denominator can be written as

$$\sum_j \pi_j \,\text{Prob}(D \mid H_j). \tag{3.55}$$

Thus the posterior probability that hypothesis H_i is true becomes

$$\text{Prob}(H_i \mid D) = \frac{\pi_i \,\text{Prob}(D \mid H_i)}{\sum_j \pi_j \,\text{Prob}(D \mid H_j)}. \tag{3.56}$$

The preferred hypothesis might then be taken as the one that maximizes $\text{Prob}(H_i \mid D)$. Since the denominator in (3.56) does not depend on i, this is the same hypothesis that maximizes the numerator of (3.56).

The Case of a Continuum of Hypotheses

Suppose now that the hypothesis test of interest refers to a parameter θ which can take any value in the interval (a, b). Given the value of θ, the probability of observed data d is $f(d|\theta)$. A prior density function $g(\theta)$ is chosen for θ. Given d, the argument of the previous section extended to the continuous case leads to a posterior density function of θ given by

$$\frac{g(\theta)f(d|\theta)}{\int_a^b g(\theta)f(d|\theta)d\theta}. \tag{3.57}$$

As an example, suppose that parameter of interest is the parameter p, the probability of success in a Bernoulli trial, and that the prior density function of p is the beta distribution (1.78), rewritten (since the random variable is now the parameter p) as

$$g(p) = \frac{\Gamma(\alpha + \beta)}{\Gamma(\alpha)\Gamma(\beta)}p^{\alpha-1}(1 - p)^{\beta-1}, \quad 0 < p < 1. \tag{3.58}$$

The choice of the parameters α and β is influenced by the prior level of certainty about the value of p. If these two parameters are both large, the density function (3.58) is tightly concentrated around the mean $\alpha/(\alpha+\beta)$, implying a high prior level of certainty that p is close to this value. Values of α and β close to 1 imply a small prior level of certainty about the value of p. Extremely small values of these two parameters imply, perhaps paradoxically, high prior probabilities that p is close either to 0 or to 1.

Suppose that in n independent Bernoulli trials there are y successes. The probability of these data is given by the binomial distribution (1.8). Following the format given in (3.57), the numerator in the posterior density function of p is the product of the prior density function (3.58) and the binomial probability (1.8), namely

$$\frac{\Gamma(\alpha + \beta)}{\Gamma(\alpha)\Gamma(\beta)} \binom{n}{y} p^{\alpha + y - 1}(1 - p)^{\beta + n - y - 1}. \tag{3.59}$$

Similarly the denominator in the posterior density function of p is

$$\int_0^1 \frac{\Gamma(\alpha + \beta)}{\Gamma(\alpha)\Gamma(\beta)} \binom{n}{y} x^{\alpha + y - 1}(1 - x)^{\beta + n - y - 1} dx \tag{3.60}$$

$$= \frac{\Gamma(\alpha + \beta)}{\Gamma(\alpha)\Gamma(\beta)} \binom{n}{y} \frac{\Gamma(\alpha + y)\,\Gamma(\beta + n - y)}{\Gamma(\alpha + \beta + n)}. \tag{3.61}$$

Thus the posterior density function of p is also a beta distribution, namely

$$f_{\text{post}}(p) = \frac{\Gamma(\alpha + \beta + n)}{\Gamma(\alpha + y)\Gamma(\beta + n - y)} p^{\alpha + y - 1}(1 - p)^{\beta + n - y - 1}. \tag{3.62}$$

By analogy with the discrete case, the hypothesized value of p could be taken as the value where this posterior density function is maximized, namely

$$\frac{\alpha + y - 1}{\alpha + \beta + n - 2}. \tag{3.63}$$

3.10   The Bayesian Approach to Estimation

There are various ways in which Bayesians estimate unknown parameters. These methods share the common feature that they are all based on the posterior distribution of that parameter. In this section we describe one such estimation method, namely that of estimating the parameter by the mean of the posterior distribution of that parameter.

In the binomial example of the previous section, the mean of the posterior distribution is given by

$$\int_0^1 p f_{\text{post}}(p) dp = \frac{\alpha + y}{\alpha + \beta + n}, \tag{3.64}$$

and this is the Bayesian estimate of p. It is a mixture of the parameters α and β in the prior distribution and the data values y and n, being heavily influenced by the latter when α and β are small and y and n are large.

Choosing α and β to be large overcomes the difficulty mentioned at the beginning of this chapter, that it is unreasonable to use the classical estimate of 1 for the probability of a head if a fair-looking coin gives three heads from three tosses. Thus if $\alpha = \beta$, the larger the value of α the closer the Bayesian estimate found from (3.64) is to 0.5. On the other hand, the larger that α and β are chosen, the more the information in the data is ignored.

From the classical perspective, an interpretation for α and β can be found through the concept of "pseudocounts." If we imagined that we observed α successes and β failures apart from the y successes and $n - y$ failures actually seen in the data, so we replace y by $y + \alpha$ and $n - y$ by $n - y + \beta$, then the classical estimator of p would be given by (3.64). The numbers α and β are then called "pseudocounts": There is a direct extension of the pseudocount concept to the multinomial case, and this extension is used in Section 6.6 for the purpose of sequence segment alignment.

3.11   Multiple Testing

Up to this point we have only discussed the testing procedure applicable for a single hypothesis. However, we often wish to test many associated hypotheses at the same time. This leads to the so-called multiple testing problem. When testing many hypotheses simultaneously, careful attention must be paid to what may be concluded from the P-values of the individual tests. As an example, suppose that a salesman claims that some substance is beneficial, and to demonstrate this, he tests 100 different null hypotheses, one for each of 100 different illnesses, each null hypothesis stating that the substance is not useful in helping to cure one of the illnesses involved and each alternative hypothesis stating that the substance is useful. Suppose that the appropriate statistical tests lead to 100 P-values, one for each disease. If a Type I error of .01 had been used for each individual test, the probability that at least one of the 100 null hypotheses will be rejected is about 0.63 if the substance is of no value for any of the illnesses. This might lead the salesman to claim that the substance is useful for those illnesses where, by chance, the null hypothesis was rejected. He may provide only the data relating to those illnesses, which taken out of context may appear convincing. This indicates the potential error in drawing conclusions from individual P-values obtained from many associated tests carried out in parallel.

One possible approach to this problem is to use a so-called "experiment-wise," or "family-wise," P-value, defined as follows. The probability of rejecting at least one null hypothesis when all null hypotheses are true is known as the *family-wise error rate* (FWER). Under this approach, the aim is to control the FWER at some value acceptable value α. To do this it is necessary to use a Type I error for each individual test that is smaller than α. One of the simplest ways of assuring an FWER of α when g tests are performed, is to use a Type I error of α/g for each individual test. This is an example of a *Bonferroni* correction. This correction applies whether the tests are independent or not, as is shown from the inequality (1.99): If the event A_j is the rejection of null hypothesis H_j, and if all null hypotheses are true, the probability of rejecting at least one null hypothesis is less than or equal to $\sum_{j=1}^{g} \alpha/g = \alpha$, from (1.99).

The Bonferroni procedure is a conservative one. A less conservative procedure for *independent tests*, with the same FWER of α, is to use $K(n, \alpha)$, defined in equation (2.141), as the Type I error for each of the n individual tests. If this is done the FWER is, from (1.111), exactly

$$1 - \prod_{j=1}^{g} \sqrt[g]{1 - \alpha} = \alpha.$$

This is the *Šidák* procedure.

We now consider a more complicated example where the multiple hypothesis testing problem arises. Suppose we measure the expression levels of a large number g of genes in a normal cell and also in a cancerous version of the same cell type, using data from cells from different individuals. From these we find the values of g different t statistics, calculated for each gene as in (3.33). The g genes are now ranked according to the absolute values of the t statistics.

Our interest is in knowing if a certain type of genes, for example genes involved in cell division, is associated with those genes with large $|t|$ values. We might test this as follows. We choose some number g_0 and assess whether cell-division genes are significantly over-represented in the genes having the g_0 highest $|t|$ values, using Fisher's exact test, described in Section 3.5. However, a non-significant result, if one occurs, might have arisen because we did not chose a suitable value of g_0. We might then decide to do this test for several choices of g_0. If we do this, we would not want to make a false claim of significance simply because we carried out a number of tests, so that we would wish to control the FWER of our procedure to an acceptable level.

There are two comments to make about the above procedure. First, Fisher's exact test requires independent counts in Table 3.5.5. To the extent that some genes act in a correlated way, this requirement might not be met in the above example. Second, the various tests corresponding to different choices of g_0 are not independent, so that even if the dependence between genes problem does not arise, a Bonferroni rather than a Šidák procedure should be used.

Both the Bonferroni and the Šidák procedures make quite stringent requirements for the rejection of any one of the null hypotheses, since when n is large both α/n and $K(n, \alpha)$ are very small. This implies a loss of power: For large n it is difficult to reject those null hypotheses that are false. Furthermore, the independence requirement necessary for the Šidák procedure is rarely satisfied for expression array data, which we consider in Chapter

13. In that chapter we consider two approaches that have been proposed to increase power. One of these controls the FWER while the other abandons this attempt and controls, instead, the so-called "false discovery rate," or FDR, discussed in Section 13.3.

The multiple testing problem arises in other contexts discussed in this book. It arises for example in BLAST calculations: See the concluding comments in Section 10.2.8. More generally the problem arises in several applications of statistics and has been discussed extensively in the statistical literature; for a good survey see Shaffer (1995), in which many further references to this problem may be found.

3.12   Combining the Results of Several Experiments

Suppose that three independent investigators, each considering the same problem and conducting the same statistical test with three independent data sets, arrive at P-values of 0.08, 0.07, and 0.06 respectively, derived from some test statistic with a continuous distribution. Although none of the individual tests is judged significant if a Type I error of 5% is used, one might feel that the results of the three tests are significant in conjunction. We now describe perhaps the most popular way in which, under the assumption that the test statistic is continuous, the individual P-values can be combined to give an overall P-value. This procedure is due to Fisher (1950).

If the null hypothesis being tested is true, any P-value derived from a test statistic with a continuous distribution has a uniform distribution in $(0, 1)$, as discussed in Section 3.4.2. The transformation theory of Section 1.15 shows that if X_1 has a uniform distribution in $(0, 1)$ and if $X_2 = -2 \log X_1$, then (see Problem 3.14) the density function of X_2 is

$$f_{X_2}(x) = \frac{1}{2}e^{-x/2}, \ x > 0. \tag{3.65}$$

Equation (1.77) shows that X_2 has a chi-square distribution with two degrees of freedom. Example 2 of Section 2.3.2 shows that the sum of independent chi-square random variables has a chi-square distribution with degrees of freedom equal to the sum of the degrees of freedom of the individual chi-square random variables. It follows that when the null hypothesis is true in each of k tests, the quantity V, defined by

$$V = -2 \sum_i \log P_i = -2 \log(P_1 P_2 \cdots P_k), \tag{3.66}$$

where $P_1, P_2, \ldots, P_k$ are the P-values of the individual tests, has a chi-square distribution with $2k$ degrees of freedom.

In the example given above, where $k = 3$ and the individual P-values are $0.08, 0.07$, and 0.06, the observed value of V is 16.00. The probability that a chi-square random variable with six degrees of freedom exceeds 16.00 is about 0.0138, so this value of V is significant at the 5% level and is almost significant at the 1% level. Therefore the tests in conjunction do provide a significant result at the 5% level.

If different sample sizes had been used in the three tests, or more generally if the results of the various tests had been thought to have different reliabilities, a generalization of this method is possible using as test statistic a weighted average of the form

$$W = -2 \sum_i \alpha_i \log P_i, \qquad (3.67)$$

where the α_i are positive weighting constants such that $\sum \alpha_i = k$. The mean and variance of W are $2k$ and $4 \sum \alpha_i^2$ respectively. It follows that W itself does not in general have a chi-square distribution, since the variance of a chi-square random variable is twice the mean. However, the quantity $kW/\sum \alpha_i^2$ has mean $2k^2/\sum \alpha_i^2$ and variance $4k^2/\sum \alpha_i^2$, twice the mean. A heuristic argument then claims that $kW/\sum \alpha_i^2$ has an approximate chi-square distribution with $2k^2/\sum \alpha_i^2$ degrees of freedom. Even though this number might well not be an integer, $kW/\sum \alpha_i^2$ can be employed for an approximate overall test procedure, using interpolation in chi-square tables.

When the test statistic has a discrete distribution, the inequality (3.23) shows that the assumption that the P-value has a continuous uniform distribution in $[0,1]$ leads to a conservative testing procedure. This implies that for a combined test, the assumption that the statistic V defined in (3.66) has a chi-square distribution with $2k$ degrees of freedom when the null hypothesis is true leads to a conservative testing procedure.

The Fisher approach described above for combining the results of independent individual tests is a heuristic one, but does have some optimality properties, as shown by Littell and Folks (1973). Other approaches are discussed by Berk and Cohen (1979), Mudholkar and George (1979), Rosenthal (1978) and Scholz (1982).

Problems

3.1 Suppose that each of n iid random variables has a normal distribution with mean μ, variance 1. Under the null hypothesis $\mu = 0$, while under the alternative hypothesis $\mu > 0$. Use normal distribution tables to find the P-value associated with an observed value of 4.20 for the maximum of the observed values of these random variables when $n = $ (i) 1,000, (ii) 10,000, (iii) 100,000, (iv) 1,000,000.

3.2 If $\hat{p}$ is the proportion of successes in n independent Bernoulli trials each having probability p of success, show that the mean value of $\hat{p}(1 - \hat{p})/n$ is $p(1 - p)/n - p(1 - p)/n^2$. Comment on the form of this expression when $n = 1$.

3.3 Suppose that $g(Y)$ is an unbiased estimator of p^{-1} in the binomial distribution (1.8). Then it follows that

$$\sum_{y=0}^{n} g(y)\, p \binom{n}{y} p^y (1 - p)^{n-y} = 1.$$

The left-hand side is a Taylor series in p. Use the result of Appendix B.13 to show that this equation cannot be solved for $g(y)$ uniformly for a continuum of values of p. This conclusion shows that there can be no unbiased estimator of p^{-1}.

3.4 *Continuation.* Find an unbiased estimator of p^2 and an unbiased estimator of p^3. What functions of p do you think admit unbiased estimation?

3.5 Prove equation (3.14).

3.6 Put $Y_i = np_i(1 + \delta_i)$ in the expression (3.41), where δ_i can be taken as being small. Use the expression (B.25) to approximate $\log(1 + \delta_i)$, as well as the fact that $\sum_i np_i\delta_i = 0$, to show that the numerical values of the expressions (3.40) and and (3.41) will usually be close.

3.7 Derive the density function (3.29) of the T statistic defined in (3.28), as follows.

T can be written as

$$T = \sqrt{n - 1} \left(\frac{(\bar{X} - \mu_0)\sqrt{n}}{\sigma} \right) \left(\frac{(n - 1)S^2}{\sigma^2} \right)^{-1/2}.$$

When the null hypothesis is true, $U = (\bar{X} - \mu_0)\sqrt{n}/\sigma$ has the standard normal distribution, and you may assume that $V = (n - 1)S^2/\sigma^2$ has a chi-square distribution with $n - 1$ degrees of freedom and is independent of $(\bar{X} - \mu_0)\sqrt{n}/\sigma$. (The proof of these facts is beyond the level of this book.) The joint density function of U and V is therefore the product of their separate density functions. From this the joint density function of T and random variable $W = V$ can be found by transformation techniques. The density function of T is then found by integrating out W.

3.8 (This example illustrates a use of indicator functions in statistics.) The test statistic in the Wilcoxon signed-rank test of Section 3.8.3 can be written as $\sum_{j=1}^{n} jI_j$, where under the null hypothesis $I_j = 1$ with probability

$1/2$, $I_j = 0$ with probability $1/2$. Use this fact to show that the null hypothesis variance of this statistic is $n(n+1)(2n+1)/24$.

3.9 (This example also illustrates a use of indicator functions in statistics.) The sum of the ranks randomly allocated to group 1 in the Mann–Whitney test of Section 3.8.2 can be written as $\sum_{j=1}^{m+n} j I_j$, where $I_j = 1$ if the jth-ranked observation is allocated to group 1, and $I_j = 0$ otherwise. Prove $E(I_j) = m/(m+n)$ and $E(I_j I_k) = m(m-1)/(m+n)(m+n-1)$, $(j \neq k)$, and use these results to show that the variance of the sum of the ranks randomly allocated to group 1 is $mn(m+n+1)/12$.

3.10 Suppose that $X_1, X_2, \ldots, X_n$ are $\mathrm{NID}(\mu, \sigma^2)$. Suppose that the prior distribution of μ is normal with mean μ_0 and variance σ_0^2. Show that the posterior distribution of μ is normal, with mean which is a linear combination of $\bar{X}$ and μ_0. Show that the variance of this distribution is the reciprocal of the sum of the reciprocals of σ_0^2 and the variance of $\bar{X}$.

3.11 Show that when the Šidák value (2.141) is used the experiment-wise Type I error is exactly α.

3.12 Show that in the permutation two-sample test using the equal variance t statistic, use of t as the test statistic is equivalent to use of $\bar{x}_1 - \bar{x}_2$ in the one-sided case.

3.13 *Continuation.* Find the mean and variance under permutation of the average $\bar{x}_1$ of the "observations" in the first group. (The answer is a function of $x_{11}, x_{12}, \ldots, x_{1m}$ and $x_{21}, x_{22}, \ldots, x_{2n}$.)

3.14 Prove the conclusion given in (3.65).

4

Stochastic Processes (i): Poisson Processes and Markov Chains

4.1 The Homogeneous Poisson Process and the Poisson Distribution

In this section we state the fundamental properties that define a Poisson process, and from these properties we derive the Poisson distribution, introduced in Section 1.3.7.

Suppose that a sequence of events occurs during some time interval. These events form a homogeneous Poisson process if the following two conditions are met:

(1) The occurrence of any event in the time interval (a, b) is independent of the occurrence of any event in the time interval (c, d), where (a, b) and (c, d) do not overlap.

(2) There is a constant $\lambda > 0$ such that for any sufficiently small time interval $(t, t + h)$, $h > 0$, the probability that one event occurs in $(t, t + h)$ is independent of t, and is $\lambda h + o(h)$ (the $o(h)$ notation is discussed in Appendix B.8), and the probability that more than one event occurs in the interval $(t, t + h)$ is $o(h)$.

Condition 2 has two implications. The first is *time homogeneity*: The probability of an event in the time interval $(t, t + h)$ is independent of t. Second, this condition means that the probability of an event occurring in a small time interval is (up to a small-order term) proportional to the length of the interval (with fixed proportionality constant λ). Thus the probability of no events in the interval $(t, t + h)$ is

$$1 - \lambda h + o(h), \tag{4.1}$$

and the probability of one *or more* events in the interval $(t, t + h)$ is

$$\lambda h + o(h).$$

The two conditions listed above are often taken as defining "randomness" of the occurrences of the events in question. Various naturally occurring phenomena follow, or very nearly follow, these two conditions. Continuing the example in Section 1.13, suppose a cellular protein degrades spontaneously, and the quantity of this protein in the cell is maintained at a constant level by the continual generation of new proteins at approximately the degradation rate. The number of proteins that degrade in any given time interval approximately satisfies conditions 1 and 2. The justification that condition 1 can be assumed in the model is that the number of proteins in the cell is essentially constant and that the spontaneous nature of the degradation process makes the independence assumption reasonable. The discussion leading to (1.116) and that following equation (2.87) makes condition 2 reasonable for spontaneously-occurring phenomena. This condition also follows from the same logic discussed in the example on page 10 that when np is small, the probability of at least one success in n Bernoulli trials is approximately np. For a precise treatment of this issue, see Feller (1968).

We now show that under conditions 1 and 2, the number N of events that occur up to any arbitrary time t has a Poisson distribution with parameter λt.

At time 0 the value of N is necessarily 0, and at any later time t, the possible values of N are 0, 1, 2, 3, ... We denote the probability that $N = j$ at any given time t by $P_j(t)$. Note that this is a departure from our standard notational convention, which would be $P_N(j)$ with t an implicit parameter. This notational change is made because the main interest here is in assessing how $P_j(t)$ behaves as a function of j and t.

The event that $N = 0$ at time $t + h$ occurs only if no events occur in $(0, t)$ and also no events occur in $(t, t + h)$. Thus for small h,

$$P_0(t + h) = P_0(t)(1 - \lambda h + o(h)) = P_0(t)(1 - \lambda h) + o(h). \tag{4.2}$$

The first equality follows from conditions 1 and 2.

The event that $N = 1$ at time $t + h$ can occur in two ways. The first is that $N = 1$ at time t and that no event occurs in the time interval $(t, t+h)$, the second is that $N = 0$ at time t and that exactly one event occurs in the time interval $(t, t + h)$. This gives

$$P_1(t + h) = P_0(t)(\lambda h) + P_1(t)(1 - \lambda h) + o(h), \tag{4.3}$$

where the term $o(h)$ is the sum of two terms, both of which are $o(h)$. Finally, for $j = 2, 3, \ldots$, the event that $N = j$ at time $t + h$ can occur in three different ways. The first is that $N = j$ at time t and that no event occurs in the time interval $(t, t+h)$. The second is that $N = j - 1$ at time t and that exactly one event occurs in $(t, t+h)$. The final possibility is that $N \leq j - 2$ at time t and that two or more events occur in $(t, t+h)$. Thus for $j = 2, 3, \ldots,$

$$P_j(t + h) = P_{j-1}(t)(\lambda h) + P_j(t)(1 - \lambda h) + o(h). \tag{4.4}$$

Equations (4.3) and (4.4) look identical, and the difference between them relates only to terms of order $o(h)$. Therefore, we can take (4.4) to hold for all $j \geq 1$. Subtracting $P_j(t)$ $(j \geq 1)$ from both sides of equation (4.4) and $P_0(t)$ from both sides of equation (4.2), and then dividing through by h, we get

$$\frac{P_0(t + h) - P_0(t)}{h} = -\frac{P_0(t)(\lambda h) + o(h)}{h}$$

$$\frac{P_j(t + h) - P_j(t)}{h} = \frac{P_{j-1}(t)(\lambda h) - P_j(t)(\lambda h) + o(h)}{h},$$

$j = 1, 2, 3, \ldots$. Letting $h \to 0$, we get

$$\frac{d}{dt} P_0(t) = -\lambda P_0(t), \tag{4.5}$$

and

$$\frac{d}{dt} P_j(t) = \lambda P_{j-1}(t) - \lambda P_j(t), \quad j = 1, 2, 3, \ldots. \tag{4.6}$$

The $P_j(t)$ are subject to the conditions

$$P_0(0) = 1, \quad P_j(0) = 0, \quad j = 1, 2, 3, \ldots. \tag{4.7}$$

Equation (4.5) is one of the most fundamental of differential equations, and has the solution

$$P_0(t) = Ce^{-\lambda t}. \tag{4.8}$$

The condition $P_0(0) = 1$ implies $C = 1$, leading to

$$P_0(t) = e^{-\lambda t}. \tag{4.9}$$

Given this, we now show that the set of equations (4.6) has the solution

$$P_j(t) = \frac{e^{-\lambda t}(\lambda t)^j}{j!}, \quad j = 1, 2, 3, \ldots. \tag{4.10}$$

The method for solving these equations follows the induction procedure described in Section B.18. Equation (4.9) shows (4.10) is true for $j = 0$. It must next be shown that the assumption that (4.10) holds for $j - 1$ implies that it holds for j. Assuming that (4.10) is true for $j - 1$, equation (4.6) gives

$$\frac{d}{dt} P_j(t) = \frac{\lambda e^{-\lambda t}(\lambda t)^{j-1}}{(j - 1)!} - \lambda P_j(t).$$

From this,

$$e^{\lambda t} \left(\frac{d}{dt} P_j(t) + \lambda P_j(t) \right) = \frac{\lambda(\lambda t)^{j-1}}{(j-1)!}.$$

This equation may be rewritten as

$$\frac{d}{dt} \left(P_j(t) e^{\lambda t} \right) = \frac{\lambda(\lambda t)^{j-1}}{(j-1)!},$$

and integration of both sides of this equation gives

$$P_j(t) e^{\lambda t} = \frac{(\lambda t)^j}{j!} + C,$$

for some constant C. From (4.7) it follows that $C = 0$. Thus

$$P_j(t) = \frac{e^{-\lambda t}(\lambda t)^j}{j!}, \quad j = 0, 1, 2, \dots. \tag{4.11}$$

This completes the induction, showing that at time t the random variable N has a Poisson distribution with parameter λt.

Conditions 1 and 2 are often taken as giving a mathematical definition of the concept of "randomness," and since many calculations in bioinformatics, some of which are described later in this book, make the randomness assumption, the Poisson distribution arises often.

4.2   The Poisson and the Binomial Distributions

An informal statement concerning the way in which the Poisson distribution arises as a limiting case of the binomial was made in Section 1.3.7. A more formally correct version of this statement is as follows. If in the binomial distribution (1.8) we let $n \to +\infty$, $p \to 0$, with the product np held constant at λ, then for any y, the binomial probability in (1.8) approaches the Poisson probability in (1.22). This may be proved by writing the binomial probability (1.8) as

$$\frac{1}{y!}(np)((n-1)p) \cdots ((n-y+1)p) \left(1 - \frac{\lambda}{n} \right)^n \left(1 - \frac{\lambda}{n} \right)^{-y}. \tag{4.12}$$

Fix y and λ and write $p = \lambda/n$. Then as $n \to \infty$, each term in the above product has a finite limit as $n \to +\infty$: Terms of the form $(n-i)\lambda/n$ approach λ for any i, and

$$\left(1 - \frac{\lambda}{n} \right)^n \to e^{-\lambda},$$

(see (B.3)), and finally,

$$\left(1 - \frac{\lambda}{n} \right)^{-y} \to 1.$$

Therefore, the expression in (4.12) approaches

$$\lambda^y e^{-\lambda}/y! \tag{4.13}$$

as $n \to +\infty$, and this is the Poisson probability (1.22).

4.3 The Poisson and the Gamma Distributions

There is an intimate connection, implied by equation (4.9), between the Poisson distribution and the exponential distribution. The (random) time until the first event occurs in a Poisson process with parameter λ is given by the exponential distribution with parameter λ. To see this, let $F(t)$ be the probability that the first event occurs before time t. Then the density function for the time until the first occurrence is the derivative $\frac{d}{dt} F(t)$. From (4.9), $F(t) = 1 - P_0(t) = 1 - e^{-\lambda t}$. Therefore, $\frac{d}{dt} F(t) = \lambda e^{-\lambda t}$. This is the exponential distribution (1.66), with notation changed from x to t.

It can also be shown that the distribution of the time between successive events is given by the exponential distribution. Thus the (random) time until the kth event occurs is the sum of k independent exponentially distributed times. The material surrounding (2.23) shows that this sum has the gamma distribution (1.75). Let t_0 be some fixed value of t. Then if the time until the kth event occurs exceeds t_0, the number of events occurring before time t_0 is less than k, and conversely. This means that the probability that $k - 1$ or fewer events occur before time t_0 must be identical to the probability that the time until the kth event occurs exceeds t_0. In other words it must be true that

$$e^{-\lambda t_0}\left(1 + (\lambda t_0) + \frac{(\lambda t_0)^2}{2!} + \cdots + \frac{(\lambda t_0)^{k-1}}{(k-1)!}\right) = \frac{\lambda^k}{\Gamma(k)} \int_{t_0}^{+\infty} x^{k-1} e^{-\lambda x} dx.$$

$$\tag{4.14}$$

This equation can also be established by repeated integration by parts of the right-hand side.

4.4 The Pure-Birth Process

In the derivation of the Poisson distribution (4.10) it was assumed that the probability of an event in any time interval $(t, t + h)$ is independent of the number of events that have occurred up to time t. In several cases of biological interest this is not a reasonable assumption, and the concept of an event is not quite appropriate. Instead, some random variable is considered whose initial value is k (usually $k \geq 1$). In the pure-birth process it is assumed that, given that the value of the random variable at time t is j, the probability that it increases to $j + 1$ in a short time interval $(t, t + h)$ is $\lambda_j h$, where (as with the Poisson process) we ignore terms of order $o(h)$.

(The Poisson case arises when $\lambda_j = \lambda$, that is λ_j is independent of j.) Following the same procedures as those carried out above for the Poisson process, we arrive at the infinite set of differential equations

$$\frac{d}{dt}P_k(t) = -\lambda_k P_k(t), \tag{4.15}$$

$$\frac{d}{dt}P_j(t) = \lambda_{j-1}P_{j-1}(t) - \lambda_j P_j(t), \quad j = k+1, k+2, k+3, \ldots. \tag{4.16}$$

for the probability that the random variable of interest takes the value j at time t. We now outline two examples of this process, and observe that in neither case does the value of the random variable at time t follow the Poisson distribution.

The Yule process. In the Yule process it is assumed that $\lambda_j = j\lambda$, for some constant λ. The motivation for this choice is that in some populations it is reasonable to assume that if the current size of the population is j, the probability that it increases to size $j+1$ in a short time interval is essentially proportional to j. It is easy to see that in this case the solution of equations (4.15) and (4.16) is

$$P_j(t) = \binom{j-1}{j-k}e^{-k\lambda t}\left(1 - e^{\lambda t}\right)^{j-k}, \quad j = k, k+1, \ldots. \tag{4.17}$$

The polymerase chain reaction (PCR). The polymerase chain reaction is a very important method used widely to amplify a comparatively short sequence of DNA. Here we model the length of the product, or "amplicon," of this reaction, considering a simplified version of the process described by Velikanov and Krapal (1999).

A primer, of initial length k (usually about $20 - 30$ base pairs), is used to initiate the reaction. The product of the reaction is formed by sequential additions of single base pair units to the primer. This addition forms a pure-birth process, and the function λ_j in the pure birth process is assumed here, for simplicity, to be of the form $m - j$. (This assumption involves a re-scaling of the time axis: Velikanov and Krapal (1999) provide a more general treatment.) The form of this function implies that, once the length of the product reaches the value m, no further increase in its length is possible. Equation (4.16) now becomes

$$\frac{d}{dt}P_j(t) = (m-j-1)P_{j-1}(t) - (m-j)P_j(t), \quad j = k+1, k+2, \ldots \tag{4.18}$$

The joint solution of this equation and equation (4.15), subject to the condition $P_k(0) = 1$, is

$$P_j(t) = \binom{m-k}{j-k}\left(1 - e^{-t}\right)^{j-k}e^{-(m-j)t}, \quad j = k, k+1, \ldots \tag{4.19}$$

The length of the additional material formed from the primer is $j - k$, and writing $i = j - k, n = m - k$, equation (4.19) can be written as

$$P_i(t) = \binom{n}{i}\left(1 - e^{-t}\right)^i e^{-(n-i)t}, \quad i = 0, 1, 2, \ldots, n. \qquad (4.20)$$

It is interesting that this solution for i and the solution to the Yule process are both in the form of the binomial distribution. Equation (4.20), the moments of the binomial distribution as given in Table 1.1 and the linearity properties of means and variances listed in Sections 1.4 and 1.5 show, for example, that at time t, $E(j) = k + (m-k)(1-e^{-t})$ and that $\mathrm{Var}(j) = (m-k)(e^{-t} - e^{-2t})$. This variance assumes its maximum value when $t = \log 2$.

4.5 Introduction to Finite Markov Chains

In this section we give a brief outline of the theory of a simple case of a discrete-time finite Markov chain. The focus is on material needed to discuss the construction of PAM matrices as described in Section 6.5.3. Further developments of Markov chain theory suitable for other applications, in particular for the evolutionary applications discussed in Chapter 14, are given in Chapter 11.

We introduce the simple discrete-time finite Markov chain in abstract terms as follows. Consider some finite discrete set S of possible "states," labeled $\{E_1, E_2, \ldots, E_s\}$. At each of the unit time points $t = 1, 2, 3, \ldots$, a Markov chain process occupies one of these states. In each time step t to $t+1$, the process either stays in the same state or moves to some other state in S. Further, it does this in a probabilistic, or stochastic, way rather than in a deterministic way. That is, if at time t the process is in state E_j, then at time $t + 1$ it either stays in this state or moves to some other state E_k according to some well-defined probabilistic rule described in more detail below. This process follows the requirements of a simple Markov chain if it has the following properties.

(i) The *Markov* property. If at some time t the process is in state E_j, the probability that one time unit later it is in state E_k depends only on E_j, and not on the past history of the states it was in before time t. That is, the current state is all that matters in determining the probabilities for the states that the process will occupy in the future.

(ii) The *temporally homogeneous transition probabilities* property. Given that at time t the process is in state E_j, the probability that one time unit later it is in state E_k is independent of t.

More general Markov processes relax one or both requirements, but we assume throughout this chapter that the above properties hold.

The concept of "time" used above is appropriate if, for example, we consider the evolution through time of the nucleotide at a given site in some population. Aspects of this process are discussed later in this book. However, the concept of time is sometimes replaced by that of "space." As an example, we may consider a DNA sequence read from left to right. Here there would be a Markov dependence between nucleotides if the nucleotide type at some site depended in some way on the type at the site immediately to its left. Aspects of the Markov chains describing this process are also discussed later in this book. Because Markov chains are widely applicable to many different situations, it is useful to describe the properties of these chains in abstract terms rather than in the concrete terms appropriate to one specific application.

In many cases the Markov chain process describes the behavior of the value of a random variable changing through time. For example, in reading a DNA sequence from left to right this random variable might be the excess of purines over pyrimidines so far observed at any point. Because of this it is often convenient to adopt a different terminology and to say that the value of the random variable is j rather than saying that the state occupied by the process is E_j. We use both forms of expression below, and also, when no confusion should arise, we abuse terminology by using expressions like "the random variable is in state E_j."

4.6   Transition Probabilities and the Transition Probability Matrix

Suppose that at time t a Markovian random variable is in state E_j. We denote the probability that at time $t + 1$ it is in state E_k by p_{jk}, called the *transition probability* from E_j to E_k. In writing this probability in this form we are already using the two Markov assumptions described above: First, no mention is made in the notation p_{jk} of the states that the random variable was in before time t (the memoryless property), and second, t does not occur in the notation p_{jk} (the time homogeneity property).

It is convenient to group the transition probabilities p_{jk} into the so-called *transition probability matrix*, or more simply the transition matrix, of the Markov chain. We denote this matrix by P, and write it as

$$
P = \begin{array}{c} \\ (\text{from } E_1) \\ (\text{from } E_2) \\ \vdots \\ (\text{from } E_s) \end{array}
\begin{array}{ccccc} (\text{to } E_1) & (\text{to } E_2) & (\text{to } E_3) & \cdots & (\text{to } E_s) \end{array} \\
\left[\begin{array}{ccccc}
p_{11} & p_{12} & p_{13} & \cdots & p_{1s} \\
p_{21} & p_{22} & p_{23} & \cdots & p_{2s} \\
\vdots & \vdots & \vdots & \ddots & \vdots \\
p_{s1} & p_{s2} & p_{s3} & \cdots & p_{ss}
\end{array} \right]. \qquad (4.21)
$$

The rows and columns of P are in correspondence with the states E_1, $E_2, \ldots, E_s$, so these states being understood, P is usually written in the simpler form

$$
P = \begin{bmatrix}
p_{11} & p_{12} & p_{13} & \cdots & p_{1s} \\
p_{21} & p_{22} & p_{23} & \cdots & p_{2s} \\
\vdots & \vdots & \vdots & \ddots & \vdots \\
p_{s1} & p_{s2} & p_{s3} & \cdots & p_{ss}
\end{bmatrix}. \tag{4.22}
$$

Any row in the matrix corresponds to the state *from* which the transition is made, and any column in the matrix corresponds to the state *to* which the transition is made. Thus the probabilities in any particular row in the transition matrix must sum to 1. However, the probabilities in any given column do not have to sum to anything in particular.

It is also assumed that there is some *initial* probability distribution for the various states in the Markov chain. That is, it is assumed that there is some probability π_i that at the initial time point the Markovian random variable is in state E_i. A particular case of such an initial distribution arises when it is known that the random variable starts in state E_i, in which case $\pi_i = 1$, $\pi_j = 0$ for $j \neq i$. In principle the initial probability distribution and the transition matrix P jointly determine all the properties the entire process. In practice, many properties are not found easily, or if found are obtained by special methods.

The probability that the Markov chain process moves from state E_i to state E_j after two steps can be found by matrix multiplication. It is this fact that makes much of Markov chain theory an application of linear algebra. The argument is as follows.

Let $p_{ij}^{(2)}$ be the probability that if the Markovian random variable is in state E_i at time t, then it is in state E_j at time $t+2$. We call this a *two-step* transition probability. Since the random variable must be in some state k at the intermediate time $t + 1$, summation over all possible states at time $t + 1$ and use of equation (1.94) gives

$$
p_{ij}^{(2)} = \sum_k p_{ik} p_{kj}.
$$

The right-hand side in this equation is the (i, j) element in the matrix P^2. Thus if the matrix $P^{(2)}$ is defined as the matrix whose (i, j) element is $p_{ij}^{(2)}$, then the (i, j) element in $P^{(2)}$ is equal to the (i, j) element in P^2. This leads to the identity

$$
P^{(2)} = P^2.
$$

Extension of this argument to an arbitrary number n of steps gives

$$
P^{(n)} = P^n. \tag{4.23}
$$

That is, the "n-step" transition probabilities are given by the entries in the nth power of P.

4.7   Markov Chains with Absorbing States

Some Markov chains have absorbing states. These can be recognized by the appearance of one or more 1's on the main diagonal of the transition matrix. If there are no 1's on the main diagonal, then there are no absorbing states. For the Markov chains with absorbing states that we consider, sooner or later some absorbing state will be entered, never thereafter to be left. The two questions we are most interested in regarding these Markov chains are:

(i) If there are two or more absorbing states, what is the probability that a specified absorbing state is the one eventually entered?

(ii) What is the mean time until one or another absorbing state is eventually entered?

We shall address these questions in detail in Chapter 11. In the remainder of this chapter we discuss only certain aspects of the theory of Markov chains with no absorbing states, focusing on the theory needed for the construction of substitution matrices, to be discussed in more detail in Chapter 6.

4.8   Markov Chains with No Absorbing States

The questions of interest about a Markov chain with no absorbing state are quite different from those asked when there are absorbing states.

In order to simplify the discussion, we assume in the remainder of this chapter that all Markov chains discussed are *finite, aperiodic*, and *irreducible*.

Finiteness means that there is a finite number of possible states. The aperiodicity assumption is that there is no state such that a return to that state is possible only at t_0, $2t_0$, $3t_0$, ... transitions later, where t_0 is an integer exceeding 1. If the transition matrix of a Markov chain with states E_1, E_2, E_3, E_4 is, for example,

$$P = \begin{bmatrix} 0 & 0 & 0.6 & 0.4 \\ 0 & 0 & 0.3 & 0.7 \\ 0.5 & 0.5 & 0 & 0 \\ 0.2 & 0.8 & 0 & 0 \end{bmatrix}, \tag{4.24}$$

then the Markov chain is periodic. If the Markovian random variable starts (at time 0) in E_1, then at time 1 it must be either in E_3 or E_4, at time 2 it must be in either E_1 or E_2, and in general it can visit only E_1 at times $2, 4, 6, \ldots$. It is therefore periodic. The aperiodicity assumption holds for essentially all applications of Markov chains in bioinformatics, and we often take aperiodicity for granted without any explicit statement being made.

The irreducibility assumption implies that any state can eventually be reached from any other state, if not in one step then after several steps. Except for the case of Markov chains with absorbing states, the irreducibility assumption also holds for essentially all applications in bioinformatics.

4.8.1 Stationary Distributions

Suppose that a Markov chain has transition matrix P and that at time t the probability that the process is in state E_j is φ_j, $j = 1, 2, \ldots, s$. This implies that the probability that at time $t + 1$ the process is in state j is $\sum_{k=1}^{s} \varphi_k p_{kj}$. Suppose that for every j these two probabilities are equal, so that

$$\varphi_j = \sum_{k=1}^{s} \varphi_k p_{kj}, \quad j = 1, 2, \ldots, s. \tag{4.25}$$

In this case the probability distribution $(\varphi_1, \varphi_2, \ldots, \varphi_s)$ is said to be *stationary*; that is, the probability that the process is in state E_j has not changed between times t and $t + 1$, and therefore will never change. Despite this, the state occupied by the process can of course change from one time point to the next.

It will be shown in Chapter 11 that for finite aperiodic irreducible Markov chains there is a unique distribution satisfying (4.25). This is then called the *stationary distribution* of the Markov chain. When we discuss stationary distributions in this book, we assume that they relate to finite aperiodic irreducible Markov chains, and thus exist and are unique.

If the row vector φ' is defined by

$$\varphi' = (\varphi_1, \varphi_2, \ldots, \varphi_s), \tag{4.26}$$

then in matrix and vector notation, the set of equations in (4.25) becomes

$$\varphi' = \varphi' P. \tag{4.27}$$

The prime here is used to indicate the transposition of the row vector into a column vector. The vector $(\varphi_1, \varphi_2, \ldots, \varphi_s)$ must also satisfy the equation $\sum_k \varphi_k = 1$. In vector notation, this is the equation

$$\varphi' \mathbf{1} = 1, \tag{4.28}$$

where $\mathbf{1} = (1, 1, \ldots, 1)'$. Equations (4.27) and (4.28) can then be used to find the stationary distribution when it exists. In this process one of the equations in (4.27) is redundant and can be omitted. An example is given in the next section.

In Chapter 11 we shall show that if the Markov chain is finite, aperiodic, and irreducible, then as n increases, $P^{(n)}$ approaches the matrix

$$\begin{bmatrix} \varphi_1 & \varphi_2 & \cdots & \varphi_s \\ \varphi_1 & \varphi_2 & \cdots & \varphi_s \\ \varphi_1 & \varphi_2 & \cdots & \varphi_s \\ \vdots & \vdots & \ddots & \vdots \\ \varphi_1 & \varphi_2 & \cdots & \varphi_s \end{bmatrix}, \tag{4.29}$$

where $(\varphi_1, \varphi_2, \ldots, \varphi_s)$ is the stationary distribution of the Markov chain.

The form of this matrix shows that no matter what the starting state was, or what was the initial probability distribution of the starting state, the probability that n time units later the process is in state j is increasingly closely approximated, as $n \to \infty$, by the value φ_j.

There is another implication, relating to long-term averages, of the calculations above. That is, if a Markov chain is observed for a very long time, then the proportion of times that it is observed to be in state E_j should be approximately φ_j, for all j.

4.8.2 Example

Consider the Markov chain with transition probability matrix given by

$$P = \begin{bmatrix} 0.6 & 0.1 & 0.2 & 0.1 \\ 0.1 & 0.7 & 0.1 & 0.1 \\ 0.2 & 0.2 & 0.5 & 0.1 \\ 0.1 & 0.3 & 0.1 & 0.5 \end{bmatrix}. \tag{4.30}$$

For this example the vector equation (4.27) consists of four separate linear equations in four unknowns. As noted above, when used jointly with (4.28) they form a redundant set of equations and any one of them can be discarded. Omission of the last equation in (4.27) leads to

$$0.6\varphi_1 + 0.1\varphi_2 + 0.2\varphi_3 + 0.1\varphi_4 = \varphi_1,$$
$$0.1\varphi_1 + 0.7\varphi_2 + 0.2\varphi_3 + 0.3\varphi_4 = \varphi_2,$$
$$0.2\varphi_1 + 0.1\varphi_2 + 0.5\varphi_3 + 0.1\varphi_4 = \varphi_3,$$
$$\varphi_1 + \varphi_2 + \varphi_3 + \varphi_4 = 1.$$

To four decimal place accuracy, these four simultaneous equations have the solution

$$\varphi' = (0.2414, 0.3851, 0.2069, 0.1667). \tag{4.31}$$

This is the stationary distribution corresponding to the matrix P given in (4.30). In informal terms, from the point of view of long-term averages, over a long time period the random variable should spend about 24.14% of the time in state E_1, about 38.51% of the time in state E_2, and so on.

The rate at which the rows in $P^{(n)}$ approach this stationary distribution can be assessed from the following values:

$$P^{(2)} = \begin{bmatrix} 0.42 & 0.20 & 0.24 & 0.14 \\ 0.16 & 0.55 & 0.15 & 0.14 \\ 0.25 & 0.29 & 0.32 & 0.14 \\ 0.16 & 0.39 & 0.15 & 0.30 \end{bmatrix}, \qquad (4.32)$$

$$P^{(4)} \cong \begin{bmatrix} 0.2908 & 0.3182 & 0.2286 & 0.1624 \\ 0.2151 & 0.4326 & 0.1899 & 0.1624 \\ 0.2538 & 0.3569 & 0.2269 & 0.1624 \\ 0.2151 & 0.4070 & 0.1899 & 0.1880 \end{bmatrix}, \qquad (4.33)$$

$$P^{(8)} \cong \begin{bmatrix} 0.24596 & 0.37787 & 0.20961 & 0.16656 \\ 0.23873 & 0.38946 & 0.20525 & 0.16656 \\ 0.24309 & 0.38223 & 0.20812 & 0.16656 \\ 0.23873 & 0.38880 & 0.20525 & 0.16721 \end{bmatrix}, \qquad (4.34)$$

$$P^{(16)} \cong \begin{bmatrix} 0.24142 & 0.38494 & 0.20692 & 0.16667 \\ 0.24135 & 0.38510 & 0.20688 & 0.16667 \\ 0.24140 & 0.38503 & 0.20691 & 0.16667 \\ 0.24135 & 0.38510 & 0.20688 & 0.16667 \end{bmatrix}. \qquad (4.35)$$

After 16 time units, the stationary distribution has, for all practical purposes, been reached. The discussion in Chapter 11 shows how the rate at which this convergence occurs can be calculated in a more informative manner.

4.9 The Graphical Representation of a Markov Chain

It is often convenient to represent a Markov chain by a directed graph. A directed graph is a set of "nodes" and a set of "edges" connecting these nodes. The edges are "directed," that is, they are marked with arrows giving each edge an orientation from one node to another.

We represent a Markov chain by identifying the states with nodes and the transition probabilities with edges. Consider, for example, the Markov chain with states E_1, E_2, and E_3 and with probability transition matrix

$$\begin{bmatrix} .2 & .1 & .7 \\ .5 & .3 & .2 \\ .6 & 0 & .4 \end{bmatrix}.$$

This Markov chain is represented by the following graph:

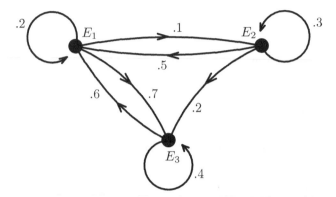

Notice that we do not draw the edge if its corresponding transition probability is known to be zero, as is the case in this example with the transition
from E_3 to E_2.

A graph helps us capture information at a glance that might not be so
apparent from the transition matrix itself. Sometimes it is also convenient
to include a *start state*; this is a dummy state that is visited only once,
at the beginning. Therefore, all transition probabilities into the start state
are zero. The transition probabilities out of the start state are given by the
initial distribution of the Markov chain. If the Markov chain starts at time
$t = 0$, we can think of the start state as being visited in time $t = -1$. We
can further have an *end state*, which stops the Markov chain when visited.

We refer to the graph structure, without any probabilities, as the *topology*
of the graph. Sometimes the topology of a model is known, but the various
probabilities are unknown.

We will use these definitions when we discuss hidden Markov models in
Chapter 12.

4.10 Modeling

There are many applications of the homogeneous Poisson process in bioinformatics. However, the two key assumptions made in the derivation of
the Poisson distribution formula (4.10), namely homogeneity and independence, do not always hold in practice. Similarly, there are many applications
of Markov chains in the literature, in particular in the evolutionary processes discussed in Chapter 14. Many of these applications also make
assumptions, specifically the two Markov assumptions stated in Section 4.5.
The modeling assumptions made in the evolutionary context are discussed
further in Section 15.9.

In the context of DNA or protein sequence analysis, where time is replaced by position along the sequence, it is very likely that neither of these two Markov assumptions is correct. The data from chromosome 22 in humans (Dunham et al. (1999)) makes it apparent that the probability that the nucleotide a is next followed by g depends to some extent on the current location in the chromosome. Further, it is likely that the Markov chain memoryless assumption does not hold: The probability that the nucleotide a is next followed by g might well depend on the nucleotide (or nucleotides) immediately preceding a. Tests for this possibility are discussed in Section 5.2: Nevertheless these tests are often not applied, and the memoryless Markov chain theory is often assumed when its applicability is uncertain.

The construction of phylogenetic trees, both by algorithmic methods and by methods involving Markov chain evolutionary models, involves many assumptions, both explicit and implicit. An example of quite different phylogenetic trees arising from different models is given in Section 15.8. A discussion of various statistical tests for appropriate evolutionary models is discussed in Section 15.9.

Given that modeling assumptions made for both Poisson processes and Markov chains often do not hold exactly, one might ask why they are made and why there is such an extensive Poisson process and Markov chain literature. Mathematical models generally make simplifying assumptions about properties of the phenomena being modelled. This concern opens up the question of why we model natural phenomena with mathematics if we cannot do so with complete accuracy. In fact, it is not necessarily desirable that we attempt to make an extremely accurate model of reality. The more closely any phenomenon is modelled, the more complicated the model becomes. If a mathematical model becomes too complicated, then solving the equations necessary to find answers to the questions we wish to ask can become intractable. Therefore, we are almost always faced with the task of finding a middle ground between tractable simple models and intractable complex models. The key point is that a model need only capture enough of the true complexity of a situation to serve our purposes, whatever they might be.

Finding this middle ground, however, is not an easy task: Being able to extract the essence of a complex reality in a simplified model that then allows a successful mathematical analysis requires some skill and experience.

In biology it is not always possible to evaluate a model's efficacy directly. Rather, a model is often tested on how well it performs its job. Sometimes benchmarks can be well defined, but often efficacy is not easy to verify empirically, and subjectivity can enter in. This is an unfortunate but usually unavoidable problem.

To illustrate this we consider the early versions of the widely used BLAST procedure discussed in more detail in Chapter 10. One of the simplifying assumptions used is that nucleotides (or amino acids) are identically and independently distributed along a DNA (or protein) sequence. Current data show that this assumption is false. Nevertheless, this simple BLAST procedure does work, in that the model captures enough of biological reality to be effective.

A further aspect of the modeling process is that we do not expect any model to be the final one used. Any given process is often initially modelled using several simplifying assumptions, and then more refined models are introduced as time goes on. Indeed, applications of the simple models often indicate those areas in which more precise modeling is needed. Various updates of the BLAST procedure exemplify this. Recent versions of BLAST remove some of the simplifying assumptions made in earlier versions and provide an example of the joint evolution of models and data analysis. Unfortunately, the mathematical theory involved in these more sophisticated versions is far more complicated than that in the simpler BLAST theory, and we shall only outline it in this book.

Not every problem we might wish to solve with a model has a happy middle ground where our assumptions find a workable balance between tractability and reality. Thus while we should be willing to accept simplifying assumptions, we should always be on the lookout for oversimplifications, especially those that are not sufficiently backed up by testing for the efficacy of the model used. Model testing is an active area of statistical research in bioinformatics, and aspects of model testing, especially in the evolutionary and phylogenetic tree contexts, are discussed further in Section 15.9.

Problems

4.1. Prove (4.14) by repeated integration by parts of the right-hand side.

4.2. Events occur in a Poisson process with parameter λ. Given that 10 events occur in the time period $[0, 2]$, what is the probability that 6 of these events occur in the time period $[0, 1]$? Given that 6 of these events occur in the time period $[0, 1]$, what is the probability that 10 of these events occur in the time period $[0, 2]$?

4.3. ("Competing Poissons.") Suppose that events occur as described in Section 4.1, but that now each event is of one of k types, labeled types $1, 2, \ldots, k$. The type of any event is independent of the type of any other event. The probability that any event is of type i is p_i. Equation (4.10) continues to give the probability that exactly j events occur in the time period $(0, t)$. Assuming this,

  (i) Find the (marginal) probability that j_i events of type i occur in the time period $(0, t)$.

(ii) Find the probability that j_i $(i = 1, \ldots, k)$ events of type i occur occur in the time period $(0, t)$.

(iii) Find the joint conditional probability that j_i $(i = 1, \ldots, k)$ events of type i occur in the time period $(0, t)$, given that j events in total occur in this time period. Relate your answer to expression (2.45).

4.4. The transition matrix of a Markov chain is

$$\begin{bmatrix} .7 & .3 \\ .4 & .6 \end{bmatrix}.$$

Find the stationary distribution of this Markov chain.

4.5. *Continuation.* If the initial probability distribution (at time 0) is $(.8, .2)$, what is the probability that at time 3 the state occupied is E_1?

4.6. The transition matrix of a Markov chain is

$$\begin{bmatrix} 1 - a & a \\ b & 1 - b \end{bmatrix}.$$

Find the stationary distribution of this Markov chain in terms of a and b, and interpret your result.

4.7. The transition matrix of a Markov chain is

$$\begin{bmatrix} 0 & \frac{1}{3} & \frac{1}{3} & \frac{1}{3} \\ \frac{1}{3} & 0 & \frac{1}{3} & \frac{1}{3} \\ \frac{1}{3} & \frac{1}{3} & 0 & \frac{1}{3} \\ \frac{1}{3} & \frac{1}{3} & \frac{1}{3} & 0 \end{bmatrix}.$$

Use induction on n (see Section B.18) to show that the probability that the Markov chain revisits the initial state at the nth transition is

$$p_{ii}^{(n)} = \frac{1}{4} + \frac{3}{4}(-\frac{1}{3})^n.$$

(This result is needed for Problem 14.7.)

4.8. Use equations (4.27) and (4.23) to show that the stationary distribution φ' satisfies the equation

$$\varphi' = \varphi' P^{(n)}, \tag{4.36}$$

for any positive integer n. For the numerical example in Section 4.8.2, use the expression for $P^{(2)}$ given in equation (4.32) and the expression for φ' given in (4.31) to check this claim for the case $n = 2$.

Why does equation (4.36) "make sense"?

4.9. Show that if the transition matrix of an irreducible, aperiodic, finite Markov chain is symmetric, then the stationary distribution is a (discrete) uniform distribution.

4.10. Show that if the transition matrix P of a Markov chain is of the circulant form

$$P = \begin{bmatrix} a_1 & a_2 & a_3 & a_4 & \cdots & a_{s-3} & a_{s-2} & a_{s-1} & a_s \\ a_s & a_1 & a_2 & a_3 & \cdots & a_{s-4} & a_{s-3} & a_{s-2} & a_{s-1} \\ a_{s-1} & a_s & a_1 & a_2 & \cdots & a_{s-5} & a_{s-4} & a_{s-3} & a_{s-2} \\ \vdots & \vdots & \vdots & \vdots & \ddots & \vdots & \vdots & \vdots & \vdots \\ a_4 & a_5 & a_6 & a_7 & \cdots & a_s & a_1 & a_2 & a_3 \\ a_3 & a_4 & a_5 & a_6 & \cdots & a_{s-1} & a_s & a_1 & a_2 \\ a_2 & a_3 & a_4 & a_5 & \cdots & a_{s-2} & a_{s-1} & a_s & a_1 \end{bmatrix}, \qquad (4.37)$$

where $a_j > 0$ for all j, then the stationary distribution is a (discrete) uniform distribution.

4.11. Suppose that the transition matrix of a Markov chain is

$$P = \begin{bmatrix} 0 & 1 & 0 & 0 & \cdots & 0 & 0 & 0 & 0 \\ q & 0 & p & 0 & \cdots & 0 & 0 & 0 & 0 \\ 0 & q & 0 & p & \cdots & 0 & 0 & 0 & 0 \\ \vdots & \vdots & \vdots & \vdots & \ddots & \vdots & \vdots & \vdots & \vdots \\ 0 & 0 & 0 & 0 & \cdots & q & 0 & p & 0 \\ 0 & 0 & 0 & 0 & \cdots & 0 & q & 0 & p \\ 0 & 0 & 0 & 0 & \cdots & 0 & 0 & 1 & 0 \end{bmatrix}. \qquad (4.38)$$

Show that this Markov chain is periodic. Despite this fact, solve equations (4.27) and (4.28) for the case $p = q$.

4.12. Suppose that a transition matrix P is of size $2s \times 2s$ and can be written in the partitioned form

$$P = \left[\begin{array}{c|c} 0 & A \\ \hline B & 0 \end{array} \right],$$

where A and B are both $s \times s$ matrices. Use this expression to find formulae for (i) $P^{(2n)}$, (ii) $P^{(2n+1)}$ in terms of the matrices A and B, and interpret your results.

4.13. Suppose that a finite Markov chain is irreducible and that there exists at least one state E_i such that $p_{ii} > 0$. Show that the Markov chain is aperiodic.

4.14. (More difficult). Any $s \times s$ matrix of non-negative numbers for which all rows sum to 1 can be regarded as the transition matrix of some Markov chain. Is it true that any such matrix can be the two-step transition matrix of some Markov chain? *Hint:* Consider the case $s = 2$. Write down the general form of a 2×2 Markov chain matrix and find when the sum of the diagonal terms is minimized.

5

The Analysis of One DNA Sequence

5.1 Shotgun Sequencing

5.1.1 Introduction

Before any analysis of a DNA sequence can take place it is first necessary to determine the actual sequence itself, at least as accurately as is reasonably possible. Unfortunately, technical considerations make it impossible to sequence very long pieces of DNA all at once. Instead, many overlapping small pieces are sequenced, each on the order of 500 bases (nucleotides). After this is done the problem arises of assembling these fragments into one long "contig." One difficulty is that the locations of the fragments within the genome and with respect to each other are not generally known. However, if enough fragments are sequenced so that there will be many overlaps between them, the fragments can be matched up and assembled. This method is called "shotgun sequencing."

It is customary to say that *n-times coverage* (or nX coverage) is obtained if, when the length of the original (long) sequence is G, the total length of the fragments sequenced is nG.

Two strategies have been used to sequence the entire human genome (International Human Genome Sequencing Consortium (2001), Venter et al. (2001)). Under one strategy the genome is partitioned into pieces whose lengths are on the order of 100,000 bases and whose locations in the genome are known. Then shotgun sequencing is performed on each piece, with high coverage, in the 8X range. The greater coverage also helps eliminate errors occurring in sequencing the fragments. A competing strategy is to adopt a "whole genome shotgun" approach, in which the entire genome is broken into small fragments, each of length approximately 500 bases. The assembly problem is much more difficult under this approach, and thus should require higher coverage. It is a matter of much debate as to which approach is to be preferred, and the human genome has been sequenced using a combination of the two methods.

In the following two sections we address several probabilistic issues arising in the reconstruction of long DNA sequences from comparatively shorter sequences, or fragments. Before proceeding it is important to note the remarks about modeling given in Section 4.10. The probabilistic models described below are simple and do not closely reflect the properties of human genome as revealed in the two references given above. References to less simplified models are given at the end of Section 5.1.3, and the calculations in the following two sections can be regarded as an introduction to these more realistic models.

5.1.2 Contigs

Figure 5.1 shows a collection of $N = 17$ fragments with their locations above the original DNA sequence. It is assumed that overlapping fragments can be recognized and used to determine a collection of "contigs" (thick black lines), of which there are 7 in the example shown. These contigs are then taken as the best possible reconstruction of the original DNA sequence. Note that Figure 5.1 is potentially misleading in that the locations and the orientations of the contigs are unknown to us.

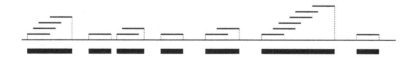

Figure 5.1.

We assume initially that there are N fragments, each of length L, and that the original full-length DNA sequence, which we call here the genome, is of length G. Therefore, the coverage a is given by

$$a = NL/G.$$

The length G is assumed to be much larger than L, so that end effects are ignored in the calculations below.

The fragments are assumed to be taken at random from the original full-length sequence, so that if end effects are ignored, the left-hand ends of the fragments are independently distributed with a common uniform distribution over $[0, G]$. This implies that any such left-hand end falls in an interval $(x, x + h)$ with probability h/G and that the number of fragments whose left-hand end falls in this interval has a binomial distribution with mean Nh/G. If N is large and h is small, the discussion of Section 4.2 shows that this distribution is approximately Poisson with mean Nh/G. We use this and other Poisson approximations throughout. The number Y of fragments whose left-hand end is located within an interval of length L to the left of a randomly-chosen point therefore has a Poisson distribution with mean a, so that the probability that at least one fragment arises in this interval is $1 - \text{Prob}(Y = 0) = 1 - e^{-a}$.

This calculation, together with other properties of homogeneous Poisson processes, is enough to provide the answer to three basic questions: What is the mean proportion of the genome covered by contigs? What is the mean number of contigs? What is the mean contig size?

The mean proportion of the genome covered by one or more fragments is the probability that a point chosen at random is covered by at least one fragment. This is the probability that at least one fragment has its left-hand end in the interval of length L immediately to the left of this point, which is $1 - e^{-a}$ as given above. Some representative values are given in Table 5.1.

a	2	4	6	8	10	12
Mean proportion of genome covered	.864665	.981684	.997521	.999665	.999955	.999994

Table 5.1. The mean proportion of the genome covered for different values of a.

It is clear from the values in Table 5.1 that, for values of a exceeding about 8, increasing the value a by increasing the value of N does not significantly increase the mean proportion of the genome covered, and in practice other methods are used to increase this proportion. Further, the simplifying assumptions made in the above calculations do not apply for several parts of the genome, as discussed below.

We now consider the mean number of contigs. Each contig has a unique rightmost fragment, so that the formula np for the mean of a binomial distribution given below (1.25) shows that the mean number of contigs is the number N of fragments multiplied by the probability that a fragment is the rightmost member of a contig. This latter probability is the probability that no other fragment has its left-hand end point on the fragment in question. From the calculation in the previous paragraph, this probability is e^{-a}. Thus

$$\text{mean number of contigs} = Ne^{-a} = Ne^{-NL/G}. \tag{5.1}$$

Table 5.2 gives some values for this mean for different values of a in the case $G = 100{,}000$, $L = 500$.

a	0.5	0.75	1	1.5	2	3	4	5	6	7
Mean number of contigs	60.7	70.8	73.6	66.9	54.1	29.9	14.7	6.7	3.0	1.3

Table 5.2. The mean number of contigs for different levels of coverage, with $G = 100,000$ and $L = 500$.

The mean number of contigs, as a function of a, increases and then decreases. The reason for this is that if there is a small number of fragments, there must be a small number of contigs, while a large number of fragments tend to form a small number of large contigs. The mean number of contigs is maximized when $N = G/L$, or equivalently when $a = 1$, corresponding to 1X coverage. In the example in Table 5.2, only one contig is expected to arise for higher than about 8X coverage. However, the expression (5.1) for the mean number of contigs becomes inaccurate in this range, since with 8X coverage, end effects (which are ignored in deriving (5.1)) become important. While the numbers G and L in this example are realistic, the assumptions of the model are simplified and unrealistic, and in practice the problem of achieving high coverage is more difficult than is implied by the above calculations. For example, the existence of repeated sequences in the junk DNA can cause fragments to appear to overlap when they do not. Furthermore, some stretches of DNA are technically much more difficult to clone and sequence than others, so that the uniform distribution assumption of fragment location is in practice not appropriate.

We next calculate the mean contig size. The mean contig size is found by considering the left-hand ends of a succession of fragments, starting with the initial left-hand fragment on a given contig. Under the Poisson approximation made, the distance from the left-hand end of the first fragment to the left-hand end of the second fragment has a geometric distribution. As discussed in Section 1.10.4, this distribution is closely approximated by the exponential distribution (1.66) with parameter $\lambda = N/G$, and we make this approximation here. This second fragment will overlap the first one if this distance is less than the length L of the first fragment. This occurs with probability

$$\int_0^L \lambda e^{-\lambda x} dx = 1 - e^{-a}.$$

A further overlap occurs if the next fragment to the right of the second fragment overlaps that second fragment. The contig is built up in this way until such an overlap fails to occur.

We think of an overlap as a "success" and a non-overlap as a "failure." The number of successively overlapping fragments, before the first non-overlap, has a geometric distribution whose mean (from Table 1.1 with $p = 1 - e^{-a}$) is $e^a - 1$. If n fragments form a contig, the total length of the contig is the length L of the final fragment together with the sum of the $n - 1$ random distances between the left-hand end of any given fragment and the left-hand end of the next fragment to the right. Equation (2.52) shows that the mean of these random distances is

$$\frac{1}{\lambda} - \frac{L}{e^a - 1}.$$

Equation (2.81), referring to the mean of a sum of a random number of random variables, shows that the mean total of these distances is

$$(e^a - 1)\left(\frac{1}{\lambda} - \frac{L}{e^a - 1}\right) = \frac{e^a - 1}{\lambda} - L.$$

Upon adding the length L of the last contig to this, the mean contig size is found to be

$$\frac{e^a - 1}{\lambda} = L\frac{e^a - 1}{a}. \tag{5.2}$$

Some examples of mean contig sizes as a function of a are given in Table 5.3 for the case $L = 500$.

a	2	4	6	8	10
Mean contig size	1,600	6,700	33,500	186,000	1,100,000

Table 5.3. The mean contig size for different values of a for the case $L = 500$.

It was assumed in the discussion above that all fragments have the same fixed length L. In reality, fragments are obtained in lengths roughly between 400 and 600 bases. Many of the above results can be generalized to the case where the length of any fragment is a random variable L with density function $f_L(\ell)$. We only consider this generalization for the problem of finding the mean proportion of the genome covered by a contig.

Let P be a given point in the genome. Ignoring end effects and terms of order $o(h)$, we find that the probability that some fragment has its left-hand end in the interval $(P - x, P - x + h)$ and overlaps P is the mean number of fragments having left-hand end point in this interval (λh) multiplied by the probability that the length of a fragment exceeds x (and thus covers the point P). Thus ignoring terms of order $o(h)$, this probability is

$$\lambda h \int_x^{+\infty} f_L(\ell)\, d\ell = \lambda h\left(1 - F_L(x)\right).$$

From this, the mean number of fragments covering the point P is

$$\lambda \int_0^{+\infty} \left(1 - F_L(x)\right) dx,$$

and from equation (1.58) this is

$$\lambda E(L) = \frac{NE(L)}{G}. \tag{5.3}$$

It follows that the probability that a random point is covered by a contig is $1 - e^{-NE(L)/G}$, and this is then also the mean proportion of the genome covered by a contig.

Further aspects of this generalization have been discussed by Lander and Waterman (1988), who, among a variety of other interesting results, show that the mean number of contigs is somewhat larger than the value $Ne^{-NE(L)/G}$, the value that might be obtained intuitively by replacing L by $E(L)$ in equation (5.1).

5.1.3   Anchored Contigs

As stated above, the locations and orientations of the contigs described in Section 5.1.2 are unknown. Short sequences in the DNA that are unique in the genome, and whose locations are known, provide markers, or "anchors," that allow us to locate and orient the fragments in the genome that happen to contain these markers. In such a case we say that the fragments are "anchored." Other concepts of an anchor are possible, but this is the simplest. For mathematical purposes an anchor can be considered as a point.

An anchored fragment is a fragment with at least one anchor on it, and an anchored contig is a collection of fragments "stapled together" by anchors. Figure 5.2 provides an example where there are 16 fragments, 9 anchors, and 6 anchored contigs, shown as thick black lines. Contigs can overlap, as

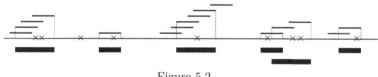

Figure 5.2.

shown in this figure, but the overlap is not recognized if there is no anchor on the overlapping section. The genome reconstruction is then carried out using anchored contigs.

We first calculate the expected proportion of the genome covered by anchored contigs in the case where all fragments are of fixed length L. The number of anchors is denoted by M, and it is assumed the location of each anchor has a uniform distribution over the genome (or part of the genome) considered, and that the locations of different anchors are independent. Essentially equivalently, we assume that the anchors are placed according to a homogeneous Poisson process with parameter M/G. Then the number of anchors in any length L of the DNA sequence has a Poisson distribution with mean $b = ML/G$. The process determining the location of the anchors is assumed to be independent of the process locating the fragments.

The mean proportion of the DNA segment covered by anchored contigs is the probability that a point P chosen at random is covered by an anchored contig. It is convenient to consider the complementary probability that P is not covered by an anchored contig. This latter probability id found in the following way.

A point can fail to be covered by an anchored contig for one of three mutually exclusive reasons. First, there might be no fragment covering this point. This event has probability e^{-a} (recall that $a = NL/G$). Second, there might be exactly one fragment covering this point (probability ae^{-a}) but no anchor on this fragment (probability e^{-b}), leading to a probability $ae^{-(a+b)}$ for this event. Finally, there might be $k \geq 2$ fragments covering this point, with no anchor on the span of these fragments. The calculations relating to this third possibility are more complicated and are carried out as follows.

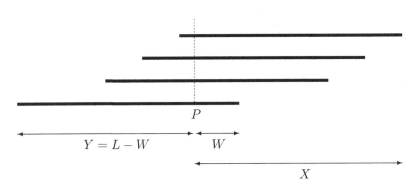

Figure 5.3.

Consider the case typified by that shown in Figure 5.3, where P is the point in question. The k ($k \geq 2$) fragments covering P have total span $X + Y$. Each of these k fragments has a "right-projection," namely the length of that part of the fragment to the right of P. Since the fragments are assumed to be placed at random, this length is uniformly distributed in $(0, L)$. The projection X is the maximum of these lengths. Similarly, each of the k fragments covering P has a "left-projection," namely the length of that part of the fragment to the left of P, also uniformly distributed in $(0, L)$. The projection Y is the maximum of these lengths. Further, $Y = L - W$, where W is the length of the smallest right-projection length.

For any fixed $k \geq 2$, the joint density function of W and X is given in (2.143) (with $x_{(1)} = W$ and $x_{(n)} = X$). With this joint density function in hand, transformation methods can be used to find the density function of the span $Z = X + Y$, given that the number of fragments crossing P is k. It can be shown (see Problem 5.1) that the density function of Z is

$$f_Z(z; k) = k(k-1)(2L - z)(z - L)^{k-2}/L^k, \quad L < z < 2L, \qquad (5.4)$$

where we use the notation $f_Z(z; k)$ to indicate the dependence of this distribution on k. The (Poisson distribution) probability that exactly k fragments cover P is $e^{-a}a^k/k!$, so that for small h, the probability of the event "$k \geq 2$ and the span is of length between z and $z+h$" is, ignoring small-order terms,

$$h \sum_{k=2}^{\infty} \frac{e^{-a}a^k}{k!} f_Z(z; k) = ha^2 e^{-2a} L^{-2}(2L - z)e^{az/L}.$$

We write the right-hand side in this expression as $h\,j(z)$. The probability that no anchor lies on a span of length z is $e^{-zb/L}$. The probability that $k \geq 2$ and that the point P is not covered by an anchored contig is then

$$\int_L^{2L} j(z)e^{-zb/L}dz = \frac{a^2}{(a-b)^2}e^{-2b} + \frac{a^2(b-a-1)}{(a-b)^2}e^{-a-b}. \qquad (5.5)$$

The total probability that the point P is not covered by an anchored contig, that is, the mean proportion of the DNA not covered by an anchored contig, is found by adding the probabilities e^{-a} (for the case $k = 0$) and $ae^{-(a+b)}$ (for the case $k = 1$) calculated above to the right-hand side in (5.5). Subtraction of the quantity so found from 1 yields

$$1 - e^{-a} - \frac{a(b^2 - ab - a)e^{-(a+b)}}{(b-a)^2} - \frac{a^2}{(b-a)^2}e^{-2b}. \qquad (5.6)$$

This is the total probability that the point P is covered by an anchored contig, or equivalently the mean proportion of the genome covered by an anchored contig, as desired.

This calculation can be checked by considering two limiting cases. First, as $b \to \infty$ the number of anchors increases indefinitely, and the probability in (5.6) approaches the value $1 - e^{-a}$ as found in Section 5.1.2, as expected. Second, as $a \to \infty$ the number of fragments increases indefinitely. In this case the probability that a point P is in an anchored contig is the probability that at least one anchor lies in the interval of length $2L$ placed symmetrically around P. This probability is $1 - e^{-2b}$, and this is the limiting value of (5.6) as $a \to \infty$.

We next calculate the mean number of anchored contigs. Each anchored contig has a unique rightmost fragment, so this mean is the mean number of rightmost fragments on an anchored contig. The key calculation needed to find this mean number of fragments is to find the probability that a given fragment C is the rightmost fragment on an anchored contig, and we now carry out this calculation.

One way in which fragment C can be such a rightmost fragment is if it is anchored, with probability $(1 - e^{-b})$, and also no other fragment overlaps it on the right, with probability e^{-a}. This event has total probability

$$(1 - e^{-b})e^{-a}. \tag{5.7}$$

The only other way in which fragment C can be the rightmost fragment on an anchored contig arises when there is at least one fragment overlapping C to the right, and that there is at least one anchor in C in the non-overlapped part of C and no anchor in C in the overlapped part of C. The probability of this event is found as follows.

If fragment C has k ($k \geq 1$) fragments overlapping it to the right, there will be some leftmost member of these fragments whose left-hand end is a distance W to the right of the left-hand end of fragment C. For the moment we take k and the length W as given. The probability of the event that there is at least one anchor on the leftmost length W of fragment C is $1 - e^{-bW/L}$ and the probability of the event that there is no anchor on the remaining rightmost length $L - W$ of fragment C is $e^{-(L-W)b/L}$. These two events are independent, so that the probability that they both occur is $\left(1 - e^{-bW/L}\right) e^{-(L-W)b/L}$.

We now continue to take k as given and find the probability density of the (random) length W. This length is the minimum of k iid uniform random variables with range $[0, L]$, each having density function $f_X(x) = \frac{1}{L}$, $0 \leq x \leq L$. The probability distribution of this minimum is given in equation (2.88) as $k(L - x)^{k-1}/L^k$, and this is the density function of W, with range $[0, L]$.

We finally observe that k is a random variable having a Poisson distribution with parameter a. The probability that fragment C is overlapped on the right by at least one other fragment, but is nevertheless the rightmost fragment on an anchored contig, is then found from the above calculations to be

$$\sum_{k=1}^{+\infty} \left(e^{-a}a^k/k!\right) \int_0^L k(L - w)^{k-1} \left(1 - e^{-wb/L}\right)(e^{-(L-w)b/L}) / L^k \, dw. \tag{5.8}$$

Now the expression $\sum_{k=1}^{+\infty} \left(e^{-a}a^k/k!\right) k(L - w)^{k-1}/L^k$ appearing in (5.8) can be simplified to

$$\frac{ae^{-a}}{L} \sum_{k=1}^{+\infty} (a(L - w)/L)^{k-1} /(k - 1)! = \frac{a}{L}e^{-a}e^{a(L-w)/L} = \frac{a}{L}e^{-aw/L}. \tag{5.9}$$

Further, the expression $\left(1 - e^{-wb/L}\right)(e^{-(L-w)b/L})$ appearing in (5.8) simplifies to $e^{-b}(e^{wb/L} - 1)$. Thus (5.8) may be written as

$$aL^{-1}e^{-b} \int_0^L e^{-aw/L}(e^{wb/L} - 1) \, dw = ae^{-b} \int_0^1 e^{-ax}(e^{bx} - 1) \, dx. \tag{5.10}$$

Evaluating this integral and adding the probability $(1 - e^{-b})e^{-a}$ given in (5.7), we arrive at the probability

$$b\frac{(e^{-a} - e^{-b})}{(b - a)}$$

that fragment C is the rightmost fragment of an anchored contig. Since there are N fragments, the mean number of anchored contigs is

$$Nb\frac{(e^{-a} - e^{-b})}{(b - a)}. \qquad (5.11)$$

As a check on this calculation, the limiting case $b \to \infty$ yields the mean number Ne^{-a} found in (5.1). The limiting case $a \to \infty$ yields a mean of Me^{-b}. This is as expected when there are infinitely many fragments, since it is the number of anchors, M, multiplied by the probability e^{-b} that no anchor arises a distance L or less to the right of any given anchor, thus preventing that anchor from being the rightmost in a contig.

The calculations for the mean contig size are more complicated and are not given here.

The calculations described above rely on several simplifying assumptions, for example that both the fragment and the anchor Poisson processes are homogeneous and that all fragments are of the same length. Calculations applying in more realistic cases, removing these and other restrictions, is discussed by Arratia et al. (1991), Schbath (1997), and Schbath et al. (2000).

5.2   Modeling DNA

On its most elementary level, the structure of DNA can be thought of as long sequences of nucleotides. These sequences are organized into coding sequences, or genes, that are separated by long intergenic regions of noncoding sequence. Most eukaryotic genes have one further level of organization: Within each gene, the coding sequences (exons) are usually interrupted by stretches of noncoding sequences (introns). During processing of DNA into messenger RNA (mRNA), these intron sequences are spliced out and thus do not appear in the final mature mRNA which is translated into protein sequence.

Intergenic regions and introns have different statistical properties from those of exons. By capturing these properties in a model we can construct statistical procedures for testing whether or not an uncharacterized piece of DNA is part of the coding region of a gene. The model is based on a set of "training" data taken from already characterized sequences. In the simplest model it is assumed that the nucleotides at the various sites are independent and identically distributed (iid). If this is the case, the difference between coding and noncoding DNA could be captured in the difference between the frequencies of the four nucleotides in the two different cases (intron vs. exon). These distributions can then be estimated from the training data.

The accuracy of this procedure depends on the accuracy of the assumptions made. Thus it is important, for example, to develop tests of the (null) hypothesis of independence. One such test is based on a Markov chain analysis. In the use of Markov chains in sequence analysis, the concept of "time" is naturally replaced with "position along the sequence." Under a Markov chain alternative hypothesis the probability that a given nucleotide is present at any site depends on the nucleotide at the preceding site. The null hypothesis of independence corresponds to a Markov chain in which all rows in the transition probability matrix are identical, so that the probabilities for the various nucleotides at any site are independent of the nucleotide at the preceding site. It is possible that neither hypothesis is correct, but it is of interest to assess whether the Markov-dependence model describes reality significantly better than the independence model and thus might raise the accuracy of our predictive procedures. Furthermore, if it does, then models with even more complex dependence may be considered.

The statistical test for Markov independence is an association test in a 4×4 table such as in Example 4 of Section 3.5. In this case $r = c = 4$, and Y_{jk} is the number of times that a nucleotide of type "j" is followed by a nucleotide of type "k" in the DNA sequence of interest. The data would then appear as shown in Table 5.4. In this case the null hypothesis of independence becomes the null hypothesis of no association in the table.

		nucleotide at site $i+1$				
		a	g	c	t	Total
	a	Y_{11}	Y_{12}	Y_{13}	Y_{14}	$y_{1.}$
nucleotide	g	Y_{21}	Y_{22}	Y_{23}	Y_{24}	$y_{2.}$
at site i	c	Y_{31}	Y_{32}	Y_{33}	Y_{34}	$y_{3.}$
	t	Y_{41}	Y_{42}	Y_{43}	Y_{44}	$y_{4.}$
	Total	$y_{.1}$	$y_{.2}$	$y_{.3}$	$y_{.4}$	y

Table 5.4. Table of numbers of times a nucleotide of one specified type follows one of another specified type.

Tests of this type show that the nucleotides at adjoining DNA sites are often dependent, and a first-order Markov model fits real data significantly better than the independence model both in introns and in exons. Higher-order homogeneous Markov models and nonhomogeneous models often provide even better fits. We discuss tests of higher-order Markov dependence in Section 11.3.

5.3 Modeling Signals in DNA

5.3.1 Introduction

Genes contain "signals" in the DNA to indicate the start and end of the transcribed region, the exon/intron boundaries, as well as other features. The cell machinery uses these signals to recognize the gene, to edit it correctly, and to translate it appropriately into protein.

A signal is a short sequence of DNA having a specific purpose, for example indicating an exon/intron boundary. If nature were kind, each such signal would consist of a unique DNA sequence that did not appear anywhere in the DNA except where it serves its specific purpose. In reality, there may be many DNA sequences that perform the same signal function; we call these "members" of that signal. Furthermore, members of signals also appear randomly in the nonfunctional DNA, making it hard to sort out the functional from nonfunctional signals.

In practice, some but not all members of a signal are known. Our aim is to use known members to assess the probability that a new uncharacterized DNA sequence is also a member of the signal. Our assumption is that the different members arose from common ancestors via stochastic processes. Therefore, it is reasonable to construct a statistical model of the data. Some signals require only simple models; others require complex ones. When the model required is complex and the data are limited, we must make carefully chosen simplifying assumptions in the model in order to utilize the data in the most efficient manner. In Chapter 12 we describe a complex model for modeling an entire gene. That application will use all of the different signal models described below.

We assume that all members of the signal of interest have the same length, which we denote by n. This assumption is not too restrictive, since the members of many signals do have the same lengths, and for those that do not we can capture portions of them with this model, which is often sufficient. To model the properties of any signal we must have a set of "training data": These are extensive data in which members of the signal are known and recognized. For example, in investigating signals in human DNA we might use a set of known members of signals taken from known human genes in public databases.

We now consider some of the basic signal models that are used in bioinformatics. For a more exhaustive discussion of this subject, see Burge (1998).

5.3.2 Weight Matrices: Independence

The test of whether a given uncharacterized DNA sequence is a member of a given signal is most easily carried out in the case where the nucleotide in any position in the signal is independent of the nucleotides in any other position in the signal.

It is therefore necessary to form a test of independence, and we consider here a test of whether the nucleotide at position a in a signal is independent of the nucleotide at position b. This can be done in various ways. It is natural to generalize the test described in connection with the data in Table 5.4, which tests for independence at adjacent sites. In this generalization "site i" is replaced by "position a in the signal" and "site $i+1$" is replaced by "position b in the signal," and Y_{jk} is interpreted as the number of times in the training data that nucleotide j occurs at position a in the signal and nucleotide k occurs at position b in the signal. This test is then performed for all pairs a and b. The individual tests are then corrected for multiple testing, either using a Bonferroni or Šidák correction, as discussed in Section 3.11.

Suppose that as a result of this or some other testing procedure independence can be assumed. A $4 \times n$ matrix is then constructed whose (i, j) entry is the proportion of cases in the training data for which the ith nucleotide occurs at the jth position of a signal. This is referred to as a *weight matrix* and is denoted by M. The set of training data is assumed to be sufficiently large that the proportions in this matrix can be taken as probabilities. An example for the case $n = 5$ is given in Table 5.5. The entries in column

		position				
		1	2	3	4	5
	a	0.33	0.34	0.19	0.20	0.21
nucleotide	g	0.22	0.27	0.23	0.24	0.21
	c	0.31	0.18	0.34	0.30	0.25
	t	0.14	0.21	0.24	0.26	0.33
	Total	1	1	1	1	1

Table 5.5. Signal probabilities.

j in this matrix give (estimated) probabilities for the four nucleotides at position j in the signal.

The matrix M defines a probability $\text{Prob}(s \mid M)$ for any sequence s of length n. Weight matrices are used in many contexts. We shall use them as a component of a gene-finding algorithm given in Chapter 12.

5.3.3 Markov Dependencies

If the nucleotides at the sites in a signal are not independent, a possibility is that there are dependencies of the first order Markov type. Under this hypothesis the first position of the signal has a probability distribution determined by overall nucleotide frequencies. Nucleotides at each subsequent position have a probability distribution depending on the nucleotide at the previous position as determined by a 4×4 Markov chain transition probability matrix whose (i, j) element is the probability of the jth nucleotide in position k, given the ith nucleotide in position $k - 1$. The elements in this matrix are estimated from the training data.

More generally, higher-order Markov dependencies such as those described in Section 11.3 may be modeled. However, as discussed in Section 11.3, these transition matrices become large as the order of the Markov dependency increases, so that for a high-order dependence the amount of data needed for satisfactory estimation might be excessive. When the set of training data is limited we must economize and capture only the most informative dependencies in our model. A method for doing this is discussed in the next section.

5.3.4   Maximal Dependence Decomposition

As discussed in Section 5.3.3, it may be impossible, due to limited data, to obtain satisfactory estimates of all dependencies in the sequences in a given signal. This motivates us to search for a method that captures those dependencies that are most informative. We now describe the approach to this problem known as *maximal dependence decomposition* (MDD) (Burge (1997)).

Suppose we wish to model a signal of length n. The first step is to find one position which has the greatest influence on the others. To do this we construct an $n \times n$ table whose (i, j) entry is the observed value of the chi-square statistic obtained from a 4×4 table such as Table 5.4 but that compares the nucleotides at a fixed position i with those at a fixed position j (instead of specifically position $i+1$). If the hypothesis of independence is true, we expect to find about one significant value out of 20 just by chance if the Type I error is 5%. Thus if only a few of the observed chi-square values are significant but not highly significant, we might conclude that the positions in the signal can be taken as independent. If more than a few are significant, or if there are values with high levels of significance, then we would conclude that the nucleotides at the various positions are not independent.

	1	2	3	4	5	total
1		34.2*	7.1	37.2*	2.8	81.3
2	34.2*		.4	72.4*	4.5	111.5
3	7.1	.4		15.3	98.3*	121.1
4	37.2*	72.4*	15.3		14.2	139.1
5	2.8	4.5	98.3*	14.2		119.8

Table 5.6.

Table 5.6 provides an example with $n = 5$ in which many of the chi-square values are significant at the 5% level (indicated by asterisks). The row with the greatest sum gives an indication of which position has the greatest influence on the other $n - 1$ positions. In the above example position 4 has the greatest row total, so we take it as the position with greatest influence and try to assess this influence further. In this case we say that we *split* on position 4.

We illustrate first by constructing a model that determines distributions for positions 1, 2, 3, and 5 conditional on position 4. This is done as follows. We divide the set of test sequences into four sets, each set being determined by the nucleotide in position 4, so that for each nucleotide x we have a set T_x consisting of those sequences where there is an x in position 4. We then calculate $p_x = n_x/d$, for each $x = a, g, c, t$, where n_x is the number of members of the training data having nucleotide x in position 4, and $d = \sum n_x$. Next, for each $x = a, g, c, t$, we calculate a 4×4 weight matrix M_x from the sequences in T_x, for positions 1, 2, 3, and 5.

The model M then consists of the distribution $\{p_a, p_g, p_c, p_t\}$ together with the four weight matrices $\{M_a, M_g, M_c, M_t\}$. For any sequence s of length 5 we calculate $\text{Prob}(s \mid M)$ as follows.

If nucleotide x occurs in position 4 of s, the weight matrix M_x is used to assign a probability p_k to positions $k = 1, 2, 3$, and 5 of s. $\text{Prob}(s \mid M)$ then equals $p_x \cdot p_1 \cdot p_2 \cdot p_3 \cdot p_5$.

In general, some or all of the sets T_x might be large enough that we can repeat the entire process recursively, splitting T_x further on one of the positions 1, 2, 3, 5. In this case, for each sufficiently large T_x we make a new table similar to Table 5.6, this time 4×4, to find the position of these four that has the greatest influence on the other three. We then decompose T_x into T_{xy} for $y = a, g, c$, and t, and continue to T_{xyz}, and so on, as long as there are enough data. A rule of thumb is to stop whenever the set $T_{xyz\cdots}$ has fewer than 100 sequences. When such a limit is reached, the remaining positions are modeled with a weight matrix.

5.4 Long Repeats

Suppose that we are interested in one specific nucleotide, say a, and ask whether there is significant evidence of long repeated sequences of this nucleotide. More specifically, we wish to test the null hypothesis that nucleotides occur at random against the alternative that there is a tendency for long repeats of a. We have discussed one test for Markov dependence of nucleotides in Section 5.2 and will discuss more detailed tests in Section 11.3. In this section we discuss a test of randomness tailored to the "repeats" alternative hypothesis.

Consider a long DNA sequence of length N, where N is assumed to be so large that end effects can be ignored in our calculations. Suppose that the null hypothesis probability that the nucleotide a occurs at any site in this sequence is some known value p. We call the occurrence of a at any site a "success" and the occurrence of any other nucleotide at any site a "failure."

Suppose that the DNA sequence is scanned from left to right, and consider the situation immediately following a failure. There will then be a sequence of successes, possibly of length 0 (if the next nucleotide is c, g, or t). Assuming that the null hypothesis is true, equation (1.15) shows that the length Y of this sequence has the geometric distribution, which is reproduced here for convenience:

$$P_Y(y) = \text{Prob}(Y = y) = (1 - p)p^y, \quad y = 0, 1, 2, \ldots. \qquad (5.12)$$

Equation (2.106) shows that if n independent such sequences are given, and $Y_{\max}$ is the length of the longest of these, then

$$\text{Prob}(Y_{\max} \geq y) = 1 - (1 - p^y)^n. \qquad (5.13)$$

If $Y_{\max}$ is used as test statistic, the P-value for any observed value $y_{\max}$ of $Y_{\max}$ can be calculated directly from equation (5.13) once an appropriate value of n is chosen. Any sequence of successes (of length 0 or more) must be preceded by a failure. Under the null hypothesis there will be approximately $(1 - p)N$ such failures, so that there will also be approximately $(1 - p)N$ sequences of successes of length 0 or more. This is the value we use for n. This implies that

$$P\text{-value} \cong 1 - (1 - p^{y_{\max}})^{(1-p)N}. \qquad (5.14)$$

If $N = 100{,}000$ and $p = \frac{1}{4}$ and the observed length $y_{\max}$ of the longest repeated sequence of a is 10, the approximate P-value calculated from (5.14) is 0.0690272. Thus, given an observed value of 10 for $Y_{\max}$, the null hypothesis would not be rejected with any conventionally used Type I error. On the other hand, if the observed longest length were 12, the P-value would be approximately 0.45%, and this would often be taken as being sufficiently small for rejection of the null hypothesis.

The discussion following the exponential approximation in (B.3) shows that the expression (5.14) is well approximated, when N is large and $(1 - p)Np^{y_{\max}} \leq 1$, by

$$P\text{-value} \cong 1 - e^{-(1-p)Np^{y_{\max}}}. \qquad (5.15)$$

In the case $N = 100{,}000$, $p = \frac{1}{4}$, the approximation (5.15) gives 0.0690275, essentially identical to the value found from (5.14). Expressions of the form (5.15) arise later in connection with BLAST.

There is one approximation used in these calculations that should be checked. It was assumed in the calculations that the number of failures is $(1-p)N$. More exactly, this number is a random variable having a binomial distribution with parameters N and $1 - p$. Thus when $N = 100{,}000$ and $1 - p = \frac{3}{4}$, a precise calculation for the null hypothesis probability that $Y_{\max} \geq 10$ is, from equations (1.53) and (1.105),

$$\sum_{j=0}^{100{,}000} \binom{100{,}000}{j} \left(\frac{3}{4}\right)^j \left(\frac{1}{4}\right)^{100{,}000-j} \left(1 - \left(1 - \left(\frac{1}{4}\right)^{10}\right)^j\right). \quad (5.16)$$

Equation (B.1) shows that this expression reduces to

$$1 - \left(1 - \frac{3}{4}\left(\frac{1}{4}\right)^{10}\right)^{100{,}000} \cong 0.0690276.$$

This agrees to seven decimal place accuracy with the approximate value found from (5.15). This close agreement is typical of the values for other cases, and we conclude that the approximation of regarding the number of sequences as fixed at its mean value is sufficiently accurate for all practical purposes.

It was shown in Table 3.4 that for the geometric distribution (the $k = 0$ case in Table 3.4), the expression in (3.51) gives a very accurate approximation for P-values, essentially identical to the exact calculation given in (2.106), provided that one uses the expressions $\mu_{\max}$ and $\sigma_{\max}^2$ for the mean and variance approximations for $Y_{\max}$ given in equations (2.118) and (2.120). These expressions are repeated here for convenience, with a slightly altered notation:

$$\mu_{\max} \approx \frac{\gamma + \log n}{-\log p} - \frac{1}{2}, \quad \sigma_{\max}^2 \approx \frac{\pi^2}{6(\log p)^2} + \frac{1}{12}. \quad (5.17)$$

Next we consider runs of any nucleotide, not only a. This case has been discussed in the disease context with repeats of triplets of nucleotides rather than single nucleotides being of interest: see for example the edition of *Brain Research Bulletin*, (Servadio et al. (2001)) devoted to this topic. The calculations for triplet repeats are a natural extension of those considered here, where we consider a collection of single nucleotide repeated sequences, some of which might be of a, some of g, some of c, and some of t.

We consider first the case when the probability that any specified nucleotide arises at any site is $1/4$. This corresponds to the case $p = 1/4$ in the expressions in (5.17). In this case Smythe (2004) has shown that, when end effects are ignored, the mean and variance of the length of the longest repeated sequence, where this sequence can be either of a, g, c or t, are given by

$$\mu_{\max} \approx \frac{\gamma + \log n}{\log 4} + \frac{1}{2}, \quad \sigma_{\max}^2 \approx \frac{\pi^2}{6(\log 4)^2} + \frac{1}{12}. \quad (5.18)$$

Thus the mean exceeds the value given in (5.17) for the case $p = 1/4$ by 1 while the variance is as given in (5.17). We discuss below why this should be so.

A more complicated result arises when the probabilities of the various nucleotide differ. Suppose that there is one nucleotide whose probability p^* of occurrence at any site exceeds that of all other nucleotides. Then when n is very large, it becomes increasingly likely that the longest repeated sequence is of this nucleotide. Smythe uses this intuition to show that the mean and variance of the longest repeat are essentially as approximated in (5.17), with now $p = p^*$.

When each nucleotide in some subset of r of the four nucleotides arises with probability p^*, and this probability exceeds the probability of any one of the $4 - r$ nucleotides not in this subset, Smythe uses a heuristic argument to claim that the mean length of the longest repeated sequence of any nucleotide is given by

$$\mu_{\max} \cong \frac{\gamma + \log n + \log r}{-\log p^*} - \frac{1}{2}. \tag{5.19}$$

In the case $r = 1$ this reduces to the value given in the previous paragraph, as expected, and in the case $r = 4$ it reduces to the value given in (5.18).

A rather different question is taken up by Karlin et al. (1983), who consider the mean and the variance of the length of the longest word, of whatever composition, that is duplicated at two (or more) positions in the sequence. Perhaps surprisingly, since this is a quite different problem from that considered above, this mean and variance are of a similar form to those given in the approximation (5.17), being

$$\mu_{\max} \cong \frac{0.6359 + 2\log n + \log(1 - P)}{-\log P} - 1, \quad \sigma_{\max}^2 \cong \frac{\pi^2}{6(\log P)^2}, \tag{5.20}$$

where P is the sum of the squares of the four nucleotide frequencies. If all nucleotides are equally frequent this mean is approximately twice that given in (5.17) when n is large, whereas the variance is close to that in (5.17).

The similarities of the various results just discussed is no coincidence. To illustrate this we show why the mean values in (5.17) and (5.18) differ by 1 and the variances are equal. In the case where a specific nucleotide is of interest, say a, we may regard the occurrence of a at any site as a success, and thus are interested in runs of successes. Suppose that there are Y_1 successes in any such run. In the case where runs of any nucleotide are of interest, we may regard a success as arising when the nucleotide at any site is the same as that at the previous site. The length of any such run may be expressed as $1 + Y_2$, the 1 arising from the initial nucleotide in the run and with Y_2 being the number of times this nucleotide then follows in succession. When all nucleotides have the same probability of arising at any site, Y_1 and Y_2 have the same probability distribution. Thus the mean of the length $1 + Y_2$ is exactly 1 more than the mean of the length Y_1, while the variances of the two lengths are identical.

The mean in (5.20) can be found by imagining the DNA sequence of length n to be copied, and then one copy compared with the other using all possible relative alignments of the two copies. A run of successes arises when exact matches of the nucleotides in the two copies occurs. There are approximately n^2 positions where such a run of successes can occur, and the fact that $\log n^2 = 2 \log n$ then explains the factor of 2 multiplying $\log n$ in (5.20).

Thus all the above results are variants of each other, and the appropriate mean and variance for any case of interest can be used, in conjunction with (3.51), to find approximate P-values for the length any observed run of "successes."

5.5 r-Scans

In this section we discuss a procedure for testing whether certain genomic features, for example genes, occur at locations that are uniformly and independently distributed in some connected segment of the genome, for example part of a chromosome arm. The tests we consider were put forward by Karlin and Macken (1991a,b), who derived a variety of so-called r-scan tests to assess heterogeneity of the location of various restriction sites and other markers in genetic sequence data. We outline their analysis in this section. An example of their tests, concerning the location of motifs, is described in Section 5.6. Various asymptotic results for r-scans not discussed here are given by Reinert et al. (2000).

The null hypothesis considered is that the genomic features of interest occur at locations that are uniformly and independently distributed in the segment of interest. We assume that the length of a gene is so short relative to this segment that we may regard the positions of the genes as points. It is also convenient to normalize lengths so that the segment is taken to be of length 1. The null hypothesis then is that the locations of a collection of n points on the unit interval $[0, 1]$ are iid uniformly distributed random on this interval. The alternative hypotheses of interest are, first, that the points tend to occur in a clumped fashion, or second, that they tend to occur in a regularly spaced fashion. The r-scan statistics of Karlin and Macken (1991a,b) described here were introduced to test these hypotheses.

We denote the locations of the n points by $X_1, X_2, \ldots, X_n$ and the corresponding order statistics by $X_{(1)}, X_{(2)}, \ldots, X_{(n)}$. These order statistics divide the interval $(0, 1)$ into $n+1$ subintervals of lengths $U_1, U_2, \ldots, U_{n+1}$, where

$$U_1 = X_{(1)}, \quad U_2 = X_{(2)} - X_{(1)}, \quad \ldots, \quad U_{n+1} = 1 - X_{(n)}. \quad (5.21)$$

The Karlin and Macken tests are based on the lengths of these subintervals. To derive the properties of these tests, it is first necessary to find the joint distribution of the U_i's when the null hypothesis is true. This is done as follows.

From (2.144), the joint density function of $X_{(1)}, X_{(2)}, \ldots, X_{(n)}$ is $n!$. The joint density function of $U_1, U_2, \ldots, U_n$ may be found from this joint density function by the transformation techniques of Section 2.13. The Jacobian matrix of the transformation from $X_{(1)}, X_{(2)}, \ldots, X_{(n)}$ to $U_1, U_2, \ldots, U_n$ is triangular, with entries 1 along the main diagonal, so that the determinant of this matrix is 1. This implies from equation (2.145) that the joint density function of $U_1, U_2, \ldots, U_n$ is

$$f_{\boldsymbol{U}_n}(u_1, u_2, \ldots, u_n) = n!, \quad u_j > 0, \quad \sum_{j=1}^{n} u_j \leq 1. \tag{5.22}$$

In this equation, and below, we use the notation $\boldsymbol{U}_i$ for the vector $(U_1, U_2, \ldots, U_i)$, $i = 1, 2, \ldots, n+1$. The joint range of $U_1, U_2, \ldots, U_n$ in (5.22) is determined by the fact that $U_j > 0$ and $\sum_{j=1}^{n} U_j \leq 1$. Since $U_{n+1} = 1 - (U_1 + U_2 + \cdots + U_n)$, the value of U_{n+1} is determined by the values of $U_1, U_2, \ldots, U_n$.

Our first aim is to test the null hypothesis against the alternative hypothesis that the points arise in a clumped fashion. The test statistic that we will initially use for this test is the length $U_{\max}$ of the maximum of $U_1, U_2, \ldots, U_{n+1}$, and we now find the null hypothesis distribution of this length.

Suppose that g of the $n+1$ lengths $U_1, U_2, \ldots, U_{n+1}$ are chosen at random, and let u be any number in $(0,1)$. We first find the probability that all of the g lengths chosen exceed u. This probability is clearly 0 if $ug > 1$, so from now on we assume that $ug < 1$.

The symmetric form of the density function (5.22) implies that the joint density function of any subset of g of the lengths $U_1, U_2, \ldots, U_{n+1}$ is independent of the subset chosen. The joint density function $f_{\boldsymbol{U}_g}(u_1, u_2, \ldots, u_g)$ of $U_1, U_2, \ldots, U_g$ is the marginal density function

$$f_{\boldsymbol{U}_g}(u_1, \ldots, u_g) = \int_0^{w_{g+1}} \int_0^{w_{g+2}} \cdots \int_0^{w_n} n! \, du_n \cdots du_{g+1}, \tag{5.23}$$

where $w_j = 1 - u_1 - u_2 - \cdots - u_{j-1}$. (These terminals arise because $U_1 + U_2 + \cdots + U_j \leq 1$ for every j.) This integration yields

$$f_{\boldsymbol{U}_g}(u_1, \ldots, u_g) = \frac{n!}{(n-g)!} (1 - u_1 - \cdots - u_g)^{n-g}. \tag{5.24}$$

The probability that each U_j, $j = 1, 2, \ldots, g$, exceeds u is then found by integration as

$$\int_u^1 \int_u^{w_2} \cdots \int_u^{w_g} \frac{n!}{(n-g)!} (1 - u_1 - \cdots - u_g)^{n-g} \, du_g \cdots du_1. \tag{5.25}$$

The value of this integral is $(1 - gu)^n$ (see Problem 5.8), so that

$$\text{Prob}(U_1 > u, \ U_2 > u, \ \ldots, \ U_g > u) = (1 - gu)^n. \tag{5.26}$$

We define h_u as the largest integer for which $h_u u < 1$. If A_i is the event $\{U_i \geq u\}$, $i = 1, 2, \ldots, n + 1$, and if $i_1, \ldots, i_g$ are g distinct indices, then from (5.26),

$$P(A_{i_1} A_{i_2} \cdots A_{i_g}) = \begin{cases} (1 - gu)^n, & g \leq h_u \\ 0, & g > h_u \end{cases}. \tag{5.27}$$

Equation (1.97) then implies that the probability that at least one of the events $A_1, A_2, \ldots, A_{n+1}$ occurs is

$$\sum_{g=1}^{h_u} (-1)^{g+1} \binom{n+1}{g} (1 - gu)^n. \tag{5.28}$$

This sum takes different forms for different values of u. When $\frac{1}{2} < u < 1$ only the first term appears in the sum; when $\frac{1}{3} < u < \frac{1}{2}$ only the first two terms appear, and so on.

The expression in (5.28) gives the probability that the maximum length exceeds u, so that it gives the P-value associated with an observed value u of $U_{\max}$.

The density function $f_{U_{\max}}(u)$ of $U_{\max}$ is found by writing (5.28) as $1 - F_{U_{\max}}(u)$, where $F_{U_{\max}}(u)$ is the cumulative distribution function of $U_{\max}$. Differentiation of this cumulative distribution function with respect to u yields

$$f_{U_{\max}}(u) = n \sum_{g=1}^{h_u} (-1)^{g+1} g \binom{n+1}{g} (1 - gu)^{n-1}. \tag{5.29}$$

As with the sum in (5.28), this density function takes different forms for different values of u.

Since this density function does not take a simple form, it would be useful to be able to approximate it using the asymptotic formula (2.131) for the density function of a maximum. Strictly speaking this should not be done, since the theory of Section 2.11.3 shows that (2.131) applies for the maximum of independent random variables whose support is of of the form $(A, +\infty)$. Neither of these requirements hold here, since $U_{\max}$ is not the maximum of independent random variables, none of which can take a value exceeding 1. Despite this, we now discuss how well use of (2.131) provides close approximations to the P-value associated with any observed value of $U_{\max}$.

To do this it is necessary first to find the mean and variance of $U_{\max}$. A long calculation which we do not give here shows that the mean of $U_{\max}$ is

$$\frac{1}{n+1} \left(\frac{1}{n+1} + \frac{1}{n} + \cdots + \frac{1}{2} + \frac{1}{1} \right) \tag{5.30}$$

and that the variance of $U_{\max}$ is approximately

$$\frac{\pi^2}{6(n+1)^2}. \tag{5.31}$$

Using these values in the approximation (2.131), we obtain, for large n,

$$\text{Prob}(U_{\max} \geq u) \sim 1 - e^{-(n+1)e^{-(n+1)u}}. \tag{5.32}$$

If $(n+1)e^{-(n+1)u}$ is small, the approximation (B.21) of Appendix B.12 shows that

$$\text{Prob}(U_{\max} \geq u) \sim (n+1)e^{-(n+1)u}. \tag{5.33}$$

In view of the various approximations involved in reaching (5.32) and (5.33), it is useful to calculate a selection of P-values exactly, using (5.28), and to compare these with the values found from (5.32) and (5.33). If $u = 0.01$ and $n+1 = 1,000$, the exact P-value found from (5.28) is 0.0482. The value found from the approximation (5.32) is 0.0444 and that found from (5.33) is 0.0454. When $u = 0.000017$ and $n+1 = 1,000,000$, the exact P-value is 0.0406. The value found from the approximation (5.32) is also 0.0406, while that found from (5.33) is 0.0414. This indicates the accuracy of (5.32), at least for large n.

The approximation (5.32) may be written in the equivalent form

$$\text{Prob}\left(U_{\max} \geq \frac{\log(n+1)+u}{n+1}\right) \sim 1 - e^{-e^{-u}}. \tag{5.34}$$

The advantage of this formulation is that it can be generalized to apply to a wide variety of "r-scan" test statistics. For example, instead of using $U_{\max}$ as test statistic, it might be thought to be more reasonable to use the maximum $R_{\max(r)}$ of the "r-fragment lengths" $R_i(r)$ $(i = 1, 2, \ldots, n-r+1)$ where

$$R_i(r) = \sum_{j=i}^{i+r-1} U_j. \tag{5.35}$$

Karlin and Macken (1991a,b) show that

$$\text{Prob}\left(R_{\max(r)} \geq \frac{\log(n+1) + (r-1)\log(\log(n+1)) + u}{n+1}\right)$$
$$\sim 1 - e^{-e^{-u}/(r-1)!}. \tag{5.36}$$

This is a direct generalization of (5.34), to which it reduces when $r = 1$. A further generalization allows the use of the kth largest of the $R_i(r)$ values as test statistic.

Suppose next that the alternative hypothesis is that the points tend to occur in a regularly spaced fashion. An initial approach might be to use $U_{\min}$ as test statistic, and to reject the null hypothesis if the observed value

u of $U_{\min}$ is too large. The case $g = n+1$ of equation (5.27) shows that for any u in $(0, (n+1)^{-1})$, the probability that the lengths of all $n+1$ intervals exceed u is $(1 - (n+1)u)^n$. This implies that

$$\text{Prob}(U_{\min} \geq u) = (1 - (n+1)u)^n, \qquad (5.37)$$

so that (5.37) gives the P-value associated with an observed value u of $U_{\min}$.

The expression (5.37) can be written as $1 - F_{U_{\min}}(u)$, where $F_{U_{\min}}(u)$ is the cumulative distribution function of $U_{\min}$. Differentiation of this expression shows that the density function of $U_{\min}$ is

$$f_{U_{\min}}(u) = n(n+1)\left(1 - (n+1)u\right)^{n-1}, \quad 0 < u < (n+1)^{-1}, \qquad (5.38)$$

which is far simpler than the expression (5.29) for $U_{\max}$.

It is possible that the points are regularly spaced except for two that are close to each other, in which case use of $U_{\min}$ will not pick up the regular spacing. Just as the procedure using $U_{\max}$ as test statistic points can be generalized to the use of the maximum of $R_i(r)$ for some predetermined value of r exceeding 1, so also generalizations using the minimum of $R_i(r)$ for some predetermined value of r exceeding 1 are available to assess whether the points tends to occur in a regularly spaced fashion. We do not however discuss the details of this here.

5.6   The Analysis of Patterns

5.6.1   Introduction

The discussion in this section and in Sections 5.7, 5.8, and 5.9 is self-contained and is not needed for the material that follows. It serves to illustrate some non-intuitive results about patterns in sequences that arise even in simple cases involving independence. It also introduces some of the statistical properties of motifs.

The concept of a word was introduced in Section 5.5. Suppose we are interested in some word and ask the following two questions of an iid DNA sequence of length N: "What is the probability distribution of the number of times that this word arises in a segment of length N?" and "What is the probability distribution of the length between one occurrence of this word and the next?" We call these two questions "number of occurrences" and "distance between occurrences" questions respectively.

There are various reasons why these questions might be asked. One is that there might be some a priori reason to suspect that the word occurs significantly more often in some DNA sequence data than would be expected if the nucleotides in the sequence were generated in an iid fashion. To test for this it is necessary to discuss probabilistic aspects of the frequency of this word under the iid assumption.

A second reason has been discussed by Bussemaker et al. (2000). Here the aim is to discover promoter signals by looking for DNA patterns common to upstream regions of genes. This is done by creating a dictionary of words of different lengths, each with an assigned probability. Any word that occurs more frequently than expected in these regions is a candidate for such a sequence. Here the analysis becomes more complicated, since there is no word that is of a priori interest, and indeed no word length that can be defined in advance as being specifically sought. Difficult statistical problems of multiple testing then arise: These are beyond the level of this book.

Examples of words that appear with unusually large frequencies at various types of sites have been investigated by Biaudet et al. (1998), Chedin et al. (1998), and Karlin et al. (1992). Leung et al. (1996) investigate under- and over-represented words in specific genomes. A general review of the theory of word occurrences is given by Reinert et al. (2000).

In some cases a specified collection of words, that is a *motif*, is of interest; these are discussed in Section 5.9.

The analysis of the frequency of a word is sometimes carried out together with an analysis of the locations of that word, as assessed by the r-scan method of Section 5.5.

The analysis of nucleotide sequences leads to a focus on an "alphabet" of size 4. We use this example throughout, although generalizations to arbitrary alphabet sizes, and thus to amino acid sequences, are straightforward.

The analysis in this chapter assumes that the nucleotide types at different sites are independent. This assumption is made to introduce some of the unexpected features of word pattern properties in a simple setting. In view of the fact that dependencies appear to exist between nucleotides at adjoining sites, the analysis is extended to the Markov-dependent case in Section 11.4.

5.6.2   Counting conventions

It is necessary to address the question of how to count the number of times that any word occurs in a DNA sequence. Suppose for example that the word of interest is *gaga*. If overlapping occurrences of *gaga* are all relevant and counted, then in the sequence

$$t\ a\ t\ g\ a\ g\ a\ g\ a\ t\ c\ c\ g\ a\ g\ a \tag{5.39}$$

this word is counted as occurring in positions 4–7, 6–9, and 13–16, a total of three times. Thus even though two of these words overlap in positions 6 and 7, both are counted. If second and higher overlapping words are not counted, the word is counted as occurring only twice, namely in positions 4–7 and 13–16. A more precise definition of the non-overlap accounting is given later, but it is already clear that the number of times that any given word is counted depends on which of the two possible overlap counting conventions is adopted. The case where overlaps are counted is discussed in Section 5.7 and the case where they are not counted is discussed in Section 5.8. Both cases are discussed in the extension of the theory to motifs, discussed in Section 5.9.

5.6.3 Notation and assumptions

In Sections 5.7 and 5.8 we consider some specified word $\mathbf{w} = w_1 w_2 \ldots w_k$ of arbitrary length k. Here each w_j is one of the four nucleotides a, g, c and t, so that for example in the word $gaga$, w_1 is g, w_2 is a, and so on. We call w_j the jth letter of $\mathbf{w}$.

Several results depend on the "self-overlapping" properties of $\mathbf{w}$. We define an indicator function ε_j to be equal to 1 if the first j letters of $\mathbf{w}$ are the same, and in the same order as, the last j letters of $\mathbf{w}$, and 0 otherwise. (With this definition, ε_k is identically 1 for a word of length k.)

We consider throughout a DNA sequence of length N, and say that $\mathbf{w}$ occurs at site n in this sequence if it occupies sites $n - k + 1, n - k + 2, \ldots, n$. The possible values of n are $k, k + 1, \ldots, N$. We assume that the nucleotide types at the various sites in the sequence are independent, with the nucleotides a, g, c, and t having respective probabilities p_a, p_g, p_c, and p_g at any site. We define $\pi(\mathbf{w})$ as the probability that $\mathbf{w}$ occupies any k consecutive sites; under the independence assumption this is simply the product of the probabilities of the various letters comprising $\mathbf{w}$. In some of the theory we also calculate the probability of some other sequence of letters: the probability of the sequence defined by the first j letters of $\mathbf{w}$, for example, that is of the sequence $w_1 w_2 \ldots w_j$, is denoted $\pi(w_1 w_2 \ldots w_j)$, with a similar definition for other sequences of letters.

When the theory is relatively complex we develop it in detail in some special case (for example $k = 4$, all nucleotides having probability $1/4$), and state without proof various general results that the special case discussed exemplifies.

The extension of the theory to consider collections of words, or motifs, requires further notation: this is introduced in Section 5.9.

5.7   Overlaps Counted

5.7.1   Number of Occurrences

We define $Y_1(N)$ as the (random) number of times that the word $\mathbf{w}$ occurs in a nucleotide sequence of length N, recalling that in this section overlaps of the word are counted. Our first aim is to find the mean and variance of $Y_1(N)$.

Define the indicator variable I_j by $I_j = 1$ if $\mathbf{w}$ occurs in position j, $I_j = 0$ if it does not. Then the total number of times $Y_1(N)$ that the $\mathbf{w}$ occurs in a sequence of length N can be represented as $I_k + I_{k+1} + \cdots + I_N$. The mean of $Y_1(N)$ can then be found by applying equation (2.69). The mean of I_j is the probability that $\mathbf{w}$ finishes in position j, namely $\pi(\mathbf{w})$. Equation (2.69) then shows that the mean number of times that $\mathbf{w}$ arises in a sequence of length N is 0 if $N < k$, while for $N \geq k$, the mean of $Y_1(N)$ is

$$E(Y_1(N)) = (N - k + 1)\pi(\mathbf{w}). \tag{5.40}$$

The variance of $Y_1(N)$ is the variance of $I_k + I_{k+1} + \cdots + I_N$, namely

$$E(I_k + I_{k+1} + \cdots + I_N)^2 - \big((N - k + 1)\pi(\mathbf{w})\big)^2.$$

Now

$$
\begin{aligned}
E(I_k + &I_{k+1} + \cdots + I_N)^2 \\
&= E(I_k^2 + I_{k+1}^2 + \cdots + I_N^2) \\
&\quad + 2(N - k) \text{ terms of the form } E(I_j I_{j+1}) \\
&\quad + 2(N - k - 1) \text{ terms of the form } E(I_j I_{j+2}) \\
&\quad + \cdots \\
&\quad + 2(N - 2k + 2) \text{ terms of the form } E(I_j I_{j+k-1}) \\
&\quad + (N - 2k + 2)(N - 2k + 1) \text{ terms of the form} \\
&\qquad E(I_j I_m), \; |m - j| > k - 1.
\end{aligned}
\tag{5.41}
$$

This calculation assumes that $N \geq 2k - 2$, the case of interest in practice, and we make this assumption from now on. The case $N < 2k - 2$ is easily handled separately.

Since the only possible values of I_j are 0 and 1, $I_j{}^2 = I_j$. Thus the first term on the right-hand side of (5.41) is $E(I_k + I_{k+1} + \cdots + I_N)$. This is the mean number of times that $\mathbf{w}$ occurs, namely $(N - k + 1)\pi(\mathbf{w})$ (from equation (5.40)).

The remaining calculations in (5.41) concern expectations of the form $E(I_j I_{j+k-i})$ for $i = 1, 2, \ldots, k - 1$. If the structure of $\mathbf{w}$ is such that this word cannot finish at both sites j and $j + k - i$, then $I_j I_{j+k-i}$ is identically zero, and hence its expected value is also identically zero. Thus to calculate the variance (5.41) it is necessary to consider, for all values of i, only those cases where $\mathbf{w}$ can finish at both positions j and $j + k - i$. These cases can arise only if the first i letters of $\mathbf{w}$ are the same, and in the same order, as the last i letters of $\mathbf{w}$. Given that $\mathbf{w}$ can finish at both positions j and $j + k - i$, the probability that it does so is the product of the probabilities of the letters in the word $w_1 \ldots w_k w_{i+1} \ldots w_k$.

Finally, when $|m - j| > k - 1$, $I_j I_m$ is zero unless $\mathbf{w}$ finishes both in position j and also in position m. These refer to non-overlapping positions, so for any such (j, m) pair the probability of this event is $(\pi(\mathbf{w}))^2$. The final term on the right-hand side of (5.41) thus contributes $(N - 2k + 2)(N - 2k + 1)(\pi(\mathbf{w}))^2$ to the variance calculation.

Addition of the various terms involved shows that, in the notation of Section 5.6.3, the variance of $Y_1(N)$ is

$$\text{Var}(Y_1(N)) = (N - k + 1)\pi(\mathbf{w}) - \left((2k - 1)N - 3k^2 + 4k - 1\right)(\pi(\mathbf{w}))^2$$

$$+ 2\sum_{j=1}^{k-1}(N - 2k + j + 1)\varepsilon_j\pi(w_1 \ldots w_k w_{j+1} \ldots w_k).$$

(5.42)

It is interesting to consider some implications of this variance formula. First, $\varepsilon_j = 0$ for $j = 1, 2, \ldots, k - 1$ when $\mathbf{w}$ is not self-overlapping. In this case the variance of $Y_1(N)$ is

$$(N - k + 1)\pi(\mathbf{w}) - \left((2k - 1)N - 3k^2 + 4k - 1\right)(\pi(\mathbf{w}))^2. \qquad (5.43)$$

If $\pi(\mathbf{w})$ is very small, this is approximately the same as the mean (5.40), suggesting that $Y_1(N)$ has an approximate Poisson distribution (see note (viii) of Section 1.5).

Next, it is of interest to compare the variances of $Y_1(N)$ for different words. To do this we consider the variances of the words *gaga*, *gggg*, *gaag*, and *gagc* for the case when all nucleotides have probability $1/4$. Under this assumption the mean of $Y_1(N)$ is $(N - 3)/256$ for all four words. However the variances of $Y_1(N)$ depend on the self-overlapping properties of the word of interest. Equation (5.42) shows that the respective variances of $Y_1(N)$ for the words *gaga*, *gggg*, *gaag*, and *gagc* are

$$\frac{281N - 895}{65536}, \ \frac{417N - 1455}{65536}, \ \frac{257N - 783}{65536}, \ \frac{249N - 735}{65536}. \qquad (5.44)$$

When $N = 1,000,000$ the variances in (5.44) are, respectively,

$$4,288, \quad 6,363, \quad 3,922, \quad \text{and} \quad 3,799. \qquad (5.45)$$

These calculations show that $Y_1(N)$ does not have a binomial distribution. If $Y_1(N)$ had (incorrectly) been taken to have a binomial distribution with index $N - 3$ and parameter $1/256$, the mean value given in equation (5.40) would continue to hold, but the variance of $Y_1(N)$ would be calculated as $(255N - 765)/65536$. This differs from the correct variances given in (5.44). More generally, the binomial variance cannot apply for any word.

The main reason why the binomial distribution is incorrect is that the number of occurrences of any self-overlapping word is more variable than the incorrect binomial variance formula suggests. For the word *gaga*, for example, this added variability arises because the occurrence of the word *gaga* at position i increases the probability of the occurrence of the word at position $i + 2$, whereas the non-occurrence of this word at position i decreases the probability of the occurrence of the word at position $i+2$. This "clumping" behavior explains the increased variance. Similar comments apply for other self-overlapping words.

5.7.2 Approximations to the Distribution of $Y_1(N)$

The calculations given above provide the mean and variance of $Y_1(N)$ for any general word **w**, and it is natural to extend these to find the complete distribution of $Y_1(N)$. However, this distribution is complicated. Exact calculations for the probability that $Y_1(N) = y$ can be found for any y using a recurrence relation (Robin and Schbath (2001)), but the calculations become very time consuming when $E(Y_1(N))$ is large (of the order 500 or more). In this case Robin and Schbath show that $Y_1(N)$ has an approximate normal distribution, with mean and variance given respectively in (5.40) and (5.42) above.

When N is large, the length k of the word **w** is 10 or more and $E(Y_1(N))$ is small to moderate, the normal distribution approximation is not accurate. For this case the occurrence of **w** might be considered as a rare event, and it might then initially be thought that $Y_1(N)$ has an approximate Poisson distribution. However, a Poisson approximation does not allow for the fact that self-overlapping words tend to occur in clumps, with the word overlapping itself one or more times in the clump.

To obtain a better approximation it is appropriate to introduce the concept of the compound Poisson distribution. The random variable Y has the compound Poisson distribution if Y is the sum of N iid non-negative integer-valued discrete random variables, where N itself is a random variable having the Poisson distribution (1.22). If $N = 0$ then $Y = 0$ (since the "empty" sum is defined to be zero). The case where $N = 0$ contributes a term $e^{-\lambda}$ to the probability that $Y = 0$, so that

$$\text{Prob}\,(Y = 0) = e^{-\lambda} + \sum_{n \geq 1} e^{-\lambda} \frac{\lambda^n}{n!} \text{Prob}(Y = 0 | N = n), \qquad (5.46)$$

$$\text{Prob}\,(Y = y) = \sum_{n \geq 1} e^{-\lambda} \frac{\lambda^n}{n!} \text{Prob}\,(Y = y | N = n), \quad y = 1, 2, \ldots \quad (5.47)$$

An important compound Poisson distribution arises when each Y_j has the "shifted geometric" distribution (1.18). In this case Y has the *Pólya–Aeppli* distribution. To find $\text{Prob}\,(Y = y | N = n)$ for this case we use the relation between the geometric distribution and the negative binomial distribution and replace $k + 1$ by n and y by $y - 1$ in (1.21) to obtain

$$\text{Prob}\,(Y = y | N = n) = \binom{y-1}{n-1} p^{y-n} (1-p)^n. \qquad (5.48)$$

From the comment below (2.80) and the pgf's of the Poisson and the shifted geometric distributions (see Problem 1.17), the pgf of the Pólya–Aeppli distribution is

$$e^{\lambda(t-1)/(1-pt)}. \qquad (5.49)$$

From this, the mean of this distribution is found to be $\lambda/(1-p)$ and the variance is found to be $\lambda(1+p)/(1-p)^2$.

Both values can also be derived from the theory of Section 2.10.1. A random variable Y_1 having the shifted geometric distribution can be thought of as $1 + Y_2$, where Y_2 has the geometric distribution. Thus from note (vi) of Section 1.4, and note (vi) of Section 1.5,

$$\text{Mean of } Y_1 = 1 + \frac{p}{1-p} = \frac{1}{1-p}, \quad \text{Variance of } Y_1 = \frac{p}{(1-p)^2}. \quad (5.50)$$

The mean of the Pólya–Aeppli distribution follows from (2.81), using the mean λ for a Poisson distribution given in Table 1.1 and the mean for the shifted geometric distribution given in equation (5.50). The value so found agrees with the value found from the pgf (5.49) of the Pólya–Aeppli distribution. The variance $\lambda(1+p)/(1-p)^2$ deriving from (5.49) agrees with the value found by applying the result of Problem 2.14.

The complete Pólya–Aeppli distribution probability distribution can be found by expanding (5.49) as a power series in t, but it is perhaps easiest to find it, for any value y of Y, directly from equations (5.46), (5.47) and (5.48). The distribution is

$$\text{Prob} \,(Y = 0) = e^{-\lambda}, \quad\quad\quad (5.51)$$

$$\text{Prob} \,(Y = y) = e^{-\lambda} \sum_{n=1}^{y} \frac{\lambda^n}{n!} \binom{y-1}{n-1} p^{y-n}(1-p)^n, \quad y = 1, 2, \ldots \quad (5.52)$$

The relevance of the Pólya–Aeppli distribution in the analysis of DNA sequences arises from the fact that, in these sequences, words tend to occur in clumps, as noted above. The number of clumps is modelled as having a Poisson distribution. Given that a clump occurs, the number of occurrences of the word in the clump is modelled as having the shifted geometric distribution with parameter p, and thus mean $1/(1-p)$, where p is the probability that the word self-overlaps. The mean number of words in the entire sequence is then given, as calculated above, by $\lambda/(1-p)$, and assuming that p is known, equating this to the value of $E(Y_1(N))$ given in (5.40) we get $\lambda = (N - k + 1)(1 - p)\pi(\mathbf{w})$. This value is then taken as the mean number of clumps.

A data-oriented approach, where parameter values are not known, is to use maximum likelihood estimation: the maximum likelihood estimates of λ and p are discussed in Example 1 of Section 8.3.

As an example of the use of the Pólya–Aeppli distribution, Robin and Schbath (2001) show that when $N = 316,000$, $k = 9$, the total variation distance $d_{\text{TV}}(P_1, P_2)$ (given in (1.23)) between the true distribution P_1 of $Y_1(N)$ and the appropriate Pólya–Aeppli distribution approximation does not exceed 0.002. For other combinations of N and k, however, the Pólya–Aeppli approximation is less satisfactory: details of the combinations of N

and k for which the approximation is satisfactory are given by Robin and Schbath (2001).

In the above calculations the nucleotide probabilities p_a, p_g, p_c and p_t have been taken as known. Robin and Schbath take up the important question of the accuracy of the calculations when these frequencies are estimated from the DNA sequence (of length N) at hand. They show that in the case where these probabilities are estimated, quite serious errors can be made by applying the above calculations.

5.7.3 *Distance Between Occurrences*

Properties of the distance between successive occurrences of **w** have been discussed by Robin and Daudin (1999), following the work of Blom (1982), Blom and Thorburn (1982), and Cowan (1991). We develop the analysis in detail only for the case $k = 4$, with all nucleotides having probability $1/4$. Formulae for the mean and variance for the distance between successive occurrences of when **w** is of arbitrary length k, and with arbitrary nucleotide probabilities, are given below in equations (5.59) and (5.60).

Given that the word **w** is of length 4, and occurred at some site i, it can next occur at site $i + y$, $(y = 1, 2, 3)$, only if $\varepsilon_{4-y} = 1$. Define Y_2 to be the distance until the next occurrence of **w** after site i, and for notational convenience, let $p(y) = p_{Y_2}(y)$ be the probability that $Y_2 = y$. Then, as explained below,

$$p(y) = \varepsilon_{4-y} 4^{-y} - \sum_{j=1}^{y-1} p(j) \varepsilon_{4+j-y} 4^{j-y} \tag{5.53}$$

for $y = 1, 2, 3$, while if $y \geq 4$,

$$p(y) = 4^{-4} - 4^{-4} \sum_{j=1}^{y-4} p(j) - \sum_{j=y-3}^{y-1} p(j) \varepsilon_{4+j-y} 4^{j-y}. \tag{5.54}$$

The reasoning behind equation (5.54) is as follows. Let E be the event that **w** arises at site $i + y$ and F be the event that **w** arises at site $i + y$ but not at any site between sites i and $i + y$. For each $j = 1, 2, \ldots, y - 1$, let A_j be the event that **w** arises at sites $i + j$ and $i + y$, and does not occur anywhere between i and $i + j$. Then

$$E = F \cup A_1 \cup \cdots \cup A_{y-1},$$

and the events on the right are all disjoint. Therefore,

$$p(y) = \text{Prob}(F) = \text{Prob}(E) - \sum_{j=1}^{y-1} \text{Prob}(A_j).$$

The event E has probability 4^{-4}. The probability of (A_j), for $j \leq y - 4$, is $p(j)4^{-4}$, since the events that **w** occurs at sites $i + j$ and $i + y$ are independent. Similar arguments show that the final term on the right-hand side of (5.54) gives the probabilities of (A_j) for $j = y - 3$, $y - 2$, and $y - 1$. These calculations lead to (5.54), and equation (5.53) is found by similar arguments.

We now find the pgf $p(t)$ of the distance between successive occurrences of $\mathbf{w}$. Multiplying throughout in equations (5.53) and (5.54) by t^y, we get, for example,

$$p(1)t = \frac{\varepsilon_3}{4}t$$
$$p(2)t^2 = \frac{\varepsilon_2}{16}t^2 \qquad\qquad -\frac{\varepsilon_3}{4}p(1)t^2$$
$$p(3)t^3 = \frac{\varepsilon_1}{64}t^3 \qquad\qquad -\frac{\varepsilon_3}{4}p(2)t^3 - \frac{\varepsilon_2}{16}p(1)t^3$$
$$p(4)t^4 = \frac{1}{256}t^4 \qquad -\frac{\varepsilon_3}{4}p(3)t^4 - \frac{\varepsilon_2}{16}p(2)t^4 - \frac{\varepsilon_1}{64}p(1)t^4$$
$$p(5)t^5 = \frac{1}{256}t^5 - \frac{1}{256}p(1)t^5 - \frac{\varepsilon_3}{4}p(4)t^5 - \frac{\varepsilon_2}{16}p(3)t^5 - \frac{\varepsilon_1}{64}p(2)t^5$$
$$p(6)t^6 = \frac{1}{256}t^6 - \frac{1}{256}\left(p(1) + p(2)\right)t^6 - \frac{\varepsilon_3}{4}p(5)t^6 - \frac{\varepsilon_2}{16}p(4)t^6 - \frac{\varepsilon_1}{64}p(3)t^6$$

We now sum these equations over all possible values of y. The sum of the terms on the left-hand sides is, by definition, $p(t)$. The sum of the terms on the right-hand side is

$$\frac{\varepsilon_3}{4}t + \frac{\varepsilon_2}{16}t^2 + \frac{\varepsilon_1}{64}t^3 + \frac{t^4}{256(1-t)} - \frac{t^4 p(t)}{256(1-t)} - \left(\frac{\varepsilon_3}{4}t + \frac{\varepsilon_2}{16}t^2 + \frac{\varepsilon_1}{64}t^3\right)p(t).$$

Equating these two sums and solving for $p(t)$, we find that

$$p(t) = \frac{t^4 + (1-t)(64\varepsilon_3 t + 16\varepsilon_2 t^2 + 4\varepsilon_1 t^3)}{t^4 + (1-t)(256 + 64\varepsilon_3 t + 16\varepsilon_2 t^2 + 4\varepsilon_1 t^3)}. \qquad (5.55)$$

Differentiation of the generating function (5.55) with respect to t, together with use of equation (1.40), shows that the mean of Y_2 is

$$E(Y_2) = 256. \qquad (5.56)$$

A second differentiation with respect to t, together with equation (1.41), shows that the variance of Y_2 is

$$\mathrm{Var}(Y_2) = 512 \sum_{j=1}^{4} \varepsilon_j 4^j - 67{,}328. \qquad (5.57)$$

While the mean of Y_2 is independent of $\mathbf{w}$, the variance depends on the self-overlapping structure of $\mathbf{w}$. For the words $gaga$, $gggg$, $gaag$, and $gagc$, for example, the variances given by formula (5.57) are, respectively,

$$71{,}936, \quad 106{,}752, \quad 65{,}792, \quad \text{and} \quad 63{,}744. \qquad (5.58)$$

The formulae (5.56) and (5.57) are readily generalized to the case where $\mathbf{w}$ is of arbitrary length k and where nucleotide probabilities are arbitrary. In this general case we find

$$E(Y_2) = (\pi(\mathbf{w}))^{-1}, \qquad (5.59)$$

$$\text{Var}(Y_2) = 2(\pi(\mathbf{w}))^{-1} \sum_{j=1}^{k} \varepsilon_j (\pi(w_1 w_2 \dots w_j))^{-1} - (2k-1)(\pi(\mathbf{w}))^{-1} - (\pi(\mathbf{w}))^{-2}.$$

$$(5.60)$$

When $\mathbf{w}$ cannot self-overlap, this variance simplifies to

$$\text{Var}(Y_2) = (\pi(\mathbf{w}))^{-2} - (2k-1)(\pi(\mathbf{w}))^{-1}. \qquad (5.61)$$

When $\pi(\mathbf{w})$ is small the variance of Y_2 is approximately the square of the mean, suggesting an approximation of the distribution of Y_2 by the exponential distribution (see Problem 1.21).

It is not a coincidence that

$$E(Y_1)E(Y_2) = N - k + 1, \qquad (5.62)$$

since we expect that the mean number of occurrences of any word, multiplied by the mean distance between words, will be equal to the length $N - k + 1$ within which $\mathbf{w}$ can occur.

It is natural to assume that the distance between consecutive occurrences of $\mathbf{w}$ has the geometric distribution, at least to a close approximation. This assumption has been made on occasion in the biological literature. The assumption is, however, incorrect, as a comparison of the pgf of the geometric distribution (given in Problem 1.17) and the pgf (5.55) shows. In some cases the error can be substantial, as we now show.

For any word of length 4, and when all nucleotides have probability 1/4 and independence holds, the mean distance between consecutive occurrences is 256. If we choose the parameters of the geometric distribution so that the mean of that distribution takes this value, then from the results of Problems 1.3 and 1.16, the variance of this geometric distribution is 65,792. This differs from three of the variances listed in (5.58), and differs substantially from the variance 106,752 for the case of the word *gggg*.

5.7.4 Beginning at the Origin

In this section we consider the number Y_3 of sites until the first occurrence of $\mathbf{w}$, starting at the origin. We develop the analysis in detail only for the case $k = 4$, with all nucleotides having probability 1/4. The the mean and variance of Y_3 for general k, and with arbitrary nucleotide probabilities, are given below in (5.69) and (5.72).

The generating function (5.55) is not appropriate for the case of the first occurrence of $\mathbf{w}$ since it was calculated allowing for overlaps of $\mathbf{w}$ with itself, whereas overlaps at the origin into "negative positions" cannot occur. This observation is confirmed by noting that if $P(y)$ is the probability that $\mathbf{w}$ first occurs at distance y from the origin, then in contrast to the values found from (5.53), $P(1) = P(2) = P(3) = 0$ and $P(4) = 1/256$.

The recurrence relation (5.54), however, continues to hold when $y \geq 5$. From these observations the generating function for the distribution of the distance until the first occurrence of **w** can be found. This is

$$\mathbb{p}(t) = \frac{t^4}{t^4 + (1-t)(256 + 64\varepsilon_3 t + 16\varepsilon_2 t^2 + 4\varepsilon_1 t^3)}. \tag{5.63}$$

This differs from the generating function (5.55). The mean and variance of the random variable Y_3, found by differentiating $\mathbb{p}(t)$, are, respectively,

$$E(Y_3) = \mu_3 = \sum_{j=1}^{4} \varepsilon_j 4^j, \tag{5.64}$$

$$\text{Var}(Y_3) = \mu_3^2 + \mu_3 - 2\sum_{j=1}^{4} j\varepsilon_j 4^j. \tag{5.65}$$

For the words *gaga, gggg, gaag*, and *gagc*, the respective means of Y_3 are

$$272, \quad 340, \quad 260, \quad \text{and} \quad 256, \tag{5.66}$$

and the respective variances of Y_3 are

$$72{,}144, \quad 113{,}436, \quad 65{,}804, \quad \text{and} \quad 63{,}744. \tag{5.67}$$

It might be surprising that both the mean and variance depend on the self-overlapping structure of **w**, even though the four words considered above are equally likely to occur in any given four consecutive sites. The values for the various means are generally close to those in (5.58), and are identical for the word *gagc*. The differences between the two sets of values emphasizes the way in which the self-overlapping property of a word affects the properties of "between word" distances and "beginning at the origin" distances. The latter distances will arise again in a different context in Section 5.8.2.

The generalizations of (5.63), (5.64), and (5.65) when **w** is of arbitrary length k and when nucleotide probabilities are arbitrary can be found by a direct generalization of the analysis leading to (5.63). This generalization can also be found by the methods of Section 5.8.2, and we show in that section that in general,

$$\mathbb{p}(t) = \frac{t^k}{t^k + (1-t)(\pi(\mathbf{w}))^{-1}\left(1 + \sum_{j=1}^{k-1} \varepsilon_j \pi(w_{j+1} w_{j+2} \cdots w_k) t^{k-j}\right)}. \tag{5.68}$$

From this,

$$E(Y_3) = (\pi(\mathbf{w}))^{-1} + (\pi(\mathbf{w}))^{-1} \sum_{j=1}^{k-1} \varepsilon_j \pi(w_{j+1} w_{j+2} \cdots w_k). \tag{5.69}$$

Equivalently,

$$E(Y_3) = (\pi(\mathbf{w}))^{-1} + \sum_{j=1}^{k-1} \varepsilon_j (\pi(w_1 w_2 \cdots w_j))^{-1}, \tag{5.70}$$

which, since $\varepsilon_k = 1$, can be written as

$$E(Y_3) = \sum_{j=1}^{k} \varepsilon_j (\pi(w_1 w_2 \ldots w_j))^{-1}, \qquad (5.71)$$

The variance of Y_3 is given by

$$\text{Var}(Y_3) = (E(Y_3))^2 + E(Y_3) - 2\sum_{j=1}^{k} j\varepsilon_j (\pi(w_1 w_2 \ldots w_j))^{-1}. \qquad (5.72)$$

For the words $\mathbf{w} = atg$ and $\mathbf{w} = ata$, for example, equation (5.70) gives, respectively,

$$E(Y_3) = (p_a p_t p_g)^{-1} \quad \text{and} \quad E(Y_3) = (p_a^2 p_t)^{-1} + p_a^{-1}. \qquad (5.73)$$

We will re-derive these results in Sections 5.8 and 11.6.2, using two further approaches.

If $\mathbf{w}$ cannot self-overlap, all values of ε_j take the value 0 except for ε_k, whose value is 1. Thus for these words, (5.70) and (5.72) give

$$E(Y_3) = (\pi(\mathbf{w}))^{-1}, \quad \text{Var }(Y_3) = (\pi(\mathbf{w}))^{-2} - (2k-1)(\pi(\mathbf{w}))^{-1}. \qquad (5.74)$$

These values agree with those in (5.59) and (5.61), as they must for a word that cannot self-overlap.

5.8   Overlaps Not Counted

5.8.1   General Comments

Let R_1 be the first occurrence of $\mathbf{w}$ in a sequence, R_2 the next occurrence that does not overlap with the first, R_3 the next that does not overlap with R_2, and so on. This gives a sequence of occurrences $R_1, R_2, R_3, \ldots$, which we refer to as the *recurrences*, rather than as the *occurrences*, of $\mathbf{w}$. R_1 called the first recurrence. Such recurrences form a special case of *recurrent events*, the investigation of the properties of which is a major area of probability theory. The definitive account is in Feller (1968), and we use some of this theory below.

There are practical reasons why overlapping words should not be counted. For example, restriction endonucleases are enzymes that cut DNA whenever a "recognition" sequence (such as *gaga*) specific to that enzyme is encountered in a DNA sequence chromosome. A property of these enzymes is that they do not cut the DNA twice in immediate succession in the sense that if a DNA sub-sequence is $\ldots attgagagaacc \ldots$, and the recognition sequence of a restriction endonuclease is *gaga*, then the DNA sequence will be cut at position 7 in the sub-sequence above but not also at position 9. This implies that cuts in the DNA occur as recurrent events.

5.8.2 Distance Between Recurrences

Let Y_4 be the random number of sites between successive recurrences of **w**. The mean $\mu = E(Y_4)$ of Y_4 is $\mu = \sum_{i=1}^{\infty} i f_i$, where f_n is the probability that the *first* recurrence of **w** occurs at site n. We find μ by considering the probability u_n that *any* recurrence of **w** occurs at at site n. A fundamental result of recurrent event theory concerns the limiting behavior of u_n as $n \to +\infty$. This result is stated as a theorem (without proof).

Theorem 5.1. $u_n \to \mu^{-1}$ as $n \to +\infty$.

This theorem allows direct calculation of μ. Suppose that **w** occurs at site n. Then **w** either occurs at site n as a recurrence (probability u_n) or at some site $n - k + j$ $(j = 1, 2, \ldots, k - 1)$ as a recurrence. Given that **w** occurs at site $n - k + j$ as a recurrence, it can also occur at site n only if the first j letters of **w** are the same as the last j letters (so that $\varepsilon_j = 1$) and the last $k - j$ letters occur at sites $n - k + j + 1, n - k + j + 2, \ldots, n$. Thus since the probability that **w** occurs at site n is $\pi(\mathbf{w})$,

$$\pi(\mathbf{w}) = u_n + \sum_{j=1}^{k-1} \varepsilon_j \pi(w_{j+1} w_{j+2} \ldots w_k) u_{n-k+j}. \tag{5.75}$$

Letting $n \to \infty$ and using the result of Theorem 5.1, we get

$$\mu = E(Y_4) = (\pi(\mathbf{w}))^{-1} + \sum_{j=1}^{k-1} \varepsilon_j \big(\pi(w_1 w_2 \ldots w_j)\big)^{-1}. \tag{5.76}$$

This equation can also be written in the form

$$\mu = E(Y_4) = (\pi(\mathbf{w}))^{-1} + (\pi(\mathbf{w}))^{-1} \sum_{j=1}^{k-1} \varepsilon_j \pi(w_{j+1} w_{j+2} \ldots w_k). \tag{5.77}$$

This form of the equation is convenient for the extension of the theory to motifs, considered in Section 5.9.

Our main observation is that the right-hand side in equation (5.76) is the same as that in equation (5.69), so that $E(Y_4) = E(Y_3)$. The reason for this identity is that the theory of the distance in counting from the origin to the first occurrence of **w** is identical to the theory of the distance between consecutive recurrences of **w**, since in the former overlaps of **w** over the origin cannot occur. This implies that that not only the mean, but indeed the entire distribution (including the variance (5.72)) of Y_4 follows immediately from the results of Section 5.7.4.

We illustrate the argument leading to (5.75) with the words *ata* and *atg*. If the word *ata* finishes at site n, either it does so as a recurrent event or it finishes at site $n-2$ as a recurrent event, with the letters *ta* then occupying sites $n - 1$ and n. This gives

$$p_a^2 p_t = u_n + p_t p_a u_{n-2}. \tag{5.78}$$

Letting $n \to \infty$ and using Theorem 5.1, we get

$$\mu = (p_a^2 p_t)^{-1} + p_a^{-1}. \tag{5.79}$$

For the word atg, a corresponding argument gives $p_a p_t p_g = u_n$ and hence $\mu = (p_a p_t p_g)^{-1}$. These results agree with those given in (5.73).

We now show how the results of this section can be used to derive the expression (5.68). Define $u_0 = 1$ and $u(t) = \sum_{j=0}^{+\infty} u_j t^j$. Multiplying throughout in equation (5.75) by t^n and summing both sides over $n = k, k+1, \ldots$, we get

$$\frac{t^k \pi(\mathbf{w})}{1-t} = \big(u(t) - 1\big) h(t), \tag{5.80}$$

where

$$h(t) = 1 + \sum_{j=1}^{k-1} \varepsilon_j \pi(w_{j+1} w_{j+2} \ldots w_k) t^{k-j}. \tag{5.81}$$

This leads to

$$u(t) = \frac{t^k (\pi(\mathbf{w})) + (1-t) h(t)}{(1-t) h(t)}. \tag{5.82}$$

Now recall that f_j is the probability that the word of interest occurs for the first time at site j. Then by definition, $\mathbb{p}(t) = \sum_{j=1}^{+\infty} f_j t^j$. Also, if the word of interest occurs at site n, it must do so either for the first time at site 1 and then recur $n - 1$ sites later, for the first time at site 2 and then recur $n - 2$ sites later, and so on, or occur for the first time at site n. This implies that

$$u_n = f_1 u_{n-1} + f_2 u_{n-2} + \cdots f_n.$$

But the left-hand side in this equation is the coefficient of t^n in $u(t) - 1$, while the right-hand side is the coefficient of t^n in $u(t) \mathbb{p}(t)$. Thus from Appendix B.13, $u(t) - 1 = u(t) \mathbb{p}(t)$. From this, $\mathbb{p}(t) = 1 - \big(u(t)\big)^{-1}$. This equation, together with the expression (5.82) for $u(t)$, leads immediately to the expression (5.68) for $\mathbb{p}(t)$.

5.8.3 Number of Recurrences

Define $Y_5(N)$ as the number of recurrences of $\mathbf{w}$ in a DNA sequence of length N. Then recurrent event theory shows that the mean $E(Y_5(N))$ of $Y_5(N)$ satisfies the asymptotic relationship

$$E(Y_5(N)) \sim \frac{N}{\mu}, \tag{5.83}$$

where μ is the mean number of sites between successive recurrences of **w**. This equation implies that $E(Y_5(N))\mu \sim N$, which resembles (5.62).

The asymptotic relation (5.83) is a general one which applies for any recurrent event, and is often sufficiently accurate for our purposes. However, it is possible to find an exact expression for $E(Y_5(N))$ when considering recurrences of words. We illustrate this with the word *gaga*, for which (5.75) becomes

$$p_a^2 p_g^2 = u_n + p_a p_g u_{n-2} \tag{5.84}$$

and for which $\mu = (1 + p_a p_g)/p_a^2 p_g^2$. Equation (5.84) is a difference equation of the form considered (in a different context) in Section 7.3, and may be solved for u_n by the methods of that section. If an indicator function I_n is defined as 1 if the word *gaga* arises at site n as a recurrent event and 0 otherwise, then $E(I_n) = u_n$ and and the mean number of recurrences of *gaga*, namely $E(\sum_{n=4}^{N} I_n)$, is $\sum_{n=4}^{N} u_n$. If terms of order $(p_a p_g)^{N/2}$ (which are extremely small) are ignored, difference equation methods (see Problem 5.13) give

$$E(Y_5(N)) = \frac{p_a^2 p_g^2}{1 + p_a p_g}\left(N - 3 + \frac{2 p_a p_g}{1 + p_a p_g}\right). \tag{5.85}$$

In general, if small terms of order c^N are ignored, (where $0 < c < 1$ and the value of c depends on the word of interest), the expression for $E(Y_5(N))$ for any word is

$$E(Y_5(N)) = \frac{1}{\mu}\left(N - k + 1 + \frac{\sum_{j=1}^{k-1}(k-j)\varepsilon_j \pi(w_{j+1} w_{j+2} \cdots w_k)}{1 + \sum_{j=1}^{k-1}\varepsilon_j \pi(w_{j+1} w_{j+2} \cdots w_k)}\right), \tag{5.86}$$

where μ is the mean number of sites between successive recurrences of the word of interest. This is identical to the expression given by Régnier (2000). The expression given in (5.83) is simpler than this, so that we often use (5.83) as a approximation to (5.86).

Equations (5.83) and (5.66) show that if all nucleotides have probability $1/4$, the respective mean number of recurrences of the words *gaga*, *gggg*, *gaag*, and *gagc* in a DNA sequence of length 1,000,000 are, approximately,

$$3676, \ 2941, \ 3846, \ \text{and} \ 3906. \tag{5.87}$$

These should be compared to the value 3906 arising for all four words when overlapping words are counted, as calculated in Section 5.7. The two means are identical for the word *gagc*, since for this word overlaps of successive occurrences are not possible. For the other three words the mean number of recurrences is less than the mean total number of occurrences, as expected.

Indicator functions can be used in a similar way to that used in arriving at the approximation (5.85) and more generally (5.86) in order to find the variance of $Y_5(N)$, using equation (5.41). If this equation is used, terms of the form $EI_j I_{j+h}$ are zero when $h < k$, since a word of length k cannot recur in both positions j and $j + h$ for these values of h. Further, since properties of further recurrences of a word after a recurrence at position j are the same as those starting at the origin, terms of the form $EI_j I_{j+h}$ for $h \geq k$ take the value $u_j u_h$, where u_j is the probability of a recurrence of the word in position j.

The complete distribution of $Y_5(N)$ is complicated. When N is large, recurrent event theory shows that the number of recurrences of **w** in a sequence of length N has an asymptotic normal distribution with mean (5.83) and variance

$$\mathrm{Var}Y_5(N) \sim \frac{N\sigma^2}{\mu^3}, \tag{5.88}$$

where μ is the mean, and σ^2 is the variance, of the number of sites between successive recurrences of the word, given in equations (5.64) and (5.65) respectively.

With $N = 1,000,000$ and when all nucleotides have probability $1/4$, (5.88) and (5.67) jointly give approximate values 3585, 2886, 3744, and 3799 for the respective variances of the number of recurrences of the words $gaga, gggg, gaag$, and $gagc$. There are two points to be made about these variance calculations. First, it is interesting to compare them with those given in (5.45) for the case where overlapping words are counted. Perhaps the most surprising comparison is that whereas the total number of occurrences of the word $gggg$ has the largest of the four variances given in (5.45), the number of recurrences of this word has the smallest variance. This fact is largely due to the high mean distance between successive recurrences. The variances for the word $gagc$ are the same in both cases, since overlaps are impossible for this word. The fact that the approximate calculation using (5.88) yields the correct value for this latter variance shows that this formula provides an excellent approximation.

Second, the Pólya–Aeppli distribution approach discussed in Section 5.7.1 assumes that the number of clumps of any word of interest, which is close to the number of recurrences of this word, has an approximate Poisson distribution. The fact that the variances calculated from (5.88) are generally close to, but nevertheless differ by about 3% from, the means listed in (5.87) shows that the Poisson approximation is fairly accurate for words of length 4.

5.9 Motifs

Many short sequences throughout DNA, for example transcription factor binding sites or splice junctions signals, serve specific functions and do not tolerate many mutations. Some mutation is generally tolerated, however, and within a species or between species there will often be several variant sequences. We try to capture this collection of variants mathematically in the concept of a motif.

For our purposes, we define a motif as a collection of m different words, no word contained within any other, in practice being rather similar to each other and often having the same length. We consider in detail here only the case where all m words in the motif have the same length k, so that with $m = 4, k = 6$ a motif might, for example, be the collection of words $tatgaa$, $tatgga$, $tatgca$, and $tatgta$. We denote the motif of interest by M, and say that M occurs at any site if any one of the words in the motif occurs at that site.

The probability that the uth word in M (which we also call word u) occurs at any site is denoted π_u. Thus the probability $\pi(M)$ that the motif M occurs at any site, that is that one of the words in the motif arises at that site, is

$$\pi(M) = \sum_{u=1}^{m} \pi_u. \tag{5.89}$$

An immediate generalization of the argument that led to equation (5.40) shows that if all occurrences of all words in M are counted, overlapping or not, the mean of $Y_6(N)$, the number of occurrences of M in the DNA sequence of length N, is given by

$$E(Y_6(N)) = (N - k + 1)\pi(M). \tag{5.90}$$

The derivation of the formula for the variance of $Y_6(N)$ follows the argument of the derivation of (5.42), which shows that this variance depends on the overlap properties of the various words in M. These can be characterized by introducing the indicator function $\varepsilon_j(u, v)$, taking the value 1 if the first j letters in word u in M are the same, and in the same order, as the last j letters of word v in M. When $\varepsilon_j(u, v) = 1$ we write $\pi_j(u, v)$ as the probability of the word (of length $2k - j$) formed by concatenating words v and u, with the last j letters of the word v being overlapped by the first j letters of the word u. The case $u = v$ is included in these definitions, and $\varepsilon_j(u, u)$ is identical to the indicator function ε_j of Section 5.6.3. Then the generalization of the variance formula (5.42) is that the variance of $Y_6(N)$ is

$$\text{Var}\,(Y_6(N)) = (N - k + 1)\pi(M) - \left((2k - 1)N - 3k^2 + 4k - 1\right)\left(\pi(M)\right)^2$$
$$+ 2\sum_{j=1}^{k-1}(N - 2k + j + 1)\sum_{u=1}^{m}\sum_{v=1}^{m}\varepsilon_j(u, v)\pi_j(u, v).$$

$$\tag{5.91}$$

When none of the words in M are self-overlapping, and none can overlap any other word in M, this variance simplifies to

$$\text{Var}\,(Y_6(N)) = (N-k+1)\pi(M) - \left((2k-1)N - 3k^2 + 4k - 1\right)\left(\pi(M)\right)^2. \tag{5.92}$$

This is a generalization of equation (5.43).

Using an argument similar to that surrounding (5.62), equation (5.90) shows that the mean of the number Y_7 of sites between successive occurrences of M is given by

$$E(Y_7) = \left(\pi(M)\right)^{-1}. \tag{5.93}$$

This equation, together with (5.59) and (5.89), shows that

$$\left(E(Y_7)\right)^{-1} = \sum_{u=1}^{m}\left(E(Y_2(u))\right)^{-1}, \tag{5.94}$$

where $E(Y_2(u))$ is the mean number of sites between successive occurrences of word u.

It was noted in Section 5.7.1 that occurrences of any self-overlapping word tend to occur in clumps. The same is true of occurrences of motifs if overlapping words in the motif are all counted. We consider the clumping behavior of motifs in Example 3 of Section 11.6.2.

We now consider the case where overlaps of words in M are not counted; that is, we consider recurrences of M. The first time that any word in M occurs is called the first recurrence of M. The next time that a word in M occurs that does not overlap with recurrence i of M is called recurrence $i + 1$ of M. If there is a recurrence of M at any site, with word u of M occurring at that site, we say that the recurrence is *determined by* word u. Our first aim is to find a formula for $\mu = E(Y_8)$, the mean number Y_8 of sites between successive recurrences of the motif.

Suppose that there has been a recurrence of the motif at a given site. We define μ_u $(u = 1, 2, \ldots, m)$ as the mean number of sites until the next recurrence of the motif that is determined by word u of the motif. The formula for μ is found by first calculating the various μ_u values. Breen et al. (1985) show that $(\mu_u)^{-1}$ is the uth element in the vector

$$W^{-1}\mathbf{1}, \tag{5.95}$$

where $\mathbf{1}$ is a (column) vector all of whose elements are 1 and the matrix W is defined in terms of the overlapping properties of the words in the motif as follows. Specifically, the uth diagonal element w_{uu} of W is given by

$$w_{uu} = \pi_u^{-1} + \pi_u^{-1}\sum_{j=1}^{k-1}\varepsilon_j(u, u)\pi_{k-j}^{(u)}, \tag{5.96}$$

where $\pi_{k-j}^{(u)}$ is the probability of the word consisting of the last $k - j$ letters of word u. (Note the similarity of this equation with equation (5.77).) The (u, v) element w_{uv} of W, $(u \neq v)$, is given by

$$w_{uv} = \pi_u^{-1}\sum_{j=1}^{k-1}\varepsilon_j(u, v)\pi_{k-j}^{(u)}. \tag{5.97}$$

Having found the various values of μ_u^{-1} by calculating (5.95), the mean number μ of sites between successive recurrences of M is found from

$$\mu^{-1} = \sum_{u=1}^{m} (\mu_u)^{-1}. \qquad (5.98)$$

This equation, together with the expression (5.95), shows that μ^{-1} is the sum of all the terms in the matrix W^{-1}.

As a simple example, suppose that M consists of two words, aaa (word 1) and ata (word 2). Then from (5.96),

$$w_{11} = p_a^{-3} + p_a^{-2} + p_a^{-1}, \; w_{22} = (p_a^2 p_t)^{-1} + p_a^{-1}, \; w_{12} = w_{21} = p_a^{-1}. \quad (5.99)$$

In the case $p_a = p_t = 1/4$, the sum of the elements in W^{-1} is 9/356 so that the mean number of sites between successive recurrences of M is $356/9 \cong 39.56$.

If there has been a recurrence of M at a given site, the expression (5.95) also shows that the mean number of sites until the next recurrence of M determined by aaa is 89 and until the next recurrence determined by ata is 71.2. For a motif consisting only of the word aaa, the mean number of sites between successive recurrences, from the theory of Section 5.8.2, is 84. The excess of 89 over 84 is accounted for by cases where the sub-sequence $\ldots ataaa \ldots$ arises, with a recurrence of the motif occurring after the letters ata in this sub-sequence. The word aaa in this sub-sequence does not then lead to a recurrence of the motif M, whereas for a motif consisting only of the word aaa it does. Thus cases of the recurrence of M determined by the word aaa arises slightly less often, or equivalently with slightly more sites between such recurrences, than recurrences of the motif consisting only of aaa. Similarly, the mean number of sites between successive recurrences of a motif consisting only of the word ata is 68, and the comparison of 71.2 and 68 is explained similarly.

These calculations are readily generalized for arbitrary values of p_a and p_t (see Problem 5.17).

Robin (2002) discusses two practical examples concerning motifs, which we now describe.

Example 1. The first example concerns the CHI (crossover hotspot initiator) motif in *Hemophilus influenzae*, which consists of the four words {*gatggtgg, gctggtgg, ggtggtgg, gttggtgg*}, and which protects this genome against certain forms of degradation. It is therefore expected to recur significantly often in this genome and also to be well spaced in the genome. In the *H. influenzae* genome (of size 1,903,356 bases), this motif occurs 223 times, with 215 occurrences being recurrences. We define a simple model as one where the nucleotides at different sites are independent, with all nucleotides having probability 1/4 at each site. Under this simple model, the sum of the elements in the matrix W^{-1} for this motif, where the matrix W is defined through (5.96) and (5.97), is 1/16,648. From the above theory, the mean number of sites between successive recurrences of the motif is thus 16,648, so that from (5.83), the mean number of recurrences of this motif is approximately $1,903,356/16648 \cong 114.32$. The observed number 215 is considerably in excess of this and leads to a highly significant z-score of about 9.5. Thus in statistical terms we reject the null hypothesis of the simple model for this motif, as would be expected given its protective function. Robin (2002) also showed by using r-scans that the motif does not arise significantly more frequently in any particular part of the genome, again as expected from its protective function.

Under the simple model assumptions made in the previous paragraph, the probability that the motif occurs at any particular site is $1/4^7$, and equation (5.90) then shows that the mean number of occurrences of the motif is $1,903,349/4^7 \cong 116.17$. This exceeds the mean number of recurrences, calculated above as 114.32, by about 1.62%. It is easy to see why this is so. By far the most likely way in which two words in the motif will overlap arises when the first word is followed by the sequence tgg, thus creating an overlap with the third word ($ggtggtgg$) of the motif. Such an overlap creates a further occurrence of the motif but not a further recurrence. The probability of the sequence tgg is 1/64 or about 1.56%, thus accounting for almost all the difference between the mean number of occurrences of the motif and the mean number of recurrences.

Example 2. The second motif considered by Robin consists of the set of all 64 six-letter self-complementary palindromes, that is of all six-letter words which are unchanged if read backwards with a replaced by g, g by a, c by t, and t by c, for example the word $acgatg$. In *Escherichia coli* various words in this motif cause breakages in the genome and thus the motif is expected to occur less often than by chance. In this case the matrix W, defined by (5.96) and (5.97), is of size 64×64. Under the simple model, the approximation (5.83) shows that the mean number of recurrences of the motif (in the *E. coli* genome of size 4,638,868) is about 68,000. The observed number, about 52,000, leads to a z-score of about -75, clearly highly significant. Therefore we would reject the simple model so far as this motif is concerned.

We conclude with some general comments. First, in the (unusual) case where no word in a motif is self-overlapping and no two words in the motif overlap with each other, $w_{uv} = 0, w_{uu} = \pi_u^{-1}$ and equations (5.94) and (5.98) reduce to the same equation. Second, recurrences of a motif form recurrent events, so that the approximation (5.83) applies for finding the mean number of recurrences in a sequence of length N, with μ found from (5.98). (This approximation was used in the CHI and palindrome motif calculations above.) Finally, more complex calculations concerning recurrences of a motif, including those for the case of Markov dependence between sites and for cases of motifs having complicated structures, are available. Aspects of the relevant theory are given by Tanushev and Arratia (1997), Régnier (2000), and Reinert et al. (2000). Because of the complexity of the various formulae, several computer programs are available for numerical calculations; see for example Nicodème (2001). It may well be that many further properties are best estimated by simulation rather than by using formulae requiring excessive computation.

Finally, we have discussed above the case where the motif of interest is known. Harder problems can arise when a motif has to be discovered. This problem is discussed, for example, by Bailey and Elkan (1994), Liu et al. (2001, 2002)), Hertz and Stormo (1999), Liu et al. (1995), Jensen and Liu (2004), and Jensen et al. (2004), using various statistical methods. This topic is beyond the scope of this book.

Problems

5.1. Use the joint density function (2.143), together with transformation techniques, to prove equation (5.4).

5.2. Show that if L and G are fixed, the mean number of contigs given in (5.1) is maximized (as a function of N) when $N = G/L$.

5.3. Use the approximation (B.21) to approximate the mean contig size (5.2) when a is small, and interpret your result.

5.4. Write the mean number of anchored contigs (5.11) as

$$\frac{abG}{L}\frac{\left(e^{-a} - e^{-b}\right)}{(b - a)}.$$

(i) Assume that G, L, and b are fixed, so that this mean number is a function of a only. Suppose that $G/L = 100,000$ and $b = 5$. Evaluate this mean for $a = 0.8, 1, 1.2, 1.4, 1.6, 1.8, 2.0$.

(ii) Now do the same calculation when $b = 10$.

(iii) For the cases $b = 5$, $b = 10$, estimate the value of a for which the mean number of contigs is maximized. How do these values compare with the case $b = +\infty$?

5.5. There are four models described below for a signal of length five: iid, weight matrix, first-order Markov, and MDD. For each of the sequences $CCGAT$ and $CATAT$ find the probability of the sequence given the model, for each of the four models (so your answer should consist of eight probabilities).

(i) iid. The probabilities of the four nucleotides are $\{p_a = .2, p_c = .1, p_g = .1, p_t = .6\}$.

(ii) Weight Matrix. The weight matrix (for the nucleotide ordering: a, c, g, t) is

$$
\begin{bmatrix}
.2 & .3 & .2 & .1 & .1 \\
.1 & .2 & .15 & .6 & .6 \\
.3 & .4 & .6 & .1 & .15 \\
.4 & .1 & .05 & .2 & .15
\end{bmatrix}.
$$

(iii) First-Order Markov. The initial distribution is $\{p_a = .2, p_c = .1, p_g = .1, p_t = .6\}$, and the transition matrix (for the nucleotide ordering a, c, g, t) is

$$
\begin{bmatrix}
.1 & .8 & .05 & .05 \\
.35 & .1 & .1 & .45 \\
.3 & .2 & .2 & .3 \\
.6 & .1 & .25 & .05
\end{bmatrix}.
$$

(iv) MDD. The first split is on position 2, with probabilities for this position $\{p_a = .2, p_c = .3, p_g = .1, p_t = .4\}$. For position 2 equal to c, g, or t we model the remaining positions with weight matrices

$$
W_c = \begin{bmatrix}
.2 & .1 & .2 & .8 \\
.5 & .1 & .2 & .1 \\
.2 & .1 & .3 & .05 \\
.1 & .7 & .3 & .05
\end{bmatrix}, \quad
W_g = \begin{bmatrix}
.4 & .1 & .2 & .2 \\
.3 & .4 & .1 & .3 \\
.2 & .1 & .3 & .2 \\
.1 & .4 & .4 & .3
\end{bmatrix}, \quad
W_t = \begin{bmatrix}
.1 & .1 & .2 & .2 \\
.6 & .6 & .4 & .35 \\
.2 & .1 & .3 & .15 \\
.1 & .2 & .1 & .3
\end{bmatrix}.
$$

If position 2 equals a, we split further the other four positions on position 1, with probabilities for this position $\{p_a = .5, p_c = .1, p_g = .1, p_t = .3\}$. The remaining three positions are modelled with weight matrices

$$
W_a = \begin{bmatrix}
.3 & .4 & .2 \\
.4 & .3 & .1 \\
.2 & .2 & .6 \\
.1 & .1 & .1
\end{bmatrix}, \quad
W_c = \begin{bmatrix}
.1 & .5 & .3 \\
.1 & .2 & .1 \\
.05 & .2 & .55 \\
.75 & .1 & .05
\end{bmatrix},
$$

$$
W_g = \begin{bmatrix}
.1 & .2 & .2 \\
.5 & .1 & .1 \\
.1 & .1 & .5 \\
.3 & .6 & .2
\end{bmatrix}, \quad
W_t = \begin{bmatrix}
.1 & .6 & .05 \\
.5 & .2 & .05 \\
.2 & .1 & .4 \\
.2 & .1 & .5
\end{bmatrix}.
$$

5.6. The expression (5.28) has an interesting consequence concerning the *random breaking of a stick* problem. Suppose that a straight stick of length 1 is randomly broken twice. Use the expression (5.28) with $u = 1/2$ to show that the probability that the three pieces formed cannot form a triangle is $3/4$.

Note: It is important to be precise about the procedure forming the breaks of the stick. The procedure assumed here is that two points where the stick is broken are chosen independently and at random, the position of each having a uniform distribution in $(0, 1)$. The breaks are then made at the two points so chosen. Other concepts of random breaking, for example that the stick is first broken at a randomly chosen point, and then one of the two pieces formed is randomly chosen and then broken at a randomly chosen point, lead to an answer different from that given above.

5.7. Derive the expression (5.26) from the integration in (5.26).

5.8. Derive the expression $(1 - gu)^n$ from the integration in (5.25).

5.9. For a sequence of DNA of length $N = 6$, find the probability that the word *gaga* occurs 0, 1, and 2 times, and hence verify the variance formula in (5.44) for this case.

5.10. What changes are needed in the variance formula (5.44) for the cases $N = 4$, $N = 5$? Find the appropriate variance formulae in these two cases.

5.11. For the case where the probability of each nucleotide at any site is $\frac{1}{4}$, find the mean number of sites until the first recurrence of the word *gagg*.

5.12. The four sets of ε_j values for the words *gaga, gggg, gaag* and *gagc* can be regarded as four vectors, namely $(0, 1, 0, 1)$, $(1, 1, 1, 1)$, $(1, 0, 0, 1)$, and $(0, 0, 0, 1)$. Is there any four-letter word having a vector of ε_j values different from any of these four? Prove your statement either by exhibiting a word with a different vector of ε_j values, or prove that there can be no such word.

5.13. Show that the solution

$$u_{2j} = \frac{p_a^2 p_g^2}{1 + p_a p_g} + \frac{p_a p_g}{1 + p_a p_g}(-p_a p_g)^j, \quad j = 1, 2, \ldots$$

satisfies equation (5.85), together with the boundary conditions $u_2 = 0$, $u_4 = p_a^2 p_g^2$. Show also that $u_3 = 0$, $u_5 = p_a^2 p_g^2$, so that $u_{2j+1} = u_{2j}$. By calculating $\sum_{i=4}^{N} u_i$, derive equation (5.85), indicating the size of the small-order terms ignored.

5.14. Prove equation (5.93).

5.15. Consider a motif made up of the words aa, at, ta, and tt, which we call words 1, 2, 3, and 4 respectively. Find the matrix M (defined by equations (5.96) and (5.97)), and thus show that $\mu_1^{-1} = p_a^2/2$, $\mu_2^{-1} = \mu_3 = p_a p_t/2$, $\mu_4^{-1} = p_t^2/2$. Use equation (5.98) to find the mean number of sites between successive recurrence of the motif, and comment on this result in the case $p_a + p_t = 1$.

5.16. Consider a motif made up of all possible words of length k. Use equation (5.94) to show that the mean number of sites between successive occurrences of this motif is 1, whatever the probabilities of the four nucleotides. Does this result make sense? What is the mean number of sites between successive recurrences of this motif?

5.17. Use the values in (5.99) to find the mean number of sites between successive recurrences of the motif $\{aaa, ata\}$ for general values of p_a and p_t.

6
The Analysis of Multiple DNA or Protein Sequences

6.1 Two Sequences: Frequency Comparisons

In Example 3 of Section 3.5 we considered the test of the hypothesis that the probabilities for the four nucleotides in a DNA sequence are equal to a set of prescribed values. In this section we consider the test of the hypothesis that the two sets of probabilities for the four nucleotides in two DNA sequences are equal, no specific claim being made as to what the probabilities are. The data used for this test are as given in Table 6.1.

	nucleotide				
	a	g	c	t	Total
sequence 1	Y_{11}	Y_{12}	Y_{13}	Y_{14}	$Y_{1.}$
sequence 2	Y_{21}	Y_{22}	Y_{23}	Y_{24}	$Y_{2.}$
Total	$Y_{.1}$	$Y_{.2}$	$Y_{.3}$	$Y_{.4}$	Y

Table 6.1. Nucleotide counts.

The test is a particular case of the two-way table test discussed in Section 3.5, and the test statistic is either (3.42) or (3.43), applied to the data of Table 6.1. When the null hypothesis that the two sequences are drawn from populations with identical nucleotide frequencies is true, both statistics have an approximate chi-square distribution with three degrees of freedom. Further, the numerical values of the two statistics are usually close. For example, with the numerical values given in Table 6.2, the statistic (3.42) and the statistic (3.43) respectively take the values 0.96416 and 0.96426. The 5% Type I error significance point for chi-square with three degrees of freedom is 7.81, so that the null hypothesis is not rejected using either statistic.

	nucleotide				
	a	g	c	t	Total
sequence 1	273	258	233	236	1000
sequence 2	281	244	246	229	1000
Total	554	502	479	465	2000

Table 6.2. Nucleotide counts: numerical example.

The statistic (3.43) when applied to the data of Table 6.1 can be written in a different form. If

$$p_{1j} = Y_{1j}/Y_{1\cdot}, \quad p_{2j} = Y_{2j}/Y_{2\cdot}, \quad p_j = Y_{\cdot j}/Y,$$

the statistic (3.43) becomes

$$2Y \left(\frac{Y_{1\cdot}}{Y} \sum p_{1j} \log p_{1j} + \frac{Y_{2\cdot}}{Y} \sum p_{2j} \log p_{2j} - \sum p_j \log p_j \right). \qquad (6.1)$$

The entropy-like statistic

$$\frac{Y_{1\cdot}}{Y} \sum p_{1j} \log_2 p_{1j} + \frac{Y_{2\cdot}}{Y} \sum p_{2j} \log_2 p_{2j} - \sum p_j \log_2 p_j \qquad (6.2)$$

has been used as a chi-square in the literature. This is not appropriate, since the statistic (6.2) uses logarithms to the base 2 and omits the factor $2Y$ in (6.1).

Bernaola-Galván et al. (2000) take up the much more difficult statistical question of testing whether the nucleotide frequencies change at some undetermined point along a DNA sequence. A DNA sequence of length N may be divided into two sub-sequences of respective lengths n and $N - n$, and the DNA frequencies in these two sub-sequences may be compared using the statistic (6.1). For any given value of n, denote this statistic by C_n. If n is varied from a small number a to a number b close to N, a collection of values $C_a, C_{a+1}, \ldots, C_b$ will be obtained. A reasonable test statistic to use for this purpose is $C_{\max}$, the maximum of $C_a, C_{a+1}, \ldots, C_b$. To assess whether the observed value of $C_{\max}$ is significant, it is necessary to find its null hypothesis distribution. This is not easy, since the values of (6.1) are highly correlated and equations such as (2.101) may not be used for this purpose. It appears that the empirical methods discussed in Section 1.16 are necessary to find approximations for this distribution. Bernaola-Galván et al. (2000) conduct simulations to find an approximation to this distribution, and claim from these that a good approximation is found by replacing N by an effective length N_{eff} (approximately $2.45 \log N$), and then using equation (2.101) with N replaced by N_{eff} and y replaced by $.84y$. However, it remains an open matter to find theoretical properties of the probability distribution of $C_{\max}$.

6.2 Alignments

As a segment of genetic material is passed on through the generations in some line of descent in a population, the sequence constituting this material will change through the process of mutation. The simplest mutations are of the form of a switch from one nucleotide to another, or in the form of an insertion or a deletion. Mutations can spread to an entire species, or nearly so, through the process of natural selection or random drift. When a switch in nucleotides spreads throughout most of a species we call it a *substitution*. (When in a population at a given site there does not exist a single nucleotide type, we say that a *polymorphism* exists at that site.) As substitutions, insertions, and deletions get passed along through two independent lines of descent, the two sequences will slowly diverge from each other. For example, the original sequence may have been

$$cggtatgcca,$$

whereas the two descendants might be

$$cgggtatccaa$$

and

$$ccctaggtccca.$$

This divergence will happen at varying rates, depending on the function of the piece of DNA in question and how well that function tolerates substitutions and other changes. For protein coding DNA, the corresponding protein sequences also evolve through time as a result of DNA sequence evolution. Individual genes typically contain stretches that change rapidly and other stretches that remain relatively constant. The latter regions are called "functional domains" since their low tolerance to change suggests that they have a critical functional role in the viability of the organism.

Many problems in bioinformatics relate to the comparison of two (or more) DNA or protein sequences. In order to compare sequences of nucleotides or amino acids, we use *alignments*. The following is an example of an alignment of the above two descendent sequences:

$$
\begin{array}{cccccccccccc}
c & g & g & g & t & a & - & - & t & c & c & a & a \\
c & c & c & - & t & a & g & g & t & c & c & c & a
\end{array}
$$

The symbol "−" is called an *indel*: it represents an assumed insertion or deletion at some point in the evolutionary history leading to the two sequences. A sequence of ℓ consecutive indels is called a *gap* of length ℓ. In the above alignment there are two gaps, one of length 1 and one of length 2.

There are many types of alignments. There are *global* alignments, in which the entire lengths of the sequences are aligned, and there are *local* alignments, which align only sub-sequences of each sequence. There are *gapped* alignments, in which indels are allowed, and there are *ungapped* alignments, in which indels are not allowed. There are *pairwise* alignments, which are alignments of two sequences, and there are *multiple* alignments, which align more than two sequences. Of course, these can be combined into, for example, "gapped global pairwise" alignments. There is a large literature on alignments, of which we shall examine a few of the basic points.

For a particular type of pairwise alignment (for instance ungapped global), there are many possible such alignments between any two sequences. Good alignments of related sequences are ones that better reflect the evolutionary relationship between them. There are several ways to discriminate between good and bad alignments. In the sections that follow we consider this problem in some detail.

6.3   Simple Tests for Significant Similarity in an Alignment

In this section we consider tests of the (null) hypothesis that the two sequences in an ungapped alignment such as those in (1.1) have been generated at random with respect to each other. We shall assess the significance of the observed similarity between the two sequences by using two "local" criteria, one requiring a quite stringent similarity criterion, the other a less stringent requirement. In Chapter 10 we consider a "local" criterion different from both of those considered here.

Exactly-Matching Sub-sequences

As the most stringent similarity criterion, we consider as similar only those sub-sequences where the elements match exactly. In the array (1.1) there are five of these: the sub-sequences '*a*' in position 3, '*c*' in position 6, '*t a g*' in positions 9–11, '*c a*' in positions 13–14, and '*t a t*' in positions 23–25. We will generically denote the length of any such sub-sequence by Y. We denote the length of the longest such sub-sequence by $Y_{\max}$, and this is the statistic we use to test for significant similarity between the two sequences. In the array (1.1) the observed value $y_{\max}$ of $Y_{\max}$ is 3. To assess in general whether the observed value $y_{\max}$ of $Y_{\max}$ is significant, we have to find out the probability distribution of $Y_{\max}$ when the null hypothesis that there is no significant similarity between the two sequences, (that is, that one is generated independently of the other), is true.

If a match is thought of as a success, the theory of Section 5.4 can be used for this problem if p is defined as the match probability $p_a^2 + p_g^2 + p_c^2 + p_t^2$, where p_a, p_g, p_c, and p_t are the respective frequencies of a, g, c, and t. With this interpretation of p, any sub-sequence of successes (of length 0 or more) must be preceded by a pair of non-matching nucleotides, (that is a failure). Under the null hypothesis there will be approximately $(1 - p)N$ such failures, so that there will also be approximately $(1 - p)N$ sub-sequences of successes in two long matched sequences each of length N. As in Section 5.4, the approximation that takes the number of sub-sequences as fixed at this value is sufficiently accurate for all practical purposes. We can now use (5.15) to provide an approximation for the P-value associated with the observed value $y_{\max}$ of the length of the longest exactly matching sub-sequence.

Well-Matching Sub-sequences

The use of the length of the longest exactly matching sub-sequence as a test statistic is probably unwise. Even if two DNA sequences have a reasonably recent common ancestor, evolutionary changes will usually cause at least a small number of differences between them. Thus it is more appropriate to use a test procedure that focuses on well-matching, rather than exactly matching, sub-sequences.

One such approach is to consider sub-sequences where k mismatches (more strictly, because of end-effects, up to k mismatches) are allowed. The length Y of any such sub-sequence is the number of trials up to but not including the $(k + 1)$th failure, so that the probability distribution of Y is the generalized geometric distribution given in equation (1.21). Because this distribution enters naturally into calculations involving well-matching sub-sequences, we now consider its cumulative distribution function at greater length than was done in Section 1.3.6.

The probability that y trials or fewer are required before failure $k + 1$ is

$$F_Y(y) = \sum_{j=k}^{y} \binom{j}{k} p^{j-k}(1-p)^{k+1}, \quad y = k, k+1, k+2, \ldots. \tag{6.3}$$

We will later wish to calculate $\text{Prob}(Y \leq y - 1)$. Replacing y by $y - 1$ in equation (6.3), we immediately obtain

$$\text{Prob}(Y \leq y - 1)$$

$$= \sum_{j=k}^{y-1} \binom{j}{k} p^{j-k}(1-p)^{k+1}, \quad y = k+1, k+2, k+3, \ldots. \tag{6.4}$$

The expression on the right-hand side of equation (6.4) can be calculated in various ways. One of these is straightforward numerical computation, which is now possible and available in most mathematical computer packages. A perhaps unexpected way of calculating this sum numerically is given in Appendix C.

We now turn to the statistical hypothesis-testing process using Y_{max} as test statistic. To use Y_{max} as test statistic we must be able to compute the P-value associated with any observed value y_{max} of Y_{max}. As an example we consider the case $N = 100,000$, $p = \frac{1}{4}$; in view of a calculation in Section 5.4, we assume that the number of matching sub-sequences, some of length 0, is fixed at the value $n = N(1 - p) = 75,000$.

The main difficulty in calculating P-values exactly is that the lengths of the sub-sequences are not independent when $k > 1$. For example, a single long run of exact matches will lead to long lengths of two or more well-matching sub-sequences overlapping this run. Unfortunately, the theory for the probability distribution of the maximum of dependent sub-sequence lengths is difficult, and we are forced to consider alternative approaches.

One possible approach is to ignore this dependence and to use values such as those given in Table 3.4, calculated assuming independence of sub-sequence lengths. This is potentially dangerous, since as we have seen, P-values change very rapidly as a function of y_{max}, and thus the approximation implied by assuming independence is not guaranteed to give accurate P-values.

Fortunately, given current computing power, the dependent case is one where a computational approach using simulations is now possible. This approach is based on the discussion in Section 1.16, and allows us to assess the magnitude of any error incurred by making the independence assumption. We now discuss this approach.

A long random DNA sequence can be generated repeatedly by a simple program, and from this an empirical probability distribution for Y_{max} can be obtained. From this, quite precise estimates of P-values for Y_{max} can be found. Some representative estimates found in this way are given in Table 6.3. A comparison of the "independence" P-values in Table 3.4 and the simulation values in Table 6.3 shows that the independence assumption does cause somewhat inaccurate P-value approximations, the inaccuracy increasing, as might be expected, as k increases.

A second approach to the approximation of P-values is based on (3.51), which requires knowledge of, or approximations for, the mean μ_{max} and the variance σ^2_{max} of Y_{max}. To assess the accuracy of this approach it is first necessary to compare the simulation values for the mean and variance of Y_{max} with the "independence" values given in Table 3.4. The simulation values are given in Table 6.4, together with their standard deviations. A comparison of the values in the two tables shows that as k increases, the "independence" mean and variance in Table 3.4 become somewhat inaccurate. Even though this inaccuracy is small, the sharp changes in P-values near the mean imply that the "independence" values for the mean and variance are possibly not sufficiently accurate for P-value approximations using (3.51) when k is large. This view is supported by the somewhat inaccurate

y	$k=0$	$k=1$	$k=2$	$k=3$	$k=4$	$k=5$
7	0.990					
8	0.683					
9	0.247	0.998				
10	0.068	0.845				
11	0.018	0.405	0.999			
12	0.005	0.133	0.875			
13	0.001	0.038	0.464			
14		0.010	0.169	0.860		
15		0.003	0.052	0.469		
16			0.005	0.179	0.823	
17			0.001	0.059	0.441	
18				0.018	0.172	0.763
19				0.006	0.058	0.395
20				0.002	0.019	0.158
21						0.056
22						0.019

Table 6.3. Approximate P-values for the length of the largest sub-sequence allowing k mismatches, in a DNA segment of length 100,000, found by simulation. Estimates have approximate standard deviation 0.001. $p = \frac{1}{4}$.

k	0	1	2	3	4	5
Mean	8.013	10.426	12.582	14.592	16.518	18.383
Variance	0.937	1.081	1.201	1.313	1.417	1.520

Table 6.4. Simulation estimates of the mean and variance of the length of the largest sub-sequence allowing k mismatches, in a DNA segment of length 100,000. Estimates of the mean have approximate standard error 0.002. Estimates of the variance have approximate standard deviation 0.01. $p = \frac{1}{4}$.

P-values, as noted above, found by assuming independence of sub-sequence lengths.

Are there approximations for the mean and variance of Y_{max} more accurate than those given in Table 3.3? Waterman (1995, page 277) claims that for large n, good approximations for the mean and variance are

$$\mu_{max} = \frac{\log n + \gamma + k \log \left(\frac{\log n}{\lambda} \right) + k \log \left(\frac{1-p}{p} \right) - \log(k!)}{\lambda} - \frac{1}{2} + r_1, \quad (6.5)$$

$$\sigma^2_{max} = \frac{\pi^2}{6\lambda^2} + \frac{1}{12} + r_2, \quad (6.6)$$

where $\lambda = (-\log p)$, $n = N(1-p)$, and, for $p = \frac{1}{4}$, $|r_1| \leq 3.45 \times 10^{-4}$ and $|r_2| \leq 2.64 \times 10^{-2}$. These approximations were first calculated assuming independence of sub-sequence lengths, and later shown not to change significantly in the dependent case. The approximations (6.5) and (6.6) are direct generalizations of the approximations (2.118) and (2.120), respectively, to the case where k is a positive integer.

If these values are to be used in conjunction with the approximation (3.51), it is important to assess their accuracy, especially if we recall the great precision necessary for calculations involving the maximum of generalized geometric random variables, as discussed in Section 3.7.2. To make this assessment, Table 6.5 shows the approximate values of the mean $\mu_{\max}$ as calculated by equation (6.5). The calculations in this table are for the values $n = 75,000$, $p = \frac{1}{4}$, $k = 0, 1, 2, 3, 4, 5$. The values in this table differ from those in Table 6.4, when $k > 0$, by more than the maximal error 3.45×10^{-4} claimed for the approximation (6.5). Presumably this arises because the asymptotic theory does not yet apply when $n = 75,000$. Thus when n is not extremely large the use of the means in Table 6.5 in conjunction with the approximation (3.51) could lead to some inaccuracies when we recall how quickly the P-values change in the neighborhood of the mean.

k	0	1	2	3	4	5
Mean	8.013	10.315	12.116	13.625	14.926	16.066

Table 6.5. Approximation for the mean of the maximum of 75,000 iid generalized geometric random variables, for various values of k, calculated from equation (6.5), $p = \frac{1}{4}$.

The corresponding approximate value for the variance $\sigma_{\max}^2$, given by equation (6.6), is 0.939. When $k > 0$, this also differs from the values in Table 6.4 by more than the maximal error 2.64×10^{-2} claimed for the approximation (6.6). This is also of concern with regard to P-value approximations using (3.51). It is thus necessary to calculate the P-value approximation given by (3.50) and (3.51), using the mean (6.5) and the variance (6.6) in the approximation, to assess the accuracy both of the approximation and of (6.5) and (6.6). Calculations corresponding to those in Table 3.4 are given in Table 6.6. It is clear that the P-value approximations based on equations (6.5) and (6.6) are somewhat in error, at least when k is 2 or more.

BLAST provides a second (and more frequently used) approach to significance testing based on "well-matching" rather than exactly matching sub-sequences. Since BLAST calculations depend on an analysis of random walks, we defer considering it until the relevant random walk theory has been described.

The analysis of patterns in DNA and protein sequences is a large and growing area of bioinformatics. The theory, however, rapidly becomes diffi-

y	$k=0$	$k=1$	$k=2$	$k=3$	$k=4$	$k=5$
7	0.990					
8	0.682					
9	0.249	0.999				
10	0.069	0.824				
11	0.018	0.353	0.995			
12	0.004	0.102	0.733			
13	0.001	0.028	0.281			
14		0.007	0.079	0.487		
15		0.002	0.020	0.154		
16			0.005	0.041	0.224	
17			0.002	0.010	0.061	
18				0.003	0.016	0.074
19				0.001	0.004	0.019
20				0.000	0.001	0.005
21						0.001
22						0.000

Table 6.6. Approximate P-values for the maximum of 75,000 iid generalized geometric random variables, for various values of k and y, calculated using equations (3.51), (6.5) and (6.6). $p = \frac{1}{4}$.

cult, and we do not pursue further topics here. Recent research results may be found, for example, in Bailey and Gribskov (1998), Jonassen, Collins, and Higgins (1995), Karlin and Brendel (1992), Neuwald and Green (1994), and Rigoutsos and Floratos (1998).

6.4   Alignment Algorithms for Two Sequences

6.4.1   Introduction

One way to discriminate between good and bad alignments is to use a scoring scheme. A simple example of a scoring scheme is

(the number of matches) − (the number of mismatches and indels). (6.7)

Scoring schemes used for aligning DNA are often not much different from this simple scheme. For protein sequences, however, a more complex scoring scheme is appropriate. Commonly used scoring schemes are developed using statistical analysis of existing data, and we discuss the statistical theory behind these scoring schemes in Section 6.5. For now, we assume that we have assigned a score to each alignment in a meaningful way that reflects the likelihood that this alignment was produced as a consequence of divergence from a common ancestor. Then we can consider the alignments with the "best" score, and we can define the score of the sequence pair to be this best score. What "best" means here depends on whether high scores in the scoring scheme are more indicative of relatedness (so the "best" score is the maximum over all alignments), or whether low scores are more indicative (so "best" is the minimum).

This mathematical framework allows a statistical analysis where we make inferences about the relatedness of the sequences. We can investigate the hypothesis that they did indeed diverge from a common ancestor by considering the probability of the observed score (or one more extreme) arising by chance, under some appropriate model of evolution. If the two sequences are judged to be related, we can use their alignment to discover common patterns in the sequences. This is useful in particular for finding functional domains. Finally, by comparing scores among several different species we can get information to help reconstruct the phylogenetic tree that relates them all.

Scores of alignments consist of two main types: *similarity scores* and *distance scores* (also commonly called *distance measures*). In similarity scores the higher the score, the more closely related are the two aligned sequences; in the distance measures the opposite is the case. In the remainder of this section we use similarity scores. These are usually computed as the sum of individual scores, one for each aligned pair of residues, together with a score for each gap. We will denote by $s(X, Y)$ the score assigned to the aligned pair consisting of the residues X and Y. This score reflects how conservative the substitution represented by the alignment of X with Y is. For example, it is much less likely that the amino acid W (tryptophan) will be substituted for V (valine) in a functional domain than it is that W will be substituted for R (arginine). (This is not only an empirically observed fact, but also makes sense in terms of the chemical properties involved.) Thus the score $s(W, V)$ assigned to an alignment of the two symbols W and V is lower than $s(W, R)$, the score assigned to an alignment of the two symbols W and R. The score assigned to a gap of length ℓ is usually a function of ℓ, which we denote by $\delta(\ell)$. It represents the cost of having a gap of length ℓ and is therefore zero or negative. The simplest gap penalty model is a *linear gap model*, where $\delta(\ell) = -\ell d$ for some non-negative constant d, called the *linear gap penalty*. Therefore, in the linear gap model, each indel in a gap is weighted in the same way, namely by a penalty of d.

Thus if the alphabet has size N ($N = 4$ for nucleotides and $N = 20$ for amino acids), a scoring scheme consists of an $N \times N$ matrix S and a gap cost function δ. The matrix S is called a *substitution matrix* and the entry in its ith row and jth column is the score of the alignment of the ith and jth symbols in the alphabet.

Example. Consider the comparison of two nucleotide sequences with a simple scoring scheme that assigns $+1$ to each match, -1 to each mismatch, and a linear gap score with $d = 2$. Then the score for the following alignment of the two sequences $cttagg$ and $catgagaa$ is $1-1+1-2+1-2+1-4 = -5$:

$$
\begin{array}{ccccccccc}
c & t & t & a & g & - & g & - & - \\
c & a & t & - & g & a & g & a & a
\end{array}
$$

One of the main aims of the statistical theory is to find for nucleotides, and more importantly for amino acids, what an optimal scoring scheme should be. This matter is taken up in detail in Section 6.5 and Chapter 10.

6.4.2 Gapped Global Comparisons and Dynamic Programming Algorithms

Suppose that we are given a scoring scheme made up of a substitution matrix and a linear gap penalty. Our aim is to find, of the possible global alignments of two sequences (with gaps allowed), the one (or those ones) with the highest score. One method in principle for doing this is to list exhaustively all possible alignments and their scores, and then note the highest-scoring alignment(s). However, when the sequences are long, this is not computationally feasible, and more efficient algorithms are needed. We describe one such algorithm below, but first we justify our assertion that the exhaustive search illustrated above is indeed not efficient, by getting a sense of how large the number of global alignments between a sequence $x = X_1 X_2 \ldots X_m$ of length m and a sequence $y = Y_1 Y_2 \ldots Y_n$ of length n is. We will denote this number by $c(m, n)$. Since there is no point in matching two deletions, no alignments of one indel with another are allowed.

Let $g(m, n)$ be the number of groups obtained by grouping together those alignments that have the same combination of aligned residue pairs ignoring the indels. Then $g(m, n) < c(m, n)$, and this provides a lower bound for $c(m, n)$. We can compute $g(m, n)$ as follows.

The number k of aligned residues for two sequences of lengths m and n is between 0 and $\min\{m, n\}$. Moreover, for each such k there are $\binom{m}{k}$ ways of choosing the residues of x that align with residues of y, and $\binom{n}{k}$ ways of choosing the residues of y that align with residues of x. So there are $\binom{m}{k}\binom{n}{k}$ alignments with k aligned residues. Therefore,

$$
g(m, n) = \sum_{k=0}^{\min\{m,n\}} \binom{m}{k}\binom{n}{k}. \tag{6.8}
$$

From the result of Problem 6.1 below, it follows that

$$
g(m, n) = \binom{m+n}{n}. \tag{6.9}
$$

In particular, when $m = n$,

$$
g(n, n) = \binom{2n}{n}.
$$

This number grows quite fast with n. Stirling's approximation (B.4), and even more directly (B.5), shows that

$$\binom{2n}{n} \sim \frac{2^{2n}}{\sqrt{\pi n}}. \tag{6.10}$$

Thus the number $c(1,000,1,000)$ of global alignments between two sequences each of length 1,000 satisfies

$$c(1,000,1,000) \geq g(1,000,1,000) \cong \frac{2^{2,000}}{\sqrt{1,000\pi}} \cong 10^{600}.$$

This shows why it is not feasible to examine all possible alignments. This motivates the search for algorithms that can compute the best score efficiently and an alignment with this score, without having to examine all possibilities. One such algorithm is the Needleman–Wunsch algorithm (1970), and we discuss a version of this procedure introduced by Gotoh (1982). These are examples of dynamic programming algorithms, and we use them to illustrate the general concept of dynamic programming.

The input consists of two sequences,

$$\boldsymbol{x} = X_1 X_2 \dots X_m \text{ and } \boldsymbol{y} = Y_1 Y_2 \dots Y_n,$$

of lengths m and n, respectively, whose elements belong to some alphabet of N symbols (for DNA or RNA sequences $N = 4$, for proteins $N = 20$). We assume that we are given a substitution matrix S and a linear gap penalty d. The output consists of the highest score over all alignments between $\boldsymbol{x}$ and $\boldsymbol{y}$ and a highest-scoring global alignment between $\boldsymbol{x}$ and $\boldsymbol{y}$.

The broad approach is to break the problem into sub-problems of the same kind and build the final solution using the solutions for the sub-problems: This is the basic idea behind any dynamic programming algorithm. In this problem we find a highest-scoring alignment using previous solutions for highest-scoring alignments of smaller sub-sequences of $\boldsymbol{x}$ and $\boldsymbol{y}$. We denote by $\boldsymbol{x}_{1,i}$ the initial segment of $\boldsymbol{x}$ given by $X_1 X_2 \cdots X_i$ and similarly we denote by $\boldsymbol{y}_{1,j}$ the initial segment of $\boldsymbol{y}$ given by $Y_1 Y_2 \cdots Y_j$. For $i = 1,2,\dots,m$ and $j = 1,2,\dots,n$, we denote by $B(i,j)$ the score of a highest-scoring alignment between $\boldsymbol{x}_{1,i}$ and $\boldsymbol{y}_{1,j}$. For $i = 1,2,\dots,m$, we denote by $B(i,0)$ the score of an alignment where $\boldsymbol{x}_{1,i}$ is aligned to a gap of length i, so $B(i,0) = -id$. Similarly, for $j = 1,2,\dots,n$, we denote by $B(0,j)$ the score of an alignment where $\boldsymbol{y}_{1,j}$ is aligned to a gap of length j, so $B(0,j) = -jd$. Finally, we initialize $B(0,0) = 0$. These calculations lead to an $(m+1) \times (n+1)$ matrix B. The entry in the last row and in the last column of B, namely $B(m,n)$, is the score of a highest-scoring alignment between our two sequences $\boldsymbol{x}$ and $\boldsymbol{y}$, and it is one of the things we want our algorithm to output.

The essence of the procedure is to fill in the elements of the matrix B recursively. We already have the values of B at $(0,0)$, $(i,0)$, and $(0,j)$, for $i = 1,2,\dots,m$ and $j = 1,2,\dots,n$. Now we proceed from top left to bottom right by noting that a highest-scoring alignment between $x_{1,i}$ and $y_{1,j}$ could terminate in one of three possible ways, namely, with

$$\frac{X_i}{Y_j}, \quad \frac{X_i}{-}, \quad \text{or} \quad \frac{-}{Y_j}.$$

In the first case, $B(i,j)$ is equal to the sum of the score for a highest-scoring alignment between $x_{1,i-1}$ and $y_{1,j-1}$ together with the extra term $s(i,j)$ to account for the match between X_i and Y_j; that is, $B(i,j) = B(i-1,j-1) + s(i,j)$. In the second case, $B(i,j)$ is equal to the sum of the score for a highest-scoring alignment between $x_{1,i-1}$ and $y_{1,j}$ together with an extra term $-d$ to account for the indel to which X_i is aligned, (i.e., $B(i,j) = B(i-1,j)-d$). Similarly, in the third case, $B(i,j) = B(i,j-1)-d$. These are all the possible options, and hence $B(i,j)$ is the highest of the three. In other words,

$$B(i,j) = \max\{B(i-1,j-1)+s(i,j), B(i-1,j)-d, B(i,j-1)-d\}. \quad (6.11)$$

In this way we recursively fill in every cell in the matrix B and determine the value of $B(m,n)$, which is the desired maximum score. The running time of this algorithm is clearly $O(mn)$. To find an alignment that has this score we must keep track, at each step of the recursion, of one of the three choices giving the value of the maximum. Although there could be more than one choice giving the maximum, if we are interested in finding only one alignment, we choose one and keep a pointer to it. Once $B(m,n)$ is obtained, by tracing back through the pointers, we can reconstruct an alignment with the highest score. We now illustrate this procedure with an example.

Example. Let $x = gaatct$ and $y = catt$, so that $m = 6$ and $n = 4$. Using the same scoring scheme as in the example in Section 6.4.1, B is given in Figure 6.1, where we have used arrows to denote where each cell came from. The best score for an alignment is given by the element in the bottom rightmost cell, which is -2. Tracing back along the bold arrows, we get the highest-scoring alignment

$$
\begin{array}{cccccc}
g & a & a & t & c & t \\
c & - & a & t & - & t
\end{array}.
$$

By making different choices of arrows in the traceback procedure we can get the following other alignments, which are also highest-scoring, (i.e., which also have a score of -2):

$$
\begin{array}{cccccc}
g & a & a & t & c & t \\
c & a & - & t & - & t
\end{array}
\quad \text{and} \quad
\begin{array}{cccccc}
g & a & a & t & c & t \\
- & c & a & t & - & t
\end{array}.
$$

We next consider modifications of the Needleman–Wunsch algorithm, which can be used to address other kinds of pairwise alignment problems.

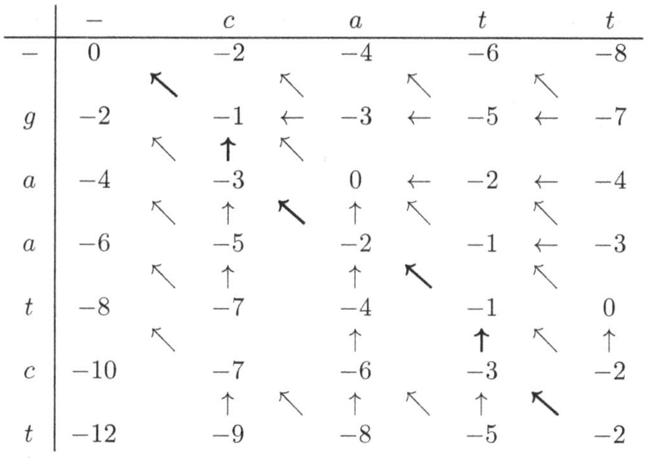

	$-$	c	a	t	t
$-$	0	-2	-4	-6	-8
g	-2	-1	-3	-5	-7
a	-4	-3	0	-2	-4
a	-6	-5	-2	-1	-3
t	-8	-7	-4	-1	0
c	-10	-7	-6	-3	-2
t	-12	-9	-8	-5	-2

Figure 6.1.

6.4.3 Fitting One Sequence into Another Using a Linear Gap Model

In this section we address the following problem: Given two sequences, a longer and a shorter one, find the sub-sequence(s) of the longer one that can be best aligned with the shorter sequence, where gaps are allowed. This procedure is relevant when one is interested in locating a specified pattern within a sequence.

Let $x = X_1 X_2 \ldots X_m$ and $y = Y_1 Y_2 \ldots Y_n$ be two sequences with $n \geq m$. For $1 \leq k \leq j \leq n$, denote by $y_{k,j}$ the sub-sequence of y given by $Y_k Y_{k+1} \ldots Y_j$. For two sequences u and v, denote by $B(u, v)$ the score of a highest-scoring (global) alignment between u and v. Our aim is to find

$$\max\{B(x, y_{k,j}) : 1 \leq k \leq j \leq n\}. \tag{6.12}$$

For each choice of k and j the running time of the Needleman–Wunsch algorithm, giving the value of $B(x, y_{k,j})$, is $O(m(j-k))$. Thus if we used this algorithm for all possible choices of k and j, and then took the maximum over all such choices, the total running time would be $O(mn^3)$, since there are $\binom{n}{2}$ possible choices for j and k. We now illustrate another approach with a better running time, namely an $O(mn)$ running time.

For $1 \leq i \leq m$ and $1 \leq j \leq n$, let $F(i, j)$ be the maximum of the scores $B(x_{1,i}, y_{k,j})$ over the values of k between 1 and j. That is, of all the possible scores for highest-scoring alignments between the initial segment of x up to x_i and the segments of y ending at y_j and beginning at some k we take $F(i, j)$ to be the greatest of such scores. The value of (6.12) is the maximum of $F(m, j)$ over all values of j between 1 and n. To find this, we initialize $F(i, 0) = -id$ for $1 \leq i \leq m$ and initialize $F(0, j) = 0$ for $0 \leq j \leq n$, since deletions of the beginning of y should clearly be without penalty. Then we fill in the matrix F recursively by

$$F(i,j) = \max\{F(i-1,j-1) + s(i,j), F(i,j-1) - d, F(i-1,j) - d\},$$

where the reasoning behind this formula is analogous to that behind (6.11). Note that there might be more than one value of j giving the maximum score. In order to recover the highest-scoring alignments of $\boldsymbol{x}$ to sub-sequences of $\boldsymbol{y}$ we can keep pointers, as in the Needleman–Wunsch algorithm.

6.4.4 Local Alignments with a Linear Gap Model

Another interesting alignment problem is to find, given two sequences, which respective sub-sequences have the highest-scoring alignment(s) (with gaps allowed). This is called a local alignment problem, and it is appropriate when one is seeking common patterns/domains in two sequences.

In the following we make the assumption that the scoring scheme we use is such that the expected (or mean) score for a random alignment is negative. If this assumption did not hold, then long matches between sub-sequences could score highly just because of their lengths, so that two long unrelated sub-sequences could give a highest-scoring alignment. Clearly, we do not want this to occur.

For $1 \le h \le i \le m$ we denote by $\boldsymbol{x}_{h,i}$ the sub-sequence of $\boldsymbol{x}$ given by $X_h X_{h+1} \ldots X_i$. With the notation as in the previous section, we want to find

$$\max\{B(\boldsymbol{x}_{h,i}, \boldsymbol{y}_{k,j}) : 1 \le h \le i \le m, 1 \le k \le j \le n\}, \tag{6.13}$$

when this is non-negative. There are $\binom{m}{2}\binom{n}{2}$ pairs of sub-sequences of $\boldsymbol{x}$ and $\boldsymbol{y}$, one for each choice of h and i among m possible values and of k and j among n possible values. Thus computing a highest-scoring alignment for each pair, using the Needleman–Wunsch algorithm, requires a total running time of $O(m^3 n^3)$. Clearly, we want to give a more efficient approach to this problem. Such an approach is provided by the Smith–Waterman algorithm (Smith and Waterman (1981)) which computes (6.13) in $O(mn)$ time. The procedure is as follows.

For each $1 \le i \le m$ and $1 \le j \le n$, we define $L(i,j)$ to be the maximum of 0 and the maximum of all possible scores for alignments between a sub-sequence of $\boldsymbol{x}$ ending at X_i and one of $\boldsymbol{y}$ ending at Y_j. That is,

$$L(i,j) = \max\{0, B(\boldsymbol{x}_{h,i}, \boldsymbol{y}_{k,j}) : 1 \le h \le i, 1 \le k \le j\}.$$

The reason we want $L(i,j) = 0$ when the max of the $B(\boldsymbol{x}_{h,i}, \boldsymbol{y}_{k,j})$'s is negative is because it is sensible to always remove the first part of an alignment if this part has a negative score, as it will just decrease the overall score of the alignment. Then the maximum of 0 and (6.13) is the maximum of $L(i,j)$ over all values of i between 1 and m and of j between 1 and n. To determine this maximum we again use dynamic programming, by initializing $L(i,0) = 0 = L(0,j)$ for $0 \leq i \leq m$ and $0 \leq j \leq n$ (since deletions at the beginning or end of our two sequences should not be penalized), and by computing

$$L(i,j) = \max\{0, L(i-1,j-1) + s(i,j), L(i-1,j) - d, L(i,j-1) - d\}.$$

We then calculate the maximum of $L(i,j)$ over all values of i and j. As in the previous maximizing procedures there might be more than one highest-scoring local alignment. To find a highest-scoring alignment, we follow the traceback procedure previously described. However, for this algorithm, we stop this process when we encounter a 0.

Figure 6.2 shows an example of an $L(i,j)$ matrix arising in locally aligning two sequences of lengths 7 and 10. In this example, the score of an

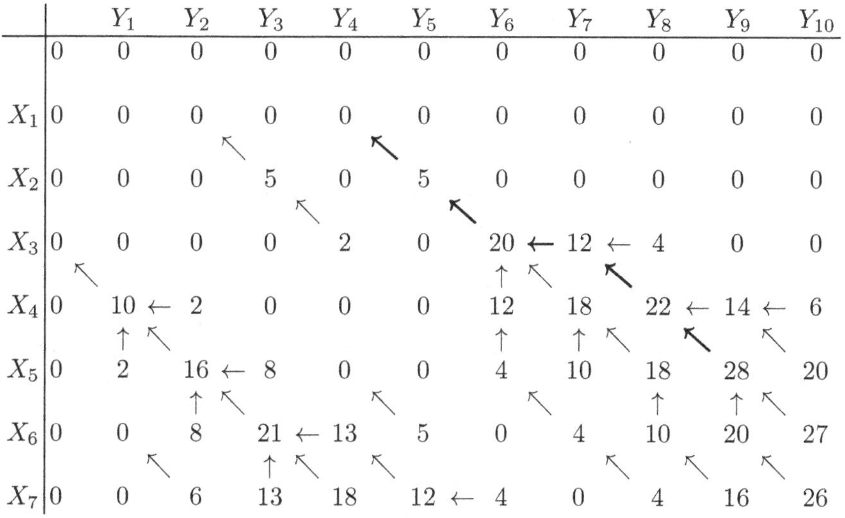

	Y_1	Y_2	Y_3	Y_4	Y_5	Y_6	Y_7	Y_8	Y_9	Y_{10}
0	0	0	0	0	0	0	0	0	0	0
X_1 0	0	0	0	0	0	0	0	0	0	0
X_2 0	0	0	5	0	5	0	0	0	0	0
X_3 0	0	0	0	2	0	20 ← 12 ← 4			0	0
X_4 0	10 ← 2		0	0	0	12	18	22 ← 14 ← 6		
X_5 0	2	16 ← 8		0	0	4	10	18	28	20
X_6 0	0	8	21 ← 13		5	0	4	10	20	27
X_7 0	0	6	13	18	12 ← 4		0	4	16	26

Figure 6.2.

optimal local alignment of the two sequences is 28, and there is only one alignment of sub-sequences giving this score, the one indicated by the bold arrows, which is

$$\begin{array}{cccccc} X_2 & X_3 & - & X_4 & X_5 \\ Y_5 & Y_6 & Y_7 & Y_8 & Y_9 \end{array}.$$

6.4.5 Other Gap Models

There are many variants and extensions of the algorithms discussed above. For example, while the linear gap model used above is appealing in its simplicity, it is often not appropriate for biological sequences, since often it is harder for a gap to open (i.e., start) than it is for it to extend. Thus it is often more appropriate not to penalize additional gap steps as much as the first one, and to use a more complicated gap cost $\delta(\ell)$. This implies that the recurrence relations need to be adjusted. For instance, in (6.11), we now have to distinguish among the alignments of $x_{1,i}$ with $y_{1,j}$ that end with X_i aligned to an indel. The score of such an alignment will also depend on how many symbols in x immediately preceding X_i are aligned to indels. Suppose that in such an alignment the last symbol preceding X_i that is not aligned to an indel (hence is aligned to Y_j) is X_k. Then the $i - k$ symbols from X_{k+1} to X_i are aligned to indels, and the score of a highest-scoring alignment is $B(k, j) + \delta(i - k)$. Similar reasoning must be applied to those alignments between $x_{1,i}$ and $y_{1,j}$ that end with Y_j aligned to an indel. Hence (6.11) must be replaced with

$$
\begin{aligned}
B(i, j) = \max\{ &B(i - 1, j - 1) + s(i, j), \\
&B(k, j) + \delta(i - k) : k = 0, 1, \ldots, i - 1, \\
&B(i, k) + \delta(j - k) : k = 0, 1, \ldots, j - 1\},
\end{aligned}
$$

and the initialization is given by $B(0, 0) = 0$, $B(i, 0) = \delta(i)$ for $i = 1, 2, \ldots, m$, and $B(0, j) = \delta(j)$ for $j = 1, 2, \ldots, n$.

Therefore, in general, finding a highest-scoring alignment between two sequences of lengths m and n takes $O(m^2 n + mn^2)$ operations, as opposed to $O(mn)$ operations needed if the gap penalty model is linear. This is because for each cell of B we now need to consider $i + j + 1$ previous cells, instead of just three.

If, however, $\delta(\ell)$ satisfies certain conditions, there are algorithms that take $O(mn)$ operations. One simple example of this is that of an *affine gap model*, in which $\delta(\ell) = -d - (\ell - 1)e$, for some (non-negative) d and e. d is called the *gap-open* penalty, and e is called the *gap-extension* penalty. Usually, e is set smaller than d. Thus all gap steps other than the first have the same cost, but each of them is penalized less than the first. We now describe a dynamic programming implementation of a global alignment algorithm for this case whose running time is also $O(mn)$.

Instead of using just one matrix B, the algorithm uses three matrices. Let x and y be the sequences we want to align. We use the same notation as above for $x_{1,i}$ and $y_{1,j}$. For $i = 1, 2, \ldots, m$ and $j = 1, 2, \ldots, n$, we denote by $S(i, j)$ the score of a highest-scoring alignment between $x_{1,i}$ and $y_{1,j}$, given that the alignment ends with X_i aligned to Y_j. We denote by $I_x(i, j)$ the score of a highest-scoring alignment between $x_{1,i}$ and $y_{1,j}$, given that the alignment ends with X_i aligned to an indel. Finally, we denote by $I_y(i, j)$ the score of a highest-scoring alignment between $x_{1,i}$ and $y_{1,j}$, given that the alignment ends with an indel aligned to Y_j. Then if we assume that a deletion will not be followed directly by an insertion, we have, for

$i = 1, 2, \ldots, m$ and $j = 1, 2, \ldots, n$,

$$S(i,j)$$
$$= \max\{S(i-1,j-1)+s(i,j), I_x(i-1,j-1)+s(i,j), I_y(i-1,j-1)+s(i,j)\},$$

where

$$I_x(i,j) = \max\{S(i-1,j) - d, I_x(i-1,j) - e\},$$

and

$$I_y(i,j) = \max\{S(i,j-1) - d, I_y(i,j-1) - e\}.$$

These recurrence relations allow us to fill in the matrices S, I_x, and I_y, once we initialize $S(0,0) = I_x(0,0) = I_y(0,0) = 0$, $S(0,j) = I_x(0,j) = -d - (j-1)e$, and $S(i,0) = I_y(i,0) = -d - (i-1)e$, for $i = 1, 2, \ldots, m$ and $j = 1, 2, \ldots, n$. The score of a highest-scoring alignment is then given by $\max\{S(m,n), I_x(m,n), I_y(m,n)\}$.

6.4.6 Limitations of the Dynamic Programming Alignment Algorithms

All the algorithms discussed above yield the exact highest score according to the given scoring scheme. However, when one has to deal with very long sequences, as would occur if one wished to align a given sequence with each of several sequences in a big database, a time complexity of $O(mn)$ might not be good enough for performing the required search in an acceptable amount of time. So various other algorithms have been developed to overcome this difficulty, BLAST being one of them. These algorithms use heuristic techniques to limit the search to a fraction of the possible alignments between two sequences in a way that attempts not to miss the high-scoring alignments. The trade-off is that one might not necessarily find the best possible score. Various aspects of BLAST will be discussed extensively in Chapter 10.

Another consideration is that of space, since memory usage can also be a limiting factor in dynamic programming. In the Needleman–Wunsch algorithm, for instance, one needs to store the $(m+1) \times (n+1)$ matrix B. If one cares only about the value of the best score, without wanting to find a highest-scoring alignment, then there is no need to store all cells of B, since the value of $B(i,j)$ depends only on entries up to one row back. Thus one can throw away rows of the matrix that are further back. However, usually one wants to find also a highest-scoring alignment, not only its score. There are methods that allow one to do this in $O(m+n)$ space, rather then $O(mn)$, and this with no more than doubling in time, so the time complexity remains at $O(mn)$. For more details see Section 2.6 in Durbin

	A	R	N	D	C	Q	E	G	H	I	L	K	M	F	P	S	T	W	Y	V
A	4	-1	-2	-2	0	-1	-1	0	-2	-1	-1	-1	-1	-2	-1	1	0	-3	-2	0
R	-1	5	0	-2	-3	1	0	-2	0	-3	-2	2	-1	-3	-2	-1	-1	-3	-2	-3
N	-2	0	6	1	-3	0	0	0	1	-3	-3	0	-2	-3	-2	1	0	-4	-2	-3
D	-2	-2	1	6	-3	0	2	-1	-1	-3	-4	-1	-3	-3	-1	0	-1	-4	-3	-3
C	0	-3	-3	-3	9	-3	-4	-3	-3	-1	-1	-3	-1	-2	-3	-1	-1	-2	-2	-1
Q	-1	1	0	0	-3	5	2	-2	0	-3	-2	1	0	-3	-1	0	-1	-2	-1	-2
E	-1	0	0	2	-4	2	5	-2	0	-3	-3	1	-2	-3	-1	0	-1	-3	-2	-2
G	0	-2	0	-1	-3	-2	-2	6	-2	-4	-4	-2	-3	-3	-2	0	-2	-2	-3	-3
H	-2	0	1	-1	-3	0	0	-2	8	-3	-3	1	-2	-1	-2	-1	-2	-2	2	-3
I	-1	-3	-3	-3	-1	-3	-3	-4	-3	4	2	-3	1	0	-3	-2	-1	-3	-1	3
L	-1	-2	-3	-4	-1	-2	-3	-4	-3	2	4	-2	2	0	-3	-2	-1	-2	-1	1
K	-1	2	0	-1	-3	1	1	-2	-1	-3	-2	5	-1	-3	-1	0	-1	-3	-2	-2
M	-1	-1	-2	-3	-1	0	-2	-3	-2	1	2	-1	5	0	-2	-1	-1	-1	-1	1
F	-2	-3	-3	-3	-2	-3	-3	-3	-1	0	0	-3	0	6	-4	-2	-2	1	3	-1
P	-1	-2	-2	-1	-3	-1	-1	-2	-2	-3	-3	-1	-2	-4	7	-1	-1	-4	-3	-2
S	1	-1	1	0	-1	0	0	0	-1	-2	-2	0	-1	-2	-1	4	1	-3	-2	-2
T	0	-1	0	-1	-1	-1	-1	-2	-2	-1	-1	-1	-1	-2	-1	1	5	-2	-2	0
W	-3	-3	-4	-4	-2	-2	-3	-2	-2	-3	-2	-3	-1	1	-4	-3	-2	11	2	-3
Y	-2	-2	-2	-3	-2	-1	-2	-3	2	-1	-1	-2	-1	3	-3	-3	-2	2	7	-1
V	0	-3	-3	-3	-1	-2	-2	-3	-3	3	1	-2	1	-1	-2	-2	0	-3	-1	4

Table 6.7. The BLOSUM62 substitution matrix

et al. (1998) or Section 9.7 in Waterman (1995).

6.5   Protein Sequences and Substitution Matrices

6.5.1   Introduction

In the study of DNA sequences, simple scoring schemes are usually effective. For protein sequences, however, some substitutions are much more likely than others. The performance of any alignment algorithm is improved when it accounts for this difference. In all cases we consider, higher scores will represent more likely substitutions.

There are two frequently used approaches to finding substitution matrices. One leads to the PAM (Accepted Point Mutation) family of matrices, and the other to the BLOSUM (BLOcks SUbstitution Matrices) family. Table 6.7 gives an example of a typical BLOSUM substitution matrix (called the BLOSUM62 matrix). In this section we discuss how these substitution matrices are derived.

```
WWYIR   CASILRKIYIYGPV   GVSRLRTAYGGRK   NRG
WFYVR   CASILRHLYHRSPA   GVGSITKIYGGRK   RNG
WYYVR   AAAVARHIYLRKTV   GVGRLRKVHGSTK   NRG
WYFIR   AASICRHLYIRSPA   GIGSFEKIYGGRR   RRG
WYYTR   AASIARKIYLRQGI   GVGGFQKIYGGRQ   RNG
WFYKR   AASVARHIYMRKQV   GVGKLNKLYGGAK   SRG
WFYKR   AASVARHIYMRKQV   GVGKLNKLYGGAK   SRG
WYYVR   TASIARRLYVRSPT   GVDALRLVYGGSK   RRG
WYYVR   TASVARRLYIRSPT   GVGALRRVYGGNK   RRG
WFYTR   AASTARHLYLRGGA   GVGSMTKIYGGRQ   RNG
WFYTR   AASTARHLYLRGGA   GVGSMTKIYGGRQ   RNG
WWYVR   AAALLRRVYIDGPV   GVNSLRTHYGGKK   DRG
```

Table 6.8. A set of four blocks from the Blocks database

Any attempt to create a scoring matrix for amino acid substitutions must start from a set of data that can be trusted. The "trusted" data are then used to determine which substitutions are more or less likely. The matrix is then derived from these data, using (as we shall see) aspects of statistical hypothesis-testing theory.

Historically, the PAM matrices were developed first (in 1978), but since the derivation of BLOSUM matrices is somewhat simpler than that for PAM matrices, we start by considering BLOSUM matrices.

6.5.2   BLOSUM Substitution Matrices

The BLOSUM approach was introduced by Henikoff and Henikoff (1992). Henikoff and Henikoff started with a set of protein sequences from public databases that had been grouped into related families. From these sequences they obtained "blocks" of aligned sequences. A block is the *ungapped* alignment of a relatively highly conserved region of a family of proteins. Methods for producing such alignments are given in Section 6.6. These alignments provide the basic data for the BLOSUM approach to constructing substitution matrices. An example of such an alignment leading to four blocks is given in Table 6.8.

Since the algorithms used to construct the aligned blocks employ substitution matrices, there is a circularity involved in the procedure if the aligned blocks are subsequently used to find substitution matrices. Henikoff and Henikoff broke this circularity as follows. They started by using a simple "unitary" substitution matrix where the score is 1 for a match, 0 for a mismatch. Then, using data from suitable groups of proteins, they constructed only those blocks that they could obtain with this simple matrix. This procedure has the effect of generating a conservative set of blocks; that is, it tends to omit blocks with low sequence identity. While this restricted the number of blocks derived, the blocks obtained were trustworthy and were not biased toward any specific scoring scheme.

Using the blocks so constructed, Henikoff and Henikoff then counted the number of occurrences of each amino acid and the number of occurrences of each pair of amino acids aligned in the same column. Consider a very simplified example, with only three amino acids, A, B, and C, and only one block:

$$
\begin{array}{cccc}
B & A & B & A \\
A & A & A & C \\
A & A & C & C \\
A & A & B & A \\
A & A & C & C \\
A & A & B & C
\end{array}
$$

In this block there are 24 amino acids observed, of which 14 are A, 4 are B, and 6 are C. Thus the observed proportions are

amino acid	proportion of times observed
A	14/24
B	4/24
C	6/24

(6.14)

There are $4 \cdot \binom{6}{2} = 60$ *aligned pairs* of amino acids in the block. These 60 pairs occur with proportions as given in the following table:

aligned pair	proportion of times observed
A to A	26/60
A to B	8/60
A to C	10/60
B to B	3/60
B to C	6/60
C to C	7/60

(6.15)

We now compare these observed proportions to the *expected* proportion of times that each amino acid pair is aligned under a random assortment of the amino acids observed, given the observed amino acid frequencies (6.14). In other words, if we choose two sequences of the same length at random with these frequencies (6.14), and put them into alignment, then the expected proportion of pairs in which A is aligned with A is $\frac{14}{24} \cdot \frac{14}{24}$, the expected proportion of pairs in which A is aligned with B is $2 \cdot \frac{14}{24} \cdot \frac{4}{24}$, and so on. (The factor of 2 in the second calculation allows for the two cases where A is in the first sequence and B in the second, and that where B is in the first sequence and A in the second.)

These fractions are now used to calculate "estimated likelihood ratios" (see Section 3.6) as shown in the following table:

aligned pair	proportion observed	proportion expected	$2\log_2\left(\dfrac{\text{proportion observed}}{\text{proportion expected}}\right)$
A to A	26/60	196/576	0.70
A to B	8/60	112/576	-1.09
A to C	10/60	168/576	-1.61
B to B	3/60	16/576	1.70
B to C	6/60	48/576	0.53
C to C	7/60	36/576	1.80

$$(6.16)$$

For each row in this table the ratio of the entries in the second and third columns is an estimate, from the data, of the ratio of the proportion of times that each amino acid combination occurs in any column to the proportion expected under random allocation of amino acids into columns. With one important qualification, which we describe later, the respective elements in the BLOSUM substitution matrix are now found by calculating twice the logarithm (to the base 2) of this ratio (as shown in the final column of the above table), and then rounding the result to the nearest integer. In this simplified example, the substitution matrix would thus be

$$
\begin{array}{c c c c}
 & A & B & C \\
A & 1 & -1 & -2 \\
B & -1 & 2 & 1 \\
C & -2 & 1 & 2
\end{array}.
$$

In general, the procedure is as follows. For each pair of amino acids x and y, first count the number of times we see x and y in the same column of an aligned block. We denote this number by n_{xy}. We then put

$$
p_{xy} = \frac{n_{xy}}{\sum_{u \leq v} n_{uv}},
$$

where we take $u \leq v$ to mean that the letter denoting u precedes the letter denoting v in the alphabet. This number p_{xy} is the estimate of the probability of a randomly chosen pair of amino acids chosen from one column of a block to be the pair x and y. Now, for each amino acid x, let p_x be the proportion of times x occurs somewhere in any block. Consider the quantity

$$
e_{xy} = \begin{cases} \frac{2 p_x p_y}{p_{xy}} & \text{if } x \neq y, \\[2mm] \frac{p_x p_y}{p_{xy}} & \text{if } x = y. \end{cases}
$$

This quantity is the ratio of the likelihood that x and y are aligned by chance, given their frequencies of occurrence in the blocks, to the proportion of times we actually observe x and y aligned in the same column in the blocks. We convert this into a score by taking -2 times its logarithm to the base 2, and rounding to the nearest integer. In this way pairs that are more likely than chance will have positive scores, and those less likely will have negative scores.

While this approach is still rudimentary, it does yield a more useful scoring scheme than the original one that merely scores 1 for a match and 0 for a mismatch. Its main shortcoming is that it overlooks an important factor that can bias the results. The substitution matrix derived will depend significantly on which sequences of each family happen to be in the database used to create the blocks. In particular, if there are many very closely related proteins in one block, and only a few others that are less closely related, then the contribution of that block will be biased toward closely related proteins. For example, suppose the data in one block are as follows:

$$
\begin{array}{cccc}
A & B & A & A \\
A & B & A & A \\
A & B & A & A \\
A & B & A & A \\
A & A & B & D \\
A & C & B & A \\
D & A & B & A \\
\end{array}
$$

The first four sequences possibly derive from closely related species and the last three from three more distant species. Since A occurs with high frequency in the first four sequences, the observed number of pairings of A with A will be higher than is appropriate if we are comparing more distantly related sequences. Ultimately, we would prefer to have sequences in each block such that any pair have roughly the same amount of "evolutionary distance" between them. The solution to this problem used by Henikoff and Henikoff is to group, or cluster, those sequences in each block that are "sufficiently close" to each other and, in effect, use the resulting cluster as a single sequence. This step requires a definition of "sufficiently close," and this is done by specifying a cut-off proportion, say 85%, and then grouping the sequences in each block into clusters in such a way that each sequence in any cluster has 85% or higher sequence identity to at least one other sequence in the cluster in that block.

We now describe how the counting is done in this case, and after the general method is described, we illustrate it with an example. The count of each amino acid is found by dividing each occurrence by the number of sequences in the cluster containing that occurrence, and summing over all occurrences. After this is done, we count aligned amino acid pairs. Here the rule we follow is that if in any block two sequences are in the same cluster, then in that block no counts are taken between amino acids in those two sequences. For any aligned amino acids in sequences in two different clusters in the same block, the count for any amino acid pair is divided by nm, where n and m are the sizes of the two clusters from which the amino acids are taken.

These weighted counts are then used in the same way as before. Consider a simple example with two blocks

$$
\begin{matrix}
B & A & B & A \\
B & A & B & C \\
A & A & C & C
\end{matrix}
$$

and

$$
\begin{matrix}
C & B & B \\
C & B & B \\
A & B & C \\
A & A & C
\end{matrix}
$$
.

Suppose the identity for clustering is taken to be .75. Thus we cluster the first two sequences in each block together. The A's are counted as follows. The first column of the first block has one A, the second column contributes two A's, since the first two sequences are clustered it has $1 + \frac{1}{2} + \frac{1}{2} = 2$ A's. The fourth column contributes $\frac{1}{2}$ A. In the second block there are three A's, since each occurrence occurs in a cluster of size one. So in total there are $13/2$ A's. Now to get the proportion of A's we must divide by 17, since each column of the first block contributes 2 to the counts of the symbols, and each column of the second block contributes 3 to the counts. So the proportion of A's is $(13/2)/17 = 13/34$. We record the proportions for all symbols in the following table:

amino acid	proportion of times observed
A	$13/34$
B	$5/17$
C	$11/34$

(6.17)

To count the A–B pairs, each occurrence in the first column of the first block contributes $\frac{1}{2}$, and in the second column of the second block the contribution is $\frac{1}{2} + \frac{1}{2} + 1$. So the total A–B count is 3. There are a total of 13 pairs in the blocks, four in the first block (each column contributes one pair, or more precisely, two half pairs) and nine in the second block. Thus the proportion of A–B pairs is $3/13$. We record the proportions for all pairs of symbols in the following table:

aligned pair	proportion of times observed
A to A	$2/13$
A to B	$3/13$
A to C	$5/26$
B to B	$1/13$
B to C	$3/13$
C to C	$3/26$

(6.18)

The procedure is then carried out as before.

A further refinement was made by Henikoff and Henikoff (1992). After obtaining a BLOSUM substitution matrix as just described, the matrix obtained is then used instead of the conservative "unitary" matrix to construct a second, less conservative, set of blocks. A new substitution matrix is then obtained from these blocks. Then the process is repeated a third time. Henikoff and Henikoff derive the final family of BLOSUM matrices from this third set of blocks, and it is these whose use is suggested.

If the .85 similarity score criterion is adopted, the final matrix is called a BLOSUM85 matrix. In general if clusters with $X\%$ identity are used, then the resulting matrix is called BLOSUMX. The BLOSUM matrices currently available on the BLAST web page at NCBI (www.ncbi.nlm.nih.gov/BLAST/) are BLOSUM45, BLOSUM62, and BLOSUM80. Note that the larger-numbered matrices correspond to more recent divergence, and the smaller-numbered matrices correspond to more distantly related sequences.

One often has prior knowledge about the evolutionary distance between the sequences of interest that helps one choose which BLOSUM matrix to use. With no information, BLOSUM62 is often used. We explore the implications of the choice of various matrices in Section 10.2.4.

A central feature of the BLOSUM substitution matrix calculation is the use of (estimated) likelihood ratios. We see in the next section that the same is true of PAM matrices. In Section 9.2.1 it is shown that use of likelihood ratios has a statistical optimality property, and this optimality property explains in part their use in the construction of both BLOSUM and PAM matrices.

6.5.3   PAM Substitution Matrices

In this section we outline the Dayhoff et al. (1978) approach to deriving the so-called PAM substitution matrices. Two essential ingredients in the construction of these matrices, as with construction of BLOSUM matrices, are the calculation of an (estimated) likelihood ratio and the use of Markov chain theory as introduced in Section 4.8. We now describe this construction in more detail.

An "accepted point mutation" is a substitution of one amino acid of a protein by another that is "accepted" by evolution, in the sense that within some given species, the mutation has not only arisen but has, over time, spread to essentially the entire species. A PAM1 transition matrix is the Markov chain matrix applying for a time period over which we expect 1% of the amino acids to undergo accepted point mutations within the species of interest.

The construction of PAM matrices starts with ungapped multiple alignments of proteins into blocks for which all pairs of sequences in any block are, as in the BLOSUM procedure, "sufficiently close" to each other. In the original construction of Dayhoff et al. (1978), the requirement was that each sequence in any block be no more than 15% different from any other sequence. This requirement resulted, for their data, in 71 blocks of aligned proteins. Imposing the requirement of close within-block similarity is aimed at minimizing the number of substitutions in the data that may have resulted from two or more consecutive substitutions at the same site. This is important because the initial goal is to create a Markov transition matrix for a short enough time period so that multiple substitutions are very unlikely to happen during this time period. We discuss later how to handle the multiple substitutions that are expected to arise over longer time periods.

The BLOSUM approach uses clustering to achieve two aims. One is to minimize biases in the databases that the sequences were taken from, since without clustering some closely related sequences may be overrepresented. The second is to account for evolutionary deviation of varying time periods. In the PAM approach, the first aim is approached by inferring a separate phylogenetic tree for the data in each aligned block of sequences, eventually using all the inferred trees in an aggregated manner to estimate a Markov chain transition matrix. The second aim is achieved by using Markov chain theory applied to this matrix.

The phylogenetic reconstruction method adopted for the data within any block in the database is the method of maximum parsimony, described in Chapter 14. This algorithm constructs trees with our sequences at the leaves, and with inferred sequences at the internal nodes, such that the total number of substitutions across the tree is minimal. Such a tree is called a *most parsimonious* tree. There are often several most parsimonious trees for any block, in which case all such trees are used and an averaging procedure is employed from the data in each tree, as described below. From now on, "tree" means one or other of the set of most parsimonious trees for one of the blocks.

For any column in any block, the data in each tree are used to obtain counts in the following manner. Suppose that two different but aligned amino acids A and B occur (in any order) in two nodes of a tree joined by a single edge. Then this edge contributes 1 to the "A–B" count. If the same amino acid A occurs aligned in two nodes of the tree joined by a single edge, then this edge contributes 2 to the "A–A" count. The counts for all "A–A" and all "A–B" amino acid pairs are then totaled over all edges of all trees in each block. If the block has n most parsimonious trees, these total counts are then divided by n. The sum of the resultant counts over all blocks is then calculated.

The following simple example demonstrates the calculations within any one block. Suppose that a given block of three sequences is

$$A\,A$$
$$A\,B \ .$$
$$B\,B$$

There are $n = 5$ most parsimonious trees leading to these three sequences at the leaves of the trees, as shown in Figure 6.3. Among these trees A is aligned with, and substituted for, B (or conversely) twice in each tree,

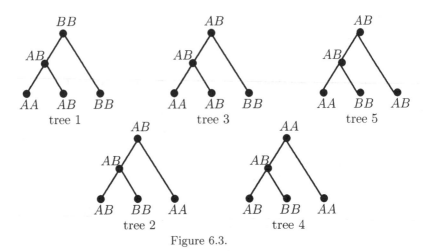

Figure 6.3.

leading to a total "A–B" count of 10. Division by the number of trees (5) in the block leads to a final contribution of 2 from this block to the A–B count.

Next, A is aligned with A two times in tree 1, three times in tree 2, three times in tree 3, four times in tree 4, and three times in tree 5, leading to a total of 15 A–A alignments. Each A–A alignment leads to a count of 2, so that the total A–A count is 30. Division by the number of trees for this block leads to a final block contribution of 6 to the overall A–A. Similar calculations show that that the contribution to the B–B count from this block is also 6.

If this were the only block in the data set, the final matrix of counts would then be

$$
\begin{array}{c|cc}
 & A & B \\
\hline
A & 6 & \\
B & 2 & 6
\end{array}
$$

In general, there will be more than one block in the data set, and if so, as indicated above, we simply add the counts from the different blocks to obtain an overall count matrix.

Suppose that the amino acids are numbered from 1 to 20 and denote the (j, k)th entry in the overall count matrix by A_{jk}. The next task is to use this count matrix to construct an estimated Markov chain transition matrix. For any j and k (not necessarily distinct), define a_{jk} by

$$
a_{jk} = \frac{A_{jk}}{\sum_m A_{jm}}. \tag{6.19}
$$

For $j \neq k$, let

$$
p_{jk} = ca_{jk}, \tag{6.20}
$$

where c is a positive scaling constant (to be determined later), and let

$$p_{jj} = 1 - \sum_{k \neq j} c a_{jk}. \tag{6.21}$$

It follows from these definitions that $\sum_k p_{jk} = 1$. If c is chosen to be sufficiently small so that each p_{jj} is non-negative, the matrix $P = \{p_{jk}\}$ then has the properties of a Markov chain transition matrix. In this matrix smaller values of c imply larger diagonal entries relative to the nondiagonal entries; however, the relative sizes of the nondiagonal entries are independent of the choice of c. In practice it will always be the case that this matrix is irreducible and aperiodic, so that it has a well-defined stationary distribution.

Although the matrix P is derived from data, and thus any probability derived from it is an estimate, we assume that the data leading to P are sufficiently extensive so that no serious error is made by thinking of P not as an estimated transition matrix but as an actual transition matrix. We thus drop the word "estimated" below in discussing the probabilities deriving from this matrix.

The value of c is now chosen so that, after one step of the Markov chain defined by the transition matrix P, the weighted expected proportion of amino acid changes is 0.01, the weights being the various amino acid frequencies. A reasonable estimate for this set of frequencies is the observed distribution found from the data in the original blocks of aligned proteins. Let p_j be the observed such frequency for the jth amino acid. Then the expected proportion of amino acids that change after one step of the Markov chain defined by the transition matrix P is

$$\sum_j p_j \sum_{k \neq j} p_{jk} = c \sum_j \sum_{k \neq j} p_j a_{jk}. \tag{6.22}$$

This implies that if c is defined by the equation

$$c = \frac{.01}{\sum_j \sum_{k \neq j} p_j a_{jk}}, \tag{6.23}$$

the "expected proportion" requirement is satisfied, and the resulting transition matrix then corresponds to an evolutionary distance of 1 PAM. This matrix is often denoted by M_1, with typical element m_{jk}, and we follow this notation here. The matrix corresponding to an evolutionary distance of n PAMs is obtained by raising M_1 to the nth power, in line with the n-step transition probability formula in equation (4.23). This matrix is denoted here by M_n and is called the PAMn matrix. As n gets larger, the matrix M_n gets closer and closer to a matrix all of whose rows are identical to the stationary distribution corresponding to the matrix M_1 (see equation (4.29)). In practice, the element common to all positions in the jth column of this matrix is often close to the background frequency p_j.

The stationary distribution of the matrix M is independent of the choice of the scaling constant c (see Problem 6.7).

The next task is to construct a substitution matrix derived from the probability transition matrix M_n. Let $m_{jk}^{(n)}$ be the (j, k) entry in the matrix M_n, for some extrinsically chosen value of n. Then $m_{jk}^{(n)}$ is the probability, after n steps of the chain defined by the matrix M_1, that the kth amino acid occurs in some specified position, given that initially the jth amino acid occurred in that position. For reasons that will be developed in Section 10.2.4, the typical entry in a PAM substitution matrix is of the form

$$C \cdot \log \left(\frac{m_{jk}^{(n)}}{p_k} \right), \tag{6.24}$$

where C is a positive constant. The choice of C is not crucial; nevertheless, this also is discussed in Section 10.2.4.

A variant of the expression (6.24) is the following. Denote the joint probability that amino acid j occurs at some nominated position at time 0 and that amino acid k occurs at this position after n steps of the Markov chain whose one-step transition matrix is M_1 by $q(j, k)$. (Note that $q(j, k)$ is a function of n, but in accordance with common practice we suppress this dependence in the notation.) Then

$$q(j, k) = p_j m_{jk}^{(n)}, \tag{6.25}$$

and (6.24) can be written as

$$C \cdot \log \left(\frac{q(j, k)}{p_j p_k} \right). \tag{6.26}$$

The choice of the value n has so far not been discussed. This matter will be taken up in Section 10.6, where the effects of an incorrect choice will be evaluated.

The BLOSUM and PAM procedures differ in one interesting respect: the larger n is for a PAM matrix, the longer is the evolutionary distance, whereas for BLOSUM matrices *smaller* values of n correspond to longer evolutionary distance.

6.5.4 A Simple Symmetric Evolutionary Matrix

In order to elucidate some properties of PAM matrices and to assess the implications of the choice of n in these matrices, it is useful to discuss a simple symmetric example, which, while it does not correspond to any PAM matrix used in practice, has properties that are found easily and that illuminate properties of PAM matrices. The model we discuss is the discrete-time analogue of the simple (and unrealistic) model considered by Bishop and Friday (1985).

Suppose that all amino acids are equally frequent (so that $p_j = 0.05$), that all are equally likely to be substituted by some other amino acid in any given time, and that all substitutions are equally likely. Then the matrix M_1 is such that its elements $\{m_{jk}\}$ are given by

$$m_{jj} = 0.99, \quad m_{jk} = 0.01/19, \quad j \neq k. \tag{6.27}$$

The value 0.99 for m_{jj} derives from the fact that we wish to mimic a PAM1 matrix, that is, a matrix for which the probability of an amino acid change in unit time is 0.01. For this simple symmetric matrix it can be shown from the spectral theory of Appendix B.19 that

$$m_{jj}^{(n)} = 0.05 + 0.95(94/95)^n, \tag{6.28}$$

$$m_{jk}^{(n)} = 0.05 - 0.05(94/95)^n, \quad j \neq k. \tag{6.29}$$

These calculations, together with equation (6.24), imply that the typical diagonal entry and the typical off-diagonal entry in the substitution matrix are, respectively,

$$C \cdot \log(1 + 19(94/95)^n), \quad C \cdot \log(1 - (94/95)^n), \tag{6.30}$$

for some positive value of C. The ratio of these is independent of C, being

$$\frac{\log(1 + 19(94/95)^n)}{\log(1 - (94/95)^n)}. \tag{6.31}$$

This leads to a substitution matrix whose diagonal elements are all

$$-\frac{\log(1 + 19(94/95)^n)}{\log(1 - (94/95)^n)} \tag{6.32}$$

and whose off-diagonal elements are all -1. When $n = 259$ (more precisely, $n = 259.0675$), the expression in (6.32) is very close to 12, corresponding to a substitution matrix whose entries are

$$S(j,j) = 12, \quad S(j,k) = -1, (j \neq k). \tag{6.33}$$

For the case $n = 259$,

$$m_{jj}^{(259)} = 0.111251, \quad m_{jk}^{(259)} = 0.046776 \quad j \neq k. \tag{6.34}$$

The definition (6.25) of $q(j,k)$ implies that for this case

$$q(j,j) = 0.0055625, \quad q(j,k) = 0.0023388, \quad j \neq k. \tag{6.35}$$

With these values the probability $20q(j,j)$ of a match at any position is 0.111251, and the probability of a mismatch is 0.888749. The mean score in the substitution matrix is then

$$12(0.111251) - 0.888749 = 0.446. \tag{6.36}$$

We discuss this calculation further in Section 10.2.4, deriving the value 0.446 there by what appears initially to be a method different from that used here.

6.6 Multiple Sequences

We may have a set of more than two related sequences, all descended from a common ancestor, and in such a case it is often desirable to put them into a multiple alignment. The definition of a multiple alignment is a straightforward generalization of a pairwise alignment. Similarly, the dynamic programming algorithms for constructing the alignments generalize in a straightforward manner. For a small number of sequences, usually fewer than 20, this works well. Problems arise, however, in applying these algorithms to many sequences. The running time for using dynamic programming to do a global multiple alignment of n sequences each of length approximately L is $O((2L)^n)$. For local alignments the situation becomes even worse.

Some algorithms have been developed to find high-scoring alignments quickly, without, however, guaranteeing to find one with the highest score. Perhaps one of the algorithms most commonly used for multiple global alignments is called CLUSTAL W (Thompson et al. (1994)).

In this section we describe a statistical method for finding ungapped local alignments, introduced by Lawrence et al. (1993). A more general algorithm allowing gapped alignments is given by Zhu et al. (1998). In Section 12.3.2 we give a different method, which constructs gapped multiple alignments.

We describe the process in terms of protein sequences. Label the amino acids in some agreed order as amino acids 1, 2, ..., 20, and suppose that these have respective "background" frequencies $p_1, p_2, \ldots, p_{20}$. Given N protein sequences, of respective lengths $L_1, L_2, \ldots, L_N$, the aim is to find N segments of length W, one in each sequence, that in some sense are most similar to each other. Here the value of W is some chosen fixed number; the choice of W is discussed below. There are $S = \prod_{j=1}^{N}(L_j - W + 1)$ possible choices for the respective locations of these N segments in the N respective sequences, and it is assumed that N and the L_j are so large that a purely algorithmic approach to finding the most similar segments by an extension of the methods discussed in Section 6.4.2 is not computationally feasible.

We describe the Lawrence et al. approach in Markov chain terms. Consider a Markov chain with S "states," each state corresponding to a choice of the locations of the N segments in the N sequences. Each state of the Markov chain is an aligned array of amino acids. This array has N rows and W columns. The aim is to find that array in which the various rows in some sense best align with each other.

The procedure consists of repeated iteration of a basic step, each step consisting of a move from one state of the Markov chain to another, that is, from one array to another. The initial array can be chosen arbitrarily or on the basis of some biological knowledge as an initial guess of the best alignment. In each step one of the various protein sequences is chosen, and the row in the array corresponding to that sequence is allowed to change in a way described below. It is convenient here to assume that the choice of this sequence is made randomly.

Before describing the way in which changes in the array are made, we illustrate in Figure 6.4 the result of the changes made in two consecutive steps of the procedure. In step 1 the third sequence happened to be chosen, so that the third row in the array was allowed to change, and in step 2 the first sequence happened to be chosen, so that the first row was allowed to change.

	position							position							position				
1	2	3	4	$\cdots$	W		1	2	3	4	$\cdots$	W		1	2	3	4	$\cdots$	W
V	Q	A	L	$\cdots$	N		V	Q	A	L	$\cdots$	N		C	A	A	N	$\cdots$	R
A	Q	B	N	$\cdots$	R		A	Q	B	N	$\cdots$	R		A	Q	B	N	$\cdots$	R
L	L	C	R	$\cdots$	N	step 1	C	Q	T	N	$\cdots$	N	step 2	C	Q	T	N	$\cdots$	N
W	R	A	A	$\cdots$	C	$\longrightarrow$	W	R	A	A	$\cdots$	C	$\longrightarrow$	W	R	A	A	$\cdots$	C
S	Q	C	C	$\cdots$	T		S	Q	C	C	$\cdots$	T		A	Q	C	C	$\cdots$	T
S	Q	T	R	$\cdots$	C		S	Q	T	R	$\cdots$	C		S	Q	T	R	$\cdots$	C
		$\vdots$							$\vdots$							$\vdots$			
G	M	C	R	$\cdots$	T		G	M	C	R	$\cdots$	T		G	M	C	R	$\cdots$	T

Figure 6.4. Full arrays of aligned segments before steps 1, 2, and 3.

We call the array just before any step is taken the "original" array and the array following this step the "new" array. This new array is also the original array for the next step. We now consider the first step in detail.

The segments in all sequences other than randomly chosen sequence 3 define a reduced array of $N-1$ segments consisting of the original array without row 3. This reduced array is the leftmost array in Figure 6.5. The

	position							position							position				
1	2	3	4	$\cdots$	W		1	2	3	4	$\cdots$	W		1	2	3	4	$\cdots$	W
V	Q	A	L	$\cdots$	N									C	A	A	N	$\cdots$	R
A	Q	B	N	$\cdots$	R		A	Q	B	N	$\cdots$	R		A	Q	B	N	$\cdots$	R
						step 1	C	Q	T	N	$\cdots$	N	step 2	C	Q	T	N	$\cdots$	N
W	R	A	A	$\cdots$	C	$\longrightarrow$	W	R	A	A	$\cdots$	C	$\longrightarrow$	C	A	A	N	$\cdots$	R
S	Q	C	C	$\cdots$	T		S	Q	C	C	$\cdots$	T							
S	Q	T	R	$\cdots$	C		S	Q	T	R	$\cdots$	C		S	Q	T	R	$\cdots$	C
		$\vdots$							$\vdots$							$\vdots$			
G	M	C	R	$\cdots$	T		G	M	C	R	$\cdots$	T		G	M	C	R	$\cdots$	T

Figure 6.5. Partial arrays of aligned segments before steps 1, 2, and 3.

reduced arrays in this figure show that sequence 3 was chosen to change in step 1 and that sequence 1 was chosen to change in step 2, and also shows that sequence 5 was chosen to change in step 3.

Suppose that in the first reduced array, amino acid j ($j = 1, 2, \ldots, 20$) occurs $c_{i,j}$ times in the ith column. From the $c_{i,j}$ values a probability estimate q_{ij} is calculated, defined by

$$q_{ij} = \frac{c_{ij} + b_j}{N - 1 + B}. \tag{6.37}$$

Here the b_j are pseudocounts, as defined in Section 3.10, and $B = \sum_j b_j$. The reason for the introduction of pseudocounts, and the actual choices of the b_j, will be discussed below. For the moment we take the b_j as given.

The aim of the first step is to replace the original segment in the third row by a new segment in a way that tends to increase the overall alignment of the N resulting segments in the new array. There are $L_3 - W$ segments of length W in sequence 3. We call that segment starting in position x in sequence 3 "segment x." Suppose that the amino acids in this segment are $x_1, x_2, \ldots, x_W$. The probability P_x of this ordered set of amino acids under the population amino acid frequencies is $P_x = p_{x_1} p_{x_2} \cdots p_{x_W}$. The estimated probability Q_x of this ordered set of amino acids using the $N - 1$ segments in the first reduced array in Figure 6.5 is taken as

$$Q_x = q_{1,x_1} q_{2,x_2} \cdots q_{W,x_W}.$$

The likelihood ratio $\text{LR}(x)$ is defined as Q_x / P_x. The numerator may be thought of as the probability of the sequence under the model reflecting only the sequences in the current alignment, while the denominator is the probability of the sequence under background frequencies. The final operation in step 1 is to replace the segment in row 3 of the original array by segment x with probability

$$\frac{\text{LR}(x)}{\sum_{m=1}^{L_3 - W} \text{LR}(m)}. \tag{6.38}$$

The reason for this choice will be discussed in Section 11.5.2. The result of this step is to produce a new array (as illustrated in the second array in Figure 6.4). The replacement procedure tends to replace the original segment in row 3 by a new segment more closely aligned with the remaining segments.

In the following step this procedure is repeated, with (in Figure 6.4) the segment from sequence 1 being randomly chosen to change. In the next step the segment from sequence 5 was randomly chosen to change, and so on. As one step follows another the N segments in the array tend to become more similar to each other, or in other words, to align better. This iterative procedure visits various possible alignments according to a random process, and thus does not systematically approach the "best" alignment. However, after many steps the best alignment should tend to arise more and more frequently and hence be recognized.

This procedure is essentially one of Gibbs sampling, and we further consider its properties in the discussion of the Gibbs sampling procedure in Section 11.5.2. In order to introduce the analysis of Section 11.5.2, we adopt here a general notation suitable for the discussion of that section. Suppose that before some specific step is taken, the current array is array s in the collection of S possible arrays, and that after this step is taken the new array is array u. Arrays s and u are identical in all rows other than the one that is changed during this step. Thus the reduced arrays obtained by eliminating the row in which arrays s and u differ lead to identical values of the q_{ij}.

Now consider the amino acids in arrays s and u in the row in which they differ. Suppose that in array s these are denoted by $s(1)$, $s(2)$,..., $s(W)$ and in array u are denoted by $u(1)$, $u(2)$,..., $u(W)$. If the transition probability from array s to array u is denoted by p_{su}, then expression (6.38) shows that

$$\frac{p_{su}}{p_{us}} = \frac{q_{1,u(1)}q_{2,u(2)}\cdots q_{W,u(W)}}{q_{1,s(1)}q_{2,s(2)}\cdots q_{W,s(W)}} \cdot \frac{p_{s(1)}p_{s(2)}\cdots p_{s(W)}}{p_{u(1)}p_{u(2)}\cdots p_{u(W)}}. \tag{6.39}$$

We return to this ratio in Section 11.5.2.

The reason for using pseudocounts in the probability estimate (6.37) is the following. In step 1 above, there might be an excellent alignment of segment x of protein sequence 3 with the $N-1$ segments in the reduced array, except that the amino acid at position i in sequence x is not represented at position i in the reduced array. If pseudocounts were not used and the probability estimate q_{ij} replaced by $c_{ij}/(N-1)$, segment x could not be chosen at this step, since in this case the probability $c_{ij}/(N-1)$ would be 0 for this segment. We therefore assume that each b_j is greater than 0, and then introduction of pseudocounts allows the choice of segment x in the above case.

The choice of the pseudocounts b_j must be made subjectively. Lawrence et al. (1993) make the reasonable suggestion of choosing b_j proportional to the background frequency p_j of amino acid j. The choice of the proportionality constant is less obvious, and Lawrence et al. claim that in practice the constant $\sqrt{N}$ works well.

So far the length W of the segments considered has been assumed to be given. The choice of this length is discussed in detail by Lawrence et al. together with further tactical questions.

It might be asked why the choice of each new segment is random. An alternative procedure in the step described above would be to replace the current segment in sequence one by that segment maximizing the ratio in (6.38). The entire operation is then fully deterministic, and runs the risk of settling on a locally- rather than a globally-best alignment. The stochastic procedure allows movement from a locally-best alignment to a globally-best alignment. However, it is not guaranteed that the procedure will escape from the neighborhood of a locally-best alignment in reasonable time, and thus may be sensitive to the choice of the initial alignment.

Problems

6.1. Prove that

$$\sum_{k=0}^{\min\{m,n\}} \binom{m}{k}\binom{n}{k} = \binom{m+n}{n} = \binom{m+n}{m}.$$

Hint: Think of having to select n objects from a set with $s = m+n$ objects, of which m are of one kind and n are of another kind.

6.2. The BLOSUM62 substitution matrix given in Table 6.7 is often used to assign a score to each pair of aligned amino acids. Use this matrix and a linear gap penalty of $d = 5$ to find all highest-scoring alignments between $x = EATGHAG$ and $y = EEAWHEAE$.

6.3. Suppose there are three symbols A, B, and C, and two blocks of data as given below. You are to construct the BLOSUM68 substitution matrix derived from these data, however in the step involving taking the log-ratios, multiply the logarithms by 8 instead of by 2 (before rounding). *Hint:* Of the six scores, one should be zero, two negative, and three positive.

The blocks are

$$
\begin{aligned}
&A\,B\,A\,C\,C\,A\\
&A\,B\,B\,C\,C\,A\\
&A\,A\,B\,C\,A\,A\\
&A\,A\,A\,C\,A\,A
\end{aligned}
$$

and

$$
\begin{aligned}
&A\,B\,A\,C\\
&A\,B\,B\,C\\
&A\,C\,C\,C\\
&C\,A\,B\,A
\end{aligned}
$$

6.4. Fit $x = cttgac$ into $y = cagtatcgtac$ with the scoring scheme of the example in Section 6.4.1.

6.5. Find the best local alignment(s) of $x = aagtatcgca$ and $y = aagttagttgg$ with the same scoring scheme as in the example in Section 6.4.1.

6.6. The matrix below has been obtained using a dynamic programming algorithm (with a linear gap model) to solve one of the following three problems relative to the two sequences

$$x = X_1 X_2 X_3 X_4$$

and

$$y = Y_1 Y_2 Y_3 Y_4 Y_5 Y_6 Y_7 Y_8 Y_9 Y_{10} Y_{11},$$

(i) global alignment, (ii) fitting x into y, (iii) local alignment.

(1) By considering the matrix it is clear which of the three problems above was the one being tackled. Which was it?

(2) What is the optimal score for the alignment sought?

(3) List all the optimal alignments (in terms of the X_i's and Y_j's) relative to the problem being tackled.

	$-$	Y_1	Y_2	Y_3	Y_4	Y_5	Y_6	Y_7	Y_8	Y_9	Y_{10}	Y_{11}
$-$	0	0	0	0	0	0	0	0	0	0	0	0
X_1	-5	-5	-5	$5\leftarrow$ 0	5	5	$5\leftarrow$ 0	5	$5\leftarrow$ 0
X_2	-10	$0\leftarrow$ -5	0	$10\leftarrow$ $5\leftarrow$ 0	0	$10\leftarrow$ 5 $\leftarrow$ 0	0
X_3	-15	-5	-5	-5	5	$5\leftarrow$ $0\leftarrow$ -5	5	$5\leftarrow$ 0 $\leftarrow$ -5
X_4	-20	-10	-10	0	0	10	$10\leftarrow$ $5\leftarrow$ 0	10	10 $\leftarrow$ 5

6.7. Prove that the stationary distribution of the matrix M, defined by equations (6.20) and (6.21), is independent of the choice of the scaling constant c.

6.8. Equations (6.28) and (6.29) can be derived in various ways. One of these is by mathematical induction (see Section B.18). It is easy to see that the equations are true for $n = 1$. Suppose that they are true for some given value n^* of n. Show that they are then true for $n = n^* + 1$, and thus complete a proof by induction that they are true for all positive integers n. *Hint:* Use the facts that

$$m_{jj}^{(n+1)} = 0.99\, m_{jj}^{(n)} + 0.01\, m_{jk}^{(n)}, \quad j \neq k,$$

$$m_{jk}^{(n+1)} = (0.01/19)\, m_{jj}^{(n)} + 0.99\, m_{jk}^{(n)} + (.18/19) m_{jr}^{(n)}, \quad j \neq k \neq r,$$

and that the latter equation can be rewritten as

$$m_{jk}^{(n+1)} = \frac{1}{19}\left(0.01\, m_{jj}^{(n)} + 18.99\, m_{jk}^{(n)}\right).$$

6.9. Use the relation (B.24) to find the limit of the ratio (6.31) as $n \to \infty$.

6.10. Show that the n-step transition probabilities $m_{jk}^{(n)}$ given in equations (6.28) and (6.29) can be found by calculating the spectral expansion of the associated matrix M.

6.11. In the Gibbs sampling method of section 6.6, devise a meaningful way to incorporate the scores from a substitution matrix into the definition of Q_x (given on page 252).

7

Stochastic Processes (ii): Random Walks

7.1 Introduction

Random walks are special cases of Markov chains, and thus can be analyzed by Markov chain methods. However, the special features of random walks allow specific methods of analysis. Some of these methods are introduced in this chapter.

Our main interest in random walk theory is that it supplies the basic probability theory behind BLAST. BLAST is a procedure that searches for high-scoring local alignments between two sequences and then tests for significance of the scores found via P-values. The P-value calculation takes into consideration, as it must, the lengths of the two sequences, since the longer the sequences, the more likely there is to be local homology simply by chance. In practice, BLAST is used to search a database consisting of a number of sequences for similarity to a single "query" sequence. P-values are calculated allowing for the size of the entire database. However, we defer consideration of this matter until Section 10.5, and consider here only the random walk theory underlying the eventual P-value calculations.

The utility of BLAST is that it is able to complete its task very quickly, even when there are many sequences in the database. The efficiency arises for two reasons. First, on the algorithmic side it uses heuristics to avoid searching through all possible ungapped local alignments, of which there is an astronomically large number. Therefore, it can fail to produce exactly the highest-scoring alignments. Its performance, however, is extremely good. Second, the calculation of the P-value uses sophisticated approximations to achieve extremely fast calculations. The calculation of these P-values is discussed in Chapter 10.

Consider the simple case of the two aligned DNA sequences given in (1.1) and repeated here for convenience:

$$
\begin{array}{c}
\downarrow \quad \downarrow \quad\quad \downarrow \quad\quad \downarrow\downarrow\downarrow \quad \downarrow\downarrow \quad\quad\quad\quad\quad\quad\quad\quad \downarrow\downarrow\downarrow \\
g\ g\ a\ g\ a\ c\ t\ g\ t\ a\ g\ a\ c\ a\ g\ c\ t\ a\ a\ t\ g\ c\ t\ a\ t\ a \quad (7.1) \\
g\ a\ a\ c\ g\ c\ c\ c\ t\ a\ g\ c\ c\ a\ c\ g\ a\ g\ c\ c\ c\ t\ t\ a\ t\ c
\end{array}
$$

Suppose we give a score of $+1$ if the two nucleotides in corresponding positions are the same and a score of -1 if they are different. As we compare the two sequences, starting from the left, the accumulated score performs a random walk. In the above example, the walk can be depicted graphically as in Figure 7.1. The filled circles in this figure relate to *ladder points*, that is to points in the walk lower than any previously reached point.

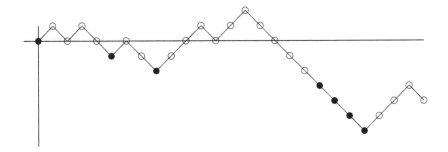

Figure 7.1.

The part of the walk from a ladder point until the highest point attained before the next ladder point is called an excursion. BLAST theory focuses on the maximum heights achieved by these excursions. In Figure 7.1 these maximum heights are, respectively, 1, 1, 4, 0, 0, 0, 3. (If the walk moves from one ladder point immediately to the next, the corresponding height is taken as 0.)

In practice, BLAST theory relates to cases that are much more complicated than this simple example. It is often applied to the comparison of two protein sequences and uses scores other than the simple scores $+1$ and -1 for matches and mismatches. These scores are described by the entries in a substitution matrix such as those given in the BLOSUM62 substitution matrix shown in Table 6.7. These scores determine the upward or downward movement of the random walk describing the score for that protein comparison. For example, if the score for any amino acid comparison is given by the appropriate entry in the BLOSUM62 substitution matrix, then for the alignment

$$
\begin{array}{c}
T\ Q\ L\ A\ A\ W\ C\ R\ M\ T\ C\ F\ E\ I\ E\ C\ K\ V \\
R\ H\ L\ D\ S\ W\ R\ R\ A\ S\ D\ D\ A\ R\ I\ E\ E\ G
\end{array} \quad (7.2)
$$

the scores are -1, 1, 5, -2, 1, 15, -4, 7, -1, 2, -4, etc..., and therefore the graph of the accumulated score goes through the points

$$(1, -1), (2, 0), (3, 5), (4, 3), (5, 4), (6, 19), (7, 15), (8, 22), \text{etc.} \qquad (7.3)$$

The graph of the accumulated score for this walk is depicted in Figure 7.2.

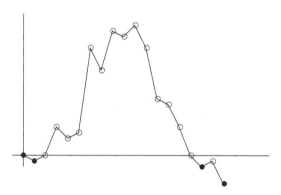

Figure 7.2.

To discuss BLAST it is necessary to consider arbitrary scoring schemes and thus aspects of the general theory of random walks. We do that in this chapter. However, we start by analyzing the simple random walk where the only possible step sizes are $+1$ and -1.

For the remainder of this chapter we consider random walks in the abstract, without reference to sequence comparisons, which we return to in Chapter 10.

7.2 The Simple Random Walk

7.2.1 Introduction

We first consider a process that starts at some arbitrary point h and moves, independently of the previous history of the process, a step down every unit time with probability q or a step up with probability p. (We use the notation "h" throughout to denote the initial position of the walk.) This process will be called a simple random walk. The walk is assumed to be restricted to the interval $[a, b]$, where a and b are integers with $a < h < b$, and stops when it reaches either a or b. We ask, "What is the probability that eventually the walk finishes at b rather than a?" and "What is the mean number of steps taken until the walk stops?"

7.3   The Difference Equation Approach

We begin by discussing the derivation of properties of the simple random walk by using the classical difference equation method, described in detail in Feller (1968). We shall find that the difference equation method does not allow ready generalizations to more complicated walks, and in the next section we shall develop the moment-generating function approach, which does allow these generalizations.

7.3.1   Absorption Probabilities

Let w_h be the probability that the simple random walk eventually finishes at, or is "absorbed" at, the point b rather than at the point a, given that the initial point is h. Then by comparing the situation just before and just after the first step of the walk, we get

$$w_h = pw_{h+1} + qw_{h-1}. \tag{7.4}$$

Furthermore,

$$w_a = 0, \quad w_b = 1. \tag{7.5}$$

Equation (7.4) is a homogeneous difference equation, with boundary conditions (7.5). A solution of (7.4) is a set of values w_h that satisfy it for all integers h. Difference equations are discussed in detail in Feller (1968), and we draw on his results here without extensive discussion because these results will be re-derived in Section 7.4. The difference equation methods are given here largely for comparison with the more powerful mgf methods.

One solution to equation (7.4) is of the form

$$w_h = e^{\theta h}$$

for some fixed constant θ. To solve for θ we substitute into (7.4) to get

$$e^{\theta h} = pe^{\theta(h+1)} + qe^{\theta(h-1)}.$$

Multiplying throughout by $e^{\theta - \theta h}$, this becomes

$$pe^{2\theta} - e^\theta + q = 0,$$

a quadratic equation in e^θ. For the case $p \neq q$, there are two distinct solutions of this equation, namely

$$e^\theta = 1, \quad e^\theta = \frac{q}{p}, \tag{7.6}$$

or equivalently

$$\theta = 0, \quad \theta = \log\left(\frac{q}{p}\right).$$

Thus when $p \neq q$, we have found two specific solutions of (7.4), namely

$$w_h = 1 \quad \text{and} \quad w_h = e^{\theta^* h},$$

where

$$\theta^* = \log\left(\frac{q}{p}\right). \tag{7.7}$$

The theory of homogeneous difference equations states that the general solution of (7.4) is any linear combination of these two solutions. That is, the general solution of (7.4) is

$$w_h = C_1 + C_2 e^{\theta^* h}, \tag{7.8}$$

for any arbitrary constants C_1 and C_2. We determine C_1 and C_2 by the boundary conditions (7.5), and this leads to

$$w_h = \frac{e^{\theta^* h} - e^{\theta^* a}}{e^{\theta^* b} - e^{\theta^* a}}. \tag{7.9}$$

This is the solution for the case $p \neq q$ only. When $p = q$ ($= \frac{1}{2}$), the two solutions in (7.6) are the same, and the above argument does not hold. In the applications of random walk theory to BLAST we never encounter the symmetric case $p = q$ or its various generalizations (which we call "zero mean" cases). Since consideration of this case usually involves extra work, we do not consider it further.

In some applications interest focuses on the probability u_h that the walk finishes at a rather than at b. The probabilities u_h satisfy the same difference equation (7.4) as w_h, but with the boundary conditions $u_a = 1$, $u_b = 0$. This leads to

$$u_h = \frac{e^{\theta^* b} - e^{\theta^* h}}{e^{\theta^* b} - e^{\theta^* a}}. \tag{7.10}$$

As we expect, $w_h + u_h = 1$.

7.3.2 Mean Number of Steps Taken Until the Walk Stops

We continue to assume that the walk starts at h and ends when one of a or b is reached, $a < h < b$. We define N as the (random) number of steps taken until the walk stops. The probability distribution of N depends on h, a, and b, and in this section we find the mean value m_h of this probability distribution, that is, the mean of N. We do not use the formula (1.24) to find this mean value; in fact, we never compute the complete probability distribution of N. Instead, we use difference equation methods similar to those in the previous section, which lead directly to the value of the mean. These calculations, too, will be repeated in the next section using moment-generating functions.

Considering the situation after one step of the walk in two different ways, we get

$$m_h - 1 = pm_{h+1} + qm_{h-1}. \tag{7.11}$$

This is an inhomogeneous difference equation. Difference equation theory shows that the general solution of (7.11) consists of any particular solution added to arbitrary multiples of solutions of the homogeneous equation

$$m_h = pm_{h+1} + qm_{h-1}. \tag{7.12}$$

A particular solution of (7.11) is $m_h = h/(q-p)$. The homogeneous equation (7.12) is the same as (7.4), and hence has the same general solution (7.8). Therefore, the general solution of (7.11) is

$$m_h = \frac{h}{q-p} + C_1 + C_2 e^{\theta^* h}. \tag{7.13}$$

The constants C_1 and C_2 are determined by boundary conditions. The particular boundary conditions $m_a = m_b = 0$ lead to

$$m_h = \frac{h-a}{q-p} - \frac{b-a}{q-p} \cdot \frac{e^{\theta^* h} - e^{\theta^* a}}{e^{\theta^* b} - e^{\theta^* a}}. \tag{7.14}$$

Using equations (7.9) and (7.10) we can rewrite this as

$$m_h = \frac{w_h(b-h) + u_h(a-h)}{p-q}. \tag{7.15}$$

We shall see later that equation (7.15) has a simple intuitive interpretation.

7.4   The Moment-Generating Function Approach

7.4.1   Introduction and Wald's identity

In this section the absorption probabilities w_h and u_h and the value of m_h are re-derived using mgf techniques. Various asymptotic results are then derived. Finally, the generalization to an arbitrary random walk will be discussed.

We start with classical result of probability theory, specialized to the case of the random walk discussed above. Suppose that the walk starts at h and continues until reaching a or b. We denote the value of the ith step by S_i ($S_i = \pm 1$). The S_i are iid random variables; let S be a random variable with their common distribution. The mgf $\mathrm{m}(\theta)$ of S is given by (7.17).

As above, we denote the (random) number of steps taken until the walk finishes by N, and let $T_N = \sum_{j=1}^{N} S_j$. T_N is either $a - h$, with probability $1 - w_h$, or $b - h$, with probability w_h. With these definitions we have the following:

Theorem 7.1 (Wald's identity).

$$E\left(\mathrm{m}(\theta)^{-N}e^{\theta T_N}\right) = 1 \tag{7.16}$$

for all θ for which the mgf is defined. (The expected value is taken with respect to the joint distribution of N and T_N.)

We do not prove this identity here, and limit the discussion to conclusions that follow from it. A derivation may be found in Karlin and Taylor (1975), page 264.

7.4.2 Absorption Probabilities

The mgf of the (random) value, either -1 or $+1$, of any single step taken in the random walk is

$$\mathrm{m}(\theta) = qe^{-\theta} + pe^{\theta}. \tag{7.17}$$

Theorem 1.1 (page 40) shows that for any random walk with non-zero mean step size and for which both positive and negative steps are possible, there exists a unique non-zero value θ^* of θ for which

$$\mathrm{m}(\theta^*) = 1. \tag{7.18}$$

The value θ^* is the same as given in equation (7.7). Using this value for θ in Wald's identity (7.16) we get

$$E\left(e^{\theta^* T_N}\right) = 1. \tag{7.19}$$

Since the only two possible values of T_N are $a - h$ (with probability $1 - w_h$) and $b - h$ (with probability w_h), equations (1.27) and (7.19) give

$$(1 - w_h)e^{\theta^*(a-h)} + w_h e^{\theta^*(b-h)} = 1. \tag{7.20}$$

From this,

$$w_h = \frac{e^{h\theta^*} - e^{a\theta^*}}{e^{b\theta^*} - e^{a\theta^*}}. \tag{7.21}$$

This is identical to the solution (7.9), found by using difference equations. The value of u_h given in (7.10) can be found similarly.

Thus use of the mgf gives us a second approach to finding absorption probabilities of the random walk. For the simple random walk considered in this section the mgf approach does not differ markedly from the approach using difference equations. However, the mgf is more readily adapted than the difference equation method for the calculations needed in the general theory for random walks (and eventually used in BLAST) discussed in Section 7.5.

7.4.3 Mean Number of Steps Until the Walk Stops

We now find m_h, the mean number of steps until the walk finishes at a or b by using Wald's equation and mgf methods. The derivative (with respect to θ) of the right-hand side of (7.16) is 0. Since there are infinitely many possible values of N, the left-hand side is an infinite sum of functions of θ. A basic theorem of calculus gives a criterion indicating when the derivative of a sum of functions can be calculated as the sum of the derivatives of the individual functions. This criterion holds in this case, so that

$$\frac{d}{d\theta} E\left(\mathrm{m}(\theta)^{-N} e^{\theta T_N}\right) = E\left(\frac{d}{d\theta}\left(\mathrm{m}(\theta)^{-N} e^{\theta T_N}\right)\right)$$

$$= E\left(-N\mathrm{m}(\theta)^{-N-1}\frac{d}{d\theta}\mathrm{m}(\theta) e^{\theta T_N} + \mathrm{m}(\theta)^{-N} T_N e^{\theta T_N}\right).$$

It follows that

$$E\left(-N\mathrm{m}(\theta)^{-N-1}\frac{d}{d\theta}\mathrm{m}(\theta) e^{\theta T_N} + \mathrm{m}(\theta)^{-N} T_N e^{\theta T_N}\right) = 0.$$

Inserting the specific value $\theta = 0$ into this equation, we get

$$E(-NE(S) + T_N) = -E(N)E(S) + E(T_N) = 0. \tag{7.22}$$

Thus

$$E(T_N) = E(S)m_h, \tag{7.23}$$

where $E(S)$ is the mean of the step size. The content of equation (7.23) is intuitively acceptable, namely that the mean value of the final total displacement of the walk, when it stops, is the mean size of each step multiplied by the mean number of steps m_h taken until the walk finishes. Equation (7.23) is a generalization of equation (2.81).

Since the eventual total displacement T_N in the walk is either $b - h$ (with probability w_h given in (7.9)) or $a - h$ (with probability u_h given in (7.10)), the mean value $E(T_N)$ of T_N is given by

$$E(T_N) = u_h(a - h) + w_h(b - h). \tag{7.24}$$

Since $E(S) = p - q$, equation (7.23) shows that

$$m_h = \frac{u_h(a - h) + w_h(b - h)}{p - q}. \tag{7.25}$$

This is the same as the expression (7.15), found by using difference equations. The present derivation is more flexible than the difference equation approach, and further, it demonstrates where the various terms in equation (7.25) come from.

7.4.4 An Asymptotic Case

The theory of BLAST concerns those walks that start at $h = 0$, where there is a lower boundary at $a = -1$ but no upper boundary, and where $E(S)$, the mean step size, is negative. Such a walk is destined eventually to reach -1. BLAST theory requires the calculation of two quantities:

(i) the probability distribution of the maximum value that the walk ever achieves before reaching -1, and

(ii) the mean number of steps until the walk eventually reaches -1.

BLAST theory generally uses the notation A to denote (ii), so for this particular walk we use this notation rather than m_0.

To discuss the maximum value that the walk ever achieves before eventually reaching -1 we install an artificial stopping boundary at the value y, where $y \geq 1$. Then if $h = 0$ and $a = -1$, equation (7.9) shows that the probability that the unrestricted walk finishes at the artificial boundary y rather than at the value a is

$$\frac{1 - e^{-\theta^*}}{e^{\theta^* y} - e^{-\theta^*}}, \tag{7.26}$$

where θ^* is defined in equation (7.7). Since θ^* is positive, the term $e^{\theta^* y}$ dominates the denominator in (7.26) when y is large, so that the probability (7.26) is asymptotic to

$$(1 - e^{-\theta^*})e^{-\theta^* y}. \tag{7.27}$$

Thus if Y is the maximum height achieved by the walk,

$$\mathrm{Prob}(Y \geq y) \sim C e^{-\theta^* y} \tag{7.28}$$

as $y \to \infty$, where

$$C = 1 - e^{-\theta^*}. \tag{7.29}$$

This asymptotic relation is in the form of the geometric-like probability displayed in equation (1.74), with the identification $\lambda = \theta^*$.

Turning to (ii), the expression for the mean number A of steps before that walk finishes at -1 or y is found by first making the substitutions $h = 0$, $a = -1$ in equation (7.25), to obtain

$$A = \frac{u_0 - y w_0}{q - p}. \tag{7.30}$$

Equation (7.21) shows that $y w_0 \to 0$ as $y \to \infty$. Since $w_0 \to 0$, $u_0 \to 1$. Taking the limit $y \to \infty$ in equation (7.30) shows that

$$A = \frac{1}{q - p}. \tag{7.31}$$

7.5 General Walks

Suppose generally that the possible step sizes in a random walk are

$$-c, -c+1, \ldots, 0, \ldots, d-1, d, \qquad (7.32)$$

and that these steps have respective probabilities

$$p_{-c}, p_{-c+1}, \ldots, p_d, \qquad (7.33)$$

some of which might be zero. We assume three conditions throughout:

(i) Both $p_{-c} > 0$ and $p_d > 0$.

(ii) This step size is of the "negative mean" type, so that

$$E(S) = \sum_{j=-c}^{d} jp_j < 0.$$

(iii) The greatest common divisor of the step sizes that have non-zero probability is 1.

The mgf of S is

$$m(\theta) = \sum_{j=-c}^{d} p_j e^{j\theta},$$

and Theorem 1.1 (page 40) states that there exists a unique positive value θ^* of θ for which

$$\sum_{j=-c}^{d} p_j e^{j\theta^*} = 1. \qquad (7.34)$$

Our aim is to obtain asymptotic results generalizing those in Section 7.4.4 applying for walks that start at 0 and have a stopping boundary at -1 and no upper boundary. To do this we impose an artificial barrier at y, where $y > 0$. The walk finishes either when it reaches the point -1 or a point less than -1, or when it reaches the point y or a point exceeding y. The possible points where the walk can finish are now

$$-c, -c+1, \ldots, -1, y, \ldots, y+d-1.$$

Let P_k be the probability that the walk finishes at the point k. From equation (7.16) (which holds also for the case of general step sizes),

$$E\left(e^{\theta^* T_N}\right) = 1,$$

where T_N is the total displacement from 0 when the walk stops. Thus

$$\sum_{k=-c}^{-1} P_k e^{k\theta^*} + \sum_{k=y}^{y+d-1} P_k e^{k\theta^*} = 1. \qquad (7.35)$$

We consider some implications of this equation in the next two examples.

Example 1. In the simple random walk, equation (7.35) becomes

$$P_{-1}e^{-\theta^*} + P_y e^{y\theta^*} = 1.$$

For this walk $P_{-1} = 1 - P_y$, and by inserting this into the above equation the value of P_y in (7.26) is obtained. Note that in this case, $P_{-1} = u_0$ and $P_y = w_0$.

Example 2. In this example we consider a walk that at each move takes *two* steps down (with probability q) or one step up (with probability p). We assume that the mean step size $p - 2q$ is negative. The mgf of the step size is $qe^{-2\theta} + pe^{\theta}$, and the unique positive value θ^* of θ for which this takes the value 1 is

$$\theta^* = \log\left(\frac{q + \sqrt{4pq + q^2}}{2p}\right). \tag{7.36}$$

With θ^* defined in equation (7.36), equation (7.35) yields

$$P_{-2}e^{-2\theta^*} + P_{-1}e^{-\theta^*} + P_y e^{y\theta^*} = 1. \tag{7.37}$$

Let $R_{-1} = \lim_{y\to+\infty} P_{-1}$ and $R_{-2} = \lim_{y\to+\infty} P_{-2}$. Then

$$P_y \sim (1 - +R_{-1}e^{-\theta^*} + R_{-2}e^{-2\theta^*})e^{-y\theta^*}. \tag{7.38}$$

This implies that if there is no upper barrier to the walk and Y is the maximum height achieved by the walk,

$$P_y = \text{Prob}(Y \geq y) \sim Ce^{-y\theta^*}, \tag{7.39}$$

where

$$C = 1 + R_{-1}e^{-\theta^*} + R_{-2}e^{-2\theta^*}. \tag{7.40}$$

Thus Y has a geometric-like distribution. We show in the next section that this conclusion is true for the general walk described above, and we shall also find expressions for the values of C and θ^* applying for any walk.

Returning to the general case, we next find a general expression for A, the mean number of steps that the walk takes before stopping at one of the points $-c, -c+1, \ldots, -1$ when no stopping boundary at y is imposed. Wald's identity (7.16) and the calculation (7.23) that follows from it make no assumption about the possible step sizes, so that (7.23) may be used in the case of a general random walk. The mean net displacement when the walk stops at one of these points is $\sum_{j=1}^{c} -jR_{-j}$, where R_{-j} is the probability that the walk finishes at $-j$. The mean step size $E(S)$ is $\sum_{j=-c}^{d} jp_j$, assumed to be negative. Thus from equation (7.23),

$$A = \frac{\sum_{j=1}^{c} j R_{-j}}{-\sum_{j=-c}^{d} j p_j}.$$
(7.41)

For BLAST calculations it is necessary to calculate the values of θ^*, C, and A for any random walk. The calculation of θ^* from equation (7.34) for any random walk is straightforward, using numerical methods if necessary. The calculation of C is far less straightforward and is considered in the following section. The calculation of A from (7.41) depends on the calculation of the R_{-j} values, a matter discussed in Section 10.2.3.

7.6 General Walks: Asymptotic Theory

7.6.1 Introduction

In this section we further develop the theory for the general random walk, focusing on the asymptotic $y \to +\infty$ case discussed in the previous section. The analysis is based on that of Karlin and Dembo (1992). Our aim is to show, for any general walk of the form described in (7.32) and (7.33) and which satisfies the three conditions following (7.33), starting at $h = 0$ and with lower boundary $a = -1$, that

$$\text{Prob}(Y \geq y) \sim C e^{-\theta^* y},$$
(7.42)

where θ^* is found from equation (7.34) and C is a constant specific to the walk in question. This implies that the maximum height achieved by the walk has a geometric-like distribution whose parameter λ is found from the mgf equation (7.34), replacing θ^* in this equation by λ. We will also find an explicit formula for the constant C in this distribution.

The value of this result lies in its generality. So long as the conditions specified above hold the geometric-like distribution (7.42) applies, whatever the specifics of the walk might be. This allows the application of the conclusion of the result to BLAST analyses with a very wide range of substitution matrices.

7.6.2 The Renewal Theorem

We start with a technical result.

Theorem 7.2 (The Renewal Theorem).
Suppose three sequences $(b_0, b_1, \dots)$, $(f_0, f_1, \dots)$, and $(u_0, u_1, \dots)$ of nonnegative constants satisfy the equation

$$u_y = b_y + (u_y f_0 + u_{y-1} f_1 + u_{y-2} f_2 + \cdots + u_1 f_{y-1} + u_0 f_y),$$
(7.43)

for all y. Suppose further that $B = \sum_i b_i < +\infty$, $\sum_i f_i = 1$, $\mu = \sum_i i f_i < +\infty$, and that the greatest common divisor of $f_0, f_1, f_2, \ldots$ is 1. Then

$$u_y \to B\mu^{-1} \text{ as } y \to +\infty. \tag{7.44}$$

The proof of this theorem is beyond the scope of this book. The interested reader is referred to Karlin and Taylor (1981).

7.6.3 Unrestricted Walks

We assume that the walk starts at 0. Suppose as in Section 7.5 that the possible step sizes in a random walk are

$$-c, -c+1, \ldots, 0, \ldots, d-1, d,$$

that these steps have respective probabilities $p_{-c}, p_{-c+1}, \ldots, p_d$, and that the three assumptions given on page 266 all hold.

In this section we consider the case where no boundaries are placed on the walk, so that the concept of stopping at a boundary point no longer applies. Because the mean step size is negative, the walk eventually drifts down to $-\infty$. Before doing so, however, the walk might visit various positive values. The first aim is to find an equation satisfied by the probabilities Q_k, where Q_k is the probability that the walk visits the positive value k before reaching any other positive value. The largest positive step size is d, so that $Q_k = 0$ for all $k > d$. It will be convenient to also define $Q_0 = 0$. Since it is possible that the walk never visits any positive value, we have $\sum_{k=1}^{d} Q_k < 1$.

Although the focus in this section is on unrestricted walks, for the moment we impose an artificial boundary at $+1$ and another at $-L$, where L is large and positive. Then equation (7.35) becomes

$$\sum_{k=-L-c+1}^{-L} Q_k(L) e^{k\theta^*} + \sum_{k=1}^{d} Q_k(L) e^{k\theta^*} = 1, \tag{7.45}$$

where $Q_k(L)$ is the probability that the walk stops at the value k and the notation recognizes the dependence of the probabilities $Q_k(L)$ on L. Since $+1$ is a boundary, if the walk stops at $k > 0$, then this k is the first positive value reached. Further,

$$\lim_{L \to \infty} Q_k(L) = Q_k, \ (1 \le k \le d).$$

Since θ^* is positive and $Q_k(L) \le 1$ for all k and L, the terms in the first sum in (7.45) all approach zero as $L \to \infty$. Thus letting $L \to \infty$ in (7.45) we have

$$\sum_{k=1}^{d} Q_k e^{k\theta^*} = 1. \tag{7.46}$$

The next aim is to find an expression for $F_{Y_{\text{unr}}}(y)$, the probability that in the unrestricted walk the maximum upward excursion is y or less, for any positive value of y. This is done as follows.

The event that in the unrestricted walk the maximum upward excursion is y or less is the union of several non-overlapping events. The first of these is the event that the maximum excursion never reaches positive values (probability $\bar{Q} = 1 - Q_1 - Q_2 - Q_3 - \cdots - Q_d$). The remaining events are that the first positive value achieved by the excursion is k, $k = 1, 2, \ldots, y$ (with probability Q_k), and then, starting from this first positive value, that the walk never achieves a further height exceeding $y - k$ (probability $F_{Y_{\text{unr}}}(y - k)$). These various possible events imply that

$$F_{Y_{\text{unr}}}(y) = \bar{Q} + \sum_{k=0}^{y} Q_k F_{Y_{\text{unr}}}(y - k) \tag{7.47}$$

(recall that $Q_0 = 0$).

This is in the form of the renewal equation (7.43), with u_k replaced by $F_{Y_{\text{unr}}}(y - k)$, b_y replaced by $\bar{Q}$, and f_k replaced by Q_k. However, we cannot immediately use the result of Theorem 7.2 to develop properties of $F_{Y_{\text{unr}}}(y)$. This is for two reasons. First, Theorem 7.2 requires that $\sum_k f_k = 1$, whereas here $\sum Q_k < 1$. Second, Theorem 7.2 requires $\sum b_y < \infty$, and this requirement does not hold when $b_y = \bar{Q}$. However, we can use Theorem 7.2 by introducing the quantity $V(y)$, defined by

$$V(y) = \left(1 - F_{Y_{\text{unr}}}(y)\right)e^{y\theta^*}, \tag{7.48}$$

where θ^* is defined by equation (7.34). From (7.48) it follows that

$$F_{Y_{\text{unr}}}(y) = 1 - V(y)e^{-y\theta^*},$$

and equation (7.47) can then be written as

$$1 - V(y)e^{-y\theta^*} = \bar{Q} + \sum_{k=0}^{y} Q_k \left(1 - V(y - k)e^{-(y-k)\theta^*}\right).$$

Elementary reorganization of this equation leads to

$$V(y) = e^{y\theta^*} \left(Q_{y+1} + Q_{y+2} + \cdots + Q_d\right) + \sum_{k=0}^{y} (Q_k e^{k\theta^*}) V(y - k) \tag{7.49}$$

when $y < d$, and

$$V(y) = \sum_{k=0}^{d} (Q_k e^{k\theta^*}) V(y - k) \tag{7.50}$$

when $y \geq d$. This is again in the form of the renewal equation, this time with $f_k = Q_k e^{k\theta^*}$, and with b_y replaced by $e^{y\theta^*}(Q_{y+1} + Q_{y+2} + \cdots + Q_d)$ if $y < d$ and $b_y = 0$ if $y \geq d$. To use Theorem 7.2 we must show that

$\sum b_y < \infty$ and that

$$\sum_{k=0}^{\infty} Q_k e^{k\theta^*} = 1.$$

The first requirement follows since $b_y = 0$ for $y \geq d$, and the second follows from equation (7.46), recalling that Q_k is zero when $k = 0$ or when $k > d$. We must find

$$B = \sum b_k = \sum_{k=0}^{d} e^{k\theta^*} \left(Q_{k+1} + Q_{k+2} + \cdots + Q_d\right).$$

If the right-hand side of this equation is multiplied by $e^{\theta^*} - 1$, the resulting expression is $\sum_{k=1}^{d} Q_k e^{k\theta^*} - (Q_1 + Q_2 + \cdots + Q_d)$, which from equation (7.34) is $1 - (Q_1 + Q_2 + \cdots + Q_d) = \bar{Q}$. Thus

$$B = \frac{\bar{Q}}{e^{\theta^*} - 1}.$$

The statement of Theorem 7.2 then implies that if

$$V = \lim_{y \to +\infty} V(y), \tag{7.51}$$

then

$$V = \frac{\bar{Q}}{(e^{\theta^*} - 1)(\sum_{k=1}^{d} k Q_k e^{k\theta^*})}. \tag{7.52}$$

7.6.4 Restricted Walks

We now consider what the unrestricted walk calculations of the previous section imply for a restricted random walk having a stopping boundary at the value -1. We again assume that the walk starts at 0. The calculations are best carried out using the complementary probabilities $F^*_{Y_{\mathrm{unr}}}(y) = 1 - F_{Y_{\mathrm{unr}}}(y)$ and $F^*_Y(y) = 1 - F_Y(y)$, the respective probabilities that the size of an excursion in the unrestricted and the restricted walks exceeds the value y. With these definitions, equation (7.48) implies that

$$F^*_{Y_{\mathrm{unr}}}(y) = V(y) e^{-y\theta^*}. \tag{7.53}$$

The definition of V in (7.51) then implies that, for large y,

$$F_{Y_{\mathrm{unr}}}(y) \sim V e^{-y\theta^*}. \tag{7.54}$$

It follows that

$$\lim_{y \to +\infty} \left(F^*_{Y_{\mathrm{unr}}}(y)\right) e^{y\theta^*} = V. \tag{7.55}$$

The size of an excursion of the unrestricted walk can exceed the value y either before or after reaching negative values. In the latter case, the first negative value reached by the walk is one of $-1, -2, \ldots, -c$. If the probability that it is $-j$ is R_{-j}, then

$$F_{Y_{\text{unr}}}^*(y) = F_Y^*(y) + \sum_{j=1}^{c} R_{-j} F_{Y_{\text{unr}}}^*(y+j). \tag{7.56}$$

Multiplying by $e^{y\theta^*}$ throughout in this equation and then letting $y \to +\infty$, we get from (7.55)

$$V = \lim_{y \to +\infty} \left(F_Y^*(y) \right) e^{y\theta^*} + V \sum_{j=1}^{c} R_{-j} e^{-j\theta^*}. \tag{7.57}$$

From this,

$$\lim_{y \to +\infty} F_Y^*(y) e^{y\theta^*} = V \left(1 - \sum_{j=1}^{c} R_{-j} e^{-j\theta^*} \right). \tag{7.58}$$

The calculation (7.52) for V then shows that

$$\lim_{y \to +\infty} F_Y^*(y) e^{y\theta^*} = \frac{\bar{Q}(1 - \sum_{j=1}^{c} R_{-j} e^{-j\theta^*})}{(e^{\theta^*} - 1)(\sum_{k=1}^{d} kQ_k e^{k\theta^*})}. \tag{7.59}$$

This implies

$$\lim_{y \to +\infty} F_Y^*(y) e^{y\theta^*} = e^{-\theta^*} C, \tag{7.60}$$

where

$$C = \frac{\bar{Q} \left(1 - \sum_{j=1}^{c} R_{-j} e^{-j\theta^*} \right)}{(1 - e^{-\theta^*})(\sum_{k=1}^{d} kQ_k e^{k\theta^*})}. \tag{7.61}$$

Recalling that $F_Y^*(y) = \text{Prob}(Y \geq y + 1)$, this equation implies that

$$\text{Prob}(Y \geq y) \sim C e^{-y\theta^*}, \tag{7.62}$$

as was to be shown.   □

This remarkable result implies that, subject to the conditions assumed, whatever the step sizes and probabilities are for the random walk, the probability distribution of Y is asymptotically the geometric-like distribution. All that is necessary to specify the distribution completely is to find the appropriate value of C, since the value of θ^* can be found from equation (7.34)).

In BLAST theory the parameter θ^*, defined in (7.34) and appearing in (7.62), is denoted by λ, so with this notation replace (7.62) by

$$\text{Prob}(Y \geq y) \sim C e^{-y\lambda}. \tag{7.63}$$

Three special cases of the calculation of C as given in (7.61) deserve attention. First, if the only possible upward step in the walk is $+1$, then $Q_k = 0$ when $k \geq 2$. Equation (7.46) then shows that $Q_1 = e^{-\theta^*}$, so that $\sum_{k=1}^{d} kQ_k e^{k\theta^*} = 1$, and $\bar{Q} = 1 - e^{-\theta^*}$. In this case (7.61) shows that

$$C = 1 - \sum_{j=1}^{c} R_{-j} e^{-j\theta^*}. \qquad (7.64)$$

This is a generalization of equation (7.40). Second, if the only possible downward step in the walk is -1, then $R_{-1} = 1$ and $R_{-j} = 0$ when $j \geq 2$. In this case (7.61) shows that

$$C = \frac{\bar{Q}}{\sum_{k=1}^{d} kQ_k e^{k\theta^*}}. \qquad (7.65)$$

Finally, exactly matching sub-sequences were considered in Section 5.4, and for these the geometric, rather than the geometric-like, distribution applies. Comparison of equations (1.70) and (1.74) shows that for this distribution, $C = 1$. Any exactly matching sub-sequence terminates at the first mismatch. In random walk terms, the possible steps sizes can be thought of as $+1$ and $-\infty$. (We could make a more precise analysis by using a step of $-L$, for L large, and then letting $L \to \infty$, but we prefer a more casual approach.) Thus we can put $R_{-\infty} = 1$ and $R_{-j} = 0$ for all finite j. Inserting these values into equation (7.64) we get $C = 1$, as required.

Karlin and Dembo (1992) give several alternative expressions for C. Of the various expressions they give, perhaps the most useful is, in our notation (which differs from theirs),

$$C = \frac{(1 - \sum_j R_{-j} e^{-j\theta^*})^2}{(1 - e^{-\theta^*})AE(Se^{\theta^* S})}, \qquad (7.66)$$

where A is given in equation (7.41) and S is the (random) size of any step in the walk. For the simple random walk of Section 7.2, $E(Se^{\theta^* S}) = q - p$, $A = 1/(q - p)$, and $R_{-1} = 1$. Inserting these values in (7.66) we obtain the value $1 - e^{-\theta^*}$ for C, which is identical to expression in (7.29).

While the formulae for C and A given above are in closed form, they still might not be useful for rapid calculation. Karlin and Dembo (1992) also give some rapidly converging series approximations for both these key parameters. Further simplifying calculations are discussed in Chapter 10.

Problems

7.1. Derive the result (7.36) by solving a cubic equation in e^{θ^*}. (Hint: It is known that one solution of the equation is $e^{\theta^*} = 1$, so that further solutions may be found by solving a quadratic equation.)

7.2. Consider the simple random walk with $h = 0$, $b = 1$ and $a = -L$, where L is positive. Use equation (7.9) to write down the probability that the walk eventually reaches 1 rather than $-L$. For the case $p < q$, show that the limiting value of this probability as $L \to \infty$ is $\frac{p}{q}$.

7.3. *Continuation.* The limiting ($L \to \infty$) probability found in Problem 7.2 is the probability in an unrestricted random walk which starts at 0 and has $p < q$, that the walk ever reaches $+1$. Show this implies that in the unrestricted case, the probability that the walk ever reaches the value y is $(\frac{p}{q})^y$, for any positive integer y.

7.4. *Continuation.* For the case of the simple random walk with $p < q$, calculate the value of V (defined in equation (7.52)). From this, calculate the right-hand side in the asymptotic expression (7.54). Compare the value found with the value given in Problem 7.3. (Recall that $F^*_{Y_{\text{unr}}}(y)$ is the probability that the unrestricted walk ever reaches a height $y + 1$ or more.)

7.5. Show that, for the simple random walk with $p < q$, the two expressions (7.64) and (7.65) are identical.

8
Statistics (ii): Classical Estimation Theory

8.1 Introduction

An introduction to classical estimation and hypothesis testing procedures
was given in Chapter 3. However, optimality aspects of the two procedures
were not addressed. In this chapter we give a brief introduction to classical
estimation theory, and in Chapter 9 we give an introduction to classical
hypothesis testing theory. In both chapters we focus on optimality aspects
of these theories. Extensive treatments of the two theories are given in
Lehmann (1991) and Lehmann (1986) respectively.

8.2 Criteria for "Good" Estimators

In this section we discuss reasonable criteria for an estimator of a parameter
to be "good". It is not always clear what makes one estimator preferable to
another, but there are certain properties of them that can be used to help
decide if one estimator is better than another. It was stated in Section 3.3.1
that one property of an estimator usually taken as desirable is that it be
unbiased, that is, that its mean value be the value of the parameter to be
estimated. For example, as shown in equation (2.74), if $X_1, X_2, \ldots, X_n$ are
iid, each having a probability distribution with mean μ, then the estimator
$\bar{X}$ is a random variable also having mean μ, so that $\bar{X}$ is an unbiased
estimator of μ.

It is not always possible to find unbiased estimators for a parameter, and even if it is possible, unbiased estimators are not always preferred to biased estimators. This is discussed further below.

A second desirable property is that of *consistency*. If $\hat{\xi}_n$ an estimator of a parameter ξ derived from n observations, then a consistent estimator is one for which, for any small ϵ, $\text{Prob}(|\hat{\xi}_n - \xi| > \epsilon) \to 0$ as $n \to +\infty$. Application of Chebyshev's inequality and the variance formula (2.74) shows that the sample average $\overline{(X)}$ of iid random variables is a consistent estimator of the mean of the probability distribution of these random variables (assuming, as we do throughout, that the variance of this distribution is finite). As the sample size increases, the estimate derived from a consistent estimator is more and more likely to be close to the parameter being estimated.

An estimator can be both unbiased and consistent, one and not the other, or neither (see Problem 8.1), so the concepts of unbiasedness and consistency concern different aspects of the properties of estimators.

A further desirable property of an unbiased estimator is that it have a low variance. One unbiased estimator of a parameter is said to be more efficient than another unbiased estimator of a parameter if it has a lower variance than the other estimator. Ideally we would like to have an unbiased estimator whose variance is lower than that of any other unbiased estimator of that parameter. In some cases such an estimator exists and can be identified, and if so, it might reasonably be thought to be the "best" estimator of that parameter.

For biased estimators a natural desirable property is that it have low mean square error (see (3.13)). In some cases the mean square error of a biased estimator is less than the mean square error (i.e., the variance) of an unbiased estimator; an example is given in Problem 8.6. In such cases the biased estimator might be preferred to the unbiased estimator.

Finally, it would also be desirable if an estimator had, at least approximately, a well-known distribution, for example a normal distribution. In this case known theory and results for this distribution can be applied when using this estimator.

Perhaps unexpectedly, there is one estimation procedure that in many practical cases achieves most, and in some cases all, of these aims, at least asymptotically as the sample size increases. This is the procedure known as *maximum likelihood estimation*.

8.3   Maximum Likelihood Estimation

8.3.1   The Discrete Case

Suppose that $Y_1, Y_2, \ldots, Y_n$ are iid discrete random variables[1] with joint probability distribution $P(Y_1; \xi) \times P(Y_2; \xi) \times \cdots \times P(Y_n; \xi)$. Here ξ is an unknown parameter that is to be estimated, and the notation indicates the dependence of the probability distribution of Y_i on ξ. We denote $(Y_1, Y_2, \ldots, Y_n)$ by $\boldsymbol{Y}$ and define the *likelihood* $L(\xi, \boldsymbol{Y})$ by

$$L(\xi, \boldsymbol{Y}) = P(Y_1; \xi) \times P(Y_2; \xi) \times \cdots \times P(Y_n; \xi). \tag{8.1}$$

While this is identical to the joint probability distribution of $\boldsymbol{Y}$, we now regard it, for a given $\boldsymbol{Y}$, as a function of ξ. The value $\hat{\xi} = \hat{\xi}(Y_1, Y_2, \ldots, Y_n)$ of ξ at which $L(\xi, \boldsymbol{Y})$ reaches a maximum (as a function of ξ) is called the maximum likelihood *estimator* of ξ. This value is a function of $Y_1, Y_2, \ldots, Y_n$, as our notation $\hat{\xi}(Y_1, Y_2, \ldots, Y_n)$ implies.

This estimator is often found by differentiating $L(\xi, \boldsymbol{Y})$ with respect to ξ and using standard derivative tests for maxima. Care must be used with this procedure, since local maxima may arise, and, as Example 5 below shows, the maximum might be reached at a boundary point.

In those cases where the maximum of $L(\xi, \boldsymbol{Y})$ is found through the differentiation procedure, it is necessary to solve the equation (in ξ)

$$\frac{d}{d\xi} L(\xi, \boldsymbol{Y}) = 0.$$

Since the logarithm is a monotonic increasing function, in practice it is equivalent to solve the equation

$$\frac{d}{d\xi} \log L(\xi, \boldsymbol{Y}) = 0. \tag{8.2}$$

Since the logarithm of $L(\xi, \boldsymbol{Y})$ is a sum rather than a product, the differentiation procedure is almost always easier when carried out in terms of the logarithm. The maximum likelihood *estimate* of ξ is the observed value of the maximum likelihood estimator, once the data have been obtained. That is, this estimate is found by replacing $\boldsymbol{Y} = (Y_1, Y_2, \ldots, Y_n)$ in the above by $\boldsymbol{y} = (y_1, y_2, \ldots, y_n)$ and $L(\xi, \boldsymbol{Y})$ by $L(\xi, \boldsymbol{y})$.

---

[1] Maximum likelihood estimators can be defined also for dependent random variables, but we assume the iid case throughout.

Example 1. Suppose that $Y_1, Y_2, \ldots, Y_n$ are iid random variables, each having a Poisson distribution with parameter λ. Then from (2.6),

$$
\begin{aligned}
L(\lambda, \boldsymbol{Y}) &= \frac{e^{-\lambda}\lambda^{Y_1}}{Y_1!} \times \frac{e^{-\lambda}\lambda^{Y_2}}{Y_2!} \times \cdots \times \frac{e^{-\lambda}\lambda^{Y_n}}{Y_n!} \\
&= \frac{e^{-n\lambda}\lambda^{\sum Y_i}}{\prod(Y_i!)},
\end{aligned}
$$

all sums and products in this example being over $i = 1, 2, \ldots, n$. Thus

$$
\log L(\lambda, \boldsymbol{Y}) = -n\lambda + \left(\sum Y_i\right)\log \lambda - \log\left(\prod(Y_i!)\right)
$$

and

$$
\frac{d}{d\lambda}\log L(\lambda, \boldsymbol{Y}) = -n + \frac{\sum Y_i}{\lambda}. \tag{8.3}
$$

This derivative is zero only when $\hat{\lambda} = \bar{Y}$, and since the derivative of the right-hand side expression in (8.3) with respect to λ is negative, this value corresponds to a maximum of $L(\lambda, \boldsymbol{Y})$. Thus

$$
\text{maximum likelihood estimator of } \lambda = \bar{Y}. \tag{8.4}
$$

From this, the maximum likelihood estimate of λ is $\bar{y}$, the observed average once the data are obtained. The mean and variance of the Poisson distribution as given in Table 1.1 together with the formulae in (2.74) show that $\bar{Y}$ is an unbiased consistent estimator of λ and has variance λ/n.

These properties of the Poisson distribution can be used in the estimation of the parameters of the Pólya–Aeppli distribution given in (5.51) and (5.52). The Pólya–Aeppli distribution is used to approximate the distribution of the number $Y_1(N)$ of times that a given word occurs in a DNA sequence of length N. It was noted in Section 5.7.1 that self-overlapping words tend to occur in clumps, with the number of clumps approximated as having a Poisson distribution with parameter λ. Estimation of λ using (8.4) is a key step in estimating various properties of the clumping behavior.

Example 2. Consider n independent Bernoulli trials, each of which results in either success (with probability p) or failure (with probability $1 - p$). If $Y_i = 1$ if trial i results in success and $Y_i = 0$ if trial i results in failure, the likelihood (8.1) is, from (1.6),

$$
L(p, \boldsymbol{Y}) = \prod_{i=1}^{n} p^{Y_i}(1-p)^{1-Y_i} = p^{Y}(1-p)^{n-Y}, \tag{8.5}
$$

where $Y = \sum_{i=1}^{n} Y_i$ is the total number of successes. Finding the maximum likelihood estimator even in this comparatively simple case is not perhaps as straightforward as might be expected. When $1 \leq Y \leq n - 1$ the solution of equation (8.2), with p replacing ξ, is $\hat{p} = Y/n = \bar{Y}$, and this may be shown to be the maximum likelihood estimator of p for these cases. However, the cases $Y = 0$, $Y = n$ must be handled differently, and we illustrate this in the case $Y = n$. Here the likelihood is p^n, and equation (8.2) gives $n/\hat{p} = 0$. There is no real solution of this equation, and this indicates that the maximum is found at a boundary point. Subject to the natural constraint $0 \leq p \leq 1$, the likelihood p^n is maximized, as a function of p, at $p = 1$, and this shows that the maximum likelihood estimator of p when $Y = n$ is $\hat{p} = 1$. This is, however, identical to $\bar{Y}$ when $Y = n$. An analogous remark applies when $Y = 0$; here again the maximum likelihood estimator may be identified with $\bar{Y}$. The estimator $\bar{Y}$ thus applies in all cases.

The fact that $\bar{Y}$ has mean p and variance $p(1-p)/n$ can be used, together with Chebyshev's inequality, to show that $\bar{Y}$ is a consistent estimator of p.

These results generalize naturally to the estimation of the parameters in the multinomial distribution.

8.3.2   The Continuous Case

We consider first the case where the random variable of interest has a density function depending on a single parameter ξ. If $(X_1, X_2, \ldots, X_n)$ is denoted by $\boldsymbol{X}$, the likelihood $L(\xi, \boldsymbol{X})$ is defined by

$$L(\xi, \boldsymbol{X}) = f_X(x_1; \xi) f_X(x_2; \xi) \cdots f_X(x_n; \xi), \tag{8.6}$$

where $f_X(x; \xi)$ is the common density function of the iid continuous random variables $X_1, X_2, \ldots, X_n$, again assumed to depend on some unknown parameter ξ. Maximum likelihood estimators and estimates are found by following a procedure essentially identical to that in the discrete case.

Example 3. Suppose that $X_1, X_2, \ldots, X_n$ are $\mathrm{NID}(\mu, \sigma^2)$ random variables and that σ^2 is known. The aim is to find the maximum likelihood estimator of μ, so we replace the generic symbol ξ by μ for this example. The likelihood $L(\mu, \boldsymbol{X})$ is, from (8.6),

$$L(\mu, \boldsymbol{X}) = \prod_{i=1}^{n} \frac{1}{\sqrt{2\pi}\sigma} e^{-\frac{(X_i - \mu)^2}{2\sigma^2}}, \tag{8.7}$$

so that the logarithm of $L(\mu, \boldsymbol{X})$ is

$$\text{constant} - \sum_{i=1}^{n} \frac{(X_i - \mu)^2}{2\sigma^2}, \tag{8.8}$$

where the constant is independent of μ. Differentiation with respect to μ leads to

$$\text{maximum likelihood estimator of } \mu = \bar{X}. \tag{8.9}$$

This estimator has mean μ, and is thus unbiased, and variance σ^2/n, and these results in conjunction with Chebyshev's inequality show that it is a consistent estimator of μ. Further, $\bar{X}$ has a normal distribution for all values of n.

Example 4. A more realistic case arises when both the mean and the variance of a normal distribution are to be estimated. There are now two unknown parameters, μ and σ^2, and the estimation procedure follows lines generalizing those described above. This procedure leads to a two-dimensional maximization problem resulting to the simultaneous equations

$$\frac{\partial \log L(\mu, \sigma^2, \boldsymbol{X})}{\partial \mu} = 0, \quad \frac{\partial \log L(\mu, \sigma^2, \boldsymbol{X})}{\partial (\sigma^2)} = 0. \tag{8.10}$$

If the solutions of 8.10) are written $\hat{\mu}$ and $\hat{\sigma}^2$, we obtain

$$-\sum_{i=1}^{n} \frac{(X_i - \hat{\mu})}{\hat{\sigma}^2} = 0, \tag{8.11}$$

$$-\frac{n}{2\hat{\sigma}^2} + \sum_{i=1}^{n} \frac{(X_i - \hat{\mu})^2}{2\hat{\sigma}^4} = 0, \tag{8.12}$$

from which

$$\hat{\mu} = \bar{X}, \quad \hat{\sigma}^2 = \sum_{i=1}^{n} \frac{(X_i - \bar{X})^2}{n}. \tag{8.13}$$

The former estimator agrees with that in (8.9) and has the same properties as that estimator. The estimator of σ^2 has bias (defined in Section 3.3.3) of order n^{-1}, and is usually replaced in practice by the unbiased estimator (3.5).

Example 5. Suppose that $X_1, X_2, \ldots, X_n$ are independent random variables, each coming from the uniform distribution

$$f_X(x) = \frac{1}{M}, \quad 0 \le x \le M. \tag{8.14}$$

The aim is to find the maximum likelihood estimator of M. In this case the likelihood is M^{-n}. The differentiation process associated with the maximization procedure shows that the maximum of the likelihood occurs at a boundary point. Inspection of the likelihood function M^{-n} shows that it is maximized when M is as small as possible, subject to the requirement $M \ge X_i, (i = 1, 2, \ldots, n$. This leads to the estimator

$$\hat{M} = X_{\max}. \tag{8.15}$$

$\hat{M}$ is not an unbiased estimator of M, as is shown (with an appropriate change of notation) by equation (2.94). An unbiased estimator of M is $U =$

$(n+1)X_{\max}/n$, and often such a simple adjustment is all that is needed to find an unbiased estimator from a maximum likelihood estimator. Equation (1.126) and the density function for $X_{\max}$ given in equation (2.137) with $i = n$, yields the density function of U as

$$f_U(u) = \frac{n^{n+1}u^{n-1}}{(n+1)^n M^n}, \quad 0 \le u \le \frac{(n+1)M}{n}. \tag{8.16}$$

8.3.3 Invariance Property

An important property of maximum likelihood estimators is that of invariance: If $\hat{\xi}$ is the maximum likelihood estimator of ξ, and if $g(\xi)$ is a monotonic function of ξ, then the maximum likelihood estimator of $g(\xi)$ is $g(\hat{\xi})$. This conclusion follows from the chain rule result

$$\frac{\partial L}{\partial \xi} = \frac{\partial L}{\partial g(\xi)} \frac{\partial g(\xi)}{\partial \xi}$$

and the fact, deriving from the monotonicity assumption, that $\partial g(\xi)/\partial \xi$ is never zero. When $g(\xi)$ is a nonlinear function of ξ, then the maximum likelihood estimator $g(\hat{\xi})$ of $g(\xi)$ is usually biased even if $\hat{\xi}$ is an unbiased estimator of ξ. However, this bias is often in practice of order of n^{-1}, where n is the sample size, and as above a simple correction often allows an unbiased estimator of $g(\xi)$ to be found.

8.3.4 Asymptotic Properties

Suppose we have an infinite sequence $X_1, X_2, \ldots$ of iid random variables whose distribution depends on a parameter ξ. We consider asymptotic properties, as n increases, of the maximum likelihood estimator $\hat{\xi}_n$ of ξ obtained from $X_1, X_2, \ldots, X_n$. It can be shown under various regularity conditions discussed below that for large n, the maximum likelihood estimator of a parameter is approximately normally distributed, that

$$E(\hat{\xi}_n) \sim \xi, \tag{8.17}$$

and that

$$\text{Var}(\hat{\xi}_n) \sim \frac{-1}{E\left(\frac{d^2}{d\xi^2} \log L\right)}. \tag{8.18}$$

Here L is given by (8.1) for discrete random variables and by (8.6) for continuous random variables.

One reason for the importance of (8.18) is that the right-hand side in this equation is the well-known "Cramér–Rao bound": Under regularity conditions similar to those discussed below, no unbiased estimator of the parameter ξ based on an iid sample of size n can have a variance smaller than this. This indicates an asymptotic optimality property of maximum likelihood estimators. For this and other reasons maximum likelihood estimators are used frequently, especially with large samples, and is the method of parameter estimation employed often in bioinformatics.

The denominator of the right-hand expression in (8.18) is always negative, so the expression itself is always positive, as is appropriate for a variance. This is seen in the following example of an application of the Cramér–Rao bound.

In Example 1 above, it was shown that, given iid random variables from a Poisson distribution with parameter λ, the maximum likelihood estimator of λ is $\bar{Y}$, and that this estimator is unbiased and has variance $\frac{\lambda}{n}$. From equation (8.3),

$$\frac{d^2}{d\lambda^2} \log L(\lambda, \boldsymbol{Y}) = -\frac{\sum Y_i}{\lambda^2},$$

and thus, since $E(\sum Y_i) = n\lambda$,

$$E\left(\frac{d^2}{d\lambda^2} \log L(\lambda, \boldsymbol{Y})\right) = -\frac{n}{\lambda}.$$

Using this value as the denominator of the right-hand side in (8.18), we find that no unbiased estimator of λ can have a variance less than λ/n. But Example 1 of Section 8.3.1 shows that this is the variance of the unbiased estimator $\bar{Y}$, so that $\bar{Y}$ is the minimum variance unbiased estimator of λ. The comments in Section 8.2 concerning criteria for "good" estimators then suggest that $\bar{Y}$ is the most desirable estimator of λ.

The regularity conditions needed for (8.17) and (8.18) to hold include the requirements that the maximum of the likelihood occurs at a point where the derivative of the likelihood is zero rather than a boundary point, that the range of the random variables not depend on the parameter being estimated, and that the parameter being estimated take values in a continuous interval of real numbers, rather than, for example, only taking integer values. The first two requirements do not hold in Example 5 above. Other examples of this are discussed in Chapter 15. To show why these regularity conditions are needed we provide a sketch of the derivation of (8.18). The proof is given for continuous random variables; the proof for discrete random variables is essentially identical.

Let X be a continuous random variable having density function $f_X(x; \xi)$ and let $X_1, X_2, \ldots, X_n$ be n iid continuous random variables each having the same density function as X. The density function $f_X(x; \xi)$ satisfies

$$\int_R f_X(x; \xi) dx = 1, \tag{8.19}$$

where R is the range of X. Differentiation throughout in equation (8.19) with respect to ξ leads to

$$\int_R \frac{df_X(x;\xi)}{d\xi} dx = 0, \qquad (8.20)$$

where the comparatively weak assumption is made that differentiation of the left-hand side in (8.19) can be performed by differentiation under the integral sign. If the range R depends on ξ, this condition usually does not hold. If $\ell(x;\xi)$ is defined by

$$\ell(x;\xi) = \frac{df_X(x;\xi)}{d\xi} \cdot \frac{1}{f_X(x;\xi)}, \qquad (8.21)$$

equation (8.20) can be written

$$\int_R \ell(x;\xi) f_X(x;\xi) dx = E\left(\ell(x;\xi)\right) = 0. \qquad (8.22)$$

Under the same assumptions as above, a further differentiation with respect to ξ yields

$$\int_R \frac{d\ell(x;\xi)}{d\xi} f_X(x;\xi) dx + \int_R \ell(x;\xi) \frac{df_X(x;\xi)}{d\xi} dx = 0,$$

or, from (8.21),

$$\int_R \frac{d\ell(x;\xi)}{d\xi} f_X(x;\xi) dx + \int_R \left(\ell(x;\xi)\right)^2 f_X(x;\xi) dx = 0.$$

This implies that

$$E\left(\frac{d\ell(X;\xi)}{d\xi}\right) = -E\left((\ell(X;\xi))^2\right). \qquad (8.23)$$

The left-hand side, namely the mean value of $d\ell(X;\xi)/d\xi$, will be denoted by $\mu(\xi)$. An extension of the argument leading to (8.23) shows that

$$E(U) = -E(V), \qquad (8.24)$$

where

$$U = \frac{d^2 \log L(\xi, \boldsymbol{X})}{d\xi^2}, \quad V = \left(\frac{d \log L(\xi, \boldsymbol{X})}{d\xi}\right)^2, \qquad (8.25)$$

where $L(\xi, \boldsymbol{X})$ is defined in (8.6). Consider now the random variable

$$n^{-1} \frac{d}{d\xi}\left(\sum_{i=1}^n \ell(X_i;\xi)\right),$$

which from (8.6) and the definition of $\ell(X_i;\xi)$ is identical to $n^{-1}U$. This average has mean value $\mu(\xi)$, and from Chebyshev's inequality and equation

(2.74) has a probability distribution closely concentrated around $\mu(\xi)$ when n is large. We therefore use the approximations

$$U/n \cong \mu(\xi), \quad V/n \cong -\mu(\xi), \tag{8.26}$$

the latter approximation following from (8.24).

Denote the maximum likelihood estimator of ξ by $\hat{\xi}$. Then the Taylor series approximation (B.29) gives

$$\left(\frac{d \log L(\xi, \boldsymbol{X})}{d\xi} \right)_{\xi=\hat{\xi}} \cong \frac{d \log L(\xi, \boldsymbol{X})}{d\xi} + (\hat{\xi} - \xi)U. \tag{8.27}$$

We assume that the maximum likelihood estimator occurs at a point where the derivative of the likelihood is zero, so that the left-hand side in this equation is 0. Thus to a first order of approximation,

$$\hat{\xi} - \xi \cong -\frac{d \log L(\xi, \boldsymbol{X})/d\xi}{U}. \tag{8.28}$$

Squaring both sides, and using the definition of V in (8.25),

$$(\hat{\xi} - \xi)^2 \cong \frac{V}{U^2}. \tag{8.29}$$

The expected value of the left-hand side is approximately the variance of $\hat{\xi}$ (and is equal to it if $E(\hat{\xi}) = \xi$). The approximations (8.26) show that the right-hand side is approximately $-\frac{1}{n\mu(\xi)}$. Thus

$$\text{variance of } \hat{\xi} \cong \frac{-1}{n\mu(\xi)}. \tag{8.30}$$

This is the desired approximation (8.18).

The above derivation of the variance of the asymptotic maximum likelihood estimate is a sketch only and can be made more rigorous by paying careful attention to the approximations made. It is given here because it emphasizes the main assumptions made in claiming that equation (8.18) applies. These are (i) that the maximum of the likelihood occurs at a point where the derivative of the likelihood is zero, (ii) that the range of the observations is independent of the parameter being estimated, (iii) that the observations are iid, (iv) that the sample size is large, and (v) since the Taylor series approximation (8.27) is used, that the parameter and its maximum likelihood estimate are both real numbers taking possible values in a continuous interval. In Section 15.7 we discuss the estimation of the topology of a phylogenetic tree by maximum likelihood methods. The theory discussed above shows that one may not automatically assume optimality properties for the maximum likelihood estimator of this topology, since this topology is not a real number taking values in some interval.

8.3.5   Many Parameters

The theory above can be generalized to cover the case of several parameters. If the (i, j) element in the *information matrix* I is defined by

$$I_{ij} = - E \left(\frac{d^2}{d\xi_i d\xi_j} \log L \right),$$ (8.31)

then the asymptotic variance of the maximum likelihood estimator of the parameter ξ_i is the (i, i) term in I^{-1}, and the asymptotic covariance of the maximum likelihood estimators of the parameters ξ_i and ξ_j is the (i, j) term in I^{-1}.

8.4   Other Methods of Estimation

8.4.1   Introduction

Since maximum likelihood estimators have optimality properties, at least in large samples, they are usually the estimators of choice in bioinformatics. Other methods of estimation, however, also have desirable properties, and in this section we describe two of these methods.

8.4.2   The Method of Moments

The "method of moments" is sometimes used as a convenient, albeit often asymptotically inefficient, way to estimate unknown parameters. It is used, for example, in the context of BLAST as described in Section 10.8. The method requires the estimation of as many moments of the distribution from which the observations come as there are unknown parameters in that distribution, and can be used for both discrete and continuous probability distributions.

Since we contrast this estimation procedure with the maximum likelihood method, we use the suffix "MM" to denote a method of moments estimator and the suffix "MLE" to denote a maximum likelihood estimator.

Suppose that $X_1, X_2, \ldots X_n$ are iid random variables, each having a probability distribution depending on one parameter ξ only. The mean μ of X_i is assumed to depend on ξ, and we denote it by $g(\xi)$, so that

$$\mu = g(\xi).$$ (8.32)

The estimator $\hat{\xi}_{\text{MM}}$ of ξ is given by mimicking equation (8.32) by equating the average $n^{-1} \sum X_i$ to $g(\hat{\xi}_{\text{MM}})$, so that

$$n^{-1} \sum X_i = g(\hat{\xi}_{\text{MM}}).$$ (8.33)

If possible, this equation is then solved explicitly for $\hat{\xi}_{MM}$. Here only the first moment of X, namely $g(\xi)$, is used in the procedure.

Example 1. We illustrate several of the points made above for the case of one unknown parameter by considering the estimation of the parameter k in the gamma distribution (1.75) for the case where λ is known to be 1. Equation (1.76) shows that the mean of this gamma distribution is k, and equation (8.33) immediately shows that the method of moments estimator $\hat{k}_{MM}$ of k is $n^{-1} \sum X_i$. This is an unbiased estimator of k, and equations (1.76) and (2.74) show that the variance of this estimator is k/n.

The maximum likelihood estimator is not so easily calculated. The likelihood $L(k, \boldsymbol{X})$ is

$$L = \frac{(\prod X_i)^{k-1} e^{-\sum X_i}}{(\Gamma(k))^n},$$

and the equation $d \log L / dk = 0$ reduces to

$$n^{-1} \sum \log X_i = \frac{d\Gamma(k)/dk}{\Gamma(k)}. \tag{8.34}$$

The estimator $\hat{k}_{MLE}$ given implicitly as the solution of this equation in k is biased. (No explicit expression for $\hat{k}_{MLE}$ is available.) While this estimator appears initially to differ substantially from the method of moments estimator $\hat{k}_{MM}$, the two will tend to be close for large n and k. This arises because for large n, $n^{-1} \sum \log X_i \cong \log \bar{X}$ while for large k, the right-hand side in (8.34) is approximately $\log k$ (Abramowitz and Stegun (1972)). Thus equation (8.34) becomes, approximately, $\log \bar{X} \cong \log \hat{k}_{MLE}$ or $\hat{k}_{MLE} \cong \bar{X} = \hat{k}_{MM}$. This argument also implies that the bias of the maximum likelihood estimator decreases as n and k increase.

The asymptotic variance of $\hat{k}_{MLE}$, found from (8.18), is $1/n\psi'(k)$, where $\psi'(k)$ is the so-called trigamma function. Properties and numerical values for this function are given by Abramowitz and Stegun (1972). In the case where $k = 1$, this variance is $6/(n\pi^2) \cong 0.607/n$. This is about 60% of the variance $1/n$ found above for the methods of moments estimator.

For large values of k, $\psi'(k) \cong k$, so the variances of the maximum likelihood estimator and the method of moments estimator are close, as is evident from the fact that the two estimators tend to be close for large k and n. In view of the bias of the maximum likelihood estimator, one might prefer method of moments estimation in this case.

In the case of two unknown parameters, called here ξ and ϕ, the first two moments of X are used. If the mean of X is $g_1(\xi, \phi)$ and the mean of X^2 is $g_2(\xi, \phi)$, the estimators $\hat{\xi}_{MM}$ and $\hat{\phi}_{MM}$ of ξ and ϕ are given implicitly by the simultaneous equations generalizing (8.33), namely

$$n^{-1} \sum X_i = g_1(\hat{\xi}_{MM}, \hat{\phi}_{MM}), \quad n^{-1} \sum X_i^2 = g_2(\hat{\xi}_{MM}, \hat{\phi}_{MM}). \tag{8.35}$$

If possible, these equations are then solved explicitly for $\hat{\xi}_{\text{MM}}$ and $\hat{\phi}_{\text{MM}}$.

Example 2. If in the gamma distribution (1.75) both λ and k are unknown and are to be estimated, equations (1.76), (1.31) and (8.35) show that the method of moments estimators $\hat{\alpha}$ and $\hat{k}$ are found from the equations

$$\hat{\lambda}_{\text{MM}} = \frac{n\hat{k}_{\text{MM}}}{\sum X_i}, \quad n^{-1}\sum X_i^2 = \left(\frac{\hat{k}_{\text{MM}}}{\hat{\lambda}_{\text{MM}}}\right)^2 + \frac{\hat{k}_{\text{MM}}}{\hat{\lambda}_{\text{MM}}^2}. \tag{8.36}$$

The values of $\hat{\lambda}_{\text{MM}}$ and $\hat{k}_{\text{MM}}$ are readily found from these equations (see Problem 8.9).

We now compare these estimators with the corresponding maximum likelihood estimators. The maximum likelihood estimator of k, namely $\hat{k}_{\text{MLE}}$, is independent of λ and is thus given by (8.34). The maximum likelihood estimator of λ, namely $\hat{\lambda}_{\text{MLE}}$, is $n\hat{k}_{\text{MLE}}/\sum X_i$. This equation is of the same form as the first equation in (8.36), implying that when $\hat{k}_{\text{MLE}}$ and $\hat{k}_{\text{MM}}$ are close, then $\hat{\lambda}_{\text{MLE}}$ and $\hat{\lambda}_{\text{MM}}$ are also close.

8.4.3 *Least Squares and Multiple Regression*

Another estimation procedure, which in some cases is equivalent to the maximum likelihood method, is that of least squares. We illustrate it in the context of the general linear model, with which it is most closely associated.

In describing the least squares approach it is convenient to depart from our standard convention and to use the notation Y for a random variable, whether it be discrete or continuous. Suppose first that $Y_1, Y_2, \ldots, Y_n$ are independently but not identically distributed random variables, Y_j having a probability distribution with mean of the form $\mu_j = \alpha + \beta x_j$ and variance σ^2. This model is most conveniently written in the form

$$Y_j = \alpha + \beta x_j + E_j, \tag{8.37}$$

where $E_1, E_2, \ldots, E_n$ are iid random variables with mean 0 and variance σ^2. The model is most frequently used when one wishes to estimate the way in which some random variable Y_j depends on some fixed quantity x_j. This is the *simple regression model*, and is used very widely in applied statistics.

The form of equation (8.37) explains the choice of the notation Y for the random variable involved in least squares calculations, since if the term E_j is ignored, this is the equation of a straight line in the standard cartesian form $y = mx + b$.

In the model (8.37), α and β are unknown parameters that we might wish to estimate. The least squares estimators of α and β are found as the values that minimize the sum of squares

$$\sum_{j=1}^{n} E_j^2 = \sum_{j=1}^{n} (Y_j - \alpha - \beta x_j)^2. \tag{8.38}$$

The resulting least squares estimators $\hat{\alpha}$ and $\hat{\beta}$ are given explicitly by

$$\hat{\beta} = \frac{\sum_{j=1}^{n} Y_j(x_j - \bar{x})}{\sum_{j=1}^{n}(x_j - \bar{x})^2}, \quad \hat{\alpha} = \bar{Y} - \hat{\beta}\bar{x}, \tag{8.39}$$

where $\bar{x} = (x_1 + x_2 + \cdots + x_n)/n$, $\bar{Y} = (Y_1 + Y_2 + \cdots + Y_n)/n$. The estimators $\hat{\beta}$ and $\hat{\alpha}$ are unbiased (see Problem 8.10).

The fact that an explicit expression is available for both $\hat{\alpha}$ and $\hat{\beta}$ should not pass without comment. If the mean of Y_j were not a linear function of α and β it might not be possible to find explicit expressions for $\hat{\alpha}$ and $\hat{\beta}$, and the best that can be done might be to find $\hat{\alpha}$ and $\hat{\beta}$ by a purely numerical procedure. We take up this comment again below.

Given observed values $y_1, y_2, \ldots, y_n$ of $Y_1, Y_2, \ldots, Y_n$, the estimates of β and α are, respectively,

$$\hat{\beta} = \frac{\sum_{j=1}^{n} y_j(x_j - \bar{x})}{\sum_{j=1}^{n}(x_j - \bar{x})^2}, \quad \hat{\alpha} = \bar{y} - \hat{\beta}\bar{x}, \tag{8.40}$$

Here we have abused notation and, for purposes of typographical clarity, have used the same symbol for the estimators and the estimates of α and β.

If $Y_1, Y_2, \ldots, Y_n$ are independent normal random variables, each having variance σ^2 and with Y_j having mean $\alpha + \beta x_j$, the maximum likelihood estimators of α and β are the least squares estimators (8.39) of these parameters (see Problem 8.11).

The model described above assumes that the various Y_j random variables all have the same variance. If some of the random variables have variances greatly exceeding that of the remaining random variables, the estimates of the parameters might be unduly influenced by those random variables with a large variance. In this case it might be thought desirable to minimize the weighted sum of squares $\sum_{j=1}^{n} w_j(Y_j - \alpha - \beta x_j)^2$ rather than the unweighted sum in (8.38), where w_j is a weighting factor associated with Y_j and is small for those random variables with a large variance. If we use the suffix "w" for the weighted least squares estimates of α and β, these estimates are given by

$$\hat{\beta}_w = \frac{(\sum w_j)(\sum w_j x_j y_j) - (\sum w_j y_j)(\sum w_j x_j)}{(\sum w_j)(\sum w_j x_j^2) - (\sum w_j x_j)^2},$$

$$\hat{\alpha}_w = \frac{\sum w_j y_j - \hat{\beta}_w \sum w_j x_j}{\sum w_j}, \tag{8.41}$$

all sums being over $j = 1, 2, \ldots, n$.

A further application of weighted least squares estimation is in the construction of *loess* curves, discussed in detail by Cleveland and Devlin (1988), following the earlier work of Cleveland (1979). (The word "loess" is an acronym (LOcally weighted regrESSion), and was chosen by Cleveland and Devlin (1988), because of its use in describing geological strata. The spelling "lowess" occurs often in the literature. Since here we describe the work of Cleveland and Devlin (1988), we adopt their spelling convention.)

Suppose that the relation between Y and x is nonlinear. Clearly any linear estimation procedure, weighted or unweighted, is inappropriate. On the other hand, a collection of linear estimation procedures, each one carried out over a short range of x values, might be reasonable. Further, it might be desirable that in any such local regression centered around the value x, higher weights are given to values of x_i close to x than to values further from x. With these aims in mind, Cleveland and Devlin (1988), suggest the following procedure.

We first consider some particular value of x, say x_j. We choose some number d and weighting factors $w_{j-d}, w_{j-d+1}, \ldots, w_{j+d}$ and carry out a weighted regression of $Y_{j-d}, Y_{j-d+1}, \ldots, Y_{j+d}$ on $x_{j-d}, x_{j-d+1}, \ldots, x_{j+d}$, using these weights. Cleveland and Devlin (1988) suggest values of d and forms of the weights that lead to suitable loess curves.

This procedure will lead to regression estimates $\hat{\beta}_{w,j}$ and $\hat{\alpha}_{w,j}$, for the weighted regression centered on x_j. The observed value y_j is then replaced by $y_j^* = \hat{\alpha}_{w,j} + \hat{\beta}_{w,j} x_j$, the value of Y corresponding to x_j predicted by this (short) weighted least-squares line. This entire procedure is then carried out for each value of j, with special calculations at boundary values where $j < d$ and $j > n - d$. The various values of the y_j^* so found are now joined to form a loess curve, which will generally be far smoother than the curve joining the original y_j values and provide a better fit to the data than an ordinary linear regression.

We return to loess curves in Section 13.1.3, where their use in connection with microarray analysis is discussed.

A second generalization of (8.37) arises when the mean μ_j of Y_j is of the form $\alpha + \beta_1 x_{j1} + \beta_2 x_{j2} + \cdots + \beta_k x_{jk}$, for some collection of known constants $x_{j1}, x_{j2}, \ldots, x_{jk}$, so that we write

$$Y_j = \alpha + \beta_1 x_{j1} + \beta_2 x_{j2} + \cdots + \beta_k x_{jk} + E_j, \quad j = 1, 2, \ldots, n. \qquad (8.42)$$

Here $\alpha, \beta_1, \beta_2, \ldots, \beta_k$ are unknown parameters that we wish to estimate, and in the unweighted case the $E_j, j = 1, 2, \ldots, n$ are assumed to be iid random variables with mean 0 and variance σ^2. This is the *multiple regression*, or *general linear*, model, and is important in many statistical procedures. A particular case of this model is the polynomial regression model, for which x_{ji} is of the form $(x_j)^i$.

The least squares estimators of $\alpha, \beta_1, \beta_2, \ldots, \beta_k$ are those which minimize the (unweighted) sum of squares $\sum_{j=1}^{n} E_j^2$. To find these estimators it is convenient to write the multiple regression model (8.42) in the matrix and vector form

$$\mathbf{Y} = C\boldsymbol{\beta} + \mathbf{E}, \qquad (8.43)$$

Here $\mathbf{Y} = (Y_1, Y_2, \ldots, Y_n)'$, $\mathbf{E} = (E_1, E_2, \ldots, E_n)'$, $\boldsymbol{\beta} = (\alpha, \beta_1, \beta_2, \ldots, \beta_k)'$ and C is an $n \times (k + 1)$ matrix whose first column consists of 1's and whose entry in row j, column $k+1$ is x_{jk}. It is necessary for the procedure described below to assume that $n < k+1$ and that the matrix C be of full rank. In this case least squares estimation of β, that is by minimization of $\mathbf{E'E}$, can be carried out efficiently by matrix methods and straightforward vector calculus. The conclusion is that the least squares estimator $\hat{\beta}$ of β is given by

$$\hat{\beta} = (C'C)^{-1}C'\mathbf{Y}. \qquad (8.44)$$

We make several remarks about this estimation procedure. The first is that, as noted above, the procedure assumes that the matrix C is of full rank, since the matrix $(C'C)^{-1}$ is not defined if this is not the case. In the one-way ANOVA model, for example, the full rank assumption is not justified, and special procedures are necessary. This ANOVA model is discussed in Section 9.5.3.

Second, assuming that the mean of each E_j is 0, so that (from (8.43)) the mean of $\mathbf{Y}$ is $C\beta$, the estimator (8.44) is unbiased. This can be seen from the sequence of matrix equations

$$E(\hat{\beta}) = E\big((C'C)^{-1}C'\mathbf{Y}\big) = (C'C)^{-1}C'E(\mathbf{Y}) = (C'C)^{-1}C'C\beta = \beta. \qquad (8.45)$$

Third, in the estimation procedure, no assumption is needed in establishing the conclusion in (8.45) about the properties of the vector $\mathbf{E}$, other than that each E_j has mean 0. By contrast, tests of hypotheses about various parameters in the model require further assumptions for their validity. Perhaps the most important of these is the test of the null hypothesis that some of the parameters $\beta_1, \beta_2, \ldots = \beta_k$ are zero. ANOVA is in effect an example of such a test. Standard statistical procedures for this and other tests of hypothesis require the further assumptions that each Y_j has a normal distribution, that the various Y_j are independent and that they all have the same variance.

Fourth, the fact that an explicit expression (8.44) exists for the estimator $\hat{\beta}$ implies that the properties of this estimator (for example the unbiasedness property shown in (8.45)) are comparatively easy to obtain. We return to this point below when considering properties of estimators of the parameters in the more general model (8.46).

Fifth, the linear model (8.42) is used in many applications in bioinformatics. The microarray ANOVA model of Section 13.3.7, for example, is a linear model (see (13.12)). However, there are also examples in bioinformatics where a linear model is not appropriate. Thus it might be the case that that the linear model (8.42) is replaced by a model of the form

$$Y_j = f(\alpha, \beta_1, \ldots, \beta_m; x_{j1}, x_{j2}, \ldots, x_{jm}) + E_j, \quad j = 1, 2, \ldots, n. \qquad (8.46)$$

for some nonlinear function $f(\cdot)$. For such a model it might not be possible to find explicit estimators of $\alpha, \beta_1, \ldots, \beta_m$ parallel to those given in explic-

itly in (8.44) for the linear case. In the nonlinear case numerical methods
are often needed.

The least-squares method requires minimizing the expression

$$\sum_{j=1}^{n} \left(y_j - f(\alpha, \beta_1, \ldots, \beta_m; x_{j1}, x_{j2}, \ldots, x_{jm})\right)^2,$$

where y_j is the observed value of Y_j, with respect to the unknown parameters. There are several potential problems with this procedure. First,
for sufficiently complicated functions $f(\cdot)$, the minimization process might
present computational difficulties. Second, because explicit estimators of
$\alpha, \beta_1, \ldots, \beta_m$ might not be available, the properties of the estimators are
often not easy to find. This is unfortunate since several models in bioinformatics are nonlinear. Finally, any least squares procedure is susceptible to
outliers, since the squaring operation exaggerates the effect of these. This
is noted for example by Li and Wong (2001), who consider a model of the
form

$$Y_{ij} = \alpha_i \beta_j + E_{ij}, \tag{8.47}$$

arising in some microarray models (see Section 13.1.2). Li and Wong suggest various alternatives to the standard least squares procedure that help
overcome the problems arising with the standard approach.

The linear model (8.44) is frequently used in the literature, perhaps
because of the computational and statistical problems arising with nonlinear models. Further, a linear model might serve as a first step, providing
information for the analysis of a more complicated model.

8.5 Multivariate Methods

8.5.1 Introduction

All of the estimation processes considered so far generalize easily to the multivariate case, where we estimate parameters by using the observed values
of random vectors of the form $(X_1, X_2, \ldots, X_k)'$, rather than the observed
values of (scalar) random variables. Such a vector might arise, for example, if we measure k characteristics on each individual, for example height,
weight, arm length .... The interesting case arises when the components of
this vector are correlated, as would be the case for these measurements. In
this brief section we consider aspects of maximum likelihood estimation in
the multivariate normal case.

8.5.2 Parameter Estimation

As for the case of a scalar random variable, maximum likelihood estimation
starts with the calculation of a likelihood. Given n iid random vectors

$\mathbf{X_j} = (X_{j1}, X_{j2}, \ldots, X'_{jk})$, $(j = 1, 2, \ldots, n)$, all having multivariate normal distribution (2.33) with mean vector $\boldsymbol{\mu}$ and variance–covariance matrix Σ, described in (2.33), this likelihood is

$$L(\boldsymbol{\mu}, \Sigma) = \prod_{j=1}^{n} \frac{1}{(2\pi)^{k/2}|\Sigma|^{1/2}} e^{-\frac{1}{2}(\boldsymbol{x}_j - \boldsymbol{\mu})'\Sigma^{-1}(\boldsymbol{x}_j - \boldsymbol{\mu})}, \quad -\infty < x_{jp} < +\infty. \tag{8.48}$$

From this it is found, as expected, that the maximum likelihood estimator of $\boldsymbol{\mu}$ is the vector of averages $(\bar{X}_1, \bar{X}_2, \ldots, \bar{X}_k)'$. This estimator is unbiased. The maximum likelihood estimator of Σ, denoted by $\hat{\Sigma}$, is a matrix whose pth diagonal element is $\hat{\sigma}_p^2$, defined by

$$\hat{\sigma}_p^2 = \frac{\sum_{j=1}^{n}(X_{jp} - \bar{X}_p)^2}{n}, \tag{8.49}$$

while the (p, q) element $\hat{\sigma}_{pq}$ of $\hat{\Sigma}$ is defined, following the format of (3.17), by

$$\hat{\sigma}_{pq} = \frac{\sum_{j=1}^{n}(X_{jp} - \bar{X}_p)(X_{jq} - \bar{X}_q)}{n}. \tag{8.50}$$

These estimators are biased, and as in the scalar case are usually replaced by the unbiased estimators

$$\hat{\sigma}_p^2 = \frac{\sum_{j=1}^{n}(X_{jp} - \bar{X}_p)^2}{n-1}, \tag{8.51}$$

and

$$\hat{\sigma}_{pq} = \frac{\sum_{j=1}^{n}(X_{jp} - \bar{X}_p)(X_{jq} - \bar{X}_q)}{n-1}. \tag{8.52}$$

With this convention, and given the observed values $(x_{j1}, x_{j2}, \ldots, x_{jk})'$, $(j = 1, 2, \ldots, n)$ of these random vectors, the estimate of $\boldsymbol{\mu}$ is, as expected, the vector of averages $(\bar{x}_1, \bar{x}_2, \ldots, \bar{x}_k)'$, and the estimate of Σ is a matrix S whose pth diagonal element is s_p^2, defined by

$$s_p^2 = \frac{\sum_{j=1}^{n}(x_{jp} - \bar{x}_p)^2}{n-1}, \tag{8.53}$$

while the (p, q) element s_{pq} of this matrix is defined by

$$s_{pq} = \frac{\sum_{j=1}^{n}(x_{jp} - \bar{x}_p)(x_{jq} - \bar{x}_q)}{n-1}. \tag{8.54}$$

8.5.3 Principal Components

The visualization of the information supplied by a set of k-dimensional vectors is often difficult, and the main aim of a principal components analysis is to replace k-dimensional random vectors by vectors of a small number (usually two or three) quantities with as little information loss as possible.

The concepts behind this procedure are best seen visually. Suppose that $k = 2$ and that the variance–covariance matrix Σ of X_1 and X_2 is

$$\begin{bmatrix} 9.1 & 3.0 \\ 3.0 & 1.1 \end{bmatrix}. \tag{8.55}$$

With this variance–covariance matrix, X_1 and X_2 are highly positively correlated, and a sample of date points might look as shown in Figure 8.1. If these points are projected onto the line shown in Figure 8.1, and the data are replaced by the various points of projection, little information is lost. The calculation of the line minimizing the information loss is equivalent to finding the first principal component.

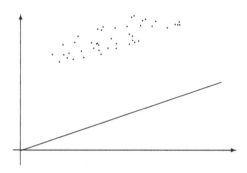

Figure 8.1.

The definition of principal components is as follows. Suppose that the elements in the random vector $\mathbf{X} = (X_1, X_2, \ldots, X_k)$ have covariance matrix Σ. We first find that linear combination $\boldsymbol{\alpha}'\mathbf{X} = \alpha_1 X_1 + \alpha_2 X_2 + \cdots + \alpha_k X_k$ of the elements in $\mathbf{X}$ that has maximum variance. Since the variance of $\boldsymbol{\alpha}'\mathbf{X}$ can be made as large as one wishes by increasing the values of the α_j, this must be done by imposing some constraint on the values of the α_j, and the constraint chosen is the normalization requirement $\boldsymbol{\alpha}'\boldsymbol{\alpha} = \sum_j \alpha_j^2 = 1$. Standard calculus procedures show that the required vector $\boldsymbol{\alpha}$ is the eigenvector of Σ corresponding to the largest eigenvalue λ_1 of Σ. This vector is called the first principal component. For the variance–covariance matrix Σ in (8.55), the largest eigenvalue is 10.1 and the eigenvector corresponding to this is

$$\begin{bmatrix} 3.0 \\ 1.0 \end{bmatrix}. \tag{8.56}$$

Given this result, we then then find that linear combination $\beta'\mathbf{X}$ which has maximum variance, subject to the normalization requirement $\beta'\beta = \sum_j \beta_j^2 = 1$ and the orthogonality requirement $\alpha'\beta = \sum_j \alpha_j \beta_j = 0$. Standard calculus procedures show that the required vector β is the eigenvector of Σ corresponding to the second largest eigenvalue λ_2 of Σ, and this vector is called the second principal component. This procedure can be carried on in a natural way to define k principal components. In the above example $k = 2$ and there are only two principal components.

In practice Σ is almost always unknown, and must be estimated from n data $\mathbf{x_j} = (x_{j1}, x_{j2}, \ldots, x_{jk})'$, $j = 1, 2, \ldots, n$, as defined above, and α is estimated by $\mathbf{a}$, the eigenvector of the matrix defined by (8.53) and (8.54). A parallel estimate $\mathbf{b}$ is made of β and of any further principal components.

We define the "total variation" in the data as

$$\sum_{j=1}^{n}\sum_{p=1}^{k}(x_{jp} - \bar{x}_p)^2.$$

The proportion of this total variation in the k random variables that is captured by the first estimated principal component is

$$\frac{\ell_1}{\sum_{p=1}^{k}\ell_p}, \tag{8.57}$$

where $\ell_1, \ell_2, \ldots, \ell_k$ are the eigenvalues of the estimated variance–covariance matrix. The eigenvalues of the true variance-covariance matrix (8.55) are 10.1 and 0.1, so the estimated proportion of the total variation explained by the first estimated principal component should be about $10.1/10.2 \approx 0.99$.

The proportion of the "total variation" explained by the second estimated principal component is

$$\frac{\ell_2}{\sum_p \ell_p}, \tag{8.58}$$

and so on. In practice, calculation of principal components is often carried out until, say, 90% of the total variation has been explained. If two principal components achieve this aim, we may calculate

$$c_{1j} = \mathbf{a}'\mathbf{x_j} \text{ and } c_{2j} = \mathbf{b}'\mathbf{x_j} \tag{8.59}$$

for $j = 1, 2, \ldots, n$, plot the resulting values on a two-dimensional plane and, together with the vectors $\mathbf{a}$ and $\mathbf{b}$, hope to obtain revealing information about the random vector $(X_1, X_2, \ldots, X_k)$. We discuss this use of principal components further in Section 13.4.

8.6 Bootstrap Methods: Estimation and Confidence Intervals

8.6.1 Introduction

An important trend in statistical inference over the last twenty years has been the introduction of computationally intensive methods. These have been made possible by the availability of convenient and greatly increased computing power, and these methods are useful in bioinformatics and computational biology. The permutation test of Section 3.8.1 provides one example of a computationally intensive procedure. Aspects of some computationally intensive "bootstrap" methods used for estimation and are outlined in this section, and a computationally intensive hypothesis bootstrap testing procedure is discussed in Section 9.8. Computationally intensive methods arise in both classical and Bayesian inference: We concentrate here on computationally intensive methods in classical inference.

8.6.2 The "Plug-in" Concept

It is convenient to start by defining "plug-in" probability distributions and statistics. Suppose that X is a random variable having an unknown probability distribution and that $x_1, x_2, \ldots, x_n$ are the observed values of n iid random variables $X_1, X_2, \ldots, X_n$ having the same probability distribution as X. These data encompass all the information that is available about this unknown distribution, and provide an empirical estimation of the probability distribution for X. This is

$$\text{Prob}_\text{C}(X = x) = \frac{m_x}{n}, \qquad (8.60)$$

where m_x is the number of values in the collection $x_1, x_2, \ldots, x_n$ equal to x and the suffix "C" indicates that this probability is conditional on the observed values in the data.

Suppose now that the distribution given in (8.60) is the actual probability distribution of X. Then from equation (1.24) the mean of X would be $\sum_{i=1}^{n} x_i/n = \bar{x}$, and from equation (1.29) the variance of X would be $\sum_{i=1}^{n} (x_i - \bar{x})^2/n$. The two respective estimates,

$$\bar{x} \quad \text{and} \quad \frac{\sum_{i=1}^{n}(x_i - \bar{x})^2}{n}, \qquad (8.61)$$

are then called the *plug-in* estimates of the mean and variance, respectively, of the actual distribution of X. The former is an unbiased estimate of the actual mean of X, but the latter is a biased estimate of the actual variance, having expected value $\sigma^2(1 - n^{-1})$ with respect to the distribution of X.

In general, a "plug-in" estimate of a parameter is calculated from the empirical distribution in the same way as the parameter itself is defined from the true, but unknown, distribution of the random variable X. Thus the observed proportion of successes in a binomial distribution is the plug-in estimate of the probability p of success, and for any probability distribution the sample median is the plug-in estimate of the median of that distribution. Plug-in estimates are clearly reasonable, but they are not necessarily unbiased, as the above example of the variance estimate, and the result of Problem 8.13, show. In both of these cases, however, the bias approaches 0 as $n \to \infty$.

8.6.3   Classical Estimation Methods

In this and the following section we describe a problem that can arise with classical estimation methods and the way in this problem is overcome by the bootstrap approach.

Suppose that $X_1, X_2, \ldots, X_n$ are iid continuous random variables, with unknown density function $f_X(x)$. We wish to estimate, for given some given probability p, value A_p such that Prob $(X_j \le A_p = p)$ and to to estimate the accuracy of this estimate. In practice, p is usually some frequently used probability such as 0.05, and it is convenient to assume that there is some integer i_0 such that $p = i_0/(n+1)$.

The estimation of A_p is based on the observed values $x_{(1)}, x_{(2)}, \ldots, x_{(n)}$ of the order statistics $X_{(1)}, X_{(2)}, \ldots, X_{(n)}$ of $X_1, X_2, \ldots, X_n$. If the cumulative distribution function corresponding to $f_X(x)$ is $F_X(x)$, then from Theorem 1.2 on page 52, the random variables U_j, defined by $U_j = F_X(X_j)$, $j = 1, 2, \ldots, n$, have uniform distributions in $(0, 1)$. Further, since $X_1, X_2, \ldots, X_n$ are iid, so are $U_1, U_2, \ldots, U_n$. Equation (2.137), applied to $U_1, U_2, \ldots, U_n$, with $L = 1$, then shows that the random variable $U_{(i_0)} = F_X(X_{(i_0)})$ has a beta distribution with parameters i_0 and $n - i_0 + 1$, and thus from equation (1.79) has mean $i_0/(n+1) = p$. That is to say, $E(F_X(X_{(i_0)})) = p$.

Equation (B.32) then shows that to a first order of approximation, $E(X_{(i_0)}) = F_X^{-1}(p)$. Since $F_X^{-1}(p) = A_p$, this implies to a first order of approximation that $X_{(i_0)}$ is an unbiased estimator of A_p. We denote this estimator $\hat{A}_p$. This addresses the first problem raised above.

The accuracy of this estimator depends on its standard deviation of $\hat{A}_p$. Since $X_{(i_0)} = F_X^{-1}(U_{(i_0)})$ and the variance of $U_{(i_0)}$ is approximately $p(1-p)/n$, equation (B.33) shows that

$$\text{standard deviation of } \hat{A}_p \cong \sqrt{\frac{p(1-p)}{n(f_X(A_p))^2}}. \qquad (8.62)$$

In the particular case $p = \frac{1}{2}$, A_p is the median M of the probability distribution of the random variable X, so that if $\hat{M}$ is the sample median, then to a close approximation,

$$\text{standard deviation of } \hat{M} \cong \sqrt{\frac{1}{4n(f_X(M))^2}}. \tag{8.63}$$

However, neither this formula nor the more general formula (8.62) is of immediate value, since they both rely on knowledge of the unknown density function $f_X(x)$. Further, no other classical approach resolves this problem when estimating A_p, as well as other parameters, when the density function $f_X(x)$ is unknown. A bootstrap approach to this problem is discussed in Example 1 of Section 8.6.6.

8.6.4 Bootstrap Estimation Methods

There is a voluminous literature on bootstrap estimation procedures, and we consider only some aspects of the procedure here. For detailed accounts see Efron and Tibshirani (1993), and for further material see Davison and Hinkley (1997), Efron (1982), Hall (1992), Manly (1997), Sprent (1998), and Chernick (1999).

Suppose we wish to estimate the mean μ of the probability distribution of some random variable X, and wish also to find a confidence interval for the mean, using the observed values $x_1, x_2, \ldots, x_n$ of n iid random variables $X_1, X_2, \ldots, X_n$ each having this probability distribution. The theory in Section 3.3.1 shows that $\bar{X}$ is an unbiased estimator of μ, and equation (3.8) shows how an approximate confidence interval for μ can be calculated. The form of this confidence interval is based on the assumption that $\bar{X}$ has an approximately normal distribution. These calculations are not applicable if one wishes to find a confidence interval for some parameter θ using an estimator that does not have an approximately normal distribution.

Suppose that $\hat{\theta} = t(X_1, X_2, \ldots, X_n)$ is an estimator of θ, based on n iid random variables $X_1, X_2, \ldots, X_n$ having a distribution depending on θ; this might or might not be the plug-in estimator. The estimate $\hat{\theta}$ corresponding to this estimator is

$$\hat{\theta} = t(x_1, x_2, \ldots, x_n), \tag{8.64}$$

where $x_1, x_2, \ldots, x_n$ are the observed values of $X_1, X_2, \ldots, X_n$. We abuse notation here and use the same symbol for the estimator and the estimate.

In the bootstrap procedure we regard the observations $x_1, x_2, \ldots, x_n$ as being given. These give the "plug-in" empirical distribution of X. In the case of discrete random variables, when the sample size n is large, each possible value of the random variable can be expected to occur in the sample with frequency close to its probability, while for continuous random variables the proportion of observations in any interval should be close to the probability that the random variable X lies in that interval. This observation forms the basis of the bootstrap procedure described below. When the sample size is small, greater caution is needed in applying bootstrap methods.

In the first step of the so-called non-parametric bootstrap estimation procedure we sample from the n observations n times *with replacement*. This is called a *bootstrap sample*. From the comments in the previous paragraph, this may be taken, when n is large, as an approximation to iid sampling from the actual distribution of X. Some of the observations in the original data might not appear in this bootstrap sample, some might appear once, some twice, and so on. From this bootstrap sample we calculate an estimate $\hat{\theta}_B$ of the parameter θ of interest, replacing $x_1, x_2, \ldots, x_n$ in (8.64) by the n bootstrap sample values. Here the suffix "B" signifies "bootstrap."

This procedure is repeated a large number R of times, leading to R "bootstrap" estimates $\hat{\theta}_{B_1}, \hat{\theta}_{B_2}, \ldots, \hat{\theta}_{B_R}$. These can be regarded as describing an empirical distribution of $\hat{\theta}$. The *bootstrap estimate* $\hat{\theta}_{B(\cdot)}$ of θ, defined by

$$\hat{\theta}_{B(\cdot)} = \frac{\hat{\theta}_{B_1} + \hat{\theta}_{B_2} + \cdots + \hat{\theta}_{B_R}}{R} \tag{8.65}$$

is simply the average of the R bootstrap estimates.

This estimate on its own conveys no information about its precision, and to obtain this information we use the bootstrap estimate of the variance of the estimator $\hat{\theta}$, namely

$$\hat{\sigma}^2_{\theta,B} = \text{estimated variance of } \hat{\theta} = \frac{(\sum_{j=1}^{R} \hat{\theta}^2_{B_j}) - R\hat{\theta}^2_{B(\cdot)}}{R-1}, \tag{8.66}$$

where R is as defined in Section 8.6.4.

Any given bootstrap sample from the original data are $x_1, x_2, \ldots, x_n$ can be written as $(Y_1)x_1, (Y_2)x_2, \ldots, (Y_n)x_n$, where by this we mean that x_j appears Y_j times in the bootstrap sample. The vector $(Y_1, Y_2, \ldots, Y_n)$ has the multinomial distribution (2.30) with $p_i = 1/n$, so that $E(Y_i) = 1$, $\text{Var}(Y_i) = (n-1)/n$, and from (2.31), the covariance between Y_i and Y_j is $-1/n$.

This representation allows various properties of bootstrap samples to be found quickly, and here we consider properties of the average of a bootstrap sample. This average can be written as $\sum_i Y_i x_i/n$. Given $x_1, x_2, \ldots, x_n$, this has (conditional) mean $\bar{x}$ and (conditional) variance

$$\frac{\sum_{i=1}^{n}(x_i - \bar{x})^2}{n^2}. \tag{8.67}$$

The unconditional mean of $\bar{X}$ is the true mean, so that the mean of any bootstrap sample is the true mean. The conditional variance (8.67) has unconditional mean $(n-1)\sigma^2/n^2$, where σ^2 is the variance of the original random variables. When n is large, this is close to the variance σ^2/n of the classical estimator $\bar{X}$.

8.6.5 Bootstrap Confidence Intervals

One major aim of bootstrap methods is to find confidence intervals when we have insufficient information about the distribution of the observations to employ classical methods, thus overcoming the problems outlined at the end of Section 8.6.3. We describe two bootstrap approaches to this problem. In general the question of the best choice of bootstrap confidence intervals is a complex one and is taken up in detail by Efron and Tibshirani (1993); see also Schenker (1985).

The first approach is based on quantile methods. If 2.5% of the R replicated bootstrap values $\hat{\theta}_{B_j}$ lie below the value A and 2.5% lie above the value B, then (A, B) can be taken as an approximate 95% confidence interval for θ. We call this a 95% *quantile* bootstrap confidence interval. The generalization of this to finding an $100\alpha\%$ confidence interval, for any value of α, is immediate.

The second, and more approximate, approach is based on the two standard deviation rule of Section 1.10.2, page 32, and should be used only if one is confident that $\hat{\theta}$ has an approximately normal distribution. This leads to an approximate 95% confidence interval for θ of the form

$$(\hat{\theta}_{B(\cdot)} - 2\hat{\sigma}_{\theta,B}, \ \hat{\theta}_{B(\cdot)} + 2\hat{\sigma}_{\theta,B}), \tag{8.68}$$

where $(\hat{\theta}_{B(\cdot)}$ is defined in (8.65) and $\hat{\sigma}_{\theta,B}$ is defined implicitly in (8.66).

8.6.6 Examples

Example 1. The problem of finding a confidence interval for the estimate of A_p was left unresolved in Section 8.6.3. The natural estimate x_{i_0} of A_p was introduced in Section 8.6.3. One can now draw R bootstrap samples and, for bootstrap replication j, compute the estimate $(\hat{A}_p)_{B_j}$ as the i_0th order statistic in that bootstrap replication. The first, or quantile, approach for a 95% bootstrap confidence interval leads to the interval (A, B), where 2.5% of the R replicated bootstrap values $(\hat{A}_p)_{B_j}$ lie below the value A and 2.5% lie above the value B

The second approach uses (8.68). The variance of the estimator X_{i_0} is calculated following the prescription in (8.66), with $\hat{\theta}$ replaced by $\hat{A}_p$ and $\hat{\theta}_{B_j}$ replaced by $(\hat{A}_p)_{B_j}$. From this an approximate 95% confidence interval is found as in (8.68).

Example 2. Methods to measure gene expression intensities in a given cell type were discussed in Section 3.5.2. Golub et al. (1999) present a data set with gene intensities measured by Affymetrix microarrays. In this data set the intensities of approximately 7000 genes are measured in 72 different experiments. As an example of the data from one such gene, Figure 8.2 shows the intensity of the gene with Genbank accession number X03934, a T-cell antigen receptor gene T3-delta, in 46 acute lymphoblastic leukemia

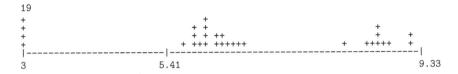

Figure 8.2. Intensities (in ln scale) of the gene with GenBank accession X03934 (T-cell antigen receptor gene T3-delta), measured in 46 acute lymphoblastic leukemia cells by Golub et al. (1999). There are 19 with intensity equal to 3. The sample mean is 5.41.

(ALL) cells. Suppose we want to estimate a 95% confidence interval for the mean of the distribution of intensities for this gene in this type of cell. By performing 5000 bootstrap estimates, we obtained a bootstrap estimate of the mean equal to 5.408, very close to the sample average of 5.407. The full range of the 5000 bootstrap estimates of the mean was (4.30, 6.86), with 95% of the samples lying between (4.77, 6.09). Using the quantile approach, we estimate the 95% confidence interval for the mean to be (4.77, 6.09). For comparison, the 95% confidence interval found using (3.9) gives (4.73, 6.09), which is virtually identical to the bootstrap interval. Therefore even for a highly irregular distribution that is very far from normal, such as that shown in Figure 8.2, the approximation (3.9) is apparently very accurate even with as few as 46 observations.

There is no simple analogue to (3.9) for estimating a confidence interval for the standard deviation of the expression intensity of this gene. The distribution in Figure 8.2 is apparently trimodal, and not of any standard type. Therefore it is not clear how to derive a theoretical confidence interval. We can however easily compute a bootstrap estimate. Using 5000 bootstrap replicates, we obtained a bootstrap estimate of the standard deviation equal to 2.28. The full range of the 5000 bootstrap estimates was (1.63, 2.69), with a 95% quantile bootstrap confidence interval of (1.97, 2.55).

More examples of applications of bootstrap methods are discussed in Chapters 13 and 15.

Problems

8.1. Show that the estimator of μ given in (8.13) is both unbiased and consistent and that the estimator of σ^2 is biased but consistent. Write down a biased estimator of μ that is not consistent and, by considering X_1 only, find an unbiased estimator of μ that is not consistent.

8.2. Show that the maximum likelihood estimators of the probabilities for the various outcomes in a multinomial distribution are the observed proportions of observations in the respective outcomes.

8.3. Replacing ξ by p in the Cramér–Rao bound (8.18), find a lower bound for the variance of any unbiased estimator of the Bernoulli parameter p, using the likelihood L defined in (8.5). Find an unbiased estimator of p whose variance reaches this lower bound.

8.4. Write down the likelihood L defined by n iid random variables X_1, $X_2, \ldots, X_n$ from the exponential distribution (1.66). By putting $1/\lambda = \xi$ in this likelihood, find the Cramér–Rao bound (8.18) for the variance of any unbiased estimator of $1/\lambda$. Find an unbiased estimator of $1/\lambda$ whose variance reaches this lower bound.

8.5. Use the Cramér–Rao bound (8.18) to find a lower bound for the variance of any unbiased estimator of the parameter λ in the Poisson distribution (1.22). Find an unbiased estimator of λ whose variance achieves this bound.

8.6. Suppose that X_1 and X_2 are independent random variables, each having the exponential distribution (1.66). An unbiased estimator of the mean $1/\lambda$ of this distribution is $\bar{X} = (X_1 + X_2)/2$, and the variance (which is also the mean square error) of this estimator is $1/(2\lambda^2)$. The aim of this and the next two questions is to investigate estimators with a smaller mean square error than that of $\bar{X}$.

 (i) Use equations (2.58) and (B.44) to show that $E(\sqrt{X_i X_j}) = \pi/4\lambda$.

 (ii) Show from (i) that the bias of $\sqrt{X_i X_j}$ as an estimator of $1/\lambda$ is $(\pi - 4)/4\lambda$.

 (iii) The variance of $\sqrt{X_i X_j}$ is, from equation (1.30), $E(X_i X_j) - \left(\pi^2/16\lambda^2\right)$. Use a result of Problem 1.21 together with equation (2.58) to show that this variance is $(16 - \pi^2)/16\pi^2$.

 (iv) Use the results of (ii) and (iii) together with equation (3.14) to show that the mean square error of $\sqrt{X_i X_j}$ as an estimator of $1/\lambda$ is $(4 - \pi)/2\lambda^2$. Show from this that the mean square error of $\sqrt{X_i X_j}$ is less than that of $\bar{X}$.

8.7. *Continuation.* A second estimator of $1/\lambda$ is $k\bar{X}$, for some constant k. Find the range of values of k for which the mean square error of this estimator is less than that of $\bar{X}$.

8.8. *Continuation.* Suppose that $X_1, \ldots, X_n$ are independent random variables, each having the exponential distribution (1.66). The variance (and mean square error) of $\bar{X}$ as an estimator of $1/\lambda$ is now $1/n\lambda^2$. Compare this mean square error with that of the estimator $\sqrt{X_1 \cdots X_n}$ for the cases

$n = 3$, $n = 4$, using the fact that $\Gamma(1.25) = 0.9064025$, $\Gamma(\frac{4}{3}) = 0.8929795$, $\Gamma(\frac{5}{3}) = 0.9027453$.

8.9. Solve the equations in (8.36) for $\hat{\lambda}$ and $\hat{k}$. Are the estimators unbiased?

8.10. Use the fact that the mean of Y_i is $\alpha + \beta x_i$, together with the fact that

$$E(\sum_i a_i Y_i) = \sum_i a_i E(Y_i)$$

for any set of constants $\{a_i\}$, to show that $\hat{\beta}$ and $\hat{\alpha}$, defined in (8.39), are unbiased estimators of β and α respectively.

8.11. Show that if $Y_1, Y_2, \ldots, Y_n$ are independent normal random variables, each having variance σ^2 and with Y_j having mean $\alpha + \beta x_j$, the maximum likelihood estimators of α and β are the least squares estimators (8.39).

8.12. Let $X_1, X_2, \ldots, X_n$ be independent continuous random variables, each having density function $f(x) = e^{-(x-\theta)}$, $\theta \le x < +\infty$. Let $X_{(1)}, X_{(2)}, \ldots, X_{(n)}$ be the corresponding order statistics. Use (2.132) to find the respective density functions of $X_{(1)}$ and $X_{(2)}$, and from this show that

$$E\left(X_{(1)}\right) = \theta + \frac{1}{n}, \quad E\left(X_{(2)}\right) = \theta + \frac{2n-1}{n(n-1)}.$$

Thus show that $X_{(1)} - (n-1)(X_{(2)} - X_{(1)})/n$ is an unbiased estimator of θ.

8.13. Use the result of Problem 2.25 to discuss the bias of the plug-in estimator of the median of the exponential distribution when n, the number of observations, is odd (so that $n = 2k + 1$ for some integer k).

8.14. Suppose that $X_1, X_2, \ldots, X_n$ are n iid random variables each having a normal distribution with mean μ, variance σ^2. In this case the median of the distribution is the same as the mean. Compare the asymptotic standard deviation calculated from (8.63) arising if the sample median is used to estimate the mean with that for estimator $\bar{X}$.

8.15. Derive the expression (8.67).

8.16. Suppose that the observed values of n continuous iid random variables are $x_1, x_2, \ldots, x_n$, where the x_j are all distinct. A bootstrap sample (with replacement) is taken from these. Show that there are $\binom{2n-1}{n}$ different possible bootstrap samples. *Hint:* A classical result of probability theory (see Feller (1968)) is that if r indistinguishable objects are placed into n distinguishable cells, the number of different occupancy arrangements is

$\binom{n+r-1}{r}$. Now put $r = n$ and make the analogy that if n_j objects are put in cell j, the value x_j arises n_j times in the bootstrap sample.

8.17. *Continuation.* Calculate the number of distinct bootstrap samples for the cases $n = 3$, $n = 4$, and $n = 5$, and display all possible samples for the case $n = 3$.

8.18. *Continuation.* Find the probability of each bootstrap sample listed in Problem 8.17 for the case $n = 3$. Use your result to find the mean number of times that x_1 appears in a bootstrap sample, and confirm your result by showing that the complete distribution of the number of times that x_1 appears in a bootstrap sample has a binomial distribution with parameter $1/3$ and index 3. Generalize your result to the case of an arbitrary value of n.

8.19. *Continuation.* Rewrite the possible samples for the case $n = 3$ in Problem 8.17 using the observed values $x_{(1)}$, $x_{(2)}$, and $x_{(3)}$ of the order statistics. From this, find the sample average and the sample median of each of the possible samples. Note: If a bootstrap sample is of the form $(x_{(i)}, x_{(i)}, x_{(j)})$ for $i \neq j$, the sample median is defined to be $x_{(i)}$.

8.20. *Continuation.* Use the result of Problem 8.18 to find the conditional mean of the bootstrap sample average and of the bootstrap sample median. (The word "conditional" emphasizes that these are functions of $x_{(1)}$, $x_{(2)}$, and $x_{(3)}$.)

8.21. *Continuation.* In the case where the three random variables have the exponential distribution (1.66), find the unconditional mean of the bootstrap sample average and of the bootstrap sample median. (For the latter, use the result of Problem 2.23.) Compare these with the corresponding means of the classical plug-in estimators.

9

Statistics (iii): Classical Hypothesis Testing Theory

9.1 Introduction

In this chapter we introduce aspects of classical hypothesis testing theory that will be useful in later chapters of this book. We consider initially the case where the sample size is fixed in advance of the experiment generating the data, and in particular does not depend on the values of the observations as they arise. We conclude with sequential tests of hypotheses, where the sample size is not fixed in advance. Sequential analysis theory forms a natural introduction to the theory of BLAST, which is discussed in Chapter 10.

9.2 Simple Fixed-Sample-Size Tests

9.2.1 The Likelihood Ratio

The five steps of classical hypothesis-testing procedures were outlined in Chapter 3. In this chapter we expand on the procedure of Step 3, leading to the choice of test statistic, and Step 4, deciding when the observed value of the test statistic should lead to rejection of the null hypothesis.

We consider first the case where both null and alternative hypotheses are *simple*, that is, where both hypotheses completely specify the probability distribution of the random variables of interest, including the numerical values of all parameters. Although the numerical value α of the Type I error is chosen by the experimenter (in Step 2 of the hypothesis-testing procedure), no specific choice is made of the numerical value β of the Type II error. However, it is desirable to have a procedure that, with the Type I error fixed, minimizes the numerical value β of the Type II error. Such a procedure would maximize the power of the test, that is, the probability $1 - \beta$ of rejecting the null hypothesis when the alternative hypothesis is true. We call such a procedure a *most powerful* test of this null hypothesis against this alternative. The methods used to find such a test are identical for both discrete and continuous random variables, so we describe in detail only the continuous random variable case.

9.2.2 The Neyman–Pearson Lemma

We consider first the simplest case, that of iid random variables $X_1, X_2, \ldots,$ X_n each having completely specified density function $f_0(x)$ under the null hypothesis H_0 and completely specified density function $f_1(x)$ under the alternative hypothesis H_1. The decision about accepting H_0 or H_1 will be based on the observed values $x_1, x_2, \ldots, x_n$ of $X_1, X_2, \ldots, X_n$.

The Neyman–Pearson lemma states that the most powerful test of H_0 against H_1 is obtained by using as test statistic the likelihood ratio, defined as

$$\mathrm{LR} = \frac{f_1(X_1)f_1(X_2)\cdots f_1(X_n)}{f_0(X_1)f_0(X_2)\cdots f_0(X_n)}. \tag{9.1}$$

The null hypothesis is rejected when the observed value of LR is greater than or equal to some constant K, where K is chosen such that

$$\mathrm{Prob}(\mathrm{LR} \geq K \text{ when } H_0 \text{ is true}) = \alpha. \tag{9.2}$$

The proof that use of LR as test statistic leads to the most powerful test is instructive, so we give it here.

Let A be the region in the joint range of $X_1, X_2, \ldots, X_n$ such that $\mathrm{LR} \geq K$, where K is defined by (9.2). Then

$$\int \cdots \int_A f_0(u_1)f_0(u_2)\cdots f_0(u_n)du_1 du_2 \cdots du_n = \alpha. \tag{9.3}$$

Let B be any other region in the joint range of $X_1, X_2, \ldots, X_n$ such that

$$\int \cdots \int_B f_0(u_1)f_0(u_2)\cdots f_0(u_n)du_1 du_2 \cdots du_n = \alpha. \tag{9.4}$$

Rejection of H_0 when the observed vector $(x_1, x_2, \ldots, x_n)$ is in B thus also leads to a test of the required Type I error. Let A and B overlap in the region C.

The power of the test is the probability that the null hypothesis is rejected when the alternative hypothesis is true. Thus the power of the test using

A is

$$\int_A \cdots \int f_1(u_1) \cdots f_1(u_n) du_1 \cdots du_n,$$

and if we define Δ by

$$\Delta = \text{power of the test using } A - \text{power of the test using } B,$$

we obtain

$$\Delta = \int_A \cdots \int f_1(u_1) \cdots f_1(u_n) du_1 \cdots du_n$$

$$- \int_B \cdots \int f_1(u_1) \cdots f_1(u_n) du_1 \cdots du_n \qquad (9.5)$$

$$= \int_{A\backslash C} \cdots \int f_1(u_1) \cdots f_1(u_n) du_1 \cdots du_n$$

$$- \int_{B\backslash C} \cdots \int f_1(u_1) \cdots f_1(u_n) du_1 \cdots du_n, \qquad (9.6)$$

where $A \backslash C$ denotes the set of points in A but not in C. Now, in $A \backslash C$,

$$f_1(u_1) \cdots f_1(u_n) \geq K f_0(u_1) \cdots f_0(u_n),$$

while in $B \backslash C$,

$$f_1(u_1) \cdots f_1(u_n) \leq K f_0(u_1) \cdots f_0(u_n).$$

Thus

$$\Delta \geq \int_{A\backslash C} \cdots \int K f_0(u_1) \cdots f_0(u_n) du_1 \cdots du_n$$

$$- \int_{B\backslash C} \cdots \int K f_0(u_1) \cdots f_0(u_n) du_1 \cdots du_n \qquad (9.7)$$

$$= \int_A \cdots \int K f_0(u_1) \cdots f_0(u_n) du_1 \cdots du_n$$

$$- \int_B \cdots \int K f_0(u_1) \cdots f_0(u_n) du_1 \cdots du_n \qquad (9.8)$$

$$= K\alpha - K\alpha = 0. \qquad (9.9)$$

This result implies that that the power of the test using A is greater than or equal to the power using B, and completes the proof of the lemma.

The lemma not only shows that LR is the appropriate test statistic (Step 3 of the hypothesis testing procedure), but also indicates which observed values of LR lead to rejection of the null hypothesis (Step 4).

It is equivalent to use any monotonic function of LR as the test statistic instead of LR itself, with a value of K appropriate to the new test statistic, and this often leads to a standard testing procedure, as the following example shows.

9.2.3 Example. Sequence matching.

In the sequence-matching example discussed in Section 3.4.1, the simple null hypothesis that the probability of a match is 0.25 was tested against the simple alternative hypothesis that this probability is 0.35. Suppose more generally that under the null hypothesis the probability of a match is p_0 and under the alternative hypothesis this probability is p_1, with $p_1 > p_0$. Then from (8.5) and (9.1) the likelihood ratio LR is

$$\frac{(p_1)^{\sum_{i=1}^n Y_i}(1-p_1)^{(n-\sum_{i=1}^n Y_i)}}{(p_0)^{\sum_{i=1}^n Y_i}(1-p_0)^{(n-\sum_{i=1}^n Y_i)}} \tag{9.10}$$

$$= \left(\frac{p_1(1-p_0)}{p_0(1-p_1)}\right)^{\sum_{i=1}^n Y_i}\left(\frac{1-p_1}{1-p_0}\right)^n. \tag{9.11}$$

The null hypothesis is rejected when the likelihood ratio is sufficiently large. However, the null hypothesis probability distribution of this ratio is not straightforward, so that the evaluation of the constant K in (9.2) is not immediate. On the other hand, the likelihood ratio is a monotonic increasing function of the number of matches $Y = \sum_{i=1}^n Y_i$, so that it is equivalent to use Y as test statistic and to reject the null hypothesis when the observed value y of Y is sufficiently large. Under the null hypothesis Y has a binomial distribution with parameter p_0 and index n, and it is straightforward to use Y as the test statistic, as described in Section 3.4.1.

The null hypothesis is rejected if $y \geq K^*$, where y is the observed value of Y and K^* is determined by the requirement

$$\text{Prob}(Y \geq K^* \,|\, p = p_0) = \alpha, \tag{9.12}$$

where α is the Type I error chosen. Provided only that $p_1 > p_0$, the calculation of the significance point K^* is not influenced by the numerical value of p_1. Thus the testing procedure in this example is the most powerful one *whatever* the alternative hypothesis might be, provided only that it defines a value of p exceeding p_0. Thus this testing procedure is *uniformly most powerful* for the (more realistic) composite alternative hypothesis $p = \text{Prob(match)} > p_0$. Despite this, if the true value of the parameter is close to p, the power even of the likelihood ratio test might be small.

9.3 Classical Hypothesis Testing: Composite Fixed-Sample-Size Tests

9.3.1 Introduction

The Neyman-Pearson testing theory described in Section 9.2.2 concerns the case of of testing whether a set of iid random variables come from one of two completely specified distributions. In most cases arising in practice, however, the random variables of interest are usually not identically distributed, at least under the alternative hypothesis. An example is the two-sample t-test, where under the alternative hypothesis an observation in the first group has a different mean from that of an observation in the second group. In this case the likelihood of the observations is not of the simple iid form such as those discussed above. For example, under the alternative hypothesis in the two sample t-test of Section 3.5 the likelihood L of the random variables $(X_{11}, X_{12}, \ldots, X_{1m}, X_{21}, X_{22}, \ldots, X_{2n})$ is, from (2.8),

$$L = \prod_{i=1}^{m} \frac{1}{\sqrt{2\pi}\sigma} e^{-(x_{1i}-\mu_1)^2/(2\sigma^2)} \prod_{i=1}^{n} \frac{1}{\sqrt{2\pi}\sigma} e^{-(x_{2i}-\mu_2)^2/(2\sigma^2)}, \qquad (9.13)$$

where μ_1 and μ_2 are different from each other.

A further generalization is that the hypotheses being tested usually do not specify completely all values of the parameters of the distribution of the observations; that is, they are usually *composite*. The two-sample t-test is a case in point. The null hypothesis states that the means μ_1 and μ_2 in (9.13) are equal, without specifying their common numerical value, and makes no specification about the variance σ^2, and the alternative hypothesis makes no specification about any of the parameters. The theory of tests involving composite hypotheses is not straightforward, especially the optimality aspects of the procedures adopted, and here we only outline those aspects of the theory relevant to the material of this book and do not address optimality issues.

9.3.2 Parameter Spaces and the Likelihood Ratio λ

The tests that we consider are tests about the numerical values of parameters. The set of values that the parameters of interest can take is called the parameter space. For example, in the normal distribution there are two parameters, the mean μ and the variance σ^2, and the parameter space is $\{-\infty < \mu < +\infty,\ 0 < \sigma^2 < +\infty\}$. The null hypothesis states that the parameters in the distribution of interest take values in some region ω of the parameter space, while the alternative hypothesis states that they take values in some region Ω of the parameter space.

In practice, ω is usually a subspace of Ω, as in the example above, and we restrict attention to this, the *nested hypotheses*, case. When the null hypothesis is nested within the alternative hypothesis, the aim of the hypothesis testing procedure is somewhat different from the aim in the unnested case. The more general alternative hypothesis can always explain the data at least as well as the narrower null hypothesis, and we are interested in whether it explains the data *significantly* better than the null hypothesis. If it does, the null hypothesis is rejected.

As stated above, there is substantial optimality theory for composite hypotheses. This theory suggests that in many cases a desirable test statistic is the ratio λ, defined by

$$\lambda = \frac{L_{\max}(\omega)}{L_{\max}(\Omega)}. \tag{9.14}$$

Here $L_{\max}(\omega)$ is the maximum of the likelihood of the observations when the parameters are confined to the region ω of the parameter space defined by the null hypothesis and $L_{\max}(\Omega)$ is the maximum of the likelihood of the observations when the parameters are confined to the region Ω defined by the alternative hypothesis. In the following section we calculate this ratio for the two-sample t-test, where the likelihood L is defined in (9.13).

The null hypothesis is rejected when λ is sufficiently small to guarantee that the test has the Type I error desired. In the equal variance two-sample t-test the choice of a Type I error leads to an explicit determination of how small λ must be for the null hypothesis to be rejected. In more complicated cases approximations are needed, as discussed in Section 9.4.

We note in passing that the numerator in the λ ratio in (9.14) refers to the null hypothesis and the denominator to the alternative hypothesis, whereas the converse is true for the likelihood ratio (9.1) for simple hypotheses. This is in the nature of a historical accident and has no significance.

As is the case for the simple hypothesis likelihood ratio LR, the composite hypothesis analogue λ is often a monotonic function of some standard test statistic whose properties are known, and when this is so it is convenient to carry out the test using that statistic.

9.3.3   Example: t-tests

The theory of the previous section leads directly to various standard statistical testing procedures. Of these perhaps the most frequently used are t-tests, and we now describe the theoretical derivation of the two-sided and the one-sided equal variance t-tests where the likelihood L is defined in (9.13). These tests generalize to the ANOVA and T^2 testing procedures, discussed in Section 9.5 and Section 9.6 respectively.

The Equal Variance Two-Sided t-test

Suppose that the random variables $X_{11}, X_{12}, \ldots, X_{1m}$ in group 1 are NID(μ_1, σ^2) and the random variables $X_{21}, X_{22}, \ldots, X_{2n}$ in group 2 are NID(μ_2, σ^2), and that X_{1i} is independent of X_{2j} for all i and j. We wish to test the null hypothesis $\mu_1 = \mu_2$ $(= \mu$, unspecified) against the alternative hypothesis that leaves the values of both μ_1 and μ_2 unspecified. The value of σ^2 is unknown and is unspecified under both hypotheses, but is assumed to be the same in the two groups considered. (This assumption might be unreasonable in many cases in practice, and the case where the variances in the two groups are different is discussed below.) The region ω is $\{\mu_1 = \mu_2,\ 0 < \sigma^2 < +\infty\}$, the region Ω is

$$\{-\infty < \mu_1 < +\infty, -\infty < \mu_2 < +\infty, 0 < \sigma^2 < +\infty\},$$

and the likelihood of these random variables is as given in (9.13). Under the null hypothesis the two sets of random variables have the same distribution. Writing μ for the mean of this (common) distribution, the maximum of (9.13) over ω occurs when

$$\hat{\mu} = \bar{\bar{X}} = \frac{X_{11} + X_{12} + \cdots + X_{1m} + X_{21} + X_{22} + \cdots + X_{2n}}{m + n},$$

and the (common) variance σ^2 is

$$\hat{\sigma}_0^2 = \frac{\sum_{i=1}^m (X_{1i} - \bar{\bar{X}})^2 + \sum_{i=1}^n (X_{2i} - \bar{\bar{X}})^2}{m + n}.$$

Inserting $\hat{\mu}$ for both μ_1 and μ_2, and $\hat{\sigma}_0^2$ for σ^2 in (9.13), we get

$$L_{\max}(\omega) = \frac{1}{(2\pi\hat{\sigma}_0^2)^{(m+n)/2}} e^{-(m+n)/2}. \tag{9.15}$$

Under the alternative hypothesis the maximum of (9.13) over Ω occurs when $\hat{\mu}_1$ is $\bar{X}_1 = (X_{11} + X_{12} + \cdots + X_{1m})/m$, $\hat{\mu}_2$ is $\bar{X}_2 = (X_{21} + X_{22} + \cdots + X_{2n})/n$, and σ^2 is

$$\hat{\sigma}_1^2 = \frac{\sum_{i=1}^m (X_{1i} - \bar{X}_1)^2 + \sum_{i=1}^n (X_{2i} - \bar{X}_2)^2}{m + n}.$$

Inserting $\hat{\mu}_1$ for μ_1, $\hat{\mu}_2$ for μ_2 and $\hat{\sigma}_1^2$ for σ^2 in (9.13), we get

$$L_{\max}(\Omega) = \frac{1}{(2\pi\hat{\sigma}_1^2)^{(m+n)/2}} e^{-(m+n)/2}. \tag{9.16}$$

Thus the test statistic λ reduces to

$$\lambda = \frac{L_{\max}(\omega)}{L_{\max}(\Omega)} = \left(\frac{\hat{\sigma}_1^2}{\hat{\sigma}_0^2}\right)^{(n+m)/2}. \tag{9.17}$$

The algebraic identity

$$\sum_{i=1}^{m}(X_{1i} - \bar{\bar{X}})^2 + \sum_{i=1}^{n}(X_{2i} - \bar{\bar{X}})^2$$

$$= \sum_{i=1}^{m}(X_{1i} - \bar{X}_1)^2 + \sum_{i=1}^{n}(X_{2i} - \bar{X}_2)^2 + \frac{mn}{m+n}(\bar{X}_1 - \bar{X}_2)^2$$

may be used to show that

$$\lambda = \left(\frac{1}{1 + t^2/(m+n-2)}\right)^{(m+n)/2}, \tag{9.18}$$

where t is defined in (3.30). It is thus equivalent by the monotonicity argument to use t^2, or alternatively $|t|$, instead of λ, as test statistic. Small values of λ correspond to large values of $|t|$, so that sufficiently large values of $|t|$ lead to rejection of the null hypothesis. The null hypothesis distribution of the statistic t is known (as the t distribution with $m + n - 2$ degrees of freedom) and significance points are widely available. Thus once a Type I error has been chosen, values of $|t|$ that are sufficiently large to lead to rejection of the null hypothesis can be determined. Thus the two-sample equal variance t-test derives from the theory of tests of composite hypotheses outlined in Section 9.3.2.

The Equal Variance One-Sided t-test

The theory for one-sided equal variance t-tests is slightly different from the theory developed for the corresponding two-sided test described above. Instead of an alternative hypothesis that leaves the values of both μ_1 and μ_2 unspecified, the alternative hypothesis in the one-sided test claims that one mean is greater than or equal to the other, for example that $\mu_1 \geq \mu_2$. In this case the region Ω is $\{\mu_1 \geq \mu_2, 0 < \sigma^2 < +\infty\}$.

If $\bar{x}_1 \geq \bar{x}_2$, all maximum likelihood estimates are as for the two-sided case and the t statistic for the two-sided case is recovered. If $\bar{x}_1 < \bar{x}_2$, the unconstrained maximum of the likelihood arises outside the region specified by the null hypothesis. There is a unique maximum of the likelihood (at $(\bar{x}_1, \bar{x}_2)$), and this implies that the maximum in the region defined by the null hypothesis arises at a boundary point of Ω. At this point the maximum likelihood estimates of both μ_1 and μ_2 are equal, both being equal to the overall average $\bar{\bar{x}}$. In this case the alternative hypothesis does not explain the observations any better than does the null hypothesis, the likelihood ratio is 1, and the null hypothesis is not rejected.

The outcome of these observations is that the null hypothesis is rejected in favor of the alternative hypothesis $\mu_1 \geq \mu_2$ only for sufficiently large *positive* values of t.

The simplicity of the two t-tests described above hides several important problems that can arise in the application of the theory in other cases. First, the ratio λ might not reduce to a function of a well-known test statistic such as t. More important, there might not even be a unique null hypothesis distribution of the statistic found by the λ ratio procedure. In the case of the above t-test, the null hypothesis distribution of the t statistic is independent of the unknown variance and also of the mean value unspecified by the null hypothesis. That is, this distribution is independent of the parameters whose values are not prescribed by the null hypothesis, so that the t statistic is a pivotal quantity. It is fortunate that this is the case, and unfortunately there are many testing procedures for which the pivotality property does not hold. If the null hypothesis defines a region in parameter space, there can be no guarantee in general that the distribution of the test statistic is the same for all points in this region, as the following example, very similar to those discussed above, shows.

The Unequal Variance Two-Sided t-test

In this example we consider the case identical to that of the two-sided t-test discussed above, except that now the observations in the first group are assumed to be $\mathrm{NID}(\mu_1, \sigma_1^2)$ and the observations in the second group are assumed to be $\mathrm{NID}(\mu_2, \sigma_2^2)$. Again we wish to test the null hypothesis $\mu_1 = \mu_2$ ($= \mu$, unspecified) against the alternative hypothesis that leaves the values of both μ_1 and μ_2 unspecified. The values of σ_1^2 and σ_2^2 are unknown, are not assumed to be identical, and are unspecified under both hypotheses. The region ω is $\{\mu_1 = \mu_2, \ 0 < \sigma_1^2, \sigma_2^2 < +\infty\}$, and the region Ω makes no constraint on the values of the parameters μ_1, μ_2, σ_1^2 and σ_2^2. The likelihood of $(X_{11}, X_{12}, \ldots, X_{1m}, X_{21}, X_{22}, \ldots, X_{2n})$ is, from (2.8),

$$\prod_{i=1}^{m} \frac{1}{\sqrt{2\pi}\sigma_1} e^{-(x_{1i}-\mu_1)^2/(2\sigma_1^2)} \prod_{i=1}^{n} \frac{1}{\sqrt{2\pi}\sigma_2} e^{-(x_{2i}-\mu_2)^2/(2\sigma_2^2)}. \tag{9.19}$$

Under the null hypothesis the two sets of random variables have the same mean, which we write μ. The likelihood then becomes

$$\prod_{i=1}^{m} \frac{1}{\sqrt{2\pi}\sigma_1} e^{-(x_{1i}-\mu)^2/(2\sigma_1^2)} \prod_{i=1}^{n} \frac{1}{\sqrt{2\pi}\sigma_2} e^{-(x_{2i}-\mu)^2/(2\sigma_2^2)}. \tag{9.20}$$

The maximum likelihood estimates $\hat{\mu}$, $\hat{\sigma}_1^2$ and $\hat{\sigma}_2^2$ satisfy the simultaneous equations

$$\frac{\sum(x_{1i} - \hat{\mu})}{\hat{\sigma}_1^2} + \frac{\sum(x_{2i} - \hat{\mu})}{\hat{\sigma}_2^2} = 0, \ \hat{\sigma}_1^2 = \frac{\sum(x_{1i} - \hat{\mu})^2}{m}, \ \hat{\sigma}_2^2 = \frac{\sum(x_{2i} - \hat{\mu})^2}{n}.$$

These lead to a cubic equation in $\hat{\mu}$. The upshot of this is that neither the λ ratio, nor any monotonic function of it, has known probability distribution when the null hypothesis is true, and also that no pivotal statistic exists. Thus the λ ratio approach does not lead to any useful testing procedure.

One might then decide to use the t statistic defined in (3.30) as a reasonable test statistic. However, as discussed in Section 3.5, the null hypothesis distribution of this statistic is unknown, since it depends on the unknown ratio σ_1^2/σ_2^2, so that the pivotality property does not hold. In practice, the heuristic procedure described in Section 3.5, employing the statistic t' defined in (3.37), is often used.

9.4   The $-2 \log \lambda$ Approximation

9.4.1   Theory

In this section we address the problems, exemplified by the final example in Section 9.3.3, that the λ ratio procedure does not lead to a test statistic whose null hypothesis is known, and that even for a statistic such as t chosen not through the λ ratio theory but as a reasonable statistic to use, the non-pivotality problem can arise.

Various approximation procedures have been proposed to address these problems in general. Perhaps the best known is based on the fact that if various regularity assumptions discussed below hold, and if the null hypothesis is true, $-2 \log \lambda$ (with λ as defined in (9.14)) has an asymptotic chi-square distribution, with degrees of freedom equal to the difference in the numbers of parameters unspecified by null and alternative hypotheses respectively. In this statement the word "asymptotic" means "as the sample size increases to infinity." This fact gives us a testing procedure that is asymptotically valid and, as such, overcomes the problem raised in the previous paragraph.

It is important to note that this statement holds only under certain restrictions. Two important requirements are, first, that the parameters involved in the test be real numbers that can take values in some interval, and second, that the maximum likelihood estimator be found from a point where the likelihood function has a turning point (and not, for example, a boundary point such as that in Example 5 of Section 8.3). A further requirement is that null and alternative hypotheses be nested, as described in Section 9.3.1. The proof of the asymptotic distribution of $-2 \log \lambda$ given, for example, in Wilks (1962, pp. 419–421) makes clear the importance of these restrictions. The approximate chi-square null hypothesis distribution of $-2 \log \lambda$ has been applied in the literature in several cases of phylogenetic tree construction where these requirements do not hold (see Section 15.9), so we outline the derivation of the result here, indicating where the requirements listed above are used.

We now outline the proof of this claim, leaving many details to be tidied up, considering for simplicity the case of a single unknown parameter θ that takes some given value θ_0 under the null hypothesis and is left unspecified under the alternative hypothesis. We assume throughout that the null hypothesis is true, that is, that $\theta = \theta_0$, and that the sample size n is sufficiently large so that we can assume that the maximum likelihood estimate $\hat{\theta}$ is close to θ_0.

The first step is to write

$$-2 \log \lambda = 2 \left(\log L(\hat{\theta}; \boldsymbol{X}) - \log L(\theta_0; \boldsymbol{X}) \right). \qquad (9.21)$$

We use the second-order Taylor series approximation

$$f(x) \cong f(a) + (x - a)f'(a) + \frac{(x - a)^2}{2} f''(a)$$

derived from (B.30) to write

$$\log L(\theta; \boldsymbol{X}) \cong \log L(\hat{\theta}; \boldsymbol{X}) + (\hat{\theta} - \theta_0) \frac{d}{d\theta} \log L(\hat{\theta}; \boldsymbol{X})$$
$$+ \frac{(\hat{\theta} - \theta_0)^2}{2} \frac{d^2}{d\theta^2} \log L(\hat{\theta}; \boldsymbol{X}). \qquad (9.22)$$

Since the maximum likelihood estimator $\hat{\theta}$ of θ is assumed to be found from a non-boundary point, the first derivative term on the right-hand side in (9.22) is zero. Then from (9.22), the right-hand side in (9.21) becomes, approximately,

$$-(\hat{\theta} - \theta_0)^2 \frac{d^2}{d\theta^2} \log L(\hat{\theta}; \boldsymbol{X}). \qquad (9.23)$$

We make the further approximation of replacing the second derivative term by its mean value, under the assumption that $\hat{\theta}$ is sufficiently close to θ_0. Then equation (8.18) shows that expression (9.23) is, approximately,

$$\frac{(\hat{\theta} - \theta_0)^2}{\text{variance of } \hat{\theta}}, \qquad (9.24)$$

and the asymptotic normality of $\hat{\theta}$ and the discussion below equation (1.77) show that this is asymptotically a chi-square random variable with 1 degree of freedom. As stated above, this sketch leaves many loose ends to be cleared up, but it is enough to demonstrate the various regularity restrictions that are assumed.

We illustrate the importance of these regularity restrictions by considering four examples, the first two of which are taken from the context of the multinomial distribution (2.30). In the second and third examples the regularity conditions do not hold (for two different reasons in the two examples), and the approximate chi-square distribution of $-2 \log \lambda$ does not apply.

9.4.2 Examples

In this section we discuss four examples of the use and potential misuse of the $-2\log\lambda$ theory.

Example 1. A frequently occurring procedure is that of testing a null hypothesis that specifies the values of the set of parameters $\{p_i\}$ in the multinomial distribution (2.30) as the set of values $\{p_{i0}\}$. The alternative hypothesis makes no restriction on the probabilities $\{p_i\}$, other than the obvious ones that they be non-negative and sum to 1. This implies that there are $k-1$ free parameters under the alternative hypothesis. In this test the null hypothesis is simple and no parameter estimation is required, and the likelihood is as given in (2.30), with p_i replaced by p_{i0}. The alternative hypothesis is composite, and under this hypothesis the maximum likelihood estimate of p_i is $\hat{p}_i = Y_i/n$, and the likelihood is as given in (2.30) with p_i replaced by $\hat{p}_i$. From this the likelihood ratio λ is given by

$$\lambda = \prod_i \left(\frac{np_{i0}}{Y_i}\right)^{Y_i}. \tag{9.25}$$

Thus $-2\log\lambda$ is identical to the statistic defined in (3.41). It can be shown (see Problem 9.1) that this statistic and the statistic defined in (3.39) are approximately equal when n is large and $Y_i/n \cong p_{i0}$. The asymptotic $-2\log\lambda$ theory described above then justifies the claim that the statistic (3.39) has an approximate chi-square distribution with $k-1$ degrees of freedom under the null hypothesis. Tables of significance points of this distribution can then be used to assess whether the null hypothesis should be rejected, given an observed value of $-2\log\lambda$.

Example 2. A more subtle testing procedure associated with the multinomial distribution (2.30) arises in the context of in situ hybridization. We describe it here to illustrate a case where the $-2\log\lambda$ theory does not apply, rather than because of its genetical relevance.

In an experimental procedure that is simplified considerably here, clones (that is, short DNA sequences) from one species tend to hybridize to sufficiently similar, or *homologous*, clones in another species. The random factors involved in the experiment, described below, imply that a statistical analysis in needed.

As a simple example, suppose that chromosome i of species A contains a proportion p_i of the DNA in the genome. If the clone from species B has no homologue in the DNA of species A and hybridization occurs, the probability that the clone hybridizes to chromosome i is taken to be p_i. If some unknown chromosome j in species A contains a single homologue of the clone from species B, and no other chromosome contains a homologue, then given that hybridization occurs, the probability of hybridization to this chromosome is assumed to be of the form $1 - \theta + p_j\theta$. The probability that hybridization, if it occurs, is to chromosome i ($i \neq j$) is taken to be $p_i \theta$. The rationale behind these formulae is that θ is an unknown parameter giving the probability of a random hybridization, so that the term $1 - \theta$ in the probability of hybridization to chromosome j is the probability of a hybridization specifically to the homologue on that chromosome. Given data on the number of hybridizations to the various chromosomes in species A, the aim is to test for a significant excess of hybridizations to one of the chromosomes in species A.

In statistical terms, the null hypothesis is that $\theta = 0$. Under the alternative hypothesis both θ and j are unknown parameters that must be estimated in the testing procedure. The details of this testing procedure are omitted here: These can be found in Ewens et al. (1992). The $-2 \log \lambda$ ratio procedure leads, asymptotically, to

$$-2 \log \lambda = (Z_{\max})^2,$$

where

$$Z_{\max} = \max_i \frac{n_i - n p_{i0}}{\sqrt{n p_{i0}(1 - p_{i0})}},$$

n_i is the number of observations in category i and $n = \sum_j n_j$. The probability distribution of the test statistic $Z_{\max}$ is *not* related to a chi-square. The reason for this is that the parameter j is discrete and the continuous real parameter theory underlying the $-2 \log \lambda$ approach does not apply.

Example 3. Suppose we wish to test the null hypothesis that the parameter M in the density function of the uniform distribution (8.14) takes some specified null hypothesis value M_0 against an alternative hypothesis that does not specify the value of M. If the null hypothesis is true, the likelihood of the observations $X_1, X_2, \ldots, X_n$ is M_0^{-n}. Under the alternative hypothesis the maximum likelihood estimator of M is $X_{\max}$ and the maximum of the likelihood is $X_{\max}^{-n}$. When the null hypothesis is true the likelihood ratio is then

$$-2 \log \lambda = 2n \log \frac{M_0}{X_{\max}}.$$

This statistic does not have an asymptotic chi-square distribution as $n \to +\infty$. This conclusion does not contradict the $-2 \log \lambda$ theory, since the maximum of the likelihood M^{-n} under the alternative hypothesis arises at the boundary point $\hat{M} = X_{\max}$.

Examples 2 and 3 illustrate the fact that the above $-2 \log \lambda$ theory is applicable only under certain restrictive circumstances. We return to this topic in Section 15.9, when considering tests of hypotheses relating to phylogenetic trees.

Example 4. It was stated in Example 4 of Section 3.5 that data often arise in the form of a two-way contingency table such as Table 3.1; examples of this were given in Sections 5.2, 5.3.4, and 6.1. We now show that the test statistic (3.43) that uses the data in Table 3.1 arises from the $-2 \log \lambda$ theory, and that often the numerical value of this statistic is close to the numerical value of the chi-square statistic (3.42).

The probability that any count falls in row j is denoted by $p_{j.}$, that it falls in column k by $p_{.k}$, and that it falls in the cell in row j, column k by p_{jk}. The null hypothesis, that there is no association between row and column classifications, implies that $p_{jk} = p_{j.}p_{.k}$. This null hypothesis defines some subspace of the parameter space corresponding to the p_{jk} values. Under this hypothesis the likelihood corresponding to the data in Table 3.1 is of the form

$$M \prod_{j=1}^{r} \prod_{k=1}^{c} (p_{j.}p_{.k})^{Y_{jk}},$$

where M is a multinomial constant. This is maximized when

$$\hat{p}_{j.} = \frac{y_{j.}}{y}, \quad \hat{p}_{.k} = \frac{y_{.k}}{y}.$$

From this,

$$L_{\max}(\omega) = M \prod_{j=1}^{r} \prod_{k=1}^{c} \left(\frac{y_{j.}}{y} \frac{y_{.k}}{y} \right)^{Y_{jk}}. \tag{9.26}$$

Under the alternative hypothesis the probability p_{jk} is unspecified. The maximum likelihood estimator of p_{jk} is Y_{jk}/y, and this leads to

$$L_{\max}(\Omega) = M \prod_{j=1}^{r} \prod_{k=1}^{c} \left(\frac{Y_{jk}}{y} \right)^{Y_{jk}}. \tag{9.27}$$

From (9.26) and (9.27)

$$-2 \log \lambda = 2 \sum_{j=1}^{r} \sum_{k=1}^{c} Y_{jk} \log \left(\frac{Y_{jk}}{E_{jk}} \right), \tag{9.28}$$

where $E_{jk} = y_{j.}y_{.k}/y$. (This notation arises because E_{jk} is often referred to as the expected number of counts in cell (j, k) when the null hypothesis is true.) The right-hand side in (9.28) is the test statistic (3.43).

We now show that the numerical value of this test statistic is often close to that of the chi-square statistic (3.42). Suppose that Y_{jk} is close to E_{jk} and write $Y_{jk} = E_{jk}(1 + \delta_{jk})$. The right-hand side in (9.28) is then

$$2 \sum_{j=1}^{r} \sum_{k=1}^{c} E_{jk}(1 + \delta_{jk}) \log(1 + \delta_{jk}).$$

From the logarithmic approximation (B.25) this is approximately

$$2 \sum_{j=1}^{r} \sum_{k=1}^{c} E_{jk}(1 + \delta_{jk}) \left(\delta_{jk} - \frac{1}{2}\delta_{jk}^2 \right).$$

The identity $\sum_{j=1}^{r} \sum_{k=1}^{c} E_{jk}\delta_{jk} = 0$ implies that if terms of order δ_{jk}^3 are ignored, this is

$$\sum_{j=1}^{r} \sum_{k=1}^{c} E_{jk}\delta_{jk}^2 = \sum_{j=1}^{r} \sum_{k=1}^{c} \frac{(Y_{jk} - E_{jk})^2}{E_{jk}},$$

and this is the chi-square statistic (3.42).

9.5   The Analysis of Variance (ANOVA)

9.5.1   Introduction

Perhaps the most frequently used hypothesis testing procedure in statistics is that of the Analysis of Variance (ANOVA). The simplest ANOVA, discussed in detail in Section 9.5.3, can be regarded as a generalization of the two-sample equal variance t-test, and we shall approach ANOVA through this generalization.

As described in Section 3.5.2, the two-sample t-test tests for equality of the means of two groups. In the notation of Section 3.5.2, the model adopted for the test is that

$$X_{ij} = \mu_i + E_{ij}, \ i = 1, 2, \tag{9.29}$$

where the X_{ij} are assumed to be independent and the E_{ij} are assu8med to be NID$(0, \sigma^2)$ random variables. The null hypothesis being tested is $\mu_1 = \mu_2$. This model is also often written in the form

$$X_{ij} = \mu + \alpha_i + E_{ij}, \ i = 1, 2. \tag{9.30}$$

In this model we can think of μ as an overall mean and α_j as a deviation from this overall mean characteristic of group j. In this form the model is *overparameterized*. There are three parameters in the model when only two are necessary. This overparameterization can be overcome by requiring that α_1 and α_2 satisfy the requirement $m\alpha_1 + n\alpha_2 = 0$, and we always assume that this requirement is imposed. It might seem to be a roundabout approach to write the model (9.29) in the form (9.30), with the condition $m\alpha_1 + n\alpha_2 = 0$ imposed, but for several reasons it is convenient to do so. The test of hypothesis is identical in the two representations of the model.

9.5.2 From t to F: Sums of Squares and the F Statistic

We start by deriving a test procedure equivalent to the two-sample two-sided t-test described in Example 2 of Section 3.5. This test is carried out by using $|t|$, the absolute value of the T statistic, as test statistic. It is equivalent to use t^2 as test statistic, since this is a monotonic function of $|t|$. Straightforward algebraic manipulation (see Problem 9.4) shows that if the square of the t statistic defined in (3.30) is written as F, then

$$F = \frac{B}{W}(m+n-2),\tag{9.31}$$

where, with the random variables involved in the procedure defined as in Section 3.5.2,

$$\bar{X}_1 = \sum_{j=1}^{m} X_{1j}/m, \quad \bar{X}_2 = \sum_{j=1}^{n} X_{2j}/n, \quad \bar{\bar{X}} = (m\bar{X}_1 + n\bar{X}_2)/(m+n),$$

$$B = \frac{mn}{m+n}(\bar{X}_1 - \bar{X}_2)^2 = m(\bar{X}_1 - \bar{\bar{X}})^2 + n(\bar{X}_2 - \bar{\bar{X}})^2\tag{9.32}$$

and

$$W = \sum_{j=1}^{m}(X_{1j} - \bar{X}_1)^2 + \sum_{j=1}^{n}(X_{2j} - \bar{X}_2)^2.\tag{9.33}$$

B is the *between group sum of squares* (more precisely called the *among group sum of squares*), and W is called the *within group sum of squares*. By assumption the random variables X_{ij} have normal distributions and thus both B and W are continuous random variables.

The sum of B and W can be shown to be

$$\sum_{i=1}^{m}(X_{1i} - \bar{\bar{X}})^2 + \sum_{i=1}^{n}(X_{2i} - \bar{\bar{X}})^2.\tag{9.34}$$

This is called the *total sum of squares*. The total number of degrees of freedom $m+n-1$ is one less than the number of random variables. Just as the total sum of squares is split up into a between group component and a within group component, so also the total number of degrees of freedom is split up into two components, 1 degree of freedom between groups and $m + n - 2$ degrees of freedom within groups.

The two main components of the statistic F are B and W. Our aim is to test for significant differences between the means of the two groups, and the component B measures this difference by the difference between the group averages $\bar{X}_1$ and $\bar{X}_2$. The observed value of $\bar{X}_1 - \bar{X}_2$ will tend to be large when the means of the two groups differ. However, the significance of any such difference must be measured relative to the variation within groups. The component W measures this variation, and is unaffected by any difference in means between the two groups. Large observed values

320 9. Statistics (iii): Classical Hypothesis Testing Theory

of F arise when the observed variation between groups is large compared with the observed variation within groups, and sufficiently large observed values of F give significant evidence that a difference exists between the two means. The ANOVA procedure makes this precise, as follows.

It can be shown that, when the null hypothesis $\mu_1 = \mu_2$ is true and when the standard ANOVA assumptions listed above are met, B/σ^2 and W/σ^2 are independent chi-square random variables having respectively 1 and $m + n - 2$ degrees of freedom. The ratio (9.31) has the F distribution (2.152) with 1 and $m+n-2$ degrees of freedom (see Problem 2.28). The test of the null hypothesis is then carried out by referring the observed value of F to tables of significance points of the F distribution (2.152) with 1 and $m + n - 2$ degrees of freedom. For any desired Type I error, the values of F that lead to rejection of the null hypothesis can be found from these tables.

This procedure demonstrates the two key steps in any ANOVA. The first is the subdivision of the observed total sum of squares into several components (in the above case the observed values of B and W), each measuring some meaningful component of variation. The second step is the comparison of these components to test some hypothesis, using for each comparison the appropriate F statistic.

The two-group comparison above generalizes immediately to a test for the equality of the means of any number of groups, and then to a hierarchy of further ANOVA tests. We now describe examples of these ANOVA tests. All of them are particular cases of the multiple regression test introduced briefly in Section 8.4.3. In illustrating these ANOVA tests we change the generic multiple regression notation Y on the left-hand side of (8.42) to X, which is standard notation in ANOVA.

9.5.3 One-way Fixed Effects ANOVA

The one-way ANOVA test is a direct generalization of the two-sample t-test to the case of an arbitrary number g of groups, in which the null hypothesis is that the means of all the groups are equal. We write the n_i observations in group i as $X_{i1}, X_{i2}, \ldots, X_{in_i}$ for $i = 1, 2, \ldots, g$. The model generalizing (9.29) is

$$X_{ij} = \mu_i + E_{ij}, \quad j = 1, 2, \ldots, n_i, \quad i = 1, 2, \ldots, g, \qquad (9.35)$$

and the model generalizing (9.30) is

$$X_{ij} = \mu + \alpha_i + E_{ij}, \quad j = 1, 2, \ldots, n_i, \quad i = 1, 2, \ldots, g. \qquad (9.36)$$

In both models the E_{ij} are assumed to be NID$(0, \sigma^2)$. The model (9.36), like the model (9.30), is overparameterized, and the overparameterization is overcome by requiring that $\sum n_i \alpha_i = 0$. This overparameterized model is often more convenient to use than is (9.35), and the test of hypothesis is the same in both models.

In some cases the assumption of independence made above is not appropriate and an analysis different from that given in this section is appropriate. This is discussed in Section 9.7.

There are several different ways in which ANOVAs can be classified. One of these concerns the distinction between a *fixed effect* model, a *mixed effects* model and a *random effect* model. For the moment we consider the properties of a fixed effects model; some of the properties of mixed and random effects models, and the distinction between the three forms of models, are discussed in Section 9.5.7. In a fixed effects model, the parameter α_i in (9.36) is considered to be a fixed (but unknown) constant characteristic of group i.

The null hypothesis in the ANOVA model (9.36) is that $\alpha_1 = \alpha_2 = \cdots = \alpha_g = 0$. The parameter μ is not involved in the test. It should be noted that despite the appearance of the word "variance" in the expression "Analysis of Variance," the hypothesis tested in this (and any) ANOVA procedure is a test about *means.* discussed later.

If the requirement $\sum n_j \alpha_j = 0$ is not imposed, the model (9.36) is overparameterized, and is a particular case of the multiple regression model (8.43). This may be seen by writing

$$\mathbf{Y} = (X_{11}, X_{12}, \ldots, X_{1n_1}, X_{21}, X_{22}, \ldots, X_{2n_2}, \ldots, X_{g1}, X_{g2}, \ldots, X_{gn_g})',$$

$$\mathbf{E} = (E_{11}, E_{12}, \ldots, E_{1n_1}, E_{21}, E_{22}, \ldots, E_{2n_2}, \ldots, E_{g1}, E_{g2}, \ldots, E_{gn_g})',$$

$$\boldsymbol{\beta} = (\mu, \alpha_1, \ldots, \alpha_g)',$$

and choosing as the matrix C in (8.43) the one-way ANOVA "design matrix," defined as in (9.37). In this matrix, all elements in the first column are 1, the first n_1 elements in the second column are 1 and the rest are 0, the elements in rows $n_1 + 1$ through n_2 of column 3 are 1 and the rest 0, ..., the elements in rows $n_1 + n_2 + \cdots + N_{g-1} + 1$ through $n_1 + n_2 + \cdots + n_g$ of column g are 1 and the rest are 0.

The matrix C so defined is not of full rank, so that the multiple regression theory of Section 8.4.3, and in particular equation (8.44), cannot be applied directly. This problem arises only because the model is overparameterized. Imposition of the requirement $\sum n_i \alpha_i = 0$ implies that (for example) α_g can be written in terms of $\alpha_1, \ldots, \alpha_{g-1}$ and this leads to a multiple regression model in which the the matrix replacing C is of full rank. This allows us to use multiple regression theory directly.

The test procedure is a direct extension of the two-group procedure described in Section 9.5.2. The total sum of squares $\sum_{i=1}^{g} \sum_{j=1}^{n_i} (X_{ij} - \bar{\bar{X}})^2$ is subdivided into the between group sum of squares B and the within group sum of squares W. These are the direct generalizations of the two-group

$$
\begin{bmatrix}
1 & 1 & 0 & \cdots & 0 \\
1 & 1 & 0 & \cdots & 0 \\
\vdots & \vdots & \vdots & \vdots & \vdots \\
1 & 1 & 0 & \cdots & 0 \\
1 & 0 & 1 & \cdots & 0 \\
1 & 0 & 1 & \cdots & 0 \\
\vdots & \vdots & \vdots & \vdots & \vdots \\
1 & 0 & 1 & \cdots & 0 \\
\vdots & \vdots & \vdots & \vdots & \vdots \\
\vdots & \vdots & \vdots & \vdots & \vdots \\
1 & 0 & 0 & \cdots & 1 \\
\vdots & \vdots & \vdots & \vdots & \vdots \\
1 & 0 & 0 & \cdots & 1
\end{bmatrix}.
\tag{9.37}
$$

values in (9.32) and (9.33), and are defined by

$$
B = \sum_{i=1}^{g} n_i (\bar{X}_i - \bar{\bar{X}})^2,
\tag{9.38}
$$

$$
W = \sum_{i=1}^{g} \sum_{j=1}^{n_i} (X_{ij} - \bar{X}_i)^2.
\tag{9.39}
$$

Here $\bar{X}_i = \sum_{j=1}^{n_i} X_{ij}/n_i$ and $\bar{\bar{X}} = \sum_{i=1}^{g} \sum_{j=1}^{n_i} X_{ij}/N$, where N is defined as $\sum_{i=1}^{g} n_i$. Just as the total sum of squares is split up into two components, so also the total degrees of freedom $N - 1$ is subdivided into two components, the between group degrees of freedom $g - 1$, and the within group degrees of freedom $N - g$.

We define the *between group mean square* as the between group sum of squares divided by the between group degrees of freedom, and the *within group mean square* as the within group sum of squares divided by the within group degrees of freedom. The test statistic F is the ratio of these two mean squares, or equivalently

$$
F = \frac{B}{W} \times \frac{N - g}{g - 1}.
\tag{9.40}
$$

The test statistic F in (9.40) is the direct generalization of that in (9.31), and has the F distribution with $g - 1, N - g$ degrees of freedom when the null hypothesis is true and the standard ANOVA assumptions are all met. The test is therefore carried out by referring the observed value of F to significance points of this F distribution.

There are various ways in which the ANOVA assumptions might not be met. The first of these arises if the random variables X_{ij} do not have a normal distribution. The procedure is fairly *robust* against non-normality, so that the F statistic has approximately the F distribution for non-normal data, provided that the non-normality is not extreme. Non-parametric alternatives generalizing the Mann–Whitney test of Section 3.8.2, and also a permutation procedure, are available if non-normality appears to be extreme. As with the t-test, the ANOVA test is fairly robust when the variances in the various groups differ, at least when the group sizes are equal.

9.5.4   The Two-Way Fixed Effects ANOVA

We now extend the one-way fixed effects ANOVAs discussed in Section 9.5.3 to two-way ANOVAs. We provide only an overview of this model, since a full description of its properties is beyond the scope of this book.

As an example of a two-way ANOVA, suppose that we wish to compare the expression levels of a gene in individuals having lung cancer. These individuals fall into a different risk classes (for example, ultrahigh, very high, intermediate, and low) and also into b different age groups. Suppose that the data consist of n individuals for each risk-class/age-group combination. Before the observations are taken, the typical expression level can be represented as X_{ijk}, where i indicates the risk class involved, $(i = 1, 2, \ldots, a)$, j indicates the age group involved, $(j = 1, 2, \ldots, b)$, while k takes the possible values $1, 2, \ldots, n$, corresponding to the n individuals in each risk class/age-group combination.

These random variables can be arranged in a two-way table, each column in the table corresponding to one risk class and one row to one age group. This explains the terminology "two-way" ANOVA. Each risk-class/age-group combination defines a "cell." The most important requirement in a two-way ANOVA is replication, that is, in this case, of obtaining more than one individual in each cell. With the data as described above we have n replications in each cell. This is a balanced design, since the value of n is the same for each cell. The theory for an unbalanced design is more complicated and is not considered here; see for example Sokal and Rohlf (1995) for a discussion of this case.

The model assumed in a two-way ANOVA is that

$$X_{ijk} = \mu + \alpha_i + \beta_j + \delta_{ij} + E_{ijk}, \ i = 1, 2, \ldots, a, \ j = 1, 2, \ldots, b, \ k = 1, 2, \ldots, n, \tag{9.41}$$

where the E_{ijk} are $\text{NID}(\mu, \sigma^2)$ random variables. Thus the mean of X_{ijk} is assumed to be of the form $\mu + \alpha_i + \beta_j + \delta_{ij}$. Here α_i is fixed parameter and is an additive contribution to the mean corresponding to risk class i, β_j is a fixed parameter and is an additive contribution to the mean corresponding to age group j, and δ_{ij} is a fixed risk-class/age-group interaction parameter whose meaning is discussed below. An interaction term of the form δ_{ij} can and should be added in any two-way ANOVA model if the possibility of interaction exists, as is presumably the case in this example. If there were no replication, so that $n = 1$, it is impossible to test for the existence of any interaction. This is the main reason why replication is important. A two-way ANOVA model that assumes that there is no interaction is of the form

$$X_{ijk} = \mu + \alpha_i + \beta_j + E_{ijk}, \quad i = 1, 2, \ldots, a, \ j = 1, 2, \ldots, b, \ k = 1, 2, \ldots, n.$$
$$(9.42)$$

We return to this model below.

As it stands the model (9.41) is overparameterized, and the overparameterization is removed by imposing the constraints $\sum_i \alpha_i = \sum_j \beta_j = 0$, $\sum_i \delta_{ij} = 0$ for all j, $\sum_j \delta_{ij} = 0$ for all i.

It should be noted that an additivity assumption is made in the model (9.41). The mean of X_{ijk} is assumed to be the sum of four terms, corresponding to an overall mean, an additive contribution from risk class i, an additive contribution from age group j, and an additive interaction term. There is no reason a priori why such an additivity assumption is appropriate, and if the additivity assumption is not justified, unreliable P-values can arise in the analysis. Authors in the bioinformatics literature often appear to be willing to assume an additive model, perhaps after a transformation of the original data (often by carrying out the analysis on the logarithms of the original data), under the assumption that the procedure is robust against mild non-additivity.

While the model (9.41) is more complicated than (9.36), it still can be represented in the general linear model form (8.42) and can be analyzed by general linear model methods.

In the two-way ANOVA with replication the total sum of squares is divided into four components. In the example considered above these are a risk-class sum of squares, an age-group sum of squares, an interaction sum of squares and a within cells, often called "residual" or "error", sum of squares. Associated with each sum of squares is a corresponding degrees of freedom and hence a corresponding mean square, defined as the sum of squares divided by the degrees of freedom. We do not give the details of these partitions here.

The mean squares are then compared using F ratios to test for significance of the various effects. The first step in the procedure is to test for a significant risk-group/age-class interaction. The F ratio used for this is the ratio of the interaction mean square and the within cells mean square. It may not be reasonable to test for significant differences between risk classes and also between age groups if a significant risk-class/age-group interaction is found. Such a situation would arise, for example, if there were two risk classes and two age groups, and the averages of the observations in the four cells are as shown in Table 9.1. Here there is no evidence of interaction, since the value of the measurement appears to increase additively from risk class 1 to risk class 2 by the same amount in the two age classes.

	risk class	
	1	2
age group 1	4	12
age group 2	7	15

Table 9.1. No evidence of interaction.

One problem with the procedure described above is that interaction effects are often hard to detect, so that the investigator might proceed to test main effects under an erroneous impression that interaction is not significant.

If there does appear to be a significant interaction, one common practice is not to test the main effects (risk classes and age groups) under the argument that if the interaction is significant, no uniform main effect statement is possible. Such a situation appears to arise, for example, if the averages of the observations in the four cells were as shown in Table 9.2. Here the value of the measurement appears to increase from risk class 1 to risk class 2 for one age group but to decrease for the other, so that it may not, for example, be said to be uniformly higher in one risk group compared to the other. On the other hand, if the averages in the four cells were as shown

	risk class	
	1	2
age group 1	4	15
age group 2	11	6

Table 9.2. An example of interaction.

in Table 9.3, there might be significant interaction, but it might still be is reasonable to conclude that there is a significant difference in the measurement between risk classes. In cases of this type, testing for main effects should be done with caution.

	risk class	
	1	2
age group 1	4	12
age group 2	3	18

Table 9.3. A second example of interaction.

In cases where there is no replication, the only way in which main effects (risk groups, age classes) can be tested arises when we are willing to assume the no interaction model (9.42). In this case we use the interaction mean square as a residual and it would appear in the denominator of the F ratios used to test for risk-class and age-group effects. However, making the assumption that there is no interaction can be dangerous.

9.5.5   Multi-Way Fixed Effects ANOVAs

It is straightforward to extend one-way and two-way fixed effects ANOVAs to multi-way ANOVAs. The models become quite complicated when interactions are allowed. For example, the overparameterized version of a three-way ANOVA model is of the form

$$X_{ijkm} = \mu + \alpha_i + \beta_j + \gamma_k + \delta_{ij} + \eta_{ik} + \nu_{jk} + \psi_{ijk} + E_{ijkm}, \qquad (9.43)$$

involving three two-way interactions and one three-way interaction. Testing is again done in a hierarchical way, starting with the three-way interaction, proceeding to two-way interactions if the three-way interaction is not significant, and then to main effects if the various two-way interactions are not significant.

9.5.6   The 2^m Design, Confounding, and Fractional Replication

A particular form of the one-way ANOVA is the 2^m factorial design. Here we consider m "factors," each taken at two "levels." There are 2^m possible combinations of levels, or groups. In this model we can test both for main effects and for interactions between main effects. In order to assess which main effects and which interactions are significant it is necessary to carry out a replicated experiment, and we suppose that there are n replications for each of the 2^m treatments.

We shall illustrate this model using the case $m = 3$, so that there are eight groups, and use as an example the case where the factors are gender (female or male), affectedness status (affected or not affected), and tissue (lung or kidney). We denote these three factors considered by the letters A, B and C, and the eight groups by $abc, ab, ac, bc, a, b, c, 1$. If a corresponds to female, b to affected and c to lung, the group "ac," for example, refers to data from lung tissue from an unaffected female and the group "b" refers to data from kidney tissue from an affected male.

We write the totals of the n observations for the various groups as $T_{abc}, T_{ab}, \ldots, T_1$. The total between group sum of squares can be subdivided into seven individual sums of squares, corresponding respectively to the three main effects (A, B and C), the three pair-wise interactions (AB, AC and BC), and one triple-wise interaction (ABC). As an example, the sum of squares for the main effect of A (gender) is

$$(T_{abc} + T_{ab} + T_{ac} + T_a - T_{bc} - T_b - T_c - T_1)^2/8n. \qquad (9.44)$$

In this expression any total involving data from a female is added in the numerator and any total involving data from a male is subtracted. Less obviously, the sum of squares for the BC (affectedness status/tissue) interaction is

$$(T_{abc} - T_{ab} - T_{ac} + T_a + T_{bc} - T_b - T_c + T_1)^2/8n. \qquad (9.45)$$

Expressions such as these provide the numerator in the various F ratios for testing for the three main effects, the three pairwise interactions, and the one triple-wise interaction. We do not pursue the analysis of this example in detail since our main interest is in confounding and partial replication procedures associated with it.

When m is about five or more, the number of groups becomes large, so that the total number of observations required, $n2^m$, is also large. In this case it is possible to reduce the number of observations by the process of *confounding*. Thus it might be claimed that, in the above example, the ABC interaction is likely to be very small or non-existent and in any event not of interest. If this claim is justified a design which loses information about this interaction would be preferred to a design that retains this information, since it reduces the amount of data to be collected. There are standard ANOVA designs that achieve this aim.

A concept closely associated with confounding is that of fractional replication. To illustrate this concept, suppose that in the example above, data are obtained only from the groups abc, a, b, and c, that is lung data from affected females, kidney data from unaffected females, kidney data from affected males, and lung data from unaffected males. Then only the totals T_{abc}, T_a, T_b, and T_c are available. The expressions (9.44) and (9.45) show that the sum of squares for both the main effect A and the BC interaction are

$$(T_{abc} + T_a - T_b - T_c)^2/4n.$$

(The denominator term involves 4 rather than 8 since now only four groups are considered.) This means that the main effect of A and the BC interaction effect cannot be distinguished from each other, and A and BC are said to be the *alias* of each other. Similarly the alias of B is AC and the alias of C is AB. The between groups sum of squares is now split up into three components, and a main effect, for example A, can be tested for significance only if it is assumed that its alias effect BC is zero.

A design involving aliasing in the context of microarrays will be discussed in Section 13.3.7.

9.5.7   Random and Mixed Effects Models

So far we have considered only fixed effects models. In the two-way ANOVA of Section 9.5.4, for example, risk classes and age groups are considered fixed. They have a significance to us and if we conducted two experiments on two different occasions we would retain these categories. Other factors In other cases an effect is thought of as "random." For example, if the data in the risk-class/age-group experiment were collected on several different essentially randomly chosen days, we have no specific interest in these various days. Nevertheless we must extract a "between days" sum of squares in the analysis, since if there is a significant day to day effect, failure to do so lowers the efficiency of the test for risk classes and age groups.

We shall see below that different procedures are used for fixed and random effects models. Because of this, it is common practice to use Greek letters to denote fixed effects and upper case Roman letters to denote random effects. As an example, suppose that in the the example of Section 9.5.4, data are collected on a risk classes on d different days, with n data values being taken on each day, and that no age group data are taken. The day is a random effect, and an appropriate model is of the form

$$X_{ik\ell} = \mu + \alpha_i + D_\ell + G_{i\ell} + E_{ik\ell}, \tag{9.46}$$

where the suffix ℓ refers to days.

It is assumed that $\sum \alpha_i = 0$, that the $E_{ik\ell}$ are NID$(0, \sigma^2,)$ that the D_ℓ are random variables corresponding to the various days and are NID$(0, \sigma_D^2)$, and that the $G_{i\ell}$ are risk-classes × days interaction random variables and are NID$(0, \sigma_{AD}^2)$. Any interaction term involving at least one random effect is necessarily random, and this notation is followed in (9.46). This is a *mixed effects* model, with one effect fixed and the other random.

The details of an ANOVA analysis depend on which effects are fixed and which are random. For example, in a two-way ANOVA described by (9.46), having one effect fixed (risk classes) and one random effect (days), the fixed effect mean square is compared to the interaction mean square for significance. The reason for this can be seen from (9.47), which displays the expected values of the three mean squares.

$$\text{between risk classes} = \sigma^2 + n\sigma_{AD}^2 + nd \sum_i \alpha_i^2/(a-1),$$

$$\text{interaction} = \sigma^2 + n\sigma_{AD}^2, \tag{9.47}$$

$$\text{residual} = \sigma^2.$$

The null hypothesis, that there is no risk-class effect, claims that the α_i are all zero, and when this null hypothesis is true the expected values of the between risk classes mean square and the interaction mean square are identical. This implies that the appropriate test statistic is the ratio of these two mean squares, and the significance of this ratio is found from F tables with $a - 1$ and $(a - 1)(b - 1)$ degrees of freedom. This is not true of the ratio of the expected values of the between risk classes mean square and the residual mean square, since this ratio is not equal to 1 under the null hypothesis.

On the other hand, in the original risk-class/age-group example, the expected values of these three mean squares are as shown in (9.48),

$$\text{between risk classes} = \sigma^2 + nb \sum_i \alpha_i^2/(a-1),$$

$$\text{interaction} = \sigma^2 + n \sum_i \sum_j \gamma_{ij}^2/\{(a-1)(b-1)\}, \qquad (9.48)$$

$$\text{residual} = \sigma^2,$$

so that when the null hypothesis is true the expected values of the between risk classes mean square and the residual mean square are identical in this model. This implies that the appropriate F ratio to test the null hypothesis is the between risk classes mean square divided by the residual mean square. The use of different denominator mean squares for different models has been emphasized in the microarray context by Churchill (2002).

The model for multi-way mixed effects ANOVAs can become quite complicated. As a simple example, and upper case Roman letters suppose that in the the example of Section 9.5.4, data are collected on d different days. An appropriate model is now

$$X_{ijk\ell} = \mu + \alpha_i + \beta_j + \delta_{ij} + D_\ell + G_{i\ell} + H_{j\ell} + P_{ij\ell} + E_{ijk\ell}, \qquad (9.49)$$

where the suffix ℓ refers to days. If there are several fixed effects and several random effects, the number of interactions becomes quite large. This often occurs in the microarray context, and often many of these interactions are simply assumed to be zero in order to simplify the analysis. We do not discuss this matter further here, and return to the question of using ANOVA models to analyze microarray data in Chapter 13.

9.6 Multivariate Methods

9.6.1 Introduction

In multivariate analysis we consider data consisting of random vectors $(X_1, X_2, \ldots, X_k)'$ rather than of (scalar) random variables. Such a vector might arise, as in Section 8.5, if we contemplate measuring k characteristics on each individual, for example height, weight, arm length, etc. The elements in the vector are usually correlated, as would be the case in this example, and indeed these correlations lie at the heart of multivariate analysis.

9.6.2 One-sample T^2 Tests

In this section we consider the multivariate generalization of the one-sample two-sided t-test of Example 1 in Section 3.5. The basic concepts and assumptions are identical to those in this t-test, the only change being that vector random variables replace scalar random variables, and the discussion below mirrors that of the t-test case.

Suppose that $\mathbf{X}_1, \mathbf{X}_2, \ldots, \mathbf{X}_n$ are independent k-dimensional random vectors, each having the multivariate normal distribution (2.33) with mean vector $\boldsymbol{\mu} = (\mu_1, \mu_2, \ldots, \mu_k)'$ and variance–covariance matrix Σ. We wish to test the null hypothesis that $\boldsymbol{\mu} = \boldsymbol{\mu}_0$, where $\boldsymbol{\mu}_0 = (\mu_{10}, \mu_{20}, \ldots, \mu_{k0})'$ is some given vector of means. The alternative hypothesis leaves $\boldsymbol{\mu}$ unspecified. The values of the elements in the matrix Σ are unspecified.

Each of the k different null hypotheses $\mu_p = \mu_{p0}$ could be tested, separately from the others, by a t-test. However, the T^2 procedure described below is to be preferred to carrying out k separate t-tests. First, the multiple testing problem discussed in Section 3.11 arises when k tests are carried out in parallel. If an experiment-wise Type I error of α is desired, each individual t-test should have a Type I error of about α/k. This problem does not arise with the T^2 test. Second, and perhaps more important, the T^2 procedure automatically exploits the correlational structure between the k different measurements. Naïve application of separate t-tests does not do this.

We first use the data to estimate $\boldsymbol{\mu}$ and Σ. The estimator of $\boldsymbol{\mu}$ is the direct parallel of the estimator in the scalar case, namely the vector of averages $\bar{\mathbf{X}}' = (\bar{X}_1, \bar{X}_2, \ldots, \bar{X}_k)'$. The typical variance estimator follows the prescription of (3.5) and the typical covariance estimator follows the prescription of (3.17). We denote the matrix formed by these estimators by $\hat{\Sigma}$. With these definitions, the test statistic used to test the null hypothesis is T^2, defined by

$$T^2 = n(\bar{\mathbf{X}} - \boldsymbol{\mu}_0)'\hat{\Sigma}^{-1}(\bar{\mathbf{X}} - \boldsymbol{\mu}_0). \tag{9.50}$$

When the null hypothesis is true, the statistic F, define by

$$F = \frac{(n-k)T^2}{(n-1)k},$$

has an F distribution with $(k, n-k)$ degrees of freedom. The observed value of F, calculated from the data, may then be used to test the null hypothesis.

We do not discuss this procedure further, since our main interest is in the two-sample case, considered in Section 9.6.3. We do however return to this one-sample case in Section 9.7.

9.6.3   Two-sample T^2 Tests

In this section we consider the multivariate generalization of the two-sample two-sided equal variance t-test of Example 2 in Section 3.5, testing now for equality of the vector of means of two groups of interest. As for the one-sample case, the basic concepts and assumptions are identical to those in the two-sample two-sided equal variance t-test, with vector random variables replacing scalar random variables, and the discussion below mirrors that of the t-test case. As in the one-sample case, we do not wish to carry out the test using k separate t-tests.

Suppose that $\mathbf{X}_{11}, \mathbf{X}_{12}, \ldots, \mathbf{X}_{1m}$ are independent random vectors, each having the multivariate normal distribution (2.33) with mean vector $\boldsymbol{\mu}_1$ and variance–covariance matrix Σ, and $\mathbf{X}_{21}, \mathbf{X}_{22}, \ldots, \mathbf{X}_{2n}$ are independent random vectors, each having the multivariate normal distribution (2.33) with mean vector $\boldsymbol{\mu}_2$ and variance–covariance matrix Σ. It is also assumed that $\mathbf{X}_{1i}$ is independent of $\mathbf{X}_{2j}$ for all i and j. The aim is to test the null hypothesis that the mean vectors in the two groups are identical, that is that $\boldsymbol{\mu}_1 = \boldsymbol{\mu}_2$ $(= \boldsymbol{\mu},$ unspecified). The alternative hypothesis leaves the values of both $\boldsymbol{\mu}_1$ and $\boldsymbol{\mu}_2$ unconstrained. The matrix Σ is unknown and is unspecified under both hypotheses, but is assumed to be the same in the two groups considered.

We first use the data to estimate $\boldsymbol{\mu}_1$, $\boldsymbol{\mu}_2$, and Σ. The estimators of $\boldsymbol{\mu}_1$ and $\boldsymbol{\mu}_2$ are the direct parallels of the estimators in the scalar case: $\boldsymbol{\mu}_i$ is estimated by the vector of averages $\bar{\mathbf{X}}_i' = (\bar{X}_{i1}, \bar{X}_{i2}, \ldots, \bar{X}_{ik})'$, $(i = 1, 2.)$

The estimation of the matrix Σ generalizes the estimation of the variance of a scalar random variable as given in (3.31). The (p, q) element $\hat{\sigma}_{pq}$ in $\hat{\Sigma}$ is defined by

$$\hat{\sigma}_{pq} = \frac{\sum_{j=1}^{m}(X_{1jp} - \bar{X}_{1p})(X_{1jq} - \bar{X}_{1q}) + \sum_{j=1}^{n}(X_{2jp} - \bar{X}_{2p})(X_{2jq} - \bar{X}_{2q})}{m + n - 2},$$

(9.51)

where X_{ijp} is the pth element in the vector $\mathbf{X}_{ij}$ and $\bar{X}_{ip}$ is the pth element in the vector $\bar{\mathbf{X}}_i$, $i = 1, 2$, and likewise for q. For the case $p = q$, this is the estimate $\hat{\sigma}_p^2$ of the variance of the pth measurement, and follows the format of (3.31).

With these definitions, the T^2 statistic is

$$T^2 = \frac{mn}{m + n}(\bar{\mathbf{X}}_1 - \bar{\mathbf{X}}_2)'\hat{\Sigma}^{-1}(\bar{\mathbf{X}}_1 - \bar{\mathbf{X}}_2).$$

(9.52)

When the various distributional assumptions of the random vectors $\mathbf{X}_{ij}$ made above hold and the null hypothesis $\boldsymbol{\mu}_1 = \boldsymbol{\mu}_2$ is true,

$$\frac{m + n - k - 1}{k(m + n - 2)} T^2$$

has an F distribution with k and $m + n - k - 1$ degrees of freedom.

Given the observed vectors $\mathbf{x}_{11}, \mathbf{x}_{12}, \ldots, \mathbf{x}_{1m}$ and $\mathbf{x}_{21}, \mathbf{x}_{22}, \ldots, \mathbf{x}_{2n}$, the observed value of T is found by replacing $\bar{\mathbf{X}}_j$ by $\bar{\mathbf{x}}_j$ and $\hat{\Sigma}$ by S, where the pq element in S is found by replacing X_{ijp} by x_{ijp} and X_{ijq} by x_{ijq} in (9.51). The resulting value is then tested using F tables. Clearly the calculation of the observed value of T assumes that the inverse matrix S^{-1} exists. In almost all classical applications of multivariate analysis this is the case.

9.6.4   Optimal Linear Functions: Discriminant Functions

A procedure associated with the T^2 test is that of discriminant analysis. Suppose that we consider the expression levels of k different genes in two risk classes of a particular tumor. The aim of a discriminant analysis is to find that linear combination of the expression levels in the various genes that best discriminates between the two risk classes. In abstract terms, the aim is to find that linear combination $\alpha_1 X_1 + \alpha_2 X_2 + \cdots + \alpha_k X_k = \boldsymbol{\alpha}'\mathbf{X}$ of the elements in the vector $(X_1, X_2, \ldots, X_k)$ that best discriminates between the two groups. This is a question of estimation, and thus might more naturally be considered in Chapter 8. However, the discriminant function procedure is closely associated with the t and the T^2 test – for example, the test of whether significant discrimination can be made between two groups reduces to a T^2 test – so we consider it, albeit briefly, here.

So far we have not defined the expression "best discriminates." We consider the data vectors $\mathbf{x}_{11}, \mathbf{x}_{12}, \ldots, \mathbf{x}_{1m}$ and $\mathbf{x}_{21}, \mathbf{x}_{22}, \ldots, \mathbf{x}_{2n}$ discussed in Section 9.6.3. The vector $\mathbf{x}_{ij}$ can be replaced by the linear combination $w_{ij} = \sum_k \alpha_k x_{ijk}$. From this it is possible to calculate a t statistic, with the values $w_{11}, w_{12}, \ldots, w_{1m}$ in the first group and $w_{21}, w_{22}, \ldots, w_{2n}$ in the second group. The combination $\boldsymbol{\alpha}'\mathbf{X}$ that best discriminates between the two original groups of vectors is defined as that which maximizes the square of the t statistic computed from the various w_{ij} values. The vector $\mathbf{a}$ achieving this aim is given (see Problem 9.5) by

$$\mathbf{a} = S^{-1}(\bar{\mathbf{x}}_1 - \bar{\mathbf{x}}_2), \tag{9.53}$$

where S is as defined above. This calculation also assumes that S^{-1} exists.

The linear combination with the elements of $\mathbf{a}$ as coefficients is called the (estimated) discriminant function. In using the word "best" above, we mean only best in the sense of maximizing the square of a t statistic, as described. In other contexts a different interpretation of "best" might be appropriate. We take up the matter of discriminant analysis again in Section 13.3.8, in the context of microarrays.

9.7   ANOVA: the Repeated Measures Case

In some cases the assumption of independence of the observations in an ANOVA, made throughout in Section 9.5, is not justified. For example, the heights of a plant measured at different time points are not independent, especially if the time points are close to each other. In this section we consider aspects of the "repeated measures" analysis designed to handle this case, using the "plants" example. Repeated measures designs can be quite complex and different assumptions can lead to different analyses. In particular, a distinction should be made between "univariate" and "multivariate" approaches to the analysis of repeated measures data. A discussion of the different assumptions made under these two approaches is given by von Ende (2001). Here we follow the multivariate approach, which is more appropriate to the use of repeated measures in microarray analysis. Many issues arise within multivariate repeated measures designs, so our brief account is necessarily a superficial one. For a more detailed account of these designs, see Davis (2003). Computational aspects are discussed, for example, in the SAS Institute SAS/STAT (1999) manual.

We consider first the one-sample case. Suppose that we plan to measure the height of n plants at k successive times points $t_1, t_2, \ldots, t_k$. We denote the (at this stage, random) heights of plant i at these time points by $X_{i1}, X_{i2} \ldots, X_{ik}$.

We assume that for each plant these heights have a multivariate normal distribution with unknown mean vector $\boldsymbol{\mu} = (\mu_1, \mu_2, \ldots \mu_k)'$ and unknown variance–covariance matrix Σ. One straightforward (null) hypothesis is that the plants are not growing, implying that $\mu_1 = \mu_2 = \cdots = \mu_k$. Because of the dependence structure, we test this null hypothesis by multivariate methods, specifically by using a T^2 statistic similar to that in Section 9.6.2. This null hypothesis cannot be tested directly by the methods of that section, since the common value of the μ_j is not specified by this null hypothesis, but a simple amendment using differences allows us to use the methods of Section 9.6.2. There are many equivalent ways in which this can be done. As one approach, if we define W_{ij} by $W_{ij} = X_{ij} - X_{i1}$, $j = 2, 3, \ldots, k$, the null hypothesis claims that the mean of W_{ij} is 0 for all i and j. We then carry out a T^2 test of the form of that in Section 9.6.2, using the random variables W_{ij}, whose null hypothesis mean vector $\boldsymbol{\mu}_0$ is $\mathbf{0}'$. Carrying out the test using the W_{ij} values instead of the X_{ij} values requires the estimation of the $k - 1$ means of the W_j random variables as well as an estimate of the variances of and the covariance between these random variables. This is done by standard methods.

We do not give the details of this test here, and have discussed it to serve as an introduction to tests of null hypotheses that might be of more interesting. One such null hypothesis is that the mean height of a plant is a linear function of time, so that μ_j is of the regression form $\alpha + \beta t_j$, for some unknown parameters α and β. However, the regression methods of Section 8.4.3 are not appropriate for the analysis, since those methods in this context would assume the heights of a plant at different times to be independent. Details of the appropriate repeated measures procedures for the test of this and other hypotheses are given by Davis (2003).

We turn now to two-sample repeated measures tests. Suppose that we have two groups of plants, m in the first group and n in the second. As above, the heights of all plants are measured at k successive time points $t_1, t_2, \ldots, t_k$. The vector of mean heights at these time points for plants in the first group is denoted $\boldsymbol{\mu}_1 = (\mu_{11}, \mu_{12}, \ldots, \mu_{1k})'$ and that for plants in the second group is denoted $\boldsymbol{\mu}_2 = (\mu_{21}, \mu_{22}, \ldots, \mu_{2k})'$. One null hypothesis of interest is that $\boldsymbol{\mu}_1 = \boldsymbol{\mu}_2$. This test is carried out directly using the two-sample T^2 test of Section 9.6.3, under the assumption that the variance–covariance matrix Σ is the same in both groups.

This test does not capture the time sequence implicit in the repeated measures data, since the T^2 test of Section 9.6.3 is invariant to any re-ordering of the k random variables considered. However, in the repeated measures context, where this ordering relates to the successive time points, is of central importance. We therefore think of it as a preliminary to tests that do take this order into account.

A perhaps more interesting null hypothesis is that the *profiles* of the means are the same in the two groups. This null hypothesis allows μ_{1j} to differ from $\mu_{2j}, j = 1, 2, \ldots, k$, but claims that the values of $\mu_{1j} - \mu_{2j}, j = 1, 2, \ldots, k$ although possibly non-zero, are the same for all values of j. Standard repeated measures designs are available to test this (less restrictive) null hypothesis. Rejection of this null hypothesis implies an interaction in plant height between time and group membership.

This test also does not capture the time sequence implicit in the repeated measures data. A test that does capture this time sequence is the test of the null hypothesis that μ_{ij} is of the form $\alpha_i + \beta t_j$, implying that the mean growth rates for the two groups of plants are equal, and the same at all time points. This null hypothesis claims that not only are the values of $\mu_{1j} - \mu_{2j}$ the same for all values of j, but also that $\mu_{i(j+1)} - \mu_{ij}$ is independent of j.

9.8   Bootstrap Methods: the Two-sample t-test

Bootstrap estimation methods were introduced in Section 8.6. In this section we consider the bootstrap alternative to the two-sample t-test. The bootstrap procedure has similarities to, and significant differences from, the permutation procedure discussed in Section 3.8.2, and we start by discussing these similarities and differences.

In both procedures we start with data in two groups, and from these calculate the numerical value of some statistic. The data are then rearranged in some way, and the value of the statistic calculated from the rearranged data. This rearrangement procedure is replicated a large number R of times, and the original value of the test statistic is declared to be significant if it is a sufficiently extreme member of all the rearranged values. However the details of the rearrangement procedure are different in the permutation and the bootstrap procedures.

A permutation procedure giving an alternative test to the two-sample t-test was described in Section 3.8.1, where it was noted that is that it relies on the assumption that the variances of the random variables in the two groups are equal. Perhaps the main aim of the bootstrap procedure is to overcome this problem.

A permutation can be thought of as sampling without replacement. Once an observation has been chosen by the permutation procedure to go into one group, it cannot be chosen again. This implies that each observation is placed exactly once into one or the other group, and may well go into the group different from that in which it actually arose. The bootstrap procedure differs from this in two respects. In each of the R bootstrap replications, a random sample of m values is taken *with replacement* from the data values in the first group; these form the "data" in the first group for the replication in question. Similarly, a random sample of n values is taken *with replacement* from the data values in the second group; and these form the "data" in the second group for replication in question. Thus in contrast to the permutation procedure, not only is sampling carried out with replacement, but also any observation always remains within its original group.

The value of some statistic, for example the t statistic, is then computed from each bootstrap sample. The observed value of t is judged to be significant with Type I error α if its value lies among the $100\alpha\%$ most extreme bootstrap t values.

The fact that in the bootstrap procedure sampling is *with* replacement, rather than without replacement, in line with the general bootstrap concepts described in Section 8.6.4, implies that there is no automatic monotonic relation between the t statistic (3.30) and $\bar{x}_1 - \bar{x}_2$. In this respect the bootstrap procedure differs from that of the permutation procedure. The choice of test statistic for the bootstrap procedure therefore requires some discussion. Efron and Tibshirani (1993) claim that use of the t statistic is preferable to use of the difference $\bar{x}_1 - \bar{x}_2$.

A key component of both the permutation and the bootstrap procedures is to use the replications to estimate the null hypothesis distribution of the test statistic t. The randomization method of the permutation test does this automatically, but in the bootstrap procedure an additional step is necessary. The null hypothesis is that the means of the two groups are equal, and to estimate the null hypothesis distribution of t, before the replications are carried out all of the original observations are adjusted in the bootstrap procedure so that the original averages in the two groups are equal. This is perhaps most easily achieved by subtracting from each observation the average of the group it is in.

The following is an important difference between the permutation and the bootstrap procedure. The permutation method applies only to the case where the null hypothesis claims identical probability distributions for the observations in the two groups, since only under this assumption is the symmetry invoked by this procedure justified. This implies in particular an assumption of equal variances in the two groups. By contrast, the bootstrap method uses the plug-in principle and probabilities estimated as in equation (8.60), and thus tests the more general null hypothesis that the two group means are equal without making the assumption of equal variances. It thus extends the range of possibilities for testing for equal means in two groups.

9.9   Sequential Analysis

9.9.1   The Sequential Probability Ratio Test

Sequential analysis is a statistical hypothesis testing procedure in which the sample size is not fixed in advance, but depends on the outcomes of the successive observations taken. Several important results in BLAST theory are borrowed from sequential analysis, and there are many parallels between sequential analysis theory and the statistical theory of BLAST, so we consider sequential analysis in some detail. We focus on the case of discrete random variables, since that provides the most appropriate comparison with BLAST.

We consider some random variable Y having probability distribution $P(y; \xi)$. In its simplest and most frequently occurring form, hypothesis testing in sequential analysis usually relates to tests about the numerical value of the parameter ξ, the form of the probability distribution of Y being known. Suppose that the null hypothesis specifies a value ξ_0 for the parameter ξ and the alternative specifies a value ξ_1. Thus both hypotheses are simple. Because the sample size is not chosen in advance, we can specify not only the desired numerical values α of the Type I error but also the desired numerical value β of the Type II error.

The sequential analysis procedure prescribes that we take observations one by one and calculate, after each observation is taken, the discrete analogue of the likelihood ratio (9.1). Sampling stops, and the alternative hypothesis is accepted, whenever this ratio becomes sufficiently large, that is, reaches or exceeds some value B. Similarly, sampling stops, and the null hypothesis is accepted, whenever it becomes sufficiently small, that is, reaches or is less than some value A. Sampling continues while

$$A < \frac{P(y_1; \xi_1)P(y_2; \xi_1) \cdots P(y_n; \xi_1)}{P(y_1; \xi_0)P(y_2; \xi_0) \cdots P(y_n; \xi_0)} < B. \tag{9.54}$$

Our aim is to choose A and B such that the Type I and Type II errors are α and β, respectively. Because of the discrete nature of the process, boundary overshoot almost always occurs. In this case the desired Type I and II errors α and β are approximately achieved by the choice $A = \beta/(1 - \alpha)$ and $B = (1 - \beta)/\alpha$.

It is often convenient to take logarithms throughout in (9.54), so that with $A = \beta/(1 - \alpha)$ and $B = (1 - \beta)/\alpha$, sampling continues as long as

$$\log \frac{\beta}{1 - \alpha} < \sum_i \log \frac{P(y_i; \xi_1)}{P(y_i; \xi_0)} < \log \frac{1 - \beta}{\alpha}. \tag{9.55}$$

Sampling stops and the null hypothesis is accepted if the lower inequality is eventually broken, and stops and the alternative hypothesis is accepted if the upper inequality is eventually broken.

In Section 1.14.3 we defined the support, or score, $S_{1,0}(y)$ of the observed value y of a random variable Y for the probability distribution claimed by the alternative hypothesis over the probability distribution claimed by the null hypothesis. In the present notation this may be written

$$S_{1,0}(y) = \log \frac{P(y; \xi_1)}{P(y; \xi_0)}. \tag{9.56}$$

The inequalities (9.55) can then be rewritten as

$$\log \frac{\beta}{1 - \alpha} < \sum_i S_{1,0}(y_i) < \log \frac{1 - \beta}{\alpha}, \tag{9.57}$$

the expression $\sum_i S_{1,0}(y_i)$ being the accumulated support provided by the observations for the alternative hypothesis over the null hypothesis.

For continuous random variables the probability $P(y; \xi)$ in (9.54) is replaced by a density function $f(x; \xi)$. We do not discuss the continuous random variable case further, since the theory and procedures are similar to those of the discrete case.

Example. Sequence matching. Suppose that in the sequence-matching test (page 307) the null and alternative hypotheses respectively specify the values $p_0 = 0.25$ and $p_1 = 0.35$ for p, and that both the Type I error and the Type II error are chosen to be 0.01. If Y_i is defined to be 1 when there is a match at site i and is defined to be 0 when there is not, equation (9.55) shows that sampling, that is moving from one position in the matched sequences to the next, is continued while

$$\log \frac{1}{99} < \sum_i S_{1,0}(Y_i) < \log 99, \tag{9.58}$$

where

$$S_{1,0}(Y_i) = \log \frac{(0.35)^{Y_i}(0.65)^{(1-Y_i)}}{(0.25)^{Y_i}(0.75)^{(1-Y_i)}} \tag{9.59}$$

is the support offered by Y_i in favor of the alternative hypothesis. In terms of the support concept, the procedure leads to the acceptance of the null hypothesis when the accumulated level of support for the alternative hypothesis reaches $\log(1/99) = -4.595$, so that the accumulated level of support for the null hypothesis reaches 4.595. The alternative hypothesis is accepted when the accumulated level of support for that hypothesis reaches $\log(99) = 4.595$.

Using (9.59), the inequalities (9.58) can be written

$$-9.581 < \sum_i (Y_i - 0.2984) < 9.581. \tag{9.60}$$

The procedure defined by (9.60) is a random walk in which the step size is either 0.7016 (for a match) or -0.2984 (for a mismatch). This illustrates the close link between sequential analysis and random walks.

It is important to note that, unlike the fixed-sample- size test discussed in Section 3.4.1, the testing procedure depends on the specifics of the alternative hypothesis. Thus if the alternative hypothesis had been $p = 0.30$ rather than $p = 0.35$, the inequalities (9.60) would be replaced by

$$-18.284 < \sum_i (Y_i - 0.2745) < 18.284. \tag{9.61}$$

This describes a random walk in which the step size is either 0.6255 (for a match) or -0.2745 (for a mismatch). Thus the step sizes of this walk differ from those of the walk described by (9.60), as do the boundaries of the walk. Thus while the sequential procedure has the advantage of allowing the Type II error to be specified, it has the disadvantage of not testing against a composite alternative hypothesis. However, whatever the true value of the parameter being tested might be, we can find the probability that the null hypothesis is rejected. This leads to a discussion of the power function of a sequential test.

9.9.2   The Power Function for a Sequential Test

Suppose that in the sequential test described in Section 9.9.1 that the true value of the parameter of interest is ξ. We might wish to know the probability that alternative hypothesis is eventually accepted, given that the parameter takes this value. This probability is the power $\mathcal{P}(\xi)$ of the test. An approximate formula for $\mathcal{P}(\xi)$, derived by ignoring the possibility of boundary overshoot, is (Wilks (1962))

$$\mathcal{P}(\xi) \cong \frac{1 - \left(\frac{\beta}{1-\alpha}\right)^{\theta^*}}{\left(\frac{1-\beta}{\alpha}\right)^{\theta^*} - \left(\frac{\beta}{1-\alpha}\right)^{\theta^*}}, \tag{9.62}$$

where, except in one specific case discussed below, θ^* is the unique non-zero solution for θ of the equation

$$\sum_{y \in R} P(y; \xi) \left(\frac{P(y; \xi_1)}{P(y; \xi_0)} \right)^{\theta} = 1, \tag{9.63}$$

where R is the range of values of Y. Equivalently, and again except in one specific case, θ^* is the unique non-zero solution for θ in the equation

$$\sum_{y \in R} P(y; \xi) e^{\theta S_{1,0}(y)} = 1, \tag{9.64}$$

where $S_{1,0}(y)$ is defined in equation (9.56).

The value of θ^* depends on the value of ξ. Two cases are of particular interest, namely $\xi = \xi_0$ and $\xi = \xi_1$. Inspection of equation (9.63) shows that when $\xi = \xi_0$ the value of θ^* is 1. Insertion of this value in (9.62) gives $P(\xi_0) \cong \alpha$, as required. When $\xi = \xi_1$ the value of θ^* is -1, and insertion of this value in (9.62) gives $P(\xi_1) \cong 1 - \beta$, again as required.

The left-hand side in (9.64) is the moment-generating function of $S_{1,0}(Y)$ if ξ is the true value of the parameter. Since $S_{1,0}(Y)$ can take both positive and negative values, the proof of Theorem 1.1 on page 40 shows that if the mean of $S_{1,0}(Y)$ is non-zero, there is a unique non-zero value θ^* of θ solving (9.64). When the mean of $S_{1,0}(Y)$ is zero there is a double root of (9.64) at 0, and a special analysis is needed for this case. However, this one isolated value is not of any significance to us.

The power calculation is interesting in its own right, since power calculations in statistical tests are frequently made. We shall see in Section 9.9.3 that the power calculation is also needed to find the mean sample size in the sequential test.

The parameter θ^*, first introduced in the literature in the context of sequential analysis, is identical to that discussed in Chapter 7 in describing properties of random walks and also identical to the parameter θ^* that will be used in Chapter 10 to discuss BLAST theory.

The sequence-matching power function. In the sequence-matching example the parameter of interest is the probability p of a match, and we replace the generic parameter notation ξ used above by the Bernoulli notation p. Suppose that the null hypothesis value of p is 0.25 and the alternative hypothesis value is 0.35. Since the random variable of interest Y can take only two values ($+1$ for a match, 0 otherwise), the expression (9.63) simplifies to

$$p \left(\frac{0.35}{0.25} \right)^{\theta} + (1-p) \left(\frac{0.65}{0.75} \right)^{\theta} = 1. \tag{9.65}$$

This may be solved for θ^* for any chosen value of p and the value found inserted in the right-hand side of equation (9.62). As an example, if $p = 0.3$

(that is, the probability of a match is halfway between the values claimed by the null and alternative hypotheses), the solution of equation (9.65) is found (by numerical methods) to be $\theta^* = -0.32$, and insertion of this value in equation (9.62) shows that the probability that the alternative hypothesis is accepted is about 0.537. Although 0.3 is halfway between the null and alternative hypothesis values 0.25 and 0.35, this probability is not 0.5.

In the discussion above it was in effect assumed that the functional form of the probability distribution of the random variables in the sequential process is known, and that all that is in question is the value of some parameter in that distribution. Thus in equation (9.63) it is assumed that the form of the true probability distribution $P(y;\xi)$ is same as that defined by the null and alternative hypotheses.

This restriction is unnecessary for power calculations. It might be the case that the true probability distribution is of a different form than that assumed under both null and alternative hypotheses. If the probability distribution of the random variable Y is $Q(y)$, the equation defining θ^* is

$$\sum_{y \in R} Q(y) \left(\frac{P(y;\xi_1)}{P(y;\xi_0)} \right)^{\theta^*} = 1. \tag{9.66}$$

Once θ^* is found from this equation, equation (9.62) may be used to find the probability that the alternative hypothesis is accepted.

9.9.3 The Mean Sample Size

The sample size in a sequential test is the (random) number of observations taken until one or the other hypothesis is accepted. An approximation to the mean sample size may be found by ignoring the possibility of boundary overshoot. The approximation is found by using methods essentially identical to those used in Section 7.4.3 in finding the mean number of steps until a random walk stops. That is, two expressions are calculated for the mean value of the final sum $\sum_i S_{1,0}(Y_i)$, one of which involves the mean sample size, and by equating the two expressions we can solve for the value of the mean sample size.

The first expression is found as follows. The sequential procedure continues so long as the inequalities (9.55) hold. If boundary overshoot is ignored, the final value of $\sum_i S_{1,0}(y_i)$ is either $\log \beta/(1-\alpha)$ (with probability $1 - \mathcal{P}(\xi)$) or $\log(1-\beta)/\alpha$ (with probability $\mathcal{P}(\xi)$), where $\mathcal{P}(\xi)$ is given by equation (9.62). The mean of the final value of $\sum_i S_{1,0}(Y_i)$ is thus, from equation (1.24),

$$(1 - \mathcal{P}(\xi)) \log \left(\frac{\beta}{1-\alpha} \right) + \mathcal{P}(\xi) \log \left(\frac{1-\beta}{\alpha} \right). \tag{9.67}$$

The second expression in effect uses the random walk equation (7.23). The term m_h on the right-hand side of equation (7.23) becomes, in the sequential case, the mean sample size. The term "$E(S)$" on the right-hand side of equation (7.23) is, in the sequential case,

$$E\left(S_{1,0}(Y_i)\right) = E\left(\log \frac{P(Y_i;\xi_1)}{P(Y_i;\xi_0)}\right) = \sum_{y \in R} P(y;\xi) \log \frac{P(y;\xi_1)}{P(y;\xi_0)}. \qquad (9.68)$$

Assuming this mean to be non-zero, it follows by equating the two expressions for the mean of the final value of $\sum_i S_{1,0}(Y_i)$ that the mean sample size is

$$\frac{(1 - \mathcal{P}(\xi)) \log(\frac{\beta}{1-\alpha}) + \mathcal{P}(\xi) \log(\frac{1-\beta}{\alpha})}{\sum_{y \in R} P(y;\xi) \log \frac{P(y;\xi_1)}{P(y;\xi_0)}}. \qquad (9.69)$$

Both numerator and denominator in this expression depend on the value of $\mathcal{P}(\xi)$, and hence on the parameter θ^*, so that the mean sample size itself also depends on the value of this parameter.

The mean of $S_{1,0}(Y_i)$ is zero for only one value of the parameter ξ, so we do not pursue the mean sample size formula for this case.

A generalization of (9.69), namely

$$\frac{(1 - \mathcal{P}(\xi)) \log(\frac{\beta}{1-\alpha}) + \mathcal{P}(\xi) \log(\frac{1-\beta}{\alpha})}{\sum_{y \in R} Q(y) \log \frac{P(y;\xi_1)}{P(y;\xi_0)}}, \qquad (9.70)$$

applies in the case where the probability distribution $Q(y)$ of Y is of a different form from that assumed in null and alternative hypotheses. A calculation similar to this is relevant to the theory of BLAST, as discussed in Section 10.6.2.

The sequence-matching mean sample size. In the sequence-matching example where the null hypothesis value of p is 0.25 and the alternative hypothesis value of p is 0.35, and the Type I and Type II errors are both 0.01, the expression (9.69) reduces to

$$\frac{9.190\mathcal{P}(p) - 4.595}{p \log \frac{7}{5} + (1 - p) \log \frac{13}{15}}. \qquad (9.71)$$

When the null hypothesis is true, so that $p = 0.25$ and the probability $\mathcal{P}(0.25)$ that the alternative hypothesis is accepted is 0.01, the mean sample size is 194. When $p = 0.30$ and the probability $\mathcal{P}(0.30)$ that the alternative hypothesis is accepted is 0.537 (calculated above), the mean sample size is found from (9.71) to be 441. When the alternative hypothesis is true, $p = 0.35$, $\mathcal{P}(0.35) = 0.99$, and the mean sample size is 182.

One of the motivations for the development of sequential methods was to find a procedure that led to a decision to accept some hypothesis with a smaller mean sample size than that needed for a fixed-sample-size test with the same Type I and Type II errors. It is thus interesting to note that in the sequence-matching example, the sample size of the fixed-size test having the same Type I and II errors as the sequential test is about 450.

It sometimes causes concern that there is a value of p for which the denominator in the expression (9.71) is zero. However, this is the value of p for which the double root of equation (9.63) arises and for which a different analysis is needed. Since this refers to one isolated value of p, this is of no importance.

The mean sample size, as with the power function, depends on the specific choice made under the alternative hypothesis for the probability p of a match at any site.

9.9.4 The Effect of Boundary Overshoot

The sequential analysis theory described above assumes that there is no boundary overshoot. In practice, there will almost always be boundary overshoot, and in this case the *actual* Type I and Type II errors of the test differ from the nominal values α and β that were used to define the boundaries in the testing procedure. Random walk theory can be used to assess how significant the effects of boundary overshoot are.

To illustrate this, consider the sequential test of the null hypothesis that the Bernoulli parameter p is 0.45 against the alternative hypothesis that it is 0.55, with nominal Type I and Type II errors of 5%. Sampling continues while

$$-14.67 < \sum_i (2y_i - 1) < 14.67,$$

where $y_i = 1$ if trial i results in success and $y_i = 0$ if trial i results in failure. Defining $u_i = 2y_i - 1$, this means that sampling continues while

$$-14.67 < \sum_i u_i < 14.67,$$

where $u_i = 1$ if trial i results in success and $u_i = -1$ if trial i results in failure. Thus $\sum_i u_i$ performs a simple random walk. Since this sum can take only integer values, the walk will continue in practice until $\sum_i u_i$ reaches either -15 or 15. Thus there must be boundary overshoot. Equation (7.10) can then be used to show that if $p = 0.45$ (that is, the null hypothesis is true), the actual Type I error is 4.7% rather than the nominal 5%. The actual Type II error is also 4.7%.

It can be shown that in any sequential procedure the sum of the actual Type I and Type II errors is less than the sum $\alpha + \beta$ of the nominal values, and usually (as in the above example) both are less than their respective nominal values.

The effect of a boundary overshoot can, however, be more important than it is in this example. BLAST theory recognizes this and deals with it in a novel way, as described in Chapter 10.

Problems

9.1. Suppose that the null hypothesis of Example 1 of Section 9.4.2 is true. Then one can write $y_i/n = p_i + h_i$, where h_i is a small deviation from zero, typically of order n^{-1}. The λ ratio (9.25) can then be written as

$$\lambda = \prod_i \left(\frac{p_i}{p_i + h_i} \right)^{y_i}. \tag{9.72}$$

This implies that

$$-2 \log \lambda = 2 \sum_{i=1}^{k} y_i \log \left(1 + \frac{h_i}{p_i} \right). \tag{9.73}$$

(i) Approximate the logarithmic term in (9.73) by an expression calculated from the approximation (B.25).

(ii) Write y_i as $np_i + nh_i$ in the expression obtained in (i).

(iii) Show that if terms of order h_i^3 are ignored, the expression in step (ii) reduces to the X^2 statistic (3.40).

9.2. Suppose that the null hypothesis being tested in Example 2 of Section 3.4.1 is true. Use the fact that the mean of Y_j is np_j and the variance of Y_j is $np_j(1 - p_j)$ to show that the mean value of the statistic (3.39) is exactly $k - 1$ (the mean of a chi-square random variable with $k - 1$ degrees of freedom (see Problem 1.26)).

(More difficult). Find the variance of the statistic (3.40) when the null hypothesis is true, and compare this with the chi-square distribution value $2(k - 1)$.

9.3. Develop the one-sample T statistic defined in (3.28) and the two-sample paired T statistic defined in (3.38) by using a likelihood ratio procedure parallel to that used in Example 1 of Section 9.3.1 for the the two-sample unpaired T statistic.

9.4. Prove the claim made in Section 9.5.2 that the square of the t statistic defined in (3.30) is the F statistic defined in (9.31).

9.5. Prove the claim leading to equation (9.53).

9.6. What value would the t statistic take in the bootstrap procedure described in Section 9.8 in the unlikely event that the same observation was sampled m times in the first group and n times in the second group?

9.7. Develop the sequential likelihood ratio test of the null hypothesis that the mean of a Poisson distribution is λ_0 against the alternative hypothesis that the mean of the distribution is λ_1, with Type I error α and Type II error β. Here λ_0 and λ_1 are given constants with $\lambda_1 > \lambda_0$.

9.8. *Continuation.* If the true mean of the Poisson distribution is λ, show that θ^* is the non-zero solution of the equation

$$\theta^*(\lambda_0 - \lambda_1) - \lambda + \lambda \left(\frac{\lambda_1}{\lambda_0}\right)^{\theta^*} = 0. \qquad (9.74)$$

If $\lambda_0 = 4$, $\lambda_1 = 5$, solve equation (9.74) for λ for the values $\theta^* = -1.2$, -1.0, -0.8, -0.6, -0.4, -0.2, $+0.2$, $+0.4$, $+0.6$, $+0.8$, $+1.0$, $+1.2$.

9.9. *Continuation.* Use the results of Problem 9.8 to sketch the power curve in the case $\alpha = \beta = 0.05$.

9.10. *Continuation.* Use the results of Problem 9.9 to sketch the mean sample size curve.

9.11. For the two respective cases $\xi = \xi_0$ and $\xi = \xi_1$, discuss the denominator in equation (9.69) as a relative entropy.

10
BLAST

10.1 Introduction

BLAST (Basic Local Alignment Search Tool) is a widely used method for
assessing which nucleic acid or protein sequences in a large database have
significant similarity to a given query sequence. Many of the results derived
in previous chapters, particularly those relating to the maximum of several
random variables, the geometric-like distribution, P-values, the renewal
theorem, random walks, and sequential analysis, were presented because
they are needed in the statistical theory associated with the BLAST procedure described in this chapter. For concreteness the discussion is in terms
of protein (amino acid) sequences; the analysis for nucleic acid sequences
is similar to that for protein sequences.

Currently there are two implementations of BLAST, one by NCBI (the
US National Center for Biotechnology Information) and the other at Washington University. For most of this chapter we consider a simple early
version of BLAST, leading to a readily understood statistical analysis. We
first describe Washington University's version 1.4, which was used to generate the examples of Section 10.5, and then describe various generalizations
leading to the current implementations.

10.2   The Comparison of Two Aligned Sequences

10.2.1   Introduction

We start by considering as given an ungapped global alignment of two protein sequences, both of length N, as shown, for example, in (7.2). This is done mainly as a preliminary step to the generalizations in the following sections. In particular, the generalization to finding the best among all local alignments of two sequences will be considered in Section 10.3. The further generalization to database searches will be considered in Section 10.4.

The null hypothesis to be tested is that for each aligned pair of amino acids, the two amino acids were generated by independent mechanisms, so that if amino acid j occurs at any given position in the first sequence with probability p_j and amino acid k occurs at any given position in the second sequence with probability p'_k, the null hypothesis probability that they occur together in a given aligned pair is

$$\text{null hypothesis probability of the pair } (j, k) = p_j p'_k. \qquad (10.1)$$

The theory of Chapter 9 shows that classical statistical testing theory requires the specification of an alternative hypothesis. For the moment we simply write

$$\text{alternative hypothesis probability of the ordered pair } (j, k) = q(j, k), \qquad (10.2)$$

without any particular specification of the form of the function $q(j, k)$. The choice of the form of this function is discussed at length in Section 10.2.4.

10.2.2   The BLAST Random Walk

In this section and the following sections we give the basic idea behind the statistical aspects of BLAST, considering first the case described above, of two aligned sequences, both of length N.

We number the positions in the alignment from left to right as positions $1, 2, \ldots, N$. A score $S(j, k)$ is allocated to each position where the aligned amino acid pair (j, k) is observed. The choice of the scores $S(j, k), j, k = 1, 2, \ldots 20$ is discussed in Section 10.2.4: For the moment we note only that there is a close connection between the choice of the choice of the scores $S(j, k)$ and the choice of the alternative hypothesis probabilities $q(j, k)$ given in (10.2).

The matrix $S = \{S(j, k)\}$ is the substitution matrix of the process: aspects of these matrices were discussed in Section 6.5. It is required in the theory, and is assumed throughout, that at least one element in the substitution matrix be positive and, for reasons discussed below, that the null hypothesis mean score $\sum_{j,k} p_j p'_k S(j, k)$ be negative. In order to apply the theory of Chapter 7 we also assume throughout that the greatest common divisor of the scores is 1.

An accumulated score at position i is calculated as the sum of the scores for the various amino acid comparisons at positions $1, 2, \ldots, i$. As i increases, this accumulated score undergoes a random walk, as described for example in (7.3) and Figure 7.2 for the protein sequence comparison given in (7.2). When the null hypothesis is true, the walk has negative drift and will go through a succession of increasingly negative ladder points, as defined in Section 7.1. Because the substitution matrix will usually include elements, or scores, whose values are -2 or less, the accumulated score at any ladder point will not necessarily be one less than the accumulated score at the preceding ladder point. This implies that, in random walk terms, boundary overshoot can occur. An analysis of this overshoot is needed for BLAST calculations, as outlined briefly below in Section 10.2.3.

Let $Y_1, Y_2, \ldots$ be the respective maximum heights of the excursions of this walk relative to the height of any ladder point after leaving this ladder point and before arriving at the next, or relative to the height of the last ladder point and arriving at the end of the sequence. We define $Y_{\max}$ as the maximum of these maxima: $Y_{\max}$ is in effect the test statistic used in BLAST, so it is necessary to find its null hypothesis distribution.

The various random variables $Y_1, Y_2, \ldots$ are independent, and ignoring end effects for now, can be taken as being identically distributed. The asymptotic probability distribution of any Y_i was shown in Chapter 7 to be the geometric-like distribution (7.63). The values of C and λ in this distribution depend on the substitution matrix used and the amino acid frequencies $\{p_j\}$ and $\{p'_k\}$. The probability distribution of $Y_{\max}$ then follows from the theory of Section 2.11, which, apart from C and λ, depends on the mean number of ladder points in the walk. (In the notation of Section 2.11, this is the value of n.) In the following section we discuss the computation of the central parameters C and θ^* as well as the mean number of ladder points, drawing on the random walk theory developed in Chapter 7.

The above procedure shows why it is necessary that the mean score $\sum_{j,k} p_j p'_k S(j, k)$ be negative. If this were not so the BLAST random walk would contain arbitrarily long upward excursions from ladder points and the entire testing procedure would break down.

10.2.3 Parameter Calculations

The expression for C is given in equation (7.61), and requires only notational amendments for application to BLAST. The step size is identified with a score $S(j, k)$, and the null hypothesis probability of taking a step of any size is found from the two sets of frequencies $\{p_j\}$ and $\{p'_j\}$.

The computation of λ also follows the random walk principles laid down in Chapter 7. As noted below equation (7.42), λ ($= \theta^*$) is found from an equation involving the mgf of the step size in this random walk. When the null hypothesis is true, this equation is

$$\sum_{j,k} p_j p'_k e^{\lambda S(j,k)} = 1. \tag{10.3}$$

The calculation of λ from this equation will usually require numerical methods: See Appendix B.15.

The calculation of the null hypothesis probability distribution of $Y_{\max}$ depends not only on C and λ but also on the mean number of ladder points in the BLAST walk. This mean number depends in turn on the mean distance A between ladder points. A general formula for A is given in equation (7.41) and is readily converted to the situation discussed here. However, the arguments leading to this formula do not necessarily provide an efficient general formula for finding the constants R_{-j} in equation (7.41), and we now describe two alternative approaches.

The first alternative approach uses a decomposition of paths. Consider as a simple example a walk in which the possible steps are $+1$ and -2, with respective probabilities p and $q = 1 - p$. Any ladder point reached in the walk is at a distance 1 or 2 below the previous one. The respective probabilities of these two cases are denoted by R_{-1} and $R_{-2} = 1 - R_{-1}$, as in Chapter 7.

The probability that -2 is a ladder point is the probability that the walk goes immediately to -2, together with the probability of the event that the walk first goes to $+1$, and then starting from $+1$, reaches 0 as the first point reached below $+1$ and then -2 as the first ladder point below 0. This implies that

$$R_{-2} = q + p(1 - R_{-2})R_{-2}. \tag{10.4}$$

The positive solution of this equation is

$$R_{-2} = \frac{-q + \sqrt{4pq + q^2}}{2p}. \tag{10.5}$$

From this the value of R_{-1} follows as $1 - R_{-2}$, and then the value of A follows from equation (7.41).

For general substitution matrices this method might not be effective. In such a case, Karlin and Altschul (1990) provide rapidly converging series expansions that give accurate values of A using only a few terms in the series. We assume from now on that a value of A, arrived at by one method or another, is in hand.

Since the two sequences compared are each of length N, and the mean distance between ladder points is A, the mean number of ladder points is equal for all practical purposes to N/A. While various approximations are involved with this calculation, the intuitive interpretation is clear: If, for example the length N is 1000 and the mean distance A between successive ladder points is 50, one expects about 20 ladder points in the walk involved with the comparison of the two sequences, and this is the value given by the expression N/A.

10.2.4   The Choice of a Score

So far, we have taken the score $S(j, k)$ as given, and have not discussed what might be a reasonable choice, on statistical and genetical grounds, for this score. In applications of BLAST this score, whether found by a BLOSUM or a PAM matrix, is a log likelihood ratio (as discussed briefly in Section 6.5), and we now indicate why this is appropriate.

The random walk described in Section 10.2.2 is determined by the sum of the scores $S(j, k)$ at each position during the walk. In sequential analysis one also considers the sum of scores. In sequential analysis the score used is a log likelihood ratio, arrived at through statistical optimality methods. Specifically, if the random variable Y whose probability distribution is being assessed is discrete, this is the "score" statistic $S_{1,0}(y)$, defined in equation (9.56) as the log likelihood ratio

$$S_{1,0}(y) = \log \frac{P(y; \xi_1)}{P(y; \xi_0)}.$$

Based on the comparison of the BLAST and the sequential analysis procedures, it can be argued that a suitable score to use in BLAST should also be the logarithm of a likelihood ratio. Under this argument, if the amino acid pair (j, k) is observed at any position, and if $p_j p'_k$ and $q(j, k)$ are, respectively, the null and the alternative hypothesis probabilities of this pair, the (discrete random variable) score $S(j, k)$ becomes

$$S(j, k) = \log \frac{q(j, k)}{p_j p'_k}. \tag{10.6}$$

Any score proportional to $S(j, k)$ is also reasonable.

The second argument favoring the choice (10.6) for the score associated with the pair (j, k) is more subtle (Karlin and Altschul (1990)). This argument also leads to the choice of a specific proportionality constant. Suppose some arbitrary substitution matrix is chosen, with (j, k) element $S(j, k)$. Now let $q(j, k)$ be defined implicitly by

$$S(j, k) = \lambda^{-1} \log \frac{q(j, k)}{p_j p'_k}, \tag{10.7}$$

where λ is defined in equation (10.3), and thus explicitly by

$$q(j, k) = p_j p'_k e^{\lambda S(j,k)}. \tag{10.8}$$

The right-hand side is the typical term on the left hand-side in equation (10.3). Therefore $\sum_{j,k} q(j, k) = 1$. Thus the $q(j, k)$ (which are all positive) form a probability distribution. This is not an arbitrary distribution. Karlin and Altschul (1990) and Karlin (1994) show that in practice, when the null hypothesis is true, the frequency with which the observation (j, k) arises in high-scoring excursions, where the score used is as given in equation (10.7), is asymptotically equal to $q(j, k)$. They then argue that a scoring scheme is "optimal" if the frequency of the observation (j, k) in high-scoring excursions is asymptotically equal to the "target" frequency $q(j, k)$, the frequency arising if the alternative hypothesis is true, (i.e., the frequency in the most biologically relevant alignments of conserved regions). This, then, argues for the use of $S(j, k)$ as defined in equation (10.7) as the score statistic.

These arguments lead us to adopt the following procedure. Suppose that the alternative hypothesis specifies a well-defined probability $q(j, k)$ for the amino acid pair (j, k), while the null hypothesis specifies a probability $p_j p'_k$ for this pair. Then we define the score $S(j, k)$ associated with this pair as that given by equation (10.7).

These arguments do not yet specify how to determine the most appropriate form for the $q(j, k)$'s. There are various possibilities for this. One frequently adopted choice is that deriving from the evolutionary arguments that lead to the PAMn matrix construction described in Section 6.5.3. In the notation of Section 6.5.3,

$$q(j, k) = p_j m^{(n)}_{jk}, \tag{10.9}$$

so that

$$S(j, k) = \log \frac{m^{(n)}_{jk}}{p'_k}. \tag{10.10}$$

The values of the $q(j, k)$'s for the simple symmetric model of Section 6.5.3 are given in equation (6.35) for one specific value of n. The derivation of these values emphasizes that $q(j, k)$ is a function of n, and that some extrinsic choice of a reasonable value of n must be made to use PAMn matrices in BLAST methods. We discuss aspects of this choice in Section 10.6.

The choice of $S(j, k)$ as the logarithm of a likelihood ratio can be related to the concepts of relative entropy and support discussed in Section 1.14.2. Specifically, the score defined by equation (10.7) is proportional to the support given by the observation (j, k) in favor of the alternative hypothesis over the null hypothesis. Equation (1.124) shows that when the alternative hypothesis is true, the mean H of this support is

$$H = \sum_{j,k} q(j, k) \log \frac{q(j, k)}{p_j p'_k}, \tag{10.11}$$

and this is the relative entropy defined in equation (1.119). Equation (10.7) shows that this relative entropy can be written as

$$H = \sum_{j,k} q(j, k) \lambda S(j, k) = \lambda E\left(S(j, k)\right), \tag{10.12}$$

the expected value being taken assuming that the alternative hypothesis is true.

From the discussion following (10.8), if the score $S(j,k)$ for the pair (j,k) is defined as in (10.7), the mean score in high-scoring segments is asymptotically $\sum_{j,k} q(j,k)S(j,k)$, and from (10.12) this is

$$\lambda^{-1}H. \tag{10.13}$$

This asymptotic result is used in BLAST calculations (see Section 10.3.3).

Simulations, however, show that the convergence to this asymptotic value is very slow. For the symmetric PAMn substitution matrix discussed in Section 6.5.4 with $n = 259$, and with equal amino acid frequencies, the asymptotic value $\lambda^{-1}H$ of the mean step size in high-scoring segments, found from computation of λ and H from (10.3) and (10.11), respectively, is 0.446. This is identical to the value given in equation (6.36), found from Markov chain considerations. For this example, Table 10.1 shows simulation estimates of this mean for various values of N, the length of the alignment. The slow rate of convergence to the asymptotic value 0.446 is clear. This

N	500	5,000	50,000	500,000	5,000,000	limiting value
mean step	1.021	.712	.608	.560	.533	.446

Table 10.1. Simulation values for the mean step size in maximally-scoring segments, as a function of N. Simulations performed with 10,000 to 1,000,000 repetitions.

observation will be relevant to the edge correction formula discussed in Section 10.3.3.

The value of the relative entropy H appears on BLAST printouts. However, the calculation used for these printouts is slightly different from that implied by (10.12). The value of $q(j,k)$ used to compute the score $S(j,k)$ may well be unknown, so that while the values of λ and $S(j,k)$ are known, direct computation of H as defined in (10.12) is not possible.

The BLAST printout value of H uses an indirect approach. With the values of λ, the $S(j,k)$, and the p_j and p'_k in hand, $q(j,k)$ is calculated by using equation (10.8). The printout value of H is now calculated as in (10.12), using the values of $q(j,k)$ so calculated.

10.2.5 Bounds and Approximations for the BLAST P-Value

We have seen that the test statistic used in BLAST is the maximum $Y_{\max}$ of $n \cong N/A$ random variables, each being a random upwards excursion height following a ladder point in the BLAST random walk. The theory of Section 7.6.4 shows that each upward excursion has, approximately and asymptotically, the geometric-like distribution (1.19). We use this result in this section to obtain asymptotic bounds for the null hypothesis distribution of $Y_{\max}$ and hence asymptotic bounds for a BLAST P-value. An approximation used in some BLAST implementations for this P-value will also be given.

The analysis of Section 2.11.3 shows that there exists an asymptotic distribution for the maximum of n iid continuous random variables whose density function has support of the form $(A, +\infty)$. The BLAST test statistic $Y_{\max}$ is, however, a discrete random variable, and an asymptotic distribution for the maximum of n iid discrete random variables, analogous to that for continuous random variables, is known not to exist. On the other hand it is possible to use the continuous distribution results to find asymptotic bounds for the distribution of $Y_{\max}$. The procedure is as follows.

If $X_{\max}$ is the maximum of n iid continuous random variables, and if $Y_{\max} = \lfloor X_{\max} \rfloor$ is the integer part of $X_{\max}$, then $Y_{\max}$ is a discrete random variable and

$$X_{\max} - 1 < Y_{\max} \le X_{\max}.$$

Thus for any positive integer y,

$$\text{Prob}(X_{\max} \le y) \le \text{Prob}(Y_{\max} \le y) \le \text{Prob}(X_{\max} \le y + 1). \quad (10.14)$$

Let $X_{\max}$ be the maximum of n iid random variables each having the exponential distribution (1.66), and put $Y_{\max} = \lfloor X_{\max} \rfloor$. Then the argument surrounding equations (2.115) and (2.116) shows that $Y_{\max}$ has the same distribution as the maximum of n iid random variables, each having the geometric distribution (1.69). Application of (2.130) and the bounds in (10.14) shows that to a close approximation,

$$e^{-ne^{-\lambda y}} \le \text{Prob}(Y_{\max} \le y) \le e^{-ne^{-\lambda(y+1)}}, \quad (10.15)$$

or equivalently

$$1 - e^{-ne^{-\lambda y}} \le \text{Prob}(Y_{\max} \ge y) \le 1 - e^{-ne^{-\lambda(y-1)}}, \quad (10.16)$$

for any positive integer y.

This discussion suggests how a parallel calculation for the maximum of random variables having a geometric-like distribution can be obtained. If $Y_{\max}$ is the maximum of n iid random variables, each having the geometric-like distribution given in (1.74), then a calculation analogous to that leading to (10.16) gives the approximate asymptotic inequality

$$1 - e^{-nCe^{-\lambda y}} \le \text{Prob}(Y_{\max} \ge y) \le 1 - e^{-nCe^{-\lambda(y-1)}}. \quad (10.17)$$

Whereas for the geometric distribution the upper bound in (10.15) and the lower bound in (10.16) hold even for small y, the inequalities in (10.17) are ultimately based on the *asymptotic* expression (1.74), which applies for large values of y as well as large n. They might not hold for small values of y, even when n is large.

If we now replace n by N/A for the mean number of BLAST ladder points and define a new parameter K by

$$K = \frac{C}{A}e^{-\lambda}, \tag{10.18}$$

the inequality (10.17) becomes

$$1 - e^{-NKe^{-\lambda(y-1)}} \leq \text{Prob}(Y_{\max} \geq y) \leq 1 - e^{-NKe^{-\lambda(y-2)}}. \tag{10.19}$$

If we replace y in this inequality by $x + \lambda^{-1}\log N$, we obtain

$$e^{-Ke^{-\lambda(x-1)}} \leq \text{Prob}(Y_{\max} - \lambda^{-1}\log N \leq x) \leq e^{-Ke^{-\lambda x}}, \tag{10.20}$$

or equivalently

$$1 - e^{-Ke^{-\lambda x}} \leq \text{Prob}(Y_{\max} \geq \lambda^{-1}\log N + x) \leq 1 - e^{-Ke^{-\lambda(x-1)}}. \tag{10.21}$$

Allowing for notational changes, this is identical to one of the inequalities (1.13) in Karlin and Dembo (1992). Equivalently, for any value $y_{\max}$,

$$1 - e^{-KNe^{-\lambda y_{\max}}} \leq \text{Prob}(Y_{\max} \geq y_{\max}) \leq 1 - e^{-KNe^{-\lambda(y_{\max}-1)}}. \tag{10.22}$$

These inequalities give bounds for the P-value corresponding to any observed value $y_{\max}$ of $Y_{\max}$. These bounds for a BLAST P-value are not directly relevant in practice, since in practice a BLAST search involves the comparison of a short query sequence with a large database, consisting of many fragments, and there is no a priori alignment of the query sequence with any part of the database. This fact introduces various complications which we shall take up in the following sections. Nevertheless, we shall see that the P-value approximation used in the implementation of BLAST described in Section 10.5 derives ultimately from the lower P-value bound in (10.22).

It is often difficult to calculate P-values even for relatively simple random variables, so it is remarkable that P-values can be approximated with the comparatively simple and efficient procedure described above. On the other hand, we shall see in Section 11.6.1 that while the approximation is often conservative (that is, can overestimate the true P-value), it can also be anti-conservative, that is underestimate it. This might be because the geometric-like distribution on which the bounds in (10.22) are ultimately based is an asymptotic one, and might not apply for comparatively small values of $y_{\max}$. Also, it would be more appropriate to use the conservative upper bound in (10.22) rather than the lower bound.

The calculation of the bounds in (10.22) requires calculation of both λ and K. The computation of λ via equation (10.3) is comparatively straightforward. The calculation of K from the right-hand side in equation (10.18) would require the calculation of C, A, and λ. However, an exact calculation of K is straightforward in at least two cases. The first of these arises when the largest of the $S(j,k)$ is +1, and the second when the smallest $S(j,k)$

is -1, arising for example in the simple DNA scoring scheme described in Section 7.1. Using the notation S for the size of a step in the BLAST random walk, the two respective formulae for K are

$$K = \left(e^{-\lambda} - e^{-2\lambda}\right) E\left(Se^{\lambda S}\right) \tag{10.23}$$

and

$$K = \left(e^{-\lambda} - e^{-2\lambda}\right) \frac{(E(S))^2}{E\left(Se^{\lambda S}\right)}, \tag{10.24}$$

the expectations being taken assuming that the null hypothesis (10.1) is true.

10.2.6 The Normalized and the Bit Scores

Karlin and Altschul (1993) call the expression

$$\lambda Y_{\max} - \log(NK) \tag{10.25}$$

a "normalized score," denoted here by S'. In terms of this score, the inequalities (10.20) can be written as

$$e^{-e^{\lambda}e^{-s}} \leq \mathrm{Prob}(S' \leq s) \leq e^{-e^{-s}}. \tag{10.26}$$

From the upper inequality we obtain the approximation

$$\mathrm{Prob}(S' \geq s) \cong 1 - e^{-e^{-s}}. \tag{10.27}$$

The P-value corresponding to an observed value $s' = \lambda y_{\max} - \log(NK)$ of S' is, from (10.27),

$$P\text{-value} \cong 1 - e^{-e^{-s'}}. \tag{10.28}$$

This is identical to the approximation given by the lower bound in (10.22). When s is large, (10.27) may be further approximated by

$$\mathrm{Prob}(S' \geq s) \cong e^{-s}. \tag{10.29}$$

The similarity between the approximation (10.27) and equation (2.127) is of course no coincidence, since S' is a (normalized) extreme value. The approximation (10.27) and the fact that the mean and variance of the density function whose cumulative distribution function is (2.126) are respectively γ (Euler's constant) and $\pi^2/6$ show that these are approximately mean and variance of S'. From (10.25) and the linearity property of a mean (see Section 1.4), the approximate null hypothesis mean and variance of $Y_{\max}$ are, approximately,

$$\lambda^{-1}\left(\log(KN) + \gamma\right), \quad \text{and} \quad \pi^2/(6\lambda^2) \tag{10.30}$$

respectively. The value of γ is usually much smaller than KN, and in BLAST calculations the mean is often approximated by

$$\lambda^{-1}\left(\log(KN)\right). \tag{10.31}$$

BLAST printouts or published papers record a score similar to the normalized score S', namely the "bit" score. In more recent printouts this is defined by

$$\text{bit score} = \frac{\lambda Y_{\max} - \log K}{\log 2}. \tag{10.32}$$

Previous printouts recorded a bit score defined by

$$\text{bit score} = \frac{\lambda Y_{\max}}{\log 2}. \tag{10.33}$$

The bit score (10.33) has an invariance property, since its value, and hence its probability distribution, does not change if all entries in the substitution matrix used are multiplied by the same constant, say G. This can be seen from the fact that such a multiplication changes the value of $Y_{\max}$ by a multiplicative factor of G, but at the same time (see equation (10.3)) changes the value of λ by a multiplicative factor $1/G$. Thus the bit score (10.33) remains unchanged by this multiplication.

If the expression on the right-hand side of (10.27) is used to approximate the distribution of the normalized score S', then to this level of approximation the normalized score has a distribution also having the invariance property, since the right-hand side in (10.27) is free of any parameter.

A much stronger result than this is true. Whereas the value of $Y_{\max}$ has no absolute interpretation if the substitution matrix from which it is calculated is not specified, the normalized score S' and the bit score do have such an interpretation. In the case of S' this is made clear by approximations such as (10.27): Here the right-hand side is free of any parameters, so that the (approximate) distribution of S' is known whatever the details of the substitution matrix. If N is given, the same can be said for the bit score (10.32).

10.2.7 The Number of High-Scoring Excursions

In this section we define and discuss the quantity E', whose calculation leads ultimately to the quantity "Expect" found on BLAST printouts. Throughout the discussion we ignore edge effects: These are discussed in detail in Section 10.3.3.

Consider excursions from a ladder point in the random walk described by the comparison of the two sequences. We have seen that under the null hypothesis, for each such excursion, the maximum height Y has a geometric-like distribution whose parameters can be calculated. Denoting as above the maximum heights of the excursions from the various ladder points by $Y_1, Y_2, \ldots$, the relation (1.74) shows that the probability that any Y_i takes a value v or larger is approximately $Ce^{-\lambda v}$, where C and λ are those appropriate to the walk in question. Since to a close approximation the number of excursions can be taken to be N/A, as discussed in Section 10.2.3, the mean number of excursions reaching a height v or more is approximately $\frac{NC}{A}e^{-\lambda v}$. In the standard BLAST calculations discussed in the printout in Section 10.5, this mean is replaced by the approximating value

$$NKe^{-\lambda v}, \tag{10.34}$$

where K is given by (10.18). In Section 10.4 we shall trace back the printout P-value calculation to the expected value expression (10.34).

Since $Y_1, Y_2, \ldots$ are iid random variables, the number of excursions having a height v or more has a binomial distribution with mean given by (10.34). The theory developed in Section 4.2 shows that when v is large, the number of excursions reaching a height greater than or equal to v, that is, the number of high-scoring segment pairs (HSPs) with a score v or more, has, using the Poisson approximation to the binomial, a Poisson distribution with mean given in (10.34). (A more sophisticated analysis, based on equations such as (2.81) that allow for the fact that the number of ladder points is a random variable, arrives at the same conclusion.) Thus, to test for significance, the actual number of such excursions achieving a score exceeding v can be compared with the tail probability of this Poisson distribution.

The expected value of the number of excursions corresponding to the observed maximal score $y_{\max}$ is found by replacing the arbitrary number v in equation (10.34) by $y_{\max}$. This expected value is denoted by E', so that

$$E' = NKe^{-\lambda y_{\max}}. \tag{10.35}$$

The relation between E' and the normalized score S' defined in (10.25) is

$$S' = -\log E', \tag{10.36}$$

and the relation between E' and the P-value approximation is found from (10.28) as

$$P\text{-value} \cong 1 - e^{-E'}, \quad E' = -\log(1 - P\text{-value}). \tag{10.37}$$

It follows from the approximation (B.21) that the approximate P-value is very close to E' when E' is small.

Similar calculations may be made for any high-scoring excursion.

10.2.8 The Karlin–Altschul Sum Statistic

Focusing on the value of $Y_{\max}$ loses the information provided by the heights of the second-largest, third-largest, etc., excursions in the random walk. In this section we discuss a statistic that uses information from these other excursions.

Consider the r largest excursion heights, that is, the r largest Y_i values, assuming that there are at least r ladder points. It is convenient to use a notation that is different from the notation for order statistics used in Chapter 2, and assume that $Y_1(= Y_{\max}) \geq Y_2 \geq \cdots \geq Y_r$. By analogy with the definition in equation (10.25) we can compute r normalized scores $S_1', S_2', \ldots, S_r'$ from $Y_1, Y_2, \ldots, Y_r$, where

$$S_i' = \lambda Y_i - \log(NK). \qquad (10.38)$$

Note that $S_1' = S'$ as defined in equation (10.25).

Karlin and Altschul (1993) show that to a close approximation, the null hypothesis joint density function $f_S(s_1, \ldots, s_r)$ of $S = (S_1', \ldots, S_r')$ is

$$f_S(s_1, \ldots, s_r) = \exp\left(-e^{-s_r} - \sum_{k=1}^{r} s_k\right). \qquad (10.39)$$

We can use any reasonable function of $S_1', S_2', \ldots, S_r'$ as test statistic. Transformation methods such as those introduced in Chapter 2 can then be used to find the distribution of this test statistic, and this in turn allows the computation of a P-value, and also an E, or Expect, value, corresponding to any observed value of this statistic.

The specific statistic suggested by Karlin and Altschul (1993) is the sum $T_r = S_1' + \cdots + S_r'$ of the normalized scores. This is called the Karlin–Altschul sum statistic. Using transformation methods such as those described in Section 2.13, Karlin and Altschul use the joint density function (10.39) to calculate the null hypothesis density function $f(t)$ of T_r. The resulting expression is

$$f_{T_r}(t) = \frac{e^{-t}}{r!(r-2)!} \int_0^{+\infty} y^{(r-2)} \exp(-e^{(y-t)/r}) dy. \qquad (10.40)$$

As an exercise in transformation theory we confirm this calculation for the case $r = 2$ in Appendix D. When t is sufficiently large, this density function can be used to find the approximate expression

$$\text{Prob}(T_r \geq t) \cong \frac{e^{-t} t^{r-1}}{r!(r-1)!}. \qquad (10.41)$$

In the case $r = 1$, this is the approximation given in equation (10.29). The approximation (10.41) is sufficiently accurate when $t > r(r+1)$, and popular implementations of BLAST use (10.41) when this inequality holds.

If t is the observed value of T_r, the right-hand side in (10.41) then provides the approximate P-value corresponding to this observed value. This is used as a component of the eventual BLAST printout P-value.

Karlin and Altschul (1993) provide an example (see their Table 1) in which the observed values of the highest two normalized scores are $s_1' = 4.4$ and $s_2' = 2.5$. Using the value $r = 1$ in the approximation (10.41), the P-value corresponding to the highest normalized score 4.4 is $e^{-4.4} = 0.012$. Using the value $r = 2$, the P-value corresponding to the sum 6.9 of the highest two normalized scores is calculated from (10.41) as $\frac{6.9}{2} e^{-6.9} = 0.0035$, and these calculations confirm those given by Karlin and Altschul. For further aspects of these calculations, and of the calculations in their Table 2, see Problems 10.13 and 10.14.

A further aspect of the use of a test statistic based on T_r is that of consistent ordering. We say that r HSPs, HSP1, HSP2, ..., HSPr, between two sequences are *consistently ordered* if whenever the midpoint in the first sequence in HSPi comes before the midpoint in the first sequence in HSPj, then the same is true for the midpoints of the second sequence. More generally, one might require that the sequences in the different HSPs not overlap, or overlap no more than some fixed proportion (in popular implementations of BLAST, the default value of this proportion is 0.125). When consistent ordering is required, the P-value calculations must be amended. In the case where overlaps are unrestricted, this requirement cuts down the search space by a factor of $r!$, the number of orderings of the r HSPs. This implies that the P-value calculated from (10.41) should be divided by $r!$. A simple approximation (Karlin and Altschul (1993)) is that P-value calculations are amended by replacing t, the observed value of T_r, by $t + \log(r!)$ in the right-hand side of (10.41), or by a corresponding amendment to the calculations using (10.40). The popular implementations of BLAST use this approach, and furthermore allow the degree to which the HSPs overlap to be restricted. Restricting overlaps should require a further adjustment of the P-value. This is not apparently done by the popular implementations. However, an adjustment is made to the "edge correction" factor discussed below, which may or may not account for this (see Section 10.3.3).

A further complication introduced by the use of the sum statistic in BLAST is that of multiple testing. In practice, the value of r is not fixed in advance and is allowed to vary. Thus the problem of multiple testing, discussed in Section 3.11, arises. We delay discussion of the way in which this problem is handled in BLAST calculations until Section 10.3.4.

In BLAST printouts the notation r is replaced by N. We have used r here because N is used to denote a sequence length. Further, r is the notation used in the fundamental paper of Karlin and Altschul (1993).

10.3   The Comparison of Two Unaligned Sequences

10.3.1   Introduction

The theory of Section 10.2 considered calculations relevant to a fixed ungapped alignment in the comparison of two sequences each of length N. In this section we consider a more general question. We are given two sequences of lengths N_1 and N_2, but we are not given any specific alignment between them. The goal is to find the significance of high-scoring segment pairs between all possible (ungapped) local alignments. The highest-scoring pair is called the maximal-scoring segment pair (MSP).

10.3.2   Theoretical and Empirical Background

BLAST considers all ungapped alignments determined by all possible relative positions of the two sequences. For each relative position, the alignment is extended as far as possible in either direction, giving a total of $N_1 + N_2 - 1$ ungapped alignments. Figure 10.1 shows the first five alignments between two sequences of length 11 and 9 respectively.

```
sequence 1    . . . . . . . . . . .
sequence 2                  . . . . . . . . .

sequence 1    . . . . . . . . . . .
sequence 2               . . . . . . . . .

sequence 1    . . . . . . . . . . .
sequence 2            . . . . . . . . .

sequence 1    . . . . . . . . . . .
sequence 2         . . . . . . . . .

sequence 1    . . . . . . . . . . .
sequence 2       . . . . . . . . .
```

Figure 10.1.

Each such alignment yields a random walk similar to that considered in Section 10.2.2, giving a collection of random walks. There are $N_1 N_2$ amino acid comparisons that can be made as the two sequences take all possible positions relative to each other.

The theory for this case is far more complicated than that outlined in Section 10.2, where only one alignment occurs. Among other matters the question of the dependence of the walks arising in different alignments must be addressed. The key papers developing the theory are Dembo et al. (1994a, 1994b). The theory is too advanced for this book, and here we simply reproduce the relevant results, the most important of which is that, to a sufficient approximation, many of the conclusions of Section 10.2 carry over to the present case, with N replaced by $N_1 N_2$.

However, there are several qualifications to make about this statement. First, several conditions (given by Dembo et al. (1994a, 1994b)) need to be satisfied before the theory of Section 10.2 can be used. Second, some of the theoretical results proved apply only in the limit as both N_1 and N_2 become large. Thus the theory might not hold in the case of interest in practice, where both sequences might be of length only a few hundred or less. Thus many simulations have been carried out to assess the extent to which the theory of Section 10.2 carries through to cases of practical interest; see, for example, Altschul and Gish (1996) and Pearson (1998). A broad conclusion reached from these simulations is that the theory of Section 10.2 does carry over to a reasonable approximation if N is replaced by the product $N_1 N_2$, or by a more refined function allowing for edge effects. Thus with much but not complete theoretical and empirical support, and remembering that cases can arise that are not covered by the theory of Section 10.2, we now use that theory for the comparison of two sequences, replacing N by $N_1 N_2$ for the moment, and by a more refined expression in Section 10.3.3.

We consider first the random variable $Y_{\max}$, the maximum score achieved in the random walk comparing the sequences, using all possible ungapped local alignments between the two. This score corresponds to the MSP. Any MSP or HSP starts at a ladder point in the BLAST random walk and finishes the first time that the maximum upward excursion from this ladder point is reached. Under the heuristic adopted, $Y_{\max}$ is the maximum of a number of geometric-like random variables, whose distribution depends on the parameters λ, C, and n. The calculations for λ and C follow as in Section 10.2.3. The mean number of ladder points in this random walk corresponding to the collection of all alignments of the two sequences is approximated by

$$\frac{N_1 N_2}{A}, \tag{10.42}$$

where A is the mean distance between ladder points. This value is used throughout the following theory. The discussion at the end of Section 10.2.3 applies equally well to explain this formula. The theory discussed in Section 10.2.3 can now be used, with the value given by equation (10.42) for the mean number of ladder points, the value of C given by equation (7.61), and the value of λ given by equation (10.3).

The key formulae discussed above are now taken over to the present case with these parameter values. Thus assuming that the null hypothesis (10.1) is true, the inequalities (10.21) are replaced by

$$1 - e^{-Ke^{-\lambda x}} \leq \mathrm{Prob}(Y_{\max} > \lambda^{-1} \log(N_1 N_2) + x) \leq 1 - e^{-Ke^{-\lambda(x-1)}}, \tag{10.43}$$

and if the normalized score S' is redefined as

$$S' = \lambda Y_{\max} - \log(N_1 N_2 K), \tag{10.44}$$

the inequality (10.26) and the approximations (10.27) and (10.29) continue to hold. As a result, the right-hand side in the latter approximation also has the interpretation of an approximate P-value corresponding to the observed value s of S' as defined in (10.44).

Similarly, the expected number E' of excursions reaching a height $y_{\max}$ or more is found by replacing equation (10.35) by

$$E' = N_1 N_2 K e^{-\lambda y_{\max}}, \tag{10.45}$$

and the approximate null hypothesis mean of $Y_{\max}$ is

$$\lambda^{-1} \left(\log(N_1 N_2 K) + \gamma \right). \tag{10.46}$$

10.3.3   Edge Effects

The calculations of the preceding sections do not allow for edge effects, an important factor in the comparison of two comparatively short sequences. In this section we discuss the adjustments to the previous calculations that are used in BLAST calculations to allow for edge effects.

A high-scoring random walk excursion induced by the comparison of the two sequences might be cut short at the end of a sequence match, so that the height of high-scoring excursions, and the number of such excursions, will tend to be less than that predicted by the theory above. Whereas much of BLAST theory concerns two long sequences for which edge effects are of less importance, in practice BLAST considers databases made up of a large number of often short sequences, for which edge effects are important. Thus BLAST calculations allow for edge effects, and do this by subtracting from both N_1 and N_2 a factor depending on the mean length of any high-scoring excursion. The justification for this is largely empirical (Altschul and Gish (1996)).

Equation (10.13) shows that the mean value of the step in a high-scoring excursion asymptotically approaches the value $\lambda^{-1}H$. Given that the height achieved by a high-scoring excursion is denoted by y, equation (7.23) suggests that the mean length $E(L|y)$ of this excursion, conditional on y, is given by

$$E(L|y) = \frac{\lambda y}{H}. \tag{10.47}$$

BLAST theory then replaces N_1 and N_2 in the calculations given above respectively by $N_1' = N_1 - E(L)$, $N_2' = N_2 - E(L)$. Specifically, the normalized score (10.25) is replaced by

$$\lambda Y_{\max} - \log(N_1' N_2' K), \tag{10.48}$$

with

$$N_1' = N_1 - \frac{\lambda Y_{\max}}{H}, \quad N_2' = N_2 - \frac{\lambda Y_{\max}}{H}. \tag{10.49}$$

The expression (10.34) for the expected number of excursions scoring v or higher is correspondingly replaced by

$$N_1' N_2' K e^{-\lambda v}, \tag{10.50}$$

with $N_1' = N_1 - \lambda v/H$, $N_2' = N_2 - \lambda v/H$. Similarly, the calculation of E' given in (10.35) is replaced by

$$E' = N_1' N_2' K e^{-\lambda y_{\max}}. \tag{10.51}$$

The use of edge corrections using (10.49) assumes that the asymptotic formula (10.13) for the mean step size in a high-scoring excursion is appropriate. The simulations discussed in Section 10.2.4 show that this might not be the case, at least when N_1 and N_2 are both of order 10^2. Table

10.2 shows empirical MSP mean lengths (from simulations with 10,000 to 1,000,000 replications) and the values calculated from (10.47) for the simulation leading to the data of Table 10.1. Clearly the values calculated from (10.47) are inaccurate for anything other than very large values of N. Thus while the calculated values approach the empirical values as N increases (in line with the convergence of the mean step sizes to the asymptotic value in Table 10.1), the use of the edge correction implied by (10.49) might in practice lead to P-value estimates less than the correct values, that is, to anti-conservative tests, for anything other than very large values of N. The use of the observed value of the length of the MSP appears to give more accurate results (Altschul and Gish (1996)).

N	500	5,000	50,000	500,000	5,000,000
Empirical mean length	43.2	106.8	181.1	258.0	335.9
Calculated mean length	98.7	168.4	237.4	301.8	373.9

Table 10.2. Empirical values for the mean length of the MSP and the value found from (10.47) and empirical values of $y_{\max}$. Simulations performed with 10,000 to 1,000,000 repetitions.

In the popular implementations of BLAST the edge effect correction factor for the Karlin–Altschul sum statistic T_r is calculated as follows. First, a raw edge effect correction is calculated as $\lambda(Y_1 + Y_2 + \cdots + Y_r)/H$, generalizing the term $\lambda Y_{\max}/H$ given in (10.49). When consistent ordering is required and overlaps are restricted by a factor f, this is then multiplied by a factor $1 - (r + 1)f/r$, where f is an "overlap adjustment factor" that can be chosen by the investigator. The default value of f is 0.125, implying that overlaps between segments of up to 12.5% are allowed. The use of f is illustrated by an example in Section 10.5.2. To this the value $r - 1$ is added, leading eventually to an edge correction value $E(L)$, defined by

$$E(L) = \frac{\lambda}{H}(Y_1 + Y_2 + \cdots + Y_r)\left(1 - \frac{r+1}{r}f\right) + r - 1. \qquad (10.52)$$

While this formula is used in BLAST, there appears to be no publication justifying its validity. It could be tested empirically in the spirit of Altschul et al. (1996). The values of N_1 and N_2 in the normalized score formula (10.38) are then replaced, respectively, by

$$N_1' = N_1 - E(L), \ N_2' = N_2 - E(L). \qquad (10.53)$$

The normalized scores in (10.38) are now redefined as

$$S_i' = \lambda Y_i - \log(N_1' N_2' K), \qquad (10.54)$$

and with this new definition the sum statistic T_r is redefined as

$$T_r = S_1' + S_2' + \cdots + S_r'. \qquad (10.55)$$

The problems discussed above concerning the accuracy of the approximation (10.47) leading to the expressions for N_1' and N_2', and hence of calculations derived from S_i' and T_r, apply here also.

If the r HSPs are required to be consistently ordered, a term $\log r!$ is added to T_r (as discussed in Section 10.2.8), and if the sum so calculated is t, the P-value is then calculated as in (10.41).

10.3.4 Multiple Testing

There is no obvious choice for the value of r when the sum statistic is used in the test procedure. It is natural to consider all $r = 1, 2, 3, \ldots$, and choose the set of HSPs with lowest sum statistic P-value as the most significant, regardless of the value of r, and this is what is done in BLAST calculations. However, this procedure implies that a sequence of tests, one for each r, rather than a single test, is performed, so that the issue of multiple testing, discussed in Section 3.11, arises. Green (unpublished results) has found through simulations that ignoring the multiple testing issue leads to a significant overestimate of BLAST P-values, so that an amendment to formal P-value calculations is indeed necessary.

Unfortunately, there is no rigorous theory available to deal with this issue, and in practice it is handled in an ad hoc manner. For example, in the Washington University versions of BLAST, the P-value is adjusted when $r > 1$ by dividing the formal P-value by a factor of $(1 - \pi)\pi^{r-1}$. The parameter π has default value .5, but its value can be chosen by the user. The default value 0.5 is used in the example in Section 10.5.

When $r = 1$ the procedure is slightly different. The factor $(1 - \pi)\pi^{r-1}$ in this case is $1 - \pi$, and this implies that the value of E' given in (10.35) is divided by $1 - \pi$ to find the amended expected value E. The BLAST default value 0.5 of π implies that $E = 2E'$, so that E is calculated to be

$$E = 2N_1' N_2' K e^{-\lambda y_{\max}}. \tag{10.56}$$

The P-value corresponding to this is then found, using the analogy with (10.37), to be

$$P\text{-value} \cong 1 - e^{-E}. \tag{10.57}$$

The P-value and Expect calculations used in BLAST embody the amendments discussed above to the theoretical values (given in Section 10.2). These amendments relate to edge effect corrections, multiple testing corrections and, in the case of the sum statistic, the consistent ordering and overlap corrections. Some details of these amendments appear not to be mentioned in BLAST documentation in the popular implementations of BLAST, and only become clear by careful reading of the code.

10.4   The Comparison of a Query Sequence Against a Database

We now consider the case that is most relevant in practice. In this case we have a single "query" sequence, and we wish to search an entire database of many sequences for those with significant similarity to the query sequence. To do this, first a (heuristic) search algorithm is employed to find the high-scoring HSPs, or sets of HSPs. The P-values and Expect values of these HSPs are then approximated. These approximations are discussed in this section.

Whereas query sequence amino acid frequencies are taken from the query sequence at hand, database frequencies often are taken from some (different) published set of estimated amino acid frequencies, for example those in Robinson and Robinson (1991). These might be different again from those used to create the substitution matrix.

In approximating database P-values and Expect values, the size of the entire database must be taken into account. This raises another multiple testing problem in addition to that discussed above. What is done in practice is first to use the results of the last section to compare the query sequence to each individual database sequence, to obtain P-values for individual sequence comparisons. Then the individual sequence P-values are adjusted to account for the size of the database. If all the sequences in the database were the same size, then we could just multiply the Expect values by the number of sequences, using the linearity property of means (see (2.66)). As an approximation to this, what is done in practice is to multiply by D/N_2, where D is the total length of the database, (i.e., the sum of the lengths of all of the database sequences), and N_2 is the length of the database sequence that aligns with the query to give the HSP (or HSPs) in question. These Expect values are then converted to P-values. The details are as follows.

We consider first the case of single HSPs $(r = 1)$. Because of its linearity properties, the most useful quantity for database searches is the quantity E, defined in (10.56), and its generalizations for other HSPs. Suppose that in the database sequence of interest there is some HSP with score v. The Poisson distribution is then used to approximate the probability that in the match between query sequence and database sequence at least one HSP scores v or more. This probability is approximately

$$1 - e^{-E}. \tag{10.58}$$

Since the entire database is D/N_2 times longer than the database sequence of interest, the mean number of HSPs scoring v or more in the entire database, namely the BLAST printout quantity Expect, is given by

$$\text{Expect} = \frac{(1 - e^{-E})D}{N_2}. \tag{10.59}$$

From this value of Expect an approximate P-value is calculated from (10.37) as

$$P\text{-value} \cong 1 - e^{-\text{Expect}}. \tag{10.60}$$

This is the BLAST printout P-value for the case $r = 1$ in the implementation of BLAST discussed in Section 10.5

We shall see in Section 10.5 that the BLAST printout P-value is found by first using (10.59) to calculate "Expect" and then finding a P-value from (10.60). Once allowance is made for multiple testing, the size of the database, edge effects, and the multiple alignment situation, the value of "Expect" derives directly from the expression for E' – see the calculations in (10.62) and (10.63) – then back to (10.51) and thence to (10.34). Thus from the relation between a P-value and an expected value given in (10.37), the BLAST printout P-value traces back to the lower bound for a P-value given in (10.22), as was claimed below (10.34).

For the case $r > 1$, sum statistics for various database sequences are calculated as described in Section 10.2.8, and P-values are calculated either from (10.40) or (10.41), using all the amendments discussed above. From each such P-value a total database value of Expect is calculated using a formula generalizing that derived from (10.59), namely

$$\text{Expect} = \frac{(P\text{-value})D}{N_2}, \tag{10.61}$$

where N_2 is the length of the database sequence from which the sum is found. From this a P-value is calculated as in (10.60).

Finally, all single (i.e., $r = 1$) HSPs or summed ($r > 1$) HSPs with sufficiently low values of Expect (or, equivalently, sufficiently low P-values) are listed, and eventually printed out in increasing order of their Expect values. The value of r, given as N in the printout, is also listed.

10.5   Printouts

In this section we relate the above theory to an actual BLAST printout, describing the comparison of a query sequence with the Swiss Protein Database SWISS-PROT.

BLAST printouts give the values of λ, calculated from (10.3), of K, calculated from (10.18) amended appropriately for sequence comparisons, and H, found from the procedure described at the end of Section 10.2.4. They also list the statistics "Score" or "High Score," which in the case $r = 1$ are the values of the maximal scores $y_{\max}$ or other high-scoring HSPs. In the case of the sum statistic T_r (with $r > 1$), the score of the highest-scoring component in the sum is listed. Also listed are the "bit scores" associated with these, together with "Expect" values and P-values calculated as described in Section 10.4. We repeat that variants of these calculations are possible for different versions of BLAST and that sometimes more sophisticated calculations, taking into account factors not discussed above, are used.

10.5.1 Example

A partial printout of Example 3 from the Washington University BLAST 1.4 program follows:[1]

```
BLASTP 1.4.10MP-WashU [29-Apr-96] [Build 22:25:52 May 19 1996]


Query=  gi|557844|sp|P40582|YIV8_YEAST HYPOTHETICAL 26.8 KD PROTEIN IN HYR1
                3'REGION.
        (234 letters)

Database:  SWISS-PROT Release 34.0
           59,021 sequences; 21,210,388 total letters.


--------------------------------------------------------------------

                                                          Smallest
                                                            Sum
                                                  High  Probability
Sequences producing High-scoring Segment Pairs:   Score  P(N)      N

sp|P46429|GTS2_MANSE GLUTATHIONE S-TRANSFERASE 2 (EC 2....  53  0.010   3
sp|P46420|GTH4_MAIZE GLUTATHIONE S-TRANSFERASE IV (EC 2...  70  0.14    1
sp|P41043|GTS2_DROME GLUTATHIONE S-TRANSFERASE 2 (EC 2....  54  0.19    2
sp|P34345|YK67_CAEEL HYPOTHETICAL 28.5 KD PROTEIN C29E4...  50  0.42    2
sp|Q04522|GTH_SILCU  GLUTATHIONE S-TRANSFERASE (EC 2.5....  62  0.87    1


--------------------------------------------------------------------


>sp|P46429|GTS2_MANSE GLUTATHIONE S-TRANSFERASE 2 (EC 2.5.1.18) (CLASS-SIG).
            Length = 203

 Score = 53 (24.4 bits), Expect = 0.010, Sum P(3) = 0.010
 Identities = 10/19 (52%), Positives = 15/19 (78%)

Query:   167 ISKNNGYLVDGKLSGADIL 185
             I+KNNG+L  G+L+ AD +
Sbjct:   136 ITKNNGFLALGRLTWADFV 154

 Score = 46 (21.2 bits), Expect = 0.010, Sum P(3) = 0.010
 Identities = 8/21 (38%), Positives = 13/21 (61%)

Query:    45 PELKKIHPLGRSPLLEVQDRE 65
             PE K   P G+ P+LE+  ++
Sbjct:    39 PEFKPNTPFGQMPVLEIDGKK 59
```

---

[1]http://sapiens.wustl.edu/blast/blast/example3-14.html

```
Score = 36 (16.6 bits), Expect = 0.010, Sum P(3) = 0.010
Identities = 8/26 (30%), Positives = 12/26 (46%)

Query:   202 EDYPAISKWLKTITSEESYAASKEKA 227
             E YP   K ++T+ S    A + A
Sbjct:   173 EQYPIFKKPIETVLSNPKLKAYLDSA 198


>sp|P46420|GTH4_MAIZE GLUTATHIONE S-TRANSFERASE IV (EC 2.5.1.18) (GST-IV)
            (GST-27) (CLASS PHI).
            Length = 222

Score = 70 (32.3 bits), Expect = 0.15, P = 0.14
Identities = 17/56 (30%), Positives = 27/56 (48%)

Query:    18 RLLWLLDHLNLEYEIVPYKRDANFRAPPELKKIHPLGRSPLLEVQDRETGKKKILA 73
             R L  L+   ++YE+VP  R      PE   +P G+ P+LE  D    + + +A
Sbjct:    18 RALLALEEAGVDYELVPMSRQDGDHRRPEHLARNPFGKVPVLEDGDLTLFESRAIA 73


>sp|Q04522|GTH_SILCU GLUTATHIONE S-TRANSFERASE (EC 2.5.1.18) (CLASS-PHI).
            Length = 216

Score = 62 (28.6 bits), Expect = 2.1, P = 0.87
Identities = 15/43 (34%), Positives = 21/43 (48%)

Query:    18 RLLWLLDHLNLEYEIVPYKRDANFRAPPELKKIHPLGRSPLLE 60
             R+L   L   +LE+E VP    A    P    ++P G+ P LE
Sbjct:    15 RVLVALYEKHLEFEFVPIDMGAGGHKQPSYLALNPFGQVPALE 57


-----------------------------------------------------------------------


Matrix name    Lambda    K        H
--------------------------------------
 BLOSUM62       0.320   0.137    0.401
```

The first calculation is to check the values of "Score" (equivalently "High score") in the printout, using the BLOSUM62 matrix in Table 6.7. As an example, the score 53 for the MANSE GLUTATHIONE match is calculated as $4 + 1 + \cdots + 1$, deriving from the $I - I$ match, the $S - T$ match, $\ldots$, the $L - V$ match in the first of the three components of the match of the query and the SWISS-PROT database. Other scores are found similarly.

We next verify the calculations leading to the Maize Glutathione match sequence value of 0.15 for Expect. For this case, the printout above shows that

$$N_1 = 234, \quad N_2 = 222, \quad y_{\max} = 70.$$

Equation (10.49), in conjunction with the printout values of λ and H, gives

$$N_1' = 234 - \frac{0.32(70)}{0.401} = 178, \quad N_2' = 222 - \frac{0.32(70)}{0.401} = 166.$$

Inserting these values and the printout value of K in (10.51), we get

$$E' \cong (178)(166)(0.137)e^{-0.32(70)} \cong 7.6(10)^{-7}. \qquad (10.62)$$

Multiplying by the multiple testing factor 2 gives $E \cong 15.2(10)^{-7}$. Inserting this value in (10.59), we get

$$\text{Expect} \cong \left(1 - e^{-15.2(10)^{-7}}\right) \frac{21{,}210{,}388}{222} \cong 0.15, \qquad (10.63)$$

in agreement with the value 0.15 for Expect found in the printout.

Given this value, equation (10.60) gives an approximate P-value of 0.14, in agreement with the printout calculation. Further, equation (10.33) gives a value $0.320(70)/\log 2 \cong 32.3$ for the bit score, in agreement with the printout value.

A similar set of calculations gives, to a close approximation, the value 2.1 for Expect in the Silcu Glutathione match.

We finally consider the Manse Glutathione match, for which $r = 3$, and describe the calculations leading to the printout value 0.010 for Expect. As noted above, this value of Expect is found using a series of amendment calculations, starting with the edge correction. The expression (10.52), together with data in the printout and the default value 0.125 for f, leads to an edge correction of

$$\frac{0.32}{0.401}(53 + 46 + 36)\left(1 - \frac{4}{3}(0.125)\right) + 2 = 91.78.$$

Thus from (10.53), $N_1' = 142.2835$ and $N_2' = 111.2835$. Using these values in (10.38), the amended observed value of T_3 is computed as

$$0.32(53 + 46 + 36) - 3\log\left((0.137)(142.22)(111.22)\right) = 20.16.$$

The consistent ordering requirement holds, so we add $\log 3! = 1.79$ to this to get the value 21.95. The P-value corresponding to this is found from (10.41) to be $1.181(10)^{-8}$. Multiplying by the multiple testing factor $2^3 = 8$ yields a value of $9.448(10)^{-8}$. The value of E is essentially identical to this.

Finally, the total database Expect value is found by multiplying this by $21{,}210{,}388/203$, and this gives the value 0.010 found in the printout.

It might be a matter of concern that various somewhat arbitrary constants enter into the above calculations. This concern is reinforced by the fact that the cumulative distribution function of maximum statistics changes very sharply, as demonstrated in Table 3.4. As a result, calculated P-values are quite sensitive to the somewhat arbitrary numerical values of these constants. In practice, this concern is not important, since users of BLAST printouts seldom view a P-value even as small as 10^{-5} as interesting, and use the numerical P-values together with significant biological judgment.

10.5.2 A More Complicated Example

The way in which some BLAST outputs are formatted can be confusing.
The partial output from a BLAST search against SWISS-PROT is given
below,[2] in which only the set of HSPs between the query and one database
sequence are shown. There are 12 HSPs in total; however, since consistent
ordering is required, the smallest sum P-value comes from a set of 8 HSPs.

```
Query= gi|604369|sp|P40692|MLH1_HUMAN MUTL PROTEIN HOMOLOG 1 (DNA MISMATCH
       (756 letters)


                                                      Smallest
                                                      Sum
                                             High  Probability
Sequences producing High-scoring Segment Pairs:  Score  P(N)    N

sp|P38920|MLH1_YEAST MUTL PROTEIN HOMOLOG 1 (DNA MIS...  675  1.7e-138  8


>sp|P38920|MLH1_YEAST MUTL PROTEIN HOMOLOG 1 (DNA MISMATCH REPAIR PROTEIN.)
          Length = 769

 Score = 675 (309.6 bits), Expect = 1.7e-138, Sum P(8) = 1.7e-138
 Identities = 127/222 (57%), Positives = 170/222 (76%)

Query:   8 IRRLDETVVNRIAAGEVIQRPANAIKEMIENCLDAKSTSIQVIVKEGGLKLIQIQDNGTG 67
             I+ LD +VVN+IAAGE+I  P NA+KEM+EN +DA +T I ++VKEGG+K++QI DNG+G
Sbjct:   5 IKALDASVVNKIAAGEIIISPVNALKEMMENSIDANATMIDILVKEGGIKVLQITDNGSG 64

Query:  68 IRKEDLDIVCERFTTSKLQSFEDLASISTYGFRGEALASISHVAHVTITTKTADGKCAYR 127
             I K DL I+CERFTTSKLQ FEDL+ I TYGFRGEALASISHVA VT+TTK + +CA+R
Sbjct:  65 INKADLPILCERFTTSKLQKFEDLSQIQTYGFRGEALASISHVARVTVTTKVKEDRCAWR 124

Query: 128 ASYSDGKLKAPPKPCAGNQGTQITVEDLFYNIATRRKALKNPSEEYGKILEVVGRYSVHN 187
             SY++GK+   PKP AG  GT I VEDLF+NI +R +AL++ ++EY KIL+VVGRY++H+
Sbjct: 125 VSYAEGKMLESPKPVAGKDGTTILVEDLFFNIPSRLRALRSHNDEYSKILDVVGRYAIHS 184

Query: 188 AGISFSVKKQGETVADVRTLPNASTVDNIRSIFGNAVSRELI 229
             I FS KK G++ +   P+ + D IR++F +V+ LI
Sbjct: 185 KDIGFSCKKFGDSNYSLSVKPSYTVQDRIRTVFNKSVASNLI 226

 Score = 215 (100.6 bits), Expect = 1.7e-138, Sum P(8) = 1.7e-138
 Identities = 39/85 (45%), Positives = 58/85 (68%)

Query: 259 LLFINHRLVESTSLRKAIETVYAAYLPKNTHPFLYLSLEISPQNVDVNVHPTKHEVHFLH 318
             + FIN+RLV   LR+A+ +VY+ YLPK  PF+YL + I P  VDVNVHPTK EV FL
Sbjct: 259 IFFINNRLVTCDLLRRALNSVYSNYLPKGNRPFIYLGIVIDPAAVDVNVHPTKREVRFLS 318

Query: 319 EESILERVQQHIESKLLGSNSSRMY 343
             ++ I+E++   + ++L   ++SR +
Sbjct: 319 QDEIIEKIANQLHAELSAIDTSRTF 343
```

[2]http://blast.wustl.edu/blast/example2-14.html

```
Score = 136 (64.7 bits), Expect = 1.7e-138, Sum P(8) = 1.7e-138
Identities = 40/121 (33%), Positives = 58/121 (47%)

Query: 636 LIGLPLLIDNYVPPLEGLPIFILRLATEVNWDEEKECFESLSKECAMFYSIRKQYISEES 695
            L LPLL+ Y+P L  LP FI RL  EV+W++E+EC + + +E A+ Y       + S
Sbjct: 649 LKSLPLLLKGYIPSLVKLPFFIYRLGKEVDWEDEQECLDGILREIALLYIPDMVPKVDTS 708

Query: 696 TLSGQQSEVPGSIPNSWKWTVEHIVYKALRSHILPPKHFTEDGNILQLANLPDLYKVFERC 756
            S + E   I     +                        +++++ANLPDLYKVFERC
Sbjct: 709 DASLSEDEKAQFINRKEHISSLLEHVLFPCIKRRFLAPRHILKDVVEIANLPDLYKVFERC 769


Score = 93 (45.2 bits), Expect = 1.7e-138, Sum P(8) = 1.7e-138
Identities = 21/52 (40%), Positives = 29/52 (55%)

Query: 539 ALAQHQTKLYLLNTTKLSEELFYQILIYDFANFGVLRLSEPAPLFDLAMLAL 590
            A  QH  KL+L++   +  ELFYQI + DFANFG + L     D+ +  L
Sbjct: 549 AAIQHDLKLFLIDYGSVCYELFYQIGLTDFANFGKINLQSTNVSDDIVLYNL 600


Score = 76 (37.4 bits), Expect = 1.7e-138, Sum P(8) = 1.7e-138
Identities = 17/49 (34%), Positives = 30/49 (61%)

Query: 501 INLTSVLSLQEEINEQGHEVLREMLHNHSFVGCVNPQWALAQHQTKLYL 549
            +NLTS+ L+E++++ H  L ++ N ++VG V+ +  LA Q  L L
Sbjct: 509 VNLTSIKKLREKVDDSIHRELTDIFANLNYVGVVDEERRLAAIQHDLKL 557


Score = 42 (22.0 bits), Expect = 1.7e-138, Sum P(8) = 1.7e-138
Identities = 8/26 (30%), Positives = 16/26 (61%)

Query: 609 EYIVEFLKKKAEMLADYFSLEIDEEG 634
            E I+ +  + + ML +Y+S+E+  +G
Sbjct: 614 EKIISKIWDMSSMLNEYYSIELVNDG 639


Score = 41 (21.5 bits), Expect = 1.7e-138, Sum P(8) = 1.7e-138
Identities = 9/33 (27%), Positives = 20/33 (60%)

Query: 365 SLTSSSTSGSSDKVYAHQMVRTDSREQKLDAFL 397
            S T++++    K   +++VR D+ + K+ +FL
Sbjct: 381 SYTTANSQLRKAKRQENKLVRIDASQAKITSFL 413


Score = 39 (20.6 bits), Expect = 1.5e-21, Sum P(5) = 1.5e-21
Identities = 9/27 (33%), Positives = 14/27 (51%)

Query: 411 IVTEDKTDISSGRARQQDEEMLELPAP 437
            + T+ K D + R    + +MLE P P
Sbjct: 112 VTTKVKEDRCAWRVSYAEGKMLESPKP 138


Score = 37 (19.7 bits), Expect = 1.7e-132, Sum P(7) = 1.7e-132
Identities = 7/22 (31%), Positives = 13/22 (59%)

Query: 503 LTSVLSLQEEINEQGHEVLREM 524
            +TS LS  ++ N +G   R++
Sbjct: 409 ITSFLSSSQQFNFEGSSTKRQL 430
```

```
Score = 36 (19.3 bits), Expect = 4.2e-46, Sum P(7) = 4.2e-46
Identities = 9/40 (22%), Positives = 20/40 (50%)

Query:   14 TVVNRIAAGEVIQRPANAIKEMIENCLDAKSTSIQVIVKE 53
            TV N+  A +I    + ++++    +D K  ++  I K+
Sbjct:  215 TVFNKSVASNLITFHISKVEDLNLESVDGKVCNLNFISKK 254


Score = 34 (18.4 bits), Expect = 1.7e-138, Sum P(8) = 1.7e-138
Identities = 7/20 (35%), Positives = 12/20 (60%)

Query:  242 MNGYISNANYSVKKCIFLLF 261
            ++G + N N+  KK I  +F
Sbjct:  241 VDGKVCNLNFISKKSISPIF 260


Score = 34 (18.4 bits), Expect = 9.1e-106, Sum P(5) = 9.1e-106
Identities = 6/23 (26%), Positives = 14/23 (60%)

Query:  209 NASTVDNIRSIFGNAVSRELIEI 231
            N +++  +R    +++ REL +I
Sbjct:  510 NLTSIKKLREKVDDSIHRELTDI 532
```

Information about the 12 HSPs is summarized in Table 10.3. The HSPs are numbered 1 to 12 as they occur above. The P-values indicated for any HSP are calculated from some Karlin–Altschul sum statistic associated with the HSP. Thus these P-values do not apply to the HSP itself, but rather to the HSP in conjunction with other HSPs with which it forms a consistent set. When more than one consistent set contains an HSP, the P-value reported for any HSP is the smallest one. Consistent sets have not been given on the standard printout, so that determining which HSPs form which consistent sets has been left to the user. An option, however, has recently been implemented in the Washington University version of BLAST 2.0 that will allow the output of consistent sets.

HSP	1	2	3	4	5	6
N	8	8	8	8	8	8
P-value	1.7e-138	1.7e-138	1.7e-138	1.7e-138	1.7e-138	1.7e-138
query span	8-229	259-343	636-756	539-590	501-549	609-634
target span	5-226	259-343	649-769	549-600	509-557	614-639

HSP	7	8	9	10	11	12
N	8	5	7	7	8	5
P-value	1.7e-138	1.5e-21	1.7e-132	4.2e-46	1.7e-138	9.1e-106
query span	365-397	411-437	503-524	14-53	242-261	209-231
target span	381-413	112-138	409-430	215-254	241-260	510-532

Table 10.3.

The eight HSPs forming the most significant set are 1, 11, 2, 7, 5, 4, 6, 3, listed in their consistent order. Notice that HSP 5 overlaps HSP 4, as shown by the HSP spans. This overlap is allowed under the default option, since there is no overlap after removing the right 12.5% of residues from HSP 5 and the left 12.5% of residues from HSP 4. The first seven HSPs are consistent; the eighth is not consistent with the previous seven, and it forms the consistent set containing HSPs 8, 5, 4, 6, 3, listed in their consistent order. HSP 9 is part of the set 1, 11, 2, 9, 4, 6, 3. HSP 10 is part of the set 10, 2, 7, 5, 4, 6, 3, and HSP 12 is part of the set 1, 12, 4, 6, 3. It might not always be so easy to find consistent sets, especially when there are hundreds of HSPs and very long HSPs. Furthermore, there may be ambiguities in that a given HSP may report an $N(=r)$ of 5, yet be consistent with two different sets of 4 HSPs. In this case BLAST reports the set with lower P-value. However, it might not be clear from the printout which set this is, and it might be necessary to calculate the significance values to find it.

10.6   Minimum Significance Lengths

10.6.1   A Correct Choice of n

When sequences are distantly related, the similarities between them might be subtle. Thus we shall not be able to detect significant similarity unless a long alignment is available. On the other hand, if sequences are very similar, then a relatively short alignment is sufficient to detect significant similarity. In this section we discuss how this issue can be put on a more rigorous foundation.

If the similarity is subtle, each aligned pair will tell us less, in terms of information, than each aligned pair in more similar sequences. This will lead us to the concept of information content per position in an alignment. The theory to be developed relates to a fixed ungapped alignment of length N.

The PAMn substitution matrix has been discussed extensively above. In this section we take for granted the evolutionary model underlying these matrices. Our analysis follows that of Altschul (1991). In particular, we assume for convenience, with Altschul, that the amino acid frequencies in the two sequences compared are the same. However, in some other respects our analysis differs from his.

The analysis of Section 10.2.4 shows that an investigator using a PAMn substitution matrix in a BLAST procedure is in effect testing the alternative hypothesis that n is the correct value to use in the evolutionary process leading to the two protein sequences compared against the null hypothesis that the appropriate value of n is $+\infty$. In this section we assume that the alternative hypothesis is correct (that is, that the correct value of n has been chosen), and in effect explore aspects of the power of the testing procedure by finding the mean length of protein sequence needed before the alternative hypothesis is accepted. In the following section we explore the effects of an incorrect choice of n.

Suppose that, in formal statistical terms, we decide to adopt a testing procedure with Type I error α. Equation (10.29) shows that the value s of the normalized score statistic S' needed to meet this P-value requirement is approximately given by $s = -\log \alpha$. From equation (10.25) the corresponding value $y_{\max}$ of $Y_{\max}$ is

$$y_{\max} = \lambda^{-1} \log \left(\frac{NK}{\alpha} \right). \tag{10.64}$$

When the alternative hypothesis is true, the mean score for the comparison of the amino acids at any position is, from (10.7),

$$\sum_{j,k} q(j,k) S(j,k) = \lambda^{-1} \sum_{j,k} q(j,k) \log \frac{q(j,k)}{p_j p_k}. \tag{10.65}$$

Equation (7.23) shows that if the mean final position in a random walk is F and the mean step size is G, the mean number of steps needed to reach the final position is F/G. This then suggests that the mean sequence length needed in the maximally scoring local alignment in order to obtain significance with Type I error α is the ratio of the expressions in (10.64) and (10.65), namely

$$\frac{\log \left(\frac{NK}{\alpha} \right)}{\sum_{j,k} q(j,k) \log \frac{q(j,k)}{p_j p_k}}. \tag{10.66}$$

Altschul (1991) calls this the "minimum significance length." The expression (10.66) does not change if we change the base of both logarithms. The choice of the base 2 for these logarithms has an "intuitive appeal" (Altschul (1991)), since then various components in the resulting expression can be interpreted in terms of bits of information, as discussed in Appendix B.10. We thus make this choice in the following discussion, and write the ratio (10.66) as

$$\frac{\log_2 \left(\frac{NK}{\alpha} \right)}{\sum_{j,k} q(j,k) \log_2 \left(\frac{q(j,k)}{p_j p_k} \right)}. \tag{10.67}$$

We consider first the denominator in (10.67). This can be thought of as the mean of the relative support, in terms of bits, provided by one observation for the alternative hypothesis against the null hypothesis, given that the alternative hypothesis is true.

It follows that the numerator in (10.67) can be thought of as the mean total number of bits of information needed to claim that the two sequences are similar. The value of K is known from experience to be typically about 0.1, and α is typically 0.05 or 0.01. Thus the value of the numerator is largely determined by the length N, and to a close approximation is $\log_2 N$. Given the value $N = 1,000$, for example, this approximate numerator expression shows that about 9.97 bits of information are needed in order to claim significant similarity between the two sequences.

Our main interest, however, is not in the numerator or the denominator of (10.67), but in the ratio of the two, that is, the minimum significant length. When n is large, $q(j,k)$ is close to $p_j p_k$; the mean information per aligned pair given in the denominator is small and the minimum significant length is large. This is as expected: If null and alternative hypotheses specify quite similar probabilities for any aligned pair, many observations will in general be needed to decide between the two hypotheses. On the other hand, if n is small, the mean relative support for the alternative hypothesis provided by each aligned pair is large, and the minimum significant length is small. The limiting $(n \to 0)$ values $q(j,j) = p_j$, $q(j,k) = 0$ for $j \neq k$, together with the convention that $0 \log 0 = 0$ (see Appendix B.7), show that as $n \to 0$, the denominator in (10.67), that is, the mean support from each position in favor of the alternative hypothesis, approaches $-\sum_j p_j \log_2 p_j$. If all amino acids are equally frequent, this mean support is $\log_2 20 = 4.32$, and we can think of this as 4.32 bits of information. In practice, the actual frequencies of the observed amino acids imply that a more appropriate value is about 4.17. Thus the minimum significant length is $(\log_2 N)/4.17$. If $N = 1,000$, this is about 2.39.

When $N = 1,000$ and $n = 250$, corresponding to a PAM250 substitution matrix, the probabilities $q(j,k)$ are such that each amino acid pair provides a mean of only 0.36 bits of information, and a minimum significance length of about $\log(1000)/0.36 = 9.97/0.36 = 28$ is required on average to accept the alternative hypothesis.

10.6.2   An Incorrect Choice of n

The above calculations all assume that the correct value for n has been chosen, and thus the correct alternative hypothesis probabilities $q(j,k)$ were used. In practice it is impossible to choose a unique correct value for n when using a PAM matrix, since different species in the database will have different distances from the species corresponding to the query sequence. This matter has been addressed by Altschul (1993). To illustrate some of the points at issue we suppose that there is a unique correct value m leading to a PAMm matrix, but that some incorrect value n was chosen and a PAMn matrix used instead. What does this imply?

Suppose that with the correct choice m the probability of the ordered pair (j,k) is $r(j,k)$. The mean score is then

$$\lambda^{-1} \sum_{j,k} r(j,k) \log \frac{q(j,k)}{p_j p_k}. \tag{10.68}$$

Clearly $r(j,k) = q(j,k)$ when $n = m$, and equation (1.120) then shows that the mean score is positive. More generally, the mean score is positive if n and m are close. However, as $m \to +\infty$, $r(j,k) \to p_j p_k$, and for this value of $r(j,k)$ the mean score is negative. Thus for any choice of n there will be values of m sufficiently large compared to n so that the mean score is negative. This matter is discussed further below.

In cases where the mean score (10.68) is positive, the minimal significance length is

$$\frac{\log\left(\frac{NK}{\alpha}\right)}{\sum_{j,k} r(j,k) \log \frac{q(j,k)}{p_j p_k}}. \tag{10.69}$$

This minimal length depends on $q(j,k)$, that is, on the choice of n. This choice of n may well involve substantial extrinsic guesswork, and it is thus important to assess the implications of an incorrect choice. Altschul (1991) gives examples of the effect on the minimal significance length of using scores derived from one PAM matrix when another is appropriate.

The fact that the mean (10.68) can be negative requires some discussion. Negative means arise when m is sufficiently large compared to n, that is, when the two species being compared diverged a long time in the past relative to the time assumed by the PAM matrix used in the analysis. In this case the data are better explained by assuming no similarity between the two sequences than by assuming a close similarity between the two sequences. The more negative this mean, the more likely it is that the null hypothesis will be accepted, and in the limit $m \to +\infty$, when $r(j,k) = p_j p_k$, the probability of rejecting the null hypothesis is equal to the chosen Type I error.

As an example of this effect, if in the simple symmetric model of Section 6.5.4 the value $n = 100$ is chosen, the mean score (10.68) is negative when m is 193 or more.

These observations indicate the perils of deciding on too small a value of n. Whereas a correctly chosen small value of n leads to shorter minimal significance lengths, as discussed above, an incorrectly small choice may lead to the possibility that a real similarity between the two sequences will not be picked up. The practice sometimes adopted of using a variety of substitution matrices to overcome this problem must be viewed with some caution, particularly in the light of the multiple testing problem discussed in Section 3.11.

10.7 BLAST: A Parametric or a Non-parametric Test?

In parametric tests the test statistic is found from likelihood ratio arguments, as discussed in Chapter 9. By contrast, the test statistic in a non-parametric test is often found on reasonable but nevertheless arbitrary grounds, as was, for example, the non-parametric Mann–Whitney test statistic discussed in Section 3.8.2.

Many of the calculations and arguments used in the immediately preceding sections derive from the derivation of the score $S(j, k)$ in a substitution matrix from likelihood ratio arguments: See, for example, equations (10.6) and (10.7). In this sense the BLAST testing theory can be thought of as a parametric procedure deriving from the likelihood ratio theory in Section 9.2.1.

The assumptions made in this theory are, however, subject to debate. For example, Benner et al. (1994) claim that the time homogeneity assumption implicit in the calculations cannot be sustained, claiming, for example, that the genetic code influenced substitutions earlier in time and various chemical properties influenced substitutions more recently. Thus comparisons of distantly related species can be problematic. Even in the comparison of more closely related species, it is not clear that a uniform set of rules governs substitutions. Further, if the data in a large database come from a collection of species whose respective evolutionary divergence times might differ widely, the concept of a uniformly correct choice of n (see Section 10.6) is not meaningful.

Even if these, and similar claims are true, the statistical aspects of the BLAST procedure are still valid, in the sense that the P-value calculations are still correct. The P-value calculations take the scores in the substitution matrix as given, so that even if these scores were chosen in any more or less reasonable way, rather than from theoretical deductions using some evolutionary Markov chain and likelihood ratio theory, no problems arise with the correctness of the P-value calculations. In this sense the BLAST testing process can be thought of as a non-parametric procedure, where the choice of test statistic does not derive from a likelihood ratio or any other optimality argument but is chosen instead on commonsense grounds. On the other hand, if the various assumptions implicit in finding a substitution matrix from likelihood ratio arguments are not correct, some of the theory in the preceding sections, particularly that associated with the optimal choice of n for a PAMn matrix, needs amendment.

10.8   Gapped BLAST and PSI BLAST

10.8.1   Gapped BLAST

In this section we outline two important generalizations that have been made and are incorporated in current BLAST implementations.

The first generalization allows gaps in the sequence alignments (Altschul et al. (1997)). To outline this generalization we first recall a result from the ungapped theory, namely that in the case of two unaligned sequences of respective lengths N_1 and N_2, the approximate mean and variance of the test statistic $Y_{\max}$, given in (10.30) and (10.31), are

$$\lambda^{-1}(\log(KN_1N_2)) \quad \text{and} \quad \pi^2/(6\lambda^2) \tag{10.70}$$

respectively, and that $Y_{\max}$ has an (approximate) distribution given implicitly by (10.43).

Suppose now that gaps are allowed in the alignment of the two sequences, with some chosen linear gap penalty. In the comparison of the two sequences there will be some maximum score $Y_{\max}^{(\text{gapped})}$, the maximum score over all possible gapped alignments. The null hypothesis probability distribution of $Y_{\max}^{(\text{gapped})}$ is determined by the substitution matrix used and the gap penalty chosen. However, this null hypothesis distribution is not easy to find, and Altschul et al. (1997) follow an empirical approach to estimating it, using simulation results of Altschul and Gish (1996).

These simulations were carried out using various substitution matrices and various gap penalties. In the case of the BLOSUM62 substitution matrix, the gap penalty used was chosen to be $12 + k$ for a gap of size k. Two independent amino acid sequences, of respective lengths N_1 and N_2, were generated at random, using the amino acid probabilities given by Robinson and Robinson (1991). From these sequences the highest score, denoted here y_1, the observed value of $Y_{\max}^{(\text{gapped})}$, was found. This procedure was then repeated a large number n of times ($n = 10,000$ in their simulations), yielding n observed highest scores $y_1, y_2, \ldots, y_n$. The mean and variance of $Y_{\max}^{(\text{gapped})}$ were then estimated using $\bar{y}$ and s^2 (defined in (3.6) with a change of notation from x_i to y_i) respectively.

The approximation is then made that the distribution of $Y_{\max}^{(\text{gapped})}$ is of the same the form (10.43) as that arising in the ungapped case, with revised values for K and λ.

The method of moments procedure, discussed in Section 8.4, is used to estimate the revised values of K and λ, using the method of moments equations

$$\bar{y} = \hat{\lambda}^{-1}(\log(\hat{K}N_1N_2)), s^2 = \quad \pi^2/(6\hat{\lambda}^2), \tag{10.71}$$

derived from (8.35) and (10.70). The solution of these equations is

$$\hat{K} = (N_1N_2)^{-1}e^{\bar{y}\hat{\lambda}}, \quad \hat{\lambda} = \pi/(s\sqrt{6}). \tag{10.72}$$

This procedure was then repeated for a number of (N_1, N_2) combinations. There is no guarantee that the estimates of K and λ are independent of N_1 and N_2. The value of $\hat{\lambda}$ does however appear to be approximately independent of N_1 and N_2, being about 85% of the corresponding value in the ungapped case. The values found for $\hat{K}$ do depend on N_1 and N_2, but carrying out edge corrections as in Section 10.3.3 does appear to overcome this problem to a large extent.

There is also no guarantee that the complete distribution of $Y_{\max}^{(\text{gapped})}$ is close to that of the ungapped statistic $Y_{\max}$, even with a change of the parameters K and λ as just described. To a first approximation, however, this appears to be the case for the BLOSUM62 matrix. With this degree of empirical support, the gapped case is handled as in the ungapped case with revised values of K and λ.

Simulations with a PAM250 matrix (with a gap penalty of $15 + 3k$) lead to similar conclusions, as do simulations using the sum statistics described in Section 10.2.8. Since the BLOSUM62 and the PAM250 substitution matrices are used often, these conclusions are useful in practice. Since only very small P-values are usually of interest in a BLAST search, comparatively small inaccuracies in the above approximations are probably not important.

Gapped BLAST calculations at NCBI no longer use the Karlin-Altschul sum statistic, so the corresponding printouts do not show the N column in the BLAST printout.

The approach described above and currently implemented depends on simulation results, carried out necessarily for a restricted range of cases. However, generalizations of the theory to cover the case of gaps have recently been made: see Mott and Tribe (1999), Siegmund and Yakir (2000), Storey and Siegmund (2001), and Chan (2003). Storey and Siegmund show that if a penalty of δ is assigned to each gap in the alignment of two sequences, then (10.45) should be replaced by

$$E' = N_1 N_2 K e^{-\lambda y_{\max}} \left(1 - \frac{T}{e^{\theta * \delta} - 1} \right), \tag{10.73}$$

for a constant T whose explicit form we do not give here. The choice $\delta = +\infty$ in effect allows no gaps, and in this case (10.73) reduces to (10.45). Chan (2003) considers the case of an arbitrary non-decreasing gap penalty, and using a generalization of the mgf equation (10.3) that incorporates this penalty, finds a sharp upper bound for the P-value associated with any observed value of $y_{\max}$.

10.8.2 PSI BLAST

A second generalization is PSI (position specific iterated) BLAST. In regular BLAST a fixed substitution matrix is used to score positions in alignments, regardless of the position in the query sequence. Substitution matrices are trained on data mainly from the alignment of well conserved functional domains in protein coding genes, and the procedure relies on one matrix to provide, on average, the most meaningful scores for all positions in the query sequence simultaneously. In PSI-BLAST, the procedure using a standard substitution matrix is used as a first step. The sequences that are found are then used to derive a separate scoring scheme for each position in the query sequence. This new scoring scheme is then used to perform a second BLAST search, which can be more sensitive and thus find subtler homology than does the first. The sequences returned on the second iteration can then be used to derive a scoring scheme again, and perform a third round, which can be more sensitive than the second. This procedure can be iterated until no further iteration seems useful.

An entire substitution matrix is not derived for each position in the query sequence. Since the base in the query sequence does not change, what will be derived for each position is essentially the one row in the matrix corresponding to the particular base at that position in the query sequence. If the same base exists in two or more positions in the sequence, each position will still get its own (most likely) unique scoring scheme. This leads to the term "position specific iterated" (PSI) BLAST.

An outline of the original procedure, which is carried out in association with gapped BLAST, is as follows. The query sequence is first compared to the data base using standard BLAST methods, and all database sequence segments having a sufficiently close similarity with the query (for example having a value of "Expect" less than 0.01) are noted. Various data-trimming procedures are now carried out; for example, only one copy of closely similar database segments are retained (found by using arguments similar to those leading to the Henikoff and Henikoff procedure of Section 6.5.2).

Consider now some site in the query sequence. This site will be aligned with some collection of the remaining database segments, and in general some interval of query sequence sites around this site will also align to these segments. From this collection of sites a frequency f_i of amino acid i is calculated. These frequencies are to be used as a basis for estimating the frequency Q_i of amino acid i at this site.

The original PSI-BLAST implementation as described in Altschul et al. (1997) estimated Q_i by using pseudocounts. The pseudocount frequencies g_i are defined by

$$g_i = \sum_j f_j q(i,j)/p_j, \qquad (10.74)$$

where $q(i,j)$ is a frequency generically of the "target" form (10.8). Q_i is then defined as a linear combination

$$Q_i = \frac{\alpha f_i + \beta g_i}{\alpha + \beta}. \qquad (10.75)$$

While in standard BLAST $q(i,j) = q(j,i)$, this equality no longer occurs automatically in the iterations of PSI-BLAST. This implies that the equation $\sum_i g_i = 1$ no longer necessarily holds, so that the g_i do not necessarily form a probability distribution. Because of this problem a new form of PSI-BLAST, described in Schäffer et al. (2001) has been implemented. In this implementation, Q_i is in effect defined as

$$Q_i = \frac{\alpha f_i + \beta \sum_j f_j p(i,j)/p_j}{\alpha + \beta}, \qquad (10.76)$$

where p_i is the background frequency of amino acid i and $p(i,j)$ is the frequency with which amino acids i and j are aligned through evolutionary descent. With this definition the required equality $\sum_i Q_i = 1$ does hold.

The logarithm of the ratio Q_k/P_k is now used in a manner similar to that on the right-hand side in (10.10) to form a score to be used in the iterated PSI-BLAST process.

Another generalization, not currently implemented in BLAST, is to the case of Markov-dependent sequences, the theory for which is developed by Karlin and Dembo (1992). However, the theory for this generalization, and the full theory for the other generalizations referred to above, is beyond that appropriate for an introductory book.

10.9   Relation to Sequential Analysis

There are many similarities between the BLAST calculations given in this chapter and sequential analysis calculations discussed in Section 9.9. First and foremost, the central BLAST parameter λ $(= \theta^*)$ was first introduced into probability theory in the context of sequential analysis, being used in that theory to calculate power curves (see equations (9.62) and (9.64)), as well as mean sample size (see equation (9.69)). Second, both sequential analysis and BLAST theory center on running sums of iid random variables, and further, the random variables in both cases are either logarithms of likelihood ratios or multiples of logarithms of likelihood ratios.

It is therefore interesting to compare further the calculations deriving from (10.66) with the analogous calculation for a sequential test of hypothesis. If the alternative hypothesis is true, the mean step size in the sequential procedure defined by (9.55) is

$$\sum_y p(y; \xi_1) \log \left(\frac{p(y; \xi_1)}{p(y; \xi_0)} \right).$$

From (9.55), the accumulated sum in the sequential procedure necessary to reject the null hypothesis is $\log\left((1 - \beta)/\alpha\right)$, where α and β are the Type I and Type II errors, respectively. If these errors are both small, as is normally the case, this is close to $\log(1/\alpha)$. If we argue as in the derivation of the ratio (10.66) above, the mean number of observations needed to reject the null hypothesis when the alternative hypothesis is true, in a test with Type I error α, would be the ratio

$$\frac{\log(1/\alpha)}{\sum_y p(y; \xi_1) \log \left(\frac{p(y;\xi_1)}{p(y;\xi_0)} \right)}. \tag{10.77}$$

If we identify the observation y in a sequential test with the pair (j, k) in a sequence comparison, the denominators in the two expressions (10.66) and (10.77) are identical. The comparison between the two expressions thus concerns only their respective numerators. The numerator in (10.66) can be written as $\log(1/\alpha) + \log(N_1 N_2 K)$. The difference between the two numerators is, then, the additive factor $\log(N_1 N_2 K)$. This factor arises because in the BLAST procedure the test statistic is essentially the maximum of $N_1 N_2 / A$ iid geometric-like random variables, and the mean of such a maximum, like the mean of the maximum of n iid geometric random variables given in equation (2.118), is approximately $\log(N_1 N_2 K)$, as shown in (10.31). This comparison shows how much more stringent a test based on a maximal test statistic must be compared to one based, in the sequential procedure, on the typical value of a statistic. Once allowance for this difference is made, the similarity between the two procedures becomes apparent.

A second connection between sequential analysis and BLAST testing derives from the comparison of the denominator in the sequential analysis expression (9.70) and the denominator in the BLAST expression (10.69). In the sequential analysis case the form of the correct probability distribution $Q(y)$ of Y differs from that assumed under the null and alternative hypotheses. In the BLAST case the parallel comment might be, for example, that the elements in the substitution matrix were calculated from the evolutionary process leading to some PAM matrix, whereas some quite different evolutionary model might be appropriate.

A further connection between the BLAST and the sequential analysis testing procedures is that in both cases the step size in the testing procedure depends implicitly on some alternative hypothesis. In this respect both procedures differ from the (fixed-sample-size) test of Section 3.4.1 for the parameter p in a binomial distribution, where the testing procedure is independent of the alternative hypothesis value of p (so long as it exceeds the null hypothesis value).

Despite these connections between BLAST and the sequential testing procedure, the two procedures are rather different, and in some respects the BLAST procedure is more like the fixed sample size test. For example, the sample size is in effect fixed in advance and the test does not rely on achieving some specified Type II error.

Problems

10.1. Consider the calculation that led to equation (10.5). Use the path decomposition method to do the analogous calculation for the probability u that the generalized random walk under consideration reaches -1 as its first ladder point. Check that $u + v = 1$.

10.2. For the simple random walk of Section 7.2 the value of θ^* is given in (7.7), the value of C is $1 - e^{-\theta^*}$, and the value of A is $(q - p)^{-1}$. From this, the value of K, calculated from equation (10.18), is $(q - p)(e^{-\theta^*} - e^{-2\theta^*})$. Making the change of notation $\theta^* = \lambda$, check that both equations (10.23) and (10.24) give this value.

10.3. (This and the following problems refer to the symmetric PAM matrix discussed in Section 6.5.4.) the case $C = 1$ corresponding to the value $n = 259$ leads to a mean step size, when the alternative hypothesis is true, of 0.446 (see equation (6.36)). BLAST theory shows that this value should also be given by the expression $\lambda^{-1} H$ (see (10.13)). Use the values for $q(j, k)$ and $q(j, j)$ given in (6.35), the values $p_j = p'_k = 0.05$ in the expression (10.11) to compute H, and equation (10.3) to compute λ (in the case $S(j, j) = 12$, $S(j, k) = -1$ $(j \neq k)$), to verify this.

10.4. *Continuation.* Show mathematically that the alternative hypothesis mean size in any simple symmetric PAM model (that is, for any value of n), is equal to the value of $\lambda^{-1} H$ for that model.

10.5. *Continuation.* Use the expression (6.32) for each diagonal element in the substitution matrix for the simple symmetric PAM model and the value -1 for each off-diagonal element, together with equation (10.3), to show that in the simple symmetric PAM model the value of λ is $- \log \left(1 - \left(\frac{94}{95} \right)^n \right)$. If the value of λ is 0.320 (as in the printout of the example of Section 10.5.1, what is the corresponding value of n?

10.6. *Continuation.* Suppose that the PAM model of the "simple symmetric" example of Section 6.5.4, for which in particular $p_j = p'_k = 0.05$, leads to a substitution matrix in which $S(j, j) = 10$ $(j = 1, 2, \ldots, 20)$ and $S(j, k) = -1$ $(j \neq k)$.

 (i) Use equation (10.3) to find the associated value of λ. (This will require numerical methods.)

 (ii) From the result of (i), use equation (10.24) to find K.

 (iii) Use equation (10.8) to find the (common) values of $q(j, j)$ $(j = 1, 2, \ldots, 20)$ and the (common) values of $q(j, k)$ $(j \neq k)$.

 (iv) From the results of (ii) and (iii), find the relative entropy H defined in (10.11).

 (v) Use equations (6.28) and (6.29) to find the value of n implied by the values of $q(j, j)$ and $q(j, k)$ found in (iii) above.

 (vi) Use the value of n found in (v) in the expression (6.31) to confirm the ratio -10 for $S(j, j)/S(j, k)$.

10.7. *Continuation.* Repeat Problem 10.6 with the value 10 for $S(j, j)$ replaced by (i) 6, 8, 12, and 14, with $S(j, k) = -1$ $(j \neq k)$ and $p_j = p'_j = 0.05$

(as in Problem 10.6). Compare your values of λ, K, and H with those on BLAST printouts.

10.8. *Continuation.* Repeat Problem 10.6 with $S(j, j)$ replaced by 20 and $S(j, k)$ replaced by -2. Comment on the similarities and differences between your calculations and those of Problem 10.6.

10.9. *Continuation.* Suppose that the diagonal elements in a simple symmetric PAM matrix all take the value S and all off-diagonal elements take the value -1. If $\lambda = 0.320$ (as in the printout of Section 10.5.1), find the value of S. From this, use equation (6.32) to find the value of n (in the PAMn matrix). Also, use equation (10.24) to find the value of K, and compare this with the value in the printout of Section 10.5.1.

10.10. *Continuation.* Suppose that in the simple symmetric example of Section 6.5.4 the value $n = 50$ is chosen to calculate the simple PAM substitution matrix. Find the values of the true value m for this model for which the mean score (10.68) is negative.

10.11. Suppose that only two amino acids "X" and "Y" exist, occurring with respective frequencies 0.6 and 0.4. Suppose that a PAM matrix is used in a sequence alignment and that the match probabilities corresponding to this matrix are $q(X, X) = 0.46$, $q(X, Y) = 0.28$, $q(Y, Y) = 0.26$. Compute the mean score (10.68) in the cases (i) $r(X, X) = 0.38$, $r(X, Y) = 0.44$, $r(Y, Y) = 0.18$, (ii) $r(X, X) = 0.40$, $r(X, Y) = 0.40$, $r(Y, Y) = 0.20$, (iii) $r(X, X) = 0.42$, $r(X, Y) = 0.36$, $r(Y, Y) = 0.22$, (iv) $r(X, X) = 0.44$, $r(X, Y) = 0.32$, $r(Y, Y) = 0.24$. Comment on your answers.

10.12. Use the BLOSUM62 substitution matrix of Table 6.7 to check the score 70 given for the Maize Glutathione match in the BLAST printout of Section 10.5.

10.13. This problem refers to Table 1 of Karlin and Altschul (1993). Given the values of λ, K, and $N(= N_1 N_2)$ referred to in their paper, confirm that the three normalized scores listed can be derived from the three corresponding scores listed, using equation (10.25).

10.14. This problem refers to Table 2 of Karlin and Altschul (1993). Given the values of "Score," λ, K, and $N(= N_1 N_2)$ referred to in their paper, confirm their calculations for the various normalized scores and the various segment and sum P-values.

11
Stochastic Processes (iii): Markov Chains

11.1 Introduction

Introductory aspects of the theory of Markov chains were discussed in Chapter 4. In the present chapter further details of the theory of Markov chains will be discussed, first for Markov chains with no absorbing states and then for Markov chains with absorbing states. Our analysis of finite Markov chain theory is often oversimplified, since an examination of some of the subtleties involved in the full theory is not appropriate for bioinformatics. A recent and more complete exposition of the theory can be found in Norris (1997).

The two distinguishing Markov characteristics were introduced in Chapter 4, as were the concepts of the states of a Markov chain, transition probabilities between these states, and the concept of a transition matrix displaying these transition probabilities. It was shown that the random walk of Chapter 7 is a special example of a Markov chain. The statistical analysis associated with BLAST can also in principle be approached through Markov chain theory, as discussed below in Example 2 of Section 11.6.2. However, the special features of BLAST make the analyses of Chapter 7 and Chapter 10 more straightforward than a Markov chain analysis.

11.2 Markov Chains with No Absorbing States

11.2.1 Introduction

Finite, aperiodic irreducible Markov chains were introduced and discussed in Chapter 4. It was shown in that chapter that perhaps the most important feature of such a chain is its stationary distribution. Several properties of this distribution were introduced but no formal proof was given of them. Here we outline these proofs, derived here under the assumption, made throughout, that the Markov chain of interest is finite, aperiodic, and irreducible.

11.2.2 Convergence to the Stationary Distribution

In Chapter 4 it was shown that the n-step transition matrix $P^{(n)}$ of a Markov chain is the nth power of the single-step transition matrix, that is, that $P^{(n)} = P^n$. It was also claimed that as $n \to +\infty$, $P^{(n)}$ approaches a matrix all of whose rows are identical and all of which display the stationary distribution of the Markov chain. Computer programs such as Maple or Mathematica have subroutines for finding powers of matrices, so that these can be used to provide a straightforward computational method for finding n-step transition probabilities, for any value of n, and for finding the stationary distribution to any desired level of approximation.

However, it is appropriate to consider a mathematical, rather than a numerical, approach to finding $P^{(n)}$. Matrix theory shows why the high powers of the transition matrix approach the "stationary distribution" matrix (4.29) and, further, indicates the rate at which these high powers converge. The relevant linear algebra theory is outlined in Appendix B.19, and we now discuss its application to Markov chains.

Suppose that P is the transition matrix of a finite aperiodic irreducible Markov chain. It can be shown that this matrix has one eigenvalue λ_1 equal to 1 and that all other eigenvalues have absolute value less than 1.

We choose right and left eigenvectors r_1 and ℓ'_1 corresponding to the eigenvalue 1 by the requirements $\ell'_1 1 = 1$ and $\ell'_1 r_1 = 1$, where 1 is a column vector all of whose elements are 1. The eigenvector ℓ'_1 satisfies the equation

$$\ell'_1 = \ell'_1 P. \tag{11.1}$$

This, however, is the same equation (4.27) as that satisfied by the stationary distribution φ. It can be shown that there is a unique solution of this equation that also satisfies the normalization requirement $\ell'_1 1 = 1$, so that $\ell'_1 = \varphi'$.

Since the n-step transition probabilities are given by the elements in the matrix P^n, it follows from the identity of ℓ'_1 and φ' and the spectral expansion (B.49) that as n increases, the n-step transition matrix $P^{(n)}$

approaches the matrix

$$\mathbf{1}\varphi', \tag{11.2}$$

where φ' is the stationary distribution of the Markov chain with transition matrix P. This is a matrix all of whose rows are equal to φ'. This completes the outline of the proof of the claim made concerning expression (4.29).

It is also important to consider the rate of convergence to the stationary distribution. The spectral expansion of P^n in (B.49) shows that this rate depends on the magnitude of the nonunit eigenvalues of P, and in particular on that of the largest absolute nonunit eigenvalue(s).

11.2.3 Stationary Distributions: A Numerical Example

The eigenvalues of the transition matrix P given in (4.30) are

$$\lambda_1 = 1.0000, \quad \lambda_2 = 0.5618, \quad \lambda_3 = 0.4000, \quad \lambda_4 = 0.3382.$$

Sets of left and right eigenvectors satisfying $\ell_j' r_j = 1$ $(j = 1, 2, 3, 4)$ are

$$(0.2414, 0.3851, 0.2069, 0.1667), \quad (-.3487, 0.5643, -.2155, 0.0000),$$
$$(0.0000, .6667, 0.0000, -0.6667), \quad (0.3800, 0.2348, -0.6148, 0.0000),$$

and

$$\begin{bmatrix} 1 \\ 1 \\ 1 \\ 1 \end{bmatrix}, \quad \begin{bmatrix} -1.3088 \\ .7662 \\ -.5162 \\ .7662 \end{bmatrix}, \quad \begin{bmatrix} .25 \\ .25 \\ .25 \\ -1.25 \end{bmatrix}, \quad \begin{bmatrix} .7953 \\ .0679 \\ -1.1090 \\ 0.0679 \end{bmatrix}.$$

The matrix P^n is, then,

$$\begin{bmatrix} .2414 & .3851 & .2069 & .1667 \\ .2414 & .3851 & .2069 & .1667 \\ .2414 & .3851 & .2069 & .1667 \\ .2414 & .3851 & .2069 & .1667 \end{bmatrix} + (.5618)^n \begin{bmatrix} .4564 & -.7385 & .2820 & 0 \\ -.2672 & .4324 & -.1651 & 0 \\ .1800 & -.2913 & .1112 & 0 \\ -.2672 & .4324 & -.1651 & 0 \end{bmatrix}$$

$$+ (.4000)^n \begin{bmatrix} 0 & .1667 & 0 & -.1667 \\ 0 & .1667 & 0 & -.1667 \\ 0 & .1667 & 0 & -.1667 \\ 0 & -.8333 & 0 & .8333 \end{bmatrix} + (.3382)^n \begin{bmatrix} .3022 & .1867 & -.4890 & 0 \\ .0258 & .0159 & -.0417 & 0 \\ -.4214 & -.2604 & .6818 & 0 \\ .0258 & .0159 & -.0417 & 0 \end{bmatrix}.$$

For the values $n = 2, 4, 8, 16$ the matrices found from this expansion agree with those given in the expressions (4.32)–(4.35). However, the spectral expansion above indicates clearly the geometric rate at which the stationary distribution is reached.

11.2.4 Reversibility and Detailed Balance

The use of a substitution matrix for BLAST calculations, as well as the evolutionary considerations to be discussed in Chapter 14, indicate the importance of examining the relationship between a Markov chain running forward in time (the direction in which evolution has actually proceeded) and the corresponding "time-reversed" chain running backward in time (the direction relevant to many evolutionary inferences made on the basis of contemporary data). This issue is particularly relevant in comparing two related contemporary sequences, where one sequence is reached from the other by going backward in time to an assumed common ancestor and then forward in time to the other sequence. We now consider aspects of this relation for a finite, irreducible, and aperiodic Markov chain with stationary distribution φ.

Suppose that at time 0 the initial state of the Markov chain with transition matrix P is chosen at random in accordance with the stationary distribution φ. The discussion surrounding equation (4.25) shows that this distribution then applies at all future times. During the course of t transitions the chain will move through a succession of states, which we denote by $S(0)$, $S(1)$, ..., $S(t)$. Define $S^*(i)$ by $S^*(i) = S(t - i)$. We show below that the probability structure determining the properties of the reversed sequence of states $S^*(0)$, $S^*(1)$, ..., $S^*(t)$ is also that of a finite aperiodic irreducible Markov chain, whose typical element p_{ij}^* is found from the typical element p_{ij} of P by the equation

$$p_{ij}^* = \frac{\varphi_j p_{ji}}{\varphi_i}. \tag{11.3}$$

Furthermore, the stationary distribution of the reversed chain is also φ.

These claims are proved as follows. The fact that the reversed process has transition probabilities given by (11.3) follows from the conditional probability formula (1.104) when A_i is the event "$S^*(u) = E_i$" and A_j is the event "$S^*(u+1) = E_j$." The fact that $\sum_j p_{ij}^* = 1$ follows immediately from the stationary distribution property $\sum_j \varphi_j p_{ji} = \varphi_i$: See equation (4.26). The fact that the stationary distribution of the time-reversed process is φ follows from the equations

$$\sum_i \varphi_i p_{ij}^* = \sum_i \varphi_j p_{ji} = \varphi_j.$$

The equality of the extreme left- and right-hand expressions is the defining property of a stationary distribution: See equation (4.25). Irreducibility and aperiodicity are equally quickly demonstrated.

Suppose that a probability distribution $\boldsymbol{\lambda} = \{\lambda_j\}$ can be found such that

$$\lambda_i p_{ij} = \lambda_j p_{ji} \tag{11.4}$$

for all (i, j) pairs. Summation of both sides of equation (11.4) over all possible values of i yields

$$\sum_i \lambda_i p_{ij} = \sum_i \lambda_j p_{ji} = \lambda_j.$$

But this is the defining equation of the stationary distribution. This observation sometimes provides the most convenient way of finding the stationary distribution of an irreducible aperiodic Markov chain: If a distribution λ can be found for which equation (11.4) holds for all (i, j) pairs, then this is the stationary distribution φ of the Markov chain, which then satisfies

$$\varphi_i p_{ij} = \varphi_j p_{ji} \quad \text{for all } (i, j). \tag{11.5}$$

These are the so-called *detailed balance* equations. From equation (11.3), the detailed balance equations hold if and only if $p_{ij}^* = p_{ij}$, in which case the properties of the reversed-time Markov chain are equivalent to those of the forward-time chain. An observer watching the successive states in the reversed chain would have no way of telling whether he/she was watching the reversed or the original chain. Examples of reversible and irreversible Markov chains used in evolutionary models are given in Section 14.1.

A version of the reversibility criterion in a form more convenient for evolutionary processes was given by Tavaré (1986): see the expression (14.23). An arbitrary Markov chain is unlikely to be reversible; for example, the Markov chain with transition matrix (4.24) is not reversible, since the terms in this matrix and in the stationary distribution (4.31) do not satisfy the detailed balance equations.

The reversibility property is often assumed implicitly in comparing sequences from two different contemporary species, since to get from one species to another one must travel backward in time to a common ancestor and then forward in time to the other species. The calculations leading to PAM matrices, discussed in Section 6.5.3, consider only Markov chains going forward in time. It is thus of interest to show that the transition matrix M_1, defined in Section 6.5.3 and used to compute various PAMn substitution matrices, is reversible for all practical purposes. This is done as follows.

M_1 is a particular case of the matrix P, defined through equations (6.20) and (6.21), so it is sufficient to prove that P is reversible. It is reasonable to assume that the quantity $\sum_m A_{jm}$ appearing in the denominator of (6.19) is, for all practical purposes, proportional to the stationary probability φ_j of amino acid j. This implies that $A_{jk} = b\varphi_j a_{jk}$ for some constant b. Equation (6.20) then implies that $A_{jk} = d\varphi_j p_{jk}$ for some constant d. But the matrix $\{A_{jk}\}$ is by construction symmetric, so that $A_{jk} = A_{kj}$ for all (j, k). From this,

$$d\varphi_j p_{jk} = d\varphi_k p_{kj}.$$

Cancellation of the constant d shows that this is the reversibility requirement (11.5).

It is of considerable interest to note that the transition matrix of the PAM model is in effect the most general of all reversible 20×20 Markov chain models (Goldman (2002)).

11.3   Higher-Order Markov Dependence

11.3.1   Testing for Higher-Order Markov Dependence

The test of Markov dependence (more exactly, first-order Markov dependence) of successive nucleotides in a DNA sequence was discussed in Section 5.2. This was presented as a chi-square test, and the calculations of Example 4 of Section 9.4 show that this is an approximation to a $-2\log\lambda$ test. In this section we extend the analysis to tests of higher-order Markov dependence (defined in the following paragraph) via the $-2\log\lambda$ procedure, following the discussion by Tavaré and Giddings (1989). For concreteness we present the discussion in terms of DNA sequences.

The probability structure of the nucleotides in a DNA sequence is described by a Markov chain of order $k \geq 1$ if the probability that any nucleotide occurs at a given site depends on the nucleotides at the preceding k sites. The transition probabilities for the case $k = 3$, for example, are of the form

$$p_{h:mnr} = \text{Prob}(Y_{j+3} = a_h \mid Y_{j+2} = a_m, Y_{j+1} = a_n, Y_j = a_r),$$

where Y_q denotes the nucleotide at site q, and the a_k are specified nucleotide types. This notation shows that for general k there are 3×4^k such transition probabilities.

Suppose that the null hypothesis is that the Markov chain is of order $k - 1$, and the alternative that it is of order k, for some value of k. (The test discussed in Section 5.2 is for the case $k = 1$.) The likelihood of the data under both hypotheses may be written down in terms of arbitrary parameters of the form $p_{h:mnr}$ under both null and alternative hypotheses. The likelihoods under null and alternative hypotheses are then maximized with respect to these parameters and the $-2\log\lambda$ statistic calculated.

The number of arbitrary parameters under null and alternative hypotheses are, respectively, $3 \times 4^{k-1}$ and 3×4^k, so that the $-2\log\lambda$ test has $9 \times 4^{k-1}$ degrees of freedom. Thus once k exceeds 2, extensive data would be required to ensure that the asymptotic chi-square properties of $-2\log\lambda$ apply. Further details are given by Tavaré and Giddings (1989) and Reinert et al. (2000). In general, it has been found that even DNA in nonfunctional intergenic regions tends to have high-order dependence.

11.3.2   Testing for a Uniform Stationary Distribution

If the hypothesis of Markov dependence is accepted, a further test of interest is whether the stationary distribution of the Markov chain is uniform. We will discuss this in the case of first-order dependence; the discussion for higher-order dependence is similar.

In the first-order case, equation (4.25) shows that a necessary and sufficient condition that the stationary distribution be uniform is that the transition probabilities in each column of the transition matrix sum to 1. This implies that, for example, the elements in the fourth row of the transition matrix are determined by the elements in the first three rows.

The probability of any observed DNA sequence can be calculated under the null hypothesis that the stationary distribution is uniform. Since under this hypothesis the elements in each row and each column of the transition matrix must sum to 1, there are 9 free parameters. The probability of the observed sequence can be maximized with respect to these parameters. Under the alternative hypothesis the only constraint is that the elements in any row must sum to 1 and therefore there are 12 free parameters. The probability of the observed sequence can then be maximized with respect to these parameters. From these two maximum likelihoods a $-2\log\lambda$ statistic may be calculated. Under the null hypothesis this statistic has an asymptotic chi-square distribution with 3 degrees of freedom. This then provides a testing procedure for uniformity of the stationary distribution.

11.4   Patterns in Sequences with First-Order Markov Dependence

If the test of independence in the previous section suggests that there is Markov dependence in a DNA sequence it is of interest to extend the calculations concerning patterns in Section 5.7 to the Markov dependence case. These extensions require only a more careful bookkeeping than is needed in Section 5.7, and have been made by several authors; summaries and references are given by Robin and Daudin (1999) and Reinert et al. (2000). The discussion here is in terms of nucleotide sequences, with attention focused on a word $\mathbf{w} = w_1 w_2 \ldots w_k$ of length k.

We assume that some 4×4 transition matrix P of the Markov chain is given or postulated, describing the transition probabilities from one nucleotide to the next along the DNA sequence. The typical term in this matrix is $p_{w_i w_j}$, and the stationary probability of the letter w_i is denoted φ_{w_i}. It is assumed that for all practical purposes stationarity has been reached.

Many expressions found in Section 5.7 can be generalized immediately to the Markov case. For example, (5.40) becomes

$$E(Y_1(N)) = (N - k + 1)\varphi_{w_1} p_{w_1 w_2} p_{w_2 w_3} \cdots p_{w_{k-1} w_k}. \qquad (11.6)$$

This is identical to (5.40) if we define

$$\pi = \varphi_{w_1} p_{w_1 w_2} p_{w_2 w_3} \cdots p_{w_{k-1} w_k}, \qquad (11.7)$$

where π is the Markov chain probability that $\mathbf{w}$ occurs at any site after site $k - 1$. The variance formula (5.42) can also be generalized easily.

The distribution of the number of sites between successive occurrences of **w** can be found by extending the calculations leading to the generating function (5.55). We suppose that **w** has just occurred at some site in the sequence. We call this "appearance A." We wish to find the probability $q(y)$ that **w** *first* reappears y sites after appearance A. The argument that led to equation (5.53) in the independent case shows that this is the probability that it does appear y sites after appearance A, less the probability that it first appears at some position j sites after appearance A ($j < y$) and then also appears $y - j$ sites after this.

If $y < k$, it is impossible for **w** appear y sites after appearance A unless it can overlap with itself appropriately. If ε_j is defined as in Section 5.7.1, it is possible for **w** to reappear y sites after appearance A only if $\varepsilon_{k-y} = 1$. When this is the case, since the letter w_k occurred in the final site corresponding to appearance A, the probability that **w** appears y sites after appearance A is

$$p_{w_k w_{k-y+1}} p_{w_{k-y+1} w_{k-y+2}} \cdots p_{w_{k-1} w_k}.$$

We denote this probability by $Q(y)$, so that the probability that **w** occurs y sites after appearance A is $\varepsilon_{k-y} Q(y)$.

The probability that **w** *first* occurs j sites after appearance A and also $y - j$ sites after this is the sum over j ($j = 1, 2, \ldots, y-1$) of the probability $p(j)$ that **w** first appears j sites after appearance A, multiplied by the probability, given that it appears j sites after appearance A, that it appears $y - j$ sites after this. These arguments lead to

$$p(y) = \varepsilon_{k-y} Q(y) - \sum_{j=1}^{y-1} p(j) \varepsilon_{k-y+j} Q(y - j). \tag{11.8}$$

The case $y \geq k$ can be handled similarly.

These equations lead to the generating function for the probability distribution of the distance until **w** next appears after appearance A, which generalizes the "independence" generating function (5.55). The mean and variance of this number of sites can then be found directly from this generating function: it is found that equation (5.59) continues to hold for the mean if π is defined as in (11.7), and that only minor modifications are needed to equation (5.60) for the variance.

Similar, and indeed simpler, calculations can be made for the generating function, for the mean and for the variance of the number of sites until the first occurrence of **w** after the origin. Further details are given by Robin and Daudin (1999).

11.5 Markov Chain Monte Carlo

Markov chains can be used for a variety of calculation and optimization purposes in bioinformatics to which they are not initially clearly related. The methods used are described as "Markov Chain Monte Carlo" methods. We discuss two of these in this section, describing first the Hastings–Metropolis algorithm on which they are based.

11.5.1 The Hastings–Metropolis Algorithm

The aim of the Hastings–Metropolis algorithm is to construct an aperiodic irreducible Markov chain having some prescribed stationary distribution $\varphi' = (\varphi_1, \varphi_2, \ldots, \varphi_s)'$, where $\varphi_j > 0$, $j = 1, \ldots, s$.

We choose a set of constants $\{q_{ij}\}$ such that $q_{ij} > 0$ for all (i, j) and $\sum_j q_{ij} = 1$ for all i. We then define a_{ij} by

$$a_{ij} = \min\left(1, \frac{\varphi_j q_{ji}}{\varphi_i q_{ij}}\right) \tag{11.9}$$

and p_{ij} by

$$p_{ij} = q_{ij} a_{ij}, \quad i \neq j, \tag{11.10}$$

and $p_{ii} = 1 - \sum_{j \neq i} p_{ij}$. Since $q_{ij} > 0$ for all (i, j), this construction shows that $p_{ij} > 0$ for all (i, j) including the case $i = j$ (see Problem 11.8), so that the Markov chain defined by the $\{p_{ij}\}$ is aperiodic and irreducible.

Theorem 10.1. *The stationary distribution of the Markov chain defined by (11.10) is φ'.*

This theorem is checked by showing that the detailed balance requirements (11.5) hold. Suppose without loss of generality that $\varphi_j q_{ji}/\varphi_i q_{ij} < 1$. Then $\varphi_i q_{ij}/\varphi_j q_{ji} > 1$ and

$$a_{ij} = \frac{\varphi_j q_{ji}}{\varphi_i q_{ij}}, \quad p_{ij} = \frac{\varphi_j q_{ji}}{\varphi_i}, \quad a_{ji} = 1, \quad p_{ji} = q_{ji}. \tag{11.11}$$

From these results the detailed balance requirement $\varphi_i p_{ij} = \varphi_j p_{ji}$ follows. The special case where $\varphi_j q_{ji}/\varphi_i q_{ij} = 1$ is easily handled separately.

The above proof can be extended to the case where some q_{ij} are zero, so long as $q_{ji} > 0$ whenever $q_{ij} > 0$ and the Markov chain defined by the resultant p_{ij} is aperiodic and irreducible.

Different Markov chains may be constructed by different choices of the q_{ij}, all having the desired stationary distribution φ', and for any given application one choice will often be more useful than another.

11.5.2   Gibbs Sampling

Let Y_i, $i = 1, 2, \ldots, k$, be discrete finite random variables, $\boldsymbol{Y}$ the random vector $(Y_1, Y_2, \ldots, Y_k)'$, and $P_{\boldsymbol{Y}}(\boldsymbol{y})$ the distribution of $\boldsymbol{Y}$. Assume furthermore that $P_{\boldsymbol{Y}}(\boldsymbol{y}) > 0$ for all $\boldsymbol{y}$.

We will define a Markov chain whose states are the possible values of $\boldsymbol{Y}$. Enumerate the vectors in some order as vectors $1, 2, \ldots, s$ and identify vector j with the jth state in a Markov chain whose transition probabilities we now define. If vectors i and j differ in more than one component, put $p_{ij} = 0$. If they differ in at most one component, suppose, to be concrete, that they differ in their first component (if they differ at all). Write vector i as $(y_1, y_2, \ldots, y_k)$ and vector j as $(y_1^*, y_2, \ldots, y_k)$. Then we define

$$p_{ij} = \text{Prob}(Y_1 = y_1^* \mid Y_2 = y_2, Y_3 = y_3, \ldots, Y_k = y_k) \qquad (11.12)$$

$$= \frac{\text{Prob}(Y_1 = y_1^*, Y_2 = y_2, Y_3 = y_3, \ldots, Y_k = y_k)}{\text{Prob}(Y_2 = y_2, Y_3 = y_3, \ldots, Y_k = y_k)}, \qquad (11.13)$$

where the probabilities in both numerator and denominator are calculated using $P_{\boldsymbol{Y}}(\boldsymbol{y})$.

We claim that this Markov chain is irreducible and aperiodic, and furthermore has stationary distribution $P_{\boldsymbol{Y}}(\boldsymbol{y})$.

Aperiodicity follows from the fact that $p_{ii} > 0$, and irreducibility follows from the fact that each state in the Markov chain can be reached, after a finite number of steps, from every other state. The fact that the stationary distribution is $P_{\boldsymbol{Y}}(\boldsymbol{y})$ is proved as follows.

Define $q_{ij} = p_{ij}$. Then from (11.9) it follows that if the denominator in (11.13) is denoted by Q,

$$a_{ij} = \min\left(1, \frac{P_{\boldsymbol{Y}}(y_1^*, y_2, \ldots, y_k)P_{\boldsymbol{Y}}(y_1, y_2, \ldots, y_k)/Q}{P_{\boldsymbol{Y}}(y_1, y_2, \ldots, y_k)P_{\boldsymbol{Y}}(y_1^*, y_2, \ldots, y_k)/Q}\right) = 1.$$

This implies that $p_{ij} = q_{ij}a_{ij}$, and from Theorem 10.1, $P_{\boldsymbol{Y}}(\boldsymbol{y})$ is the stationary distribution of the Markov chain defined by the $\{p_{ij}\}$.

Example. In this example we show that the segment alignment procedure of Section 6.6 is essentially a Gibbs sampling procedure. The alignment procedure was described in Section 6.6 as a Markov chain process with S states, each state corresponding to an array of amino acids having N rows and W columns. We define $c_{ij}(s)$ as the number of times that amino acid j occurs in column i in the array corresponding to state s in this Markov chain, and put

$$q_{ij}^*(s) = \frac{c_{ij}(s) + b_j}{N + B}, \quad q_{ij}(s) = \frac{c_{ij}(s)}{N},$$

where the b_j are pseudocounts, $B = \sum b_j$, as defined in Section 6.6.

We define p_j as the background frequency of amino acid j. The relative entropy between $\{q_{ij}^*(s)\}$ and $\{p_j\}$ is

$$\sum_{i=1}^{W}\sum_{j=1}^{20} q_{ij}^*(s) \log\left(\frac{q_{ij}^*(s)}{p_j}\right). \tag{11.14}$$

States for which this relative entropy is high are those corresponding to good alignments. Our aim is therefore to find states for which this relative entropy is high.

We associate with state s in this Markov chain the probability λ_s, defined by

$$\lambda_s = \text{const} \prod_{i=1}^{W}\prod_{j=1}^{20} \left(\frac{q_{ij}^*(s)}{p_j}\right)^{c_{ij}(s)}, \tag{11.15}$$

where the constant is chosen so that $\sum_{s=1}^{S} \lambda_s = 1$.

We will say that states s and u are neighbors if either $s = u$ or if the arrays corresponding to these two states differ only in the entries in one row. This implies that in the step-by-step procedure described in Section 6.6 it is possible to move in one step from state s to state u and also from state u to state s.

If states s and u are neighbors, for any position i the respective values of the counts $c_{ij}(s)$ and $c_{ij}(u)$ either will be identical for all 20 values of j or will be identical for 18 values of j and differ by $+1$ and -1 for the remaining two values of j. The values $+1$ and -1 arise when there are different amino acids in position i in the row where states s and u differ.

For any state a, define a_r as the reduced array obtained by removing row r from state a. Then if states s and u differ in row r, $q_{ij}(s_r) = q_{ij}(u_r)$. For each (i, j) pair, the values of $q_{ij}^*(s)$ and $q_{ij}^*(u)$ are very close to this common value $q_{ij}(s_r)$, and from now on we make the approximation that both are equal to $q_{ij}(s_r)$.

Making this approximation, the ratio λ_s/λ_u becomes, in the notation adopted in equation (6.39),

$$\frac{\lambda_s}{\lambda_u} = \frac{q_{1,s(1)}q_{2,s(2)} \cdots q_{W,s(W)}}{q_{1,u(1)}q_{2,u(2)} \cdots q_{W,u(W)}} \cdot \frac{p_{u(1)}p_{u(2)} \cdots p_{u(W)}}{p_{s(1)}p_{s(2)} \cdots p_{s(W)}}. \tag{11.16}$$

Equations (6.39) and (11.16) jointly imply that with the approximation made above,

$$\frac{\lambda_s}{\lambda_u} = \frac{p_{us}}{p_{su}},$$

so that

$$\lambda_s p_{su} = \lambda_u p_{us} \tag{11.17}$$

for all neighboring states. Equation (11.17) is also true for states that are not neighbors, since in this case both sides of the equation are 0. But this

equation is identical in form to equation (11.4) and the detailed balance
equation (11.5). Thus the step procedure described in Section 6.6 is, for
all practical purposes, that of a Gibbs sampling process, and λ_s is the
stationary probability of state s in the procedure.

States with high stationary probability are visited comparatively frequently in the procedure and thus may be recognized. Since $\log \lambda_s$ is
approximately a linear function of the relative entropy (11.14), we have
achieved the aim of ensuring that states such that $\{q_{ij}^*(s)\}$ and $\{p_j\}$ have
high relative entropy are visited comparatively frequently and can thus be
identified. This is the reason for the choice of the transition probabilities
introduced in Section 6.6.

11.5.3 Simulated Annealing

The goal of the simulated annealing procedure is to find, at least approximately, the minimum of some positive function defined on an extremely
large number s of "states" $E_1, \ldots, E_s$, and to find those states for which
this function is minimized (or approximately minimized). Write the value
of this function for state E_j as $f(j)$.

To illustrate some aspects of the simulated annealing procedure, it is
useful to consider the travelling salesman problem, which is equivalent in
complexity to many problems that arise in bioinformatics. In this classical
problem, a salesman wishes to find the minimal travelling path linking n
cities, returning finally to his city of origin. The number of possible paths
is $O(n!)$, which is extremely large when n exceeds 20 or 30. In this case
there is no hope of listing the total distance along each path and thus
finding the minimum distance by exhaustive search. We may call any such
path a "state" E_j, and define $f(j)$ as the total distance travelled along the
path. Although the number of paths is very large, once a path E_j is given,
the distance $f(j)$ along that path is easily computed, and "similar" paths
should have similar total distances.

The concept of a collection of states and a function $f(j)$ corresponding
to state E_j may be generalized. Let T be a fixed positive parameter whose
value is as yet unspecified. The aim is to construct a Markov chain with
states $E_j, j = 1, 2, \ldots, s$, such that the stationary distribution probability
$\varphi(j; T)$ of the state E_j is

$$\varphi(j; T) = C \cdot \exp(-f(j)/T). \tag{11.18}$$

Here the constant C is chosen to ensure that the sum of the probabilities
in the stationary distribution is 1.

The reason for this construction is the following. Suppose that for some
state E_j, $f(j)$ is small. If a Hastings–Metropolis procedure can be used to
build a Markov chain with stationary distribution $\{\varphi(j; T)\}$, then at stationarity the chain should visit state E_j comparatively often, since from
(11.18) the stationary distribution probability for this state is compara-

tively large. Thus starting in any state and following the procedure for a large number of iterations, states with low values of $f(\cdot)$ should become recognizable.

We construct a "neighborhood" of each state, the neighborhood of state E_j being chosen as a set of states in some sense "close" to E_j, and to which the variable in the Markov chain can move, in one step, from E_j. Moves in one step to states outside the neighborhood are not allowed. We make the following four requirements about these neighborhoods.

(a) If E_j is in the neighborhood of E_k, then E_k is in the neighborhood of E_j.

(b) The number of states in the neighborhood of any given state is independent of that state. This number is denoted by N.

(c) The neighborhoods are linked in the sense that the Markov chain can eventually move, after a finite number of steps, from E_j to E_m, for all (j, m).

(d) Given that the Markov chain is in E_j, it may next move only to a state in the neighborhood of E_j.

Suppose that in the travelling salesman problem the cities to be visited are A, B, C, D, E. Any choice of a path through these cities by the salesman is equivalent to an ordering of these cities, for example ADEBC. If a neighboring path is defined by switching the order of two adjoining cities, for example in the above case to ADBEC, then requirements (a) and (b) above are satisfied.

Requirements (a)–(d) can be relaxed with care, but for simplicity we impose them as stated. Subject to these requirements, the choice of N and of the states in the neighborhood of any given state are in principle arbitrary, but in practice the usefulness of the simulated annealing algorithm depends to some extent on a wise choice of these.

Suppose that the variable in the Markov chain is currently in state E_i. The Hastings–Metropolis quantity q_{ij} is chosen in the simplest possible way, namely

$$q_{ij} = \begin{cases} N^{-1} & \text{if } E_j \text{ is in the neighborhood of } E_i, \\ q_{ij} = 0 & \text{otherwise.} \end{cases}$$

For each (i, j) for which $q_{ij} = N^{-1}$, this choice implies that the Hastings–Metropolis quantity a_{ij} is given by

$$a_{ij} = \min\left(1, \frac{\varphi(j; T)N}{\varphi(i; T)N}\right) = \min\left(1, e^{(f(i)-f(j))/T}\right). \tag{11.19}$$

Consequently, the desired Markov chain has transition probabilities p_{ij} as follows: When $i \neq j$,

$$p_{ij} = \begin{cases} N^{-1}\frac{\varphi(j;T)}{\varphi(i;T)} & \text{if } E_j \text{ is a neighbor of } E_i \text{ and } \varphi(i;T) > \varphi(j;T), \\ N^{-1} & \text{if } E_j \text{ is a neighbor of } E_i \text{ and } \varphi(i;T) < \varphi(j;T), \\ 0 & \text{if } E_j \text{ is not a neighbor of } E_i, \end{cases}$$

and

$$p_{ii} = 1 - \sum_{j \neq i} p_{ij}.$$

In the above procedure a large value of T implies that all states in the neighborhood of the present state of the Markov chain are chosen with approximately equal probability, and also that the stationary distribution of the Markov chain tends to be uniform. A small value of T implies that different states in the neighborhood of E_i tend to have rather different stationary distribution probabilities. However, too small a choice of T might lead to the procedure tending to stay near local maxima and not move to other higher maxima. Part of the art of the process is in choosing a value of T that allows comparatively rapid movement from one neighborhood to another (large T) but at the same time picks out states within neighborhoods with comparatively large stationary probabilities (small T).

Example. Waterman (1995) describes an application of the simulated annealing process to the so-called double digest problem. Restriction endonucleases were described briefly in Section 5.8.1. In the double digest procedure a length of DNA is subjected to two restriction endonucleases, both separately and together. When the DNA is subjected to the first restriction endonuclease it will be cut wherever the recognition sequence for that restriction endonuclease occurs, and a parallel comment applies for the second restriction endonuclease. When both are applied together the DNA is cut whenever either recognition sequence occurs. The data thus consist of three sets of sequence lengths, the sum of the lengths in each set being the length of the original DNA sequence, and the aim is to reconstruct, from these lengths, the original set of locations of the two recognition sequences. This problem is "NP-complete," implying in particular that no polynomial-time algorithm is known (or, conjecturally, can ever be known) for its solution. Thus heuristic methods are required, and the simulated annealing method is one such approach.

Suppose that the application of both restriction endonucleases together cuts the DNA into s segments of lengths $c(1), c(2), \ldots, c(s)$, numbered so that $c(1) \leq c(2) \leq \cdots \leq c(s)$. Suppose also that the first restriction endonuclease cuts the DNA into n segments, which we write $A_1, A_2, \ldots, A_n$, and the second cuts the DNA into m segments, which we write $B_1, B_2, \ldots, B_m$. Without mixing the A_i's with the B_i's, these segments may be placed jointly into $n!m!$ different orderings. Each ordering defines a sequence of

points on the DNA as the cut points of one of the two restriction endonucleases, and for any ordering these j cut points will divide the DNA into s
segments, some of whose lengths can be 0. These lengths are now numbered
$d_j(1), d_j(2), \ldots, d_j(s)$ in such a way that $d_j(1) \le d_j(2) \le \cdots \le d_j(s)$.

If all measurements are error-free, at least one of these $n!m!$ orderings
will produce the lengths $c(1), c(2), \ldots, c(s)$. To find these orderings, or more
generally if measurement errors are possible to find those orderings that
most closely achieve this, it is reasonable to attempt to find the values of
j that minimize a function $f(j)$ of the form

$$f(j) = \sum_{u=1}^{s} \frac{(d_j(u) - c(u))^2}{c(u)}.$$

An interesting choice of the neighborhood of any ordering for the simulated annealing procedure is the generalization of that given above for
the travelling salesman. Consider any of the $n!m!$ orderings of the segments produced by the two restriction endonucleases, for example (with
$n = 5, m = 4$) $A_5 A_2 A_1 A_4 A_3$ and $B_3 B_4 B_1 B_2$. The neighborhood of this ordering can be taken as any ordering for which at most one switch between
adjoining segments in each ordering is made, for example $A_5 A_1 A_2 A_4 A_3$ and
$B_3 B_4 B_2 B_1$. Waterman (1995) discusses more efficient choices than this in
both the double digest and the travelling salesman problems.

11.6   Markov Chains with Absorbing States

11.6.1   Theory

We now turn to Markov chains where there are absorbing states. These
were introduced and discussed briefly in Section 4.7. In this section we
discuss two questions asked about these Markov chains, namely, "If there
are two or more absorbing states, what is the probability that a specified
absorbing state is the one eventually entered?" and "What is the mean
time until an absorbing state is eventually entered?"

Let $Y(1), Y(2), \ldots$ be a sequence of random variables such that

$$\text{Prob}(Y(t+1) = j \mid Y(t) = i) = p_{ij}, \quad i, j = 1, 2, \ldots, s.$$

Thus the successive values of $Y(\cdot)$ have probabilistic behavior governed by
a Markov chain with s states $E_1, E_2, \ldots, E_s$.

We first consider the case with two absorbing states, which are assumed
to be E_1 and E_s, so that $p_{1j} = 0$ for $j \ne 1$ and $p_{sj} = 0$ for $j \ne s$. For
$i = 2, 3, \ldots, s - 1$, let w_i be the probability that eventually $Y(t) = s$, given
that $Y(0) = i$. Then a generalization of the argument that led to equation

(7.4) shows that the w_i are the solution of the simultaneous equations

$$w_i = \sum_{j=2}^{s-1} p_{ij} w_j + p_{is}, \quad i = 1, 2, 3, \ldots, s. \qquad (11.20)$$

A generalization of the argument that led to (7.11) shows that if t_i is the mean number of transitions in the Markov chain until either $Y(t) = 1$ or $Y(t) = s$, given that $Y(0) = i$, then the t_i are the solution of the simultaneous equations

$$t_i = \sum_{j=2}^{s-1} p_{ij} t_j + 1, \quad i = 2, 3, \ldots, s-1. \qquad (11.21)$$

These equations are subject to the boundary conditions $t_1 = t_s = 0$.

We next consider the case where there is only one absorbing state, which we take to be E_s. In this case the probability of eventual absorption in E_s is 1, and is thus not of interest. On the other hand, the probability that absorption in E_s takes place at or before some designated number of steps in the chain might well be of interest: See Example 2 in Section 11.6.2. The mean time until the absorbing state E_s is eventually entered is found by replacing equations (11.21) by

$$t_i = \sum_{j=1}^{s-1} p_{ij} t_j + 1, \quad i = 1, 2, \ldots, s-1, \qquad (11.22)$$

with boundary condition $t_s = 0$.

11.6.2 Examples

Example 1. Word recurrence lengths. The theory of Markov chains with absorbing states can be used to find the mean distance between successive recurrences of a word consisting of a short DNA sequence, as discussed in the "non-overlapping" analysis of Section 5.8. The Markov chain approach is due to Karlin and Brendel (1996), and we initially follow their analysis.

We assume that the nucleotides a, g, c, and t occur with respective frequencies p_a, p_g, p_c, and p_t and that successive nucleotides are independent. Suppose first that the word of interest is *atg*. A calculation in Section 5.8.2 shows that when overlaps are not counted, the mean distance between successive occurrences of this word is $1/(p_a p_t p_g)$. This result can be found using the theory of absorbing Markov chains as follows.

We consider all words consisting of three nucleotides and divide these into four "types": *atg*, *xya*, *xat*, and "*other.*" Here x and y are arbitrary nucleotides, possibly identical, and "*other*" consists of all words of length three other than those listed. These types then form the four states in a Markov chain, as described below. The DNA sequence is scanned from left to right, starting immediately following an occurrence of *atg*, and our aim is to find the mean number of nucleotide sites until the next occurrence. This is done by making atg an absorbing state of the Markov chain and seeking the mean number of sites visited until it is entered, assuming that the initial state is "*other*." The Markov chain transition matrix is then

$$
\begin{array}{c}
 \\
atg \\
xya \\
xat \\
other
\end{array}
\begin{array}{cccc}
atg & xya & xat & other \\
\left[\begin{array}{cccc}
1 & 0 & 0 & 0 \\
0 & p_a & p_t & p_c + p_g \\
p_g & p_a & 0 & p_c + p_t \\
0 & p_a & 0 & 1 - p_a
\end{array}\right].
\end{array}
\tag{11.23}
$$

Let t_2, t_3, and t_4 be the mean number of sites until atg is first observed, given respectively that the current state is xya, xat, and "*other*." Our aim is to find t_4. This Markov chain has only one absorbing state, so equations (11.22) show that t_2, t_3, and t_4 satisfy the simultaneous equations

$$
t_2 = p_a t_2 + p_t t_3 + (p_c + p_g)t_4 + 1,
$$
$$
t_3 = p_a t_2 + (p_c + p_t)t_4 + 1,
$$
$$
t_4 = p_a t_2 + (1 - p_a)t_4 + 1.
$$

Solving these equations gives $t_4 = 1/(p_a p_t p_g)$, as found in Section 5.8.2.

We next consider the word ata. A calculation in Section 5.8.2 shows that the mean number of nucleotides before this sequence first occurs is $(1 + p_a p_t)/p_a^2 p_t$. To find this result using Markov chain theory it is necessary to consider the four "types" ata, $\overline{at}a$, xat, and "*other*," which again we use as the four states in a Markov chain. Here x is an arbitrary nucleotide, and $\overline{at}a$ is any word of length three finishing with a and not starting with at. The transition matrix for this Markov chain, with ata being regarded as an absorbing state, is

$$
\begin{array}{c}
 \\
ata \\
\overline{at}a \\
xat \\
other
\end{array}
\begin{array}{cccc}
ata & \overline{at}a & xat & other \\
\left[\begin{array}{cccc}
1 & 0 & 0 & 0 \\
0 & p_a & p_t & p_c + p_g \\
p_a & 0 & 0 & 1 - p_a \\
0 & p_a & 0 & 1 - p_a
\end{array}\right].
\end{array}
\tag{11.24}
$$

If t_2, t_3, and t_4 are the mean numbers of sites until ata is first observed, given that the current state is $\overline{at}a$, xat, and "*other*," respectively, equation (11.22) shows that

$$
t_2 = p_a t_2 + p_t t_3 + (p_c + p_g)t_4 + 1,
$$
$$
t_3 = (1 - p_a)t_4 + 1,
$$
$$
t_4 = p_a t_2 + (1 - p_a)t_4 + 1.
$$

These equations yield $t_4 = (1 + p_a p_t)/(p_a^2 p_t)$, again as found in Section 5.8.2.

This approach generalizes naturally to the case where there is a Markov dependence between successive nucleotides. Some references to the recent

Markov dependent literature are given at the end of Section 5.9.

Example 2. Markov chains and the alignment of two sequences. The BLAST theory relating to two aligned sequences of equal length discussed in Section 10.2 relies on asymptotic results for geometric-like random variables. In this section we discuss an exact as opposed to an asymptotic analysis, following the work of Daudin and Mercier (2000). To simplify the discussion we consider the analysis in terms of DNA sequences. An essentially identical analysis applies for protein sequences.

The two sequences are scanned from left to right, and a score $S(i, k)$, determined from some score matrix, is allocated at any site if the two respective nucleotides at that site are nucleotides i and k. The accumulated score after the first j sites have been scanned is denoted by S_j. As in Chapter 10, the test statistic of interest is $Y_{\max}$, which is redefined here by

$$Y_{\max} = \max_{a,b}(S_b - S_a),$$

where a runs over all ladder points of the random walk defined by the successive S_j values and $b \geq a$.

The calculations in Chapter 10 provide an approximate asymptotic null hypothesis probability distribution for $Y_{\max}$, and this allows a readily computed P-value approximation for the eventual observed value $y_{\max}$ of $Y_{\max}$. By contrast, the Daudin and Mercier analysis provides, at least in principle although not necessarily easily in practice, an exact P-value calculation for the observed value $y_{\max}$ of $Y_{\max}$.

It is assumed, as in Chapter 10, that the mean score at any site is negative. The random walk defined by the accumulated score S_j, $j = 1, 2, 3, \ldots$, proceeds through a sequence of increasingly negative ladder points, as described in Chapter 10. We define $S^*(j)$ as the accumulated score at the last ladder point before site j. (The point $(0, 0)$ is classified as a ladder point.) From the accumulated score we define a new sequence $\{U_j\}$, the height of the current excursion up to site j, given by

$$U_0 = 0, \quad U_j = \max(0, S_j - S^*(j)), \tag{11.25}$$

but subject to the further requirement that if $U_j \geq s$ for some j and some number s discussed below, then U_j is immediately set to the value s and stays there thereafter, that is, that $U_j = U_{j+1} = \cdots = s$.

The successive values of the U_j are governed by a Markov chain in which the possible values of U_j are $0, 1, \ldots, s$, with the value $U_j = s$ corresponding to an absorbing state. The transition probability matrix of this Markov chain, which we denote by P, is found from the probability distribution of the steps $S(i, k)$.

Let the length of each of the aligned sequences be N. The event that $Y_{\max} \geq s$ is the event that this Markov chain enters the state $U_j = s$ at or before the Nth step in the chain. This probability can be calculated from the Nth power of P, and this allows the calculation of an exact P-value associated with an observed value s of $Y_{\max}$. Thus application of this approach does not take place until this observed value is known.

The matrix P could be quite large, and N could be extremely large. Thus the calculation of P^N could involve substantial computing, but would still be feasible for values of s and N arising in practice. Mercier and Daudin (2001) carry out these calculations for the representative examples of Karlin and Altschul (1990) to show that the upper bound for the P-value in a BLAST search given in Karlin and Altschul (1990) (and given in (10.21)) is usually quite accurate, although in two of the ten cases considered is less than the actual P-value. In these two cases the P-value is comparatively large, and so presumably this anomaly arises because the asymptotic theory on which BLAST calculations depend is not sufficiently accurate.

The calculations described above refer to a fixed alignment. The general BLAST procedure of comparing two *unaligned* sequences requires that this calculation be repeated a large number of times, which might be impractical with current computing power.

In the case $s = +\infty$ the successive values of the U_j follow a *Lindley process*. This process possesses a stationary distribution whose properties allow alternative bounds for the P-value in a BLAST search with a single fixed alignment. This approach to the theory is developed by Bacro et al. (2002)., who develop the theory of Mercier and Daudin (2001) further. In particular, they give a simple derivation of the bounds (10.21) and derive sharper bounds than these that are easy to compute. For further theoretical developments along these lines, see Daudin et al. (2001) and Daudin et al. (2003).

Example 3. Clumping behavior of motifs. The CHI motif in *H. influenzae*, consisting of the four words {*gatggtgg, gctggtgg, ggtggtgg, gttggtgg*}, was discussed in Section 5.9. Under a simple Bernoulli model, this motif has a comparatively simple clumping behavior: the probability P that the first word of the motif in any clump is overlapped by itself or some other word is independent of the first word. In this case equation (11.22) specializes to

$$t = Pt + 1. \tag{11.26}$$

It is straightforward to calculate P (see Problem 11.9), and then to calculate t from (11.26).

A slightly more complicated case arises for the motif {*aaa, ata*}. We think of a Markov chain process starting in state E_1 if the first word in a clump is *aaa* and in state E_2 if the first word in a clump is *ata*. The process continues to occupy one or other of these states while the clump continues, and enters absorbing state E_3 when the clump stops. If t_i is the mean number of words in the clump if the initial state is E_i, equations

(11.22) give

$$t_1 = p_a t_1 + p_t p_a t_2 + 1, \quad t_2 = p_a^2 t_1 + p_t p_a t_2 + 1. \tag{11.27}$$

These equations give

$$t_1 = \frac{1}{1 - p_a - p_a p_t + p_a^2 p_t - p_a^3 p_t}, \quad t_2 = \frac{1 - p_a + p_a^2}{1 - p_a - p_a p_t + p_a^2 p_t - p_a^3 p_t}.$$

Given that a clump occurs, it is natural to assume that the first word is aaa with probability $p_a/(p_a + p_t)$ and is ata with probability $p_t/(p_a + p_t)$. In this case the unconditional mean number of words in the clump is $(t_1 p_a + t_2 p_t)/(p_a + p_t)$.

11.7 Continuous-Time Markov Chains

11.7.1 Definitions

Markov processes can be in either discrete or continuous time, and in either discrete or continuous space; that is to say, there are four types of Markov processes. The Markov chains considered in Chapter 4 and so far in this chapter are in discrete time and discrete space. In this section we consider a random variable Y that takes values in some discrete space but whose values can change in continuous time. The value of Y at time t is denoted by $Y(t)$.

The "Markov" and the "time homogeneity" assumptions are defined for the continuous-time process as follows. The *Markov* assumption is that, given that $Y = i$ at any time u, the probability that $Y = j$ at any future time $u + t$ does not depend further on the values before time u. The *time homogeneity* assumption is that the conditional probability

$$\text{Prob}(Y(u + t) = j \,|\, Y(u) = i) \tag{11.28}$$

is independent of u, so we write it as $P_{ij}(t)$.

We focus on the case where the possible values of $Y(t)$ are $1, 2, \ldots, s$ and where it is assumed that the transition probability equations (11.28) take the form

$$P_{ij}(h) = q_{ij} h + o(h), \quad j \neq i, \tag{11.29}$$
$$P_{ii}(h) = 1 - q_i h + o(h), \tag{11.30}$$

as $h \to 0$, with q_i defined by

$$q_i = \sum_{j \neq i} q_{ij}. \tag{11.31}$$

We shall call the q_{ij} the *instantaneous transition rates* of the process. If q_i is independent of i, the expression (11.30) becomes identical in form to (4.1). The justification for the choice of the mathematical forms on the right-hand sides of (11.29) and (11.30) is the same as that given in Chapter 4 leading to the expression (4.1).

In Section 14.3.1 we consider evolutionary models in which equations of the form (11.29) and (11.30) are assumed. The justification for this is, in effect, that a change at a position in a population from i to j in a small time interval of length $2h$ is approximately twice the probability that this happens in a time interval of length h. Therefore, a generalization of the argument leading to the expression (4.1) leads to equations (11.29) and (11.30). In some evolutionary models that we consider it happens that q_i is independent of i, so that the Poisson process theory of Chapter 4 may be used directly for them. When q_i depends on i we may regard the process governing the number of transitions as a generalization of the Poisson process of Chapter 4.

11.7.2   Time-Dependent Solutions

By the time-homogeneity property of the process of interest,

$$P_{ij}(t + h) = \sum_k P_{ik}(t) P_{kj}(h),$$

where the sum is taken over all possible states. From this, equations (11.29)–(11.31) imply that for fixed i,

$$P_{ij}(t + h) = P_{ij}(t)(1 - q_j h) + h \sum_{k \neq j} P_{ik}(t) q_{kj} + o(h). \qquad (11.32)$$

From this equation and the assumptions (11.29)–(11.31) we arrive at the following system of differential equations:

$$\frac{d}{dt} P_{ij}(t) = -q_j P_{ij}(t) + \sum_{k \neq j} P_{ik}(t) q_{kj}, \quad j = 1, 2, \ldots, s. \qquad (11.33)$$

These are called the *forward Kolmogorov equations* of the system. They can be solved explicitly in cases where the q_{kj} take simple forms. An example is given in Problem 11.10. Some applications in the evolutionary context where these equations can be solved are given in Sections 14.3.1, 14.3.2, and 14.3.3.

11.7.3   The Stationary Distribution

A stationary distribution $\{\varphi_j\}$ has the property that if at any time t $\text{Prob}(Y(t) = j) = \varphi_j$ for all j, then for all j, $\text{Prob}(Y(u) = j) = \varphi_j$ for all $u > t$. Thus a stationary distribution, if it exists, can be found by replacing the derivatives on the left-hand sides of the system of equations (11.33) by zero and replacing $P_{ij}(t)$ and $P_{ik}(t)$ by φ_j and φ_k, respectively, to get

$$q_j \varphi_j = \sum_{k \neq j} \varphi_k q_{kj}, \quad j = 1, 2, \ldots, s. \tag{11.34}$$

This equation is used in an evolutionary context in Sections 14.3.1, 14.3.2, and 14.3.3.

11.7.4 Detailed Balance

When a stationary distribution $\{\varphi_j\}$ exists, the detailed balance conditions analogous to the discrete-time conditions (11.5) are that for all i, j, and t,

$$\varphi_i P_{ij}(t) = \varphi_j P_{ji}(t). \tag{11.35}$$

A simpler form of this criterion is that

$$\varphi_i q_{ij} = \varphi_j q_{ji}, \tag{11.36}$$

where the q_{ij} are defined in equation (11.29).

11.7.5 Exponential Holding Times

Suppose that $Y(t) = j$. Then $Y(\cdot)$ will remain at the value j for some length of time until it changes to some value other than j, and we now find the probability density function of the time until such a change occurs.

Let T be the (random) time until the value of $Y(\cdot)$ moves to some value different from j. Then

$$\text{Prob}(T \geq t + h) = \text{Prob}(T \geq t) \cdot \text{Prob}(T \geq t + h \,|\, T \geq t).$$

From the Markov and time homogeneity properties of the process, this is

$$\text{Prob}(T \geq t + h) = \text{Prob}(T \geq t) \cdot \text{Prob}(T \geq h). \tag{11.37}$$

The probability that $T \geq h$ is the probability that the random variable has not moved from the value j before time h, and if terms of order $o(h)$ are ignored, this is the probability $1 - q_j h + o(h)$ given in equation (11.30). Thus

$$\text{Prob}(T \geq t + h) = \text{Prob}(T \geq t) \cdot (1 - q_j h) + o(h). \tag{11.38}$$

Rearrangement of terms gives

$$\frac{\text{Prob}(T \geq t + h) - \text{Prob}(T \geq t)}{h} = -q_j \,\text{Prob}(T \geq t) + \frac{o(h)}{h},$$

and the limiting operation $h \to 0$ gives

$$\frac{d}{dt} \text{Prob}(T \geq t) = -q_j \,\text{Prob}(T \geq t). \tag{11.39}$$

Standard differential equation calculations show that

$$\text{Prob}(T \geq t) = C \cdot e^{-q_j t},$$

for some constant C. The case $t = 0$ shows that $C = 1$, so that

$$\text{Prob}(T \geq t) = e^{-q_j t}. \tag{11.40}$$

Allowing for a change in notation, this is identical to equation (1.67), and thus T has an exponential distribution. From (1.54), the density function of T is

$$f(t) = q_j e^{-q_j t}, \quad t > 0. \tag{11.41}$$

To summarize, if $Y(\cdot)$ has just arrived at the value j, it next moves to some other value after a random length of time having the exponential distribution given in equation (11.41). This implies that having just arrived at the value j, the mean time spent at this value before moving to some other value is q_j^{-1}.

11.7.6   The Embedded Chain

In some applications we might not be interested in the time spent by $Y(\cdot)$ at any value but only in the sequence of values that $Y(\cdot)$ assumes. In other words we are interested only in the so-called *embedded chain* of the process. This embedded chain is a discrete-time Markov chain whose transition probabilities are

$$p_{jk} = \frac{q_{jk}}{q_j}. \tag{11.42}$$

These are conditional probabilities, derived from (1.101); given that a change occurs, the probability that it is to k is given by (11.42). Markov chain theory can be used to find properties of this embedded process. These properties can provide information about the original time-dependent process. For example, if a time-dependent process has absorbing states, the probability that the process enters a specific absorbing state is the same as the corresponding probability in the embedded chain. The latter probability might be found more easily using discrete time Markov chain theory than by the continuous time theory of this section. On the other hand it is not possible to find absorption time properties for the continuous time process from the embedded chain.

Problems

11.1. Suppose that the transition matrix P of a Markov chain is given by

$$P = \begin{bmatrix} 0.6 & 0.4 \\ 0.3 & 0.7 \end{bmatrix}. \tag{11.43}$$

Use the definitions (B.45) and (B.46) to find the (two) eigenvalues and the (two pairs of) corresponding left and right eigenvectors of P.

11.2. Use your answer to Problem 11.1 to check that equation (B.48) holds.

11.3. For the matrix P given in (11.43), find P^2 by direct matrix multiplication. Then find P^2 by using the eigenvalues and eigenvectors calculated in Problem 11.1, together with the right-hand side of equation (B.49).

11.4. Find the stationary distribution of the Markov chain whose transition matrix is given in (11.43). Find P^3 and P^4 using the spectral expansion, and thus check that P^n is approaching the matrix defined through the stationary distribution.

11.5. Suppose that a reversible Markov chain with transition matrix $P = \{p_{ij}\}$ and stationary distribution φ whose typical element is φ_i. Show that for such a matrix $\varphi_i p_{ij}^n = \varphi_j p_{ji}^n$. (Hint: this equation is true for $n = 1$: see equation (11.5)). Now use induction and the fact that

$$p_{ij}^{(n+1)} = \sum_k p_{ik}^{(n)} p_{kj} = \sum_k p_{ik} p_{kj}^{(n)}.$$

Assuming that the matrix M_1 defined in Section 6.5.3 is reversible, use this result to show that the expression in (6.24) is unchanged by reversing the roles of j and k (as is desired in the context of PAM matrices).

11.6. Use equation (11.5) to show that if the transition matrix of a finite irreducible aperiodic Markov chain is symmetric, then that Markov chain is reversible.

11.7. Suppose that a finite aperiodic irreducible Markov chain has transition probability matrix P and stationary distribution φ'. Show that if k is any constant, $0 < k < 1$, then the Markov chain with transition probability matrix $P^* = kP + (1 - k)I$ also has stationary distribution φ'. What interpretation or explanation can you give for this result?

11.8. Prove the assertion made below equation (11.10), that $p_{ij} > 0$ for all (i, j) including the case $i = j$.

11.9. Calculate the value of P in equation (11.26).

11.10. A simple Markov chain has two states, called here "0" and "1," with instantaneous transition rates $q_{01} = q_{10} = \alpha$. Write down the differential equation (11.33) for this case, and show that the solution is

$$P_{ii}(t) = \frac{1}{2} + \frac{1}{2}e^{-2\alpha t}, \quad P_{ij}(t) = \frac{1}{2} - \frac{1}{2}e^{-2\alpha t}, \quad (i \neq j, i, j = 0, 1). \quad (11.44)$$

Let $t \to +\infty$ to find the stationary distribution of this Markov process, and show that this distribution satisfies (11.34).

12
Hidden Markov Models

We divide this brief account of hidden Markov models into three sections: (i) a description of the properties of these models, (ii) the three main algorithms of the models, and (iii) applications. For a more complete account of these models, see Rabiner (1989).

12.1 What is a Hidden Markov Model?

A hidden Markov model (HMM) is similar to a Markov chain, but is more general, and hence more flexible, allowing us to model phenomena that we cannot model sufficiently well with a regular Markov chain model. An HMM is a discrete-time Markov model with some extra features. The main addition is that when a state is visited by the Markov chain, the state "emits" a letter from a fixed time-independent alphabet. Letters are emitted via a time-independent, but usually state-dependent, probability distribution over the alphabet. When the HMM runs there is, first, a sequence of states visited, which we denote by $q_1, q_2, q_3, \ldots$, and second, a sequence of emitted symbols, denoted by $\mathcal{O}_1, \mathcal{O}_2, \mathcal{O}_3, \ldots$. Their generation can be visualized as a two-step process as follows:

$$\underset{q_1}{\text{initial}} \to \underset{\mathcal{O}_1}{\text{emission}} \to \underset{\text{to } q_2}{\text{transition}} \to \underset{\mathcal{O}_2}{\text{emission}} \to \underset{\text{to } q_3}{\text{transition}} \to \underset{\mathcal{O}_3}{\text{emission}} \to \cdots$$

We denote the entire sequence of q_i's by Q and the entire sequence of $\mathcal{O}_i$'s by $\mathcal{O}$, and we write "the observed sequence $\mathcal{O} = \mathcal{O}_1, \mathcal{O}_2, \ldots$" and "the state sequence $Q = q_1, q_2, \ldots$."

Often we know the sequence $\mathcal{O}$ but do not know the sequence Q. In such a case the sequence Q is called "hidden." An important feature of HMMs is that we can efficiently answer several questions about $\mathcal{O}$ and Q.

One of these questions concerns the estimation of the hidden state sequence that has the highest probability given the observed sequence. We illustrate this with a simple example. Consider the Markov chain with two states S_1 and S_2, with uniform initial distribution and transition matrix

$$\begin{bmatrix} .9 & .1 \\ .8 & .2 \end{bmatrix}.$$

Let A be an alphabet consisting only of the numbers 1 and 2. State S_1 emits a 1 or 2 with equal probability $\frac{1}{2}$, state S_2 emits a 1 with probability $\frac{1}{4}$ and a 2 with probability $\frac{3}{4}$. Suppose the observed sequence is $\mathcal{O} = 2, 2, 2$. What sequence of states $Q = q_1, q_2, q_3$ has the highest probability given $\mathcal{O}$? In other words, what is

$$\operatorname*{argmax}_{Q} \operatorname{Prob}(Q \mid \mathcal{O})?$$

There are eight possibilities for Q. Each of these can be written down and its probability calculated, and from this it is found that the answer to the above question is $Q = S_2, S_1, S_1$. The sequence Q contains more S_1's, even though S_2 is more likely to produce a 2 when visited (probability $\frac{3}{4}$) than S_1 (probability $\frac{1}{2}$). The reason is because S_1 is much more likely to be visited than S_2 ($p_{11} = .9$ and $p_{21} = .8$).

We can also calculate

$$\operatorname{Prob}(\mathcal{O}) = \sum_{Q} \operatorname{Prob}(\mathcal{O} \mid Q) \cdot \operatorname{Prob}(Q). \tag{12.1}$$

This calculation is useful in distinguishing which of several models is most likely to have produced $\mathcal{O}$.

In the above example all of these calculations can be done by hand. However, models arising in practice have many states, sometimes hundreds, and an alphabet with many symbols (often 20, one for each amino acid). In these cases, calculation of the quantities above by exhaustive methods becomes impossible even for the fastest computers. Fortunately, there are dynamic programming approaches that overcome this problem, which we discuss in detail below. Before turning to the algorithms, however, it is necessary to introduce some specific notation. An HMM will consist of the following five components:

(1) A set of N states $S_1, S_2, \ldots, S_N$.

(2) An alphabet of M distinct observation symbols $A = \{a_1, a_2, \ldots, a_M\}$.

(3) The transition probability matrix $P = (p_{ij})$, where

$$p_{ij} = \operatorname{Prob}(q_{t+1} = S_j \mid q_t = S_i)$$

(4) The emission probabilities: For each state S_i and a in A,

$$b_i(a) = \mathrm{Prob}(S_i \text{ emits symbol } a).$$

The probabilities $b_i(a)$ form the elements in an $N \times M$ matrix $B = (b_i(a))$.

(5) An initial distribution vector $\pi = (\pi_i)$, where $\pi_i = \mathrm{Prob}(q_1 = S_i)$.

Components 1 and 2 describe the structure of the model, and 3–5 describe the parameters. It is convenient to let $\lambda = (P, B, \pi)$ represent the full set of parameters. We can now describe the main algorithms.

12.2   Three Algorithms

There are three calculations that are frequently required in HMM theory. Given some observed output sequence $\mathcal{O} = \mathcal{O}_1, \mathcal{O}_2, \ldots, \mathcal{O}_T$, these are:

(i) Given the parameters λ, efficiently calculate

$$\mathrm{Prob}(\mathcal{O} \mid \lambda).$$

That is, efficiently calculate the probability of some given sequence of observed outputs.

(ii) Efficiently calculate the hidden sequence $Q = q_1, q_2, \ldots, q_T$ of states that is most likely to have occurred, given $\mathcal{O}$. That is, calculate

$$\underset{Q}{\mathrm{argmax}}\ \mathrm{Prob}(Q \mid \mathcal{O}).$$

(iii) Assuming a fixed topology of the model (i.e., a fixed graph structure of the underlying Markov chain, as defined in 4.9), find the parameters $\lambda = (P, B, \pi)$ that maximize $\mathrm{Prob}(\mathcal{O} \mid \lambda)$.

We address these problems in turn.

12.2.1   The Forward and Backward Algorithms

We first consider problem (i). The naive way of calculating $\mathrm{Prob}(\mathcal{O})$ is to use formula (12.1). This calculation involves the sum of N^T multiplications, each being a multiplication of $2T$ terms. The total number of operations is thus on the order of $2T \cdot N^T$.

Unless T is quite small, this calculation is computationally infeasible. For example, if $N = 4$, $T = 100$, the number of calculations is on the order of 10^{60}. It would take the life of the universe to make such a calculation. Fortunately, there is a much more efficient and computationally feasible procedure, called the *forward algorithm*.

The forward algorithm focuses on the calculation of the quantity

$$\alpha(t, i) = \text{Prob}(\mathcal{O}_1, \mathcal{O}_2, \mathcal{O}_3, \ldots, \mathcal{O}_t, q_t = S_i), \qquad (12.2)$$

which is the joint probability that the sequence of observations seen up to and including time t is $\mathcal{O}_1, \mathcal{O}_2, \mathcal{O}_3, \ldots, \mathcal{O}_t$, and that the state of the HMM at time t is S_i. The $\alpha(t, i)$ are called the *forwards* variables.

Once we know $\alpha(T, i)$ for all i, then $\text{Prob}(\mathcal{O})$ can be calculated as

$$\text{Prob}(\mathcal{O}) = \sum_{i=1}^{N} \alpha(T, i). \qquad (12.3)$$

We calculate the $\alpha(t, i)$'s inductively on t. The first calculation is of the *initialization* step, and uses the obvious result

$$\alpha(1, i) = \pi_i b_i(\mathcal{O}_1). \qquad (12.4)$$

Next, the equation

$$\alpha(t + 1, i) = \sum_{j=1}^{N} \text{Prob}(\mathcal{O}_1, \mathcal{O}_2, \mathcal{O}_3, \ldots, \mathcal{O}_{t+1}, q_{t+1} = S_i \text{ and } q_t = S_j)$$

leads to the *induction* step

$$\alpha(t + 1, i) = \sum_{j=1}^{N} \alpha(t, j) p_{ji} b_i(\mathcal{O}_{t+1}). \qquad (12.5)$$

This equation gives $\alpha(t+1, i)$ in terms of the $\alpha(t, j)$, so that $\alpha(t+1, i)$ can be calculated quickly once the $\alpha(t, j)$ are known. We use (12.4) to calculate $\alpha(1, i)$ for all i; then we use (12.5) to calculate $\alpha(2, i)$ for all i and again to calculate $\alpha(3, i)$ for all i, and so on, until we have obtained the $\alpha(T, i)$ for all i, needed in (12.3).

This procedure provides an algorithm for the solution to problem (i). The algorithm requires on the order of TN^2 computations, and thus is feasible in practice, even for very large models.

Before going on to problem (ii), we consider briefly the *backward* part of the forward–backward algorithm. This provides another approach to solving problem (i), but we introduced it because we will use the "backwards" variables when we discuss problem (iii).

In the above, we calculated successively $\alpha(1, \cdot), \alpha(2, \cdot), \ldots, \alpha(T, \cdot)$, that is, we calculated forward in time. In the backward algorithm we calculate another quantity backwards in time, as the name suggests. We do not use these quantities to solve (12.3): instead, we shall need them for a later calculation. The goal of the backwards algorithm is to calculate the probability $\beta(t, i)$, defined by

$$\beta(t, i) = \text{Prob}(\mathcal{O}_{t+1}, \mathcal{O}_{t+2}, \ldots, \mathcal{O}_T \mid q_t = S_i), \qquad (12.6)$$

for $1 \le t \le T - 1$, and for convenience we define $\beta(T, j)$ to be 1 for all j. We then calculate (12.6) working backwards from $t = T - 1$. The relevant equations for this procedure are

$$\beta(t - 1, i) = \sum_{j=1}^{N} p_{ij}\, b_j(\mathcal{O}_t)\beta(t, j). \tag{12.7}$$

Using these equations we can successively calculate $\beta(T - 1, i)$ for all i, $\beta(T - 2, i)$ for all $i, \ldots$, and $\beta(1, i)$ for all i.

12.2.2 The Viterbi Algorithm

Given some observed sequence $\mathcal{O} = \mathcal{O}_1, \mathcal{O}_2, \mathcal{O}_3, \ldots, \mathcal{O}_T$ of outputs, we want to compute efficiently a state sequence $Q = q_1, q_2, q_3, \ldots, q_T$ that has the highest conditional probability given $\mathcal{O}$. In other words, we want to find a Q that makes $\mathrm{Prob}(Q \,|\, \mathcal{O})$ maximal, that is, we want to calculate

$$\operatorname*{argmax}_{Q} \ \mathrm{Prob}(Q \,|\, \mathcal{O}). \tag{12.8}$$

There may be many Q's that maximize $\mathrm{Prob}(Q \,|\, \mathcal{O})$. We give an algorithm that finds one of them. It can easily be generalized to find them all. However, for our applications this generalization will not be necessary.

The Viterbi algorithm carries out the efficient computation of (12.8). The algorithm is divided into two parts. It first finds $\max_Q \mathrm{Prob}(Q \,|\, \mathcal{O})$, and then "backtracks" to find a Q that realizes this maximum. This is another dynamic programming algorithm.

First define, for arbitrary t and i,

$$\delta_t(i) = \max_{q_1, q_2, \ldots, q_{t-1}} \mathrm{Prob}(q_1, q_2, \ldots, q_{t-1}, q_t = S_i \text{ and } \mathcal{O}_1, \mathcal{O}_2, \mathcal{O}_3, \ldots, \mathcal{O}_t)$$

($\delta_1(i) = \mathrm{Prob}(q_1 = S_i \text{ and } \mathcal{O}_1)$). In words, $\delta_t(i)$ is the maximum probability of all ways to end in state S_i at time t and have observed sequence $\mathcal{O}_1$, $\mathcal{O}_2, \ldots, \mathcal{O}_t$. Then

$$\max_{Q} \ \mathrm{Prob}(Q \text{ and } \mathcal{O}) = \max_{i} \ \delta_T(i). \tag{12.9}$$

The probability in this expression is the joint probability of Q and $\mathcal{O}$, not a conditional probability. Our aim is to find a sequence Q for which the maximum conditional probability (12.8) is achieved. Since

$$\max_{Q} \ \mathrm{Prob}(Q \,|\, \mathcal{O}) = \max_{Q} \ \frac{\mathrm{Prob}(Q \text{ and } \mathcal{O})}{\mathrm{Prob}(\mathcal{O})},$$

and since the denominator on the right-hand side does not depend on Q,

$$\operatorname*{argmax}_{Q} \ \mathrm{Prob}(Q \,|\, \mathcal{O}) = \operatorname*{argmax}_{Q} \ \frac{\mathrm{Prob}(Q \text{ and } \mathcal{O})}{\mathrm{Prob}(\mathcal{O})} = \operatorname*{argmax}_{Q} \ \mathrm{Prob}(Q \text{ and } \mathcal{O}).$$

The first step is to calculate the $\delta_t(i)$'s inductively. Then we will "backtrack" and recover the sequence that gives the largest $\delta_T(i)$. The *initialization* step is

$$\delta_1(i) = \pi_i b_i(\mathcal{O}_1), \quad 1 \leq i \leq N. \tag{12.10}$$

The induction step is

$$\delta_t(j) = \max_{1 \leq i \leq N} \delta_{t-1}(i) p_{ij} b_j(\mathcal{O}_t), \quad 2 \leq t \leq T,\ 1 \leq j \leq N. \tag{12.11}$$

We recover the q_i's as follows. Define

$$\psi_T = \operatorname*{argmax}_{1 \leq i \leq N} \delta_T(i),$$

and put $q_T = S_{\psi_T}$. Then q_T is the final state in the state sequence required. The remaining q_t for $t \leq T - 1$ are found recursively by first defining

$$\psi_t = \operatorname*{argmax}_{1 \leq i \leq N} \delta_t(i) p_{i\psi_{t+1}},$$

and then putting $q_t = S_{\psi_t}$. If the argmax is not unique, we arbitrarily take one value of i giving the maximum.

12.2.3   The Estimation Algorithms

We now address problem (iii). Suppose we are given a set of observed data from an HMM for which the topology is known (by topology we mean the graph structure of the underlying Markov model). We wish to try to estimate the parameters in that HMM. The parameter space is usually far too large to allow exact calculation of a set of parameter estimates that maximizes the probability of the data. Instead, we employ algorithms that find "locally" best sets of parameters. This partial solution to the problem has proven to be useful in many applications.

The focus on local estimation means that the procedure is heuristic. Therefore, the efficacy of the procedure must evaluated empirically by using benchmarks and test sets for which there are known outcomes. This matter is discussed further below.

Some additional comments are in order. It is not necessary to assume that the data come from an HMM. Instead, it is usually more accurate to assume that the data are generated by some random process that we try to "fit" with an HMM. Sometimes it might be possible to achieve a tight fit with an HMM and sometimes it might not.

The discussion above shows that we should use the term "estimation" of parameters cautiously in this section. Our aim is to "set" parameters at values providing a good fit to data rather than to estimate parameters in the sense of Chapter 8.

We now describe the Baum–Welch method of parameter estimation. This is a difficult algorithm, so we do not provide proofs of the claims made but instead indicate the intuition behind the method.

We assume that the alphabet A and number of states N is fixed at the outset, and that the parameters π_i, p_{jk}, and $b_i(a)$ are unknown and are to be "estimated." The data we use to estimate the parameters constitute a set of observed sequences $\{\mathcal{O}^{(d)}\}$. Each observed sequence $\mathcal{O}^{(d)} = \mathcal{O}_1^{(d)}, \mathcal{O}_2^{(d)}, \ldots$ has a corresponding hidden state sequence $Q^{(d)} = q_1^{(d)}, q_2^{(d)}, \ldots$.

The procedure starts by setting the parameters π_i, p_{jk}, and $b_i(a)$ at some initial values. These can be chosen from some uniform distribution or can be chosen to incorporate prior knowledge about them. We then calculate, using these initial parameter values,

$$\overline{\pi}_i = \text{the expected proportion of times in state } S_i \text{ at} \qquad (12.12)$$
$$\text{the first time point, given } \{\mathcal{O}^{(d)}\},$$

$$\overline{p}_{jk} = \frac{E(N_{jk} \mid \{\mathcal{O}^{(d)}\})}{E(N_j \mid \{\mathcal{O}^{(d)}\})}, \qquad (12.13)$$

$$\overline{b}_i(a) = \frac{E(N_i(a) \mid \{\mathcal{O}^{(d)}\})}{E(N_i \mid \{\mathcal{O}^{(d)}\})}, \qquad (12.14)$$

where N_{jk} is the (random) number of times $q_t^{(d)} = S_j$ and $q_{t+1}^{(d)} = S_k$ for some d and t; N_i is the (random) number of times $q_t^{(d)} = S_i$ for some d and t; and $N_i(a)$ equals the (random) number of times $q_t^{(d)} = S_i$ and it emits symbol a, for some d and t. The expected values in (12.13) and (12.14) are conditional expected values, as defined in (2.59).

We show how to calculate these efficiently below. These are the "re-estimation" parameter values that then replace π_i, p_{jk}, and $b_i(a)$. These values follow the form of estimation used, for example, in equation (3.10). The algorithm proceeds by iterating this step.

It can be shown that if $\lambda = (\pi_i, p_{jk}, b_i(a))$ is replaced by $\overline{\lambda} = (\overline{\pi}_i, \overline{p}_{jk}, \overline{b}_i(a))$, then $\text{Prob}(\{\mathcal{O}^{(d)}\} \mid \overline{\lambda}) \geq \text{Prob}(\{\mathcal{O}^{(d)}\} \mid \lambda)$, with equality holding if and only if $\overline{\lambda} = \lambda$. Thus successive iterations continually increase the probability of the data, given the model. Iterations continue until either a local maximum of the probability is reached or until the change in the probability becomes negligible.

In order to discuss the calculations needed for (12.13)–(12.12), define $\xi_t^{(d)}(i, j)$ by

$$\xi_t^{(d)}(i, j) = \text{Prob}(q_t^{(d)} = S_i, q_{t+1}^{(d)} = S_j \mid \mathcal{O}^{(d)}), \qquad (12.15)$$

where $i, j = 1, \ldots, N$, and $t \geq 1$. The conditional probability formula (1.101) shows that this is equal to

$$\frac{\text{Prob}(q_t^{(d)} = S_i, q_{t+1}^{(d)} = S_j, \mathcal{O}^{(d)})}{\text{Prob}(\mathcal{O}^{(d)})}.$$

The denominator is $\text{Prob}(\mathcal{O}^{(d)})$ and is thus calculated efficiently using the methods of Section 12.2.1. The numerator is calculated efficiently by writing it in terms of the forwards and backwards variables discussed in Section 12.2.1,

$$\text{Prob}(q_t^{(d)} = S_i, q_{t+1}^{(d)} = S_j, \mathcal{O}^{(d)}) = \alpha_t(i)p_{ij}b_j(\mathcal{O}_{t+1}^{(d)})\beta_{t+1}(j). \qquad (12.16)$$

Let $I_t^{(d)}(i)$ be the indicator variables defined by

$$I_t^{(d)}(i) = \begin{cases} 1, & \text{if } q_t^{(d)} = S_i, \\ 0, & \text{otherwise.} \end{cases}$$

The number of times S_i is visited is then $\sum_d \sum_t I_t^{(d)}(i)$. The expected number of times S_i is visited, given $\{\mathcal{O}^{(d)}\}$, is then

$$\sum_d \sum_t E(I_t^{(d)}(i) \,|\, \mathcal{O}^{(d)}). \qquad (12.17)$$

Now $E(I_t^{(d)}(i) \,|\, \mathcal{O}^{(d)})$ is $\text{Prob}(q_t^{(d)} = S_i \,|\, \mathcal{O}^{(d)})$, which is

$$\sum_{j=1}^{N} \xi_t^{(d)}(i,j). \qquad (12.18)$$

Thus the expected number of times S_i is visited, given $\{\mathcal{O}^{(d)}\}$, is

$$\sum_d \sum_t \sum_{j=1}^{N} \xi_t^{(d)}(i,j).$$

Similarly, the expected number of transitions from S_i to S_j given $\{\mathcal{O}^{(d)}\}$ is

$$\sum_d \sum_t \xi_t^{(d)}(i,j).$$

These expressions give efficient formulae to calculate all the quantities in equations (12.12)–(12.14) except the numerator of (12.14). This is calculated as follows.

Define the indicator random variables $I_t^{(d)}(i,a)$ by

$$I_t^{(d)}(i,a) = \begin{cases} 1, & \text{if } q_t^{(d)} = S_i \text{ and } \mathcal{O}_t^{(d)} = a, \\ 0, & \text{otherwise.} \end{cases}$$

Then $E(I_t^{(d)}(i,a) \,|\, \mathcal{O}^{(d)})$ is the expected number of times the dth process is in state S_i at time t and emits symbol a, given $\mathcal{O}^{(d)}$. The numerator of (12.14) is equal to $\sum_d \sum_t E(I_t^{(d)}(i,a) \,|\, \mathcal{O}^{(d)})$, which is

$$\sum_d \sum_t \sum_{\mathcal{O}_t^{(d)} = a} \sum_{j=1}^{N} \xi_t^{(d)}(i,j).$$

12.3 Applications

We sketch here the applications of HMMs in several different areas of computational biology. Only a brief outline of each application is given: further details may be found in the references provided.

12.3.1 Modeling Protein Families

In this section we develop an HMM to model protein families, and we shall use the model for two purposes: to construct multiple sequence alignments and to determine the family of a query sequence. These applications were first presented in Krogh et al. (1994). In order to present the main ideas we simplify many of the details.

Figure 12.1 gives an example of the basic type of HMM we shall use. This example has "length" five; any length is possible. The underlying Markov model is presented in graphical form (as in Section 4.9). The states are

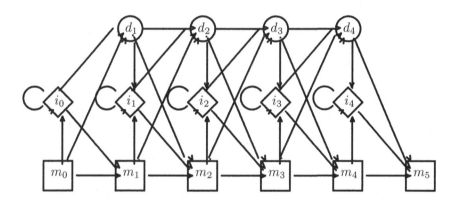

Figure 12.1. Hidden Markov model for a protein family.

the squares, diamonds, and circles labeled $m_0, m_1, \ldots, m_5, i_0, i_1, \ldots, i_4$, and $d_1, d_2, \ldots, d_4$, respectively. The squares are called the *match* states, the diamonds the *insert* states, and the circles the *delete* states. The edges not shown have transition probability zero. State m_0 is the *start* state, so that the process always starts in state m_0. A transition never moves to the left, so that as time progresses the current state gradually moves to the right, eventually ending in match state m_5, the *end* state. When this state is reached the process ends. A match or delete state is never visited more than once.

The alphabet A consists of the twenty amino acids together with one "dummy" symbol representing "delete" (denoted δ). Delete states output

δ with probability one. Each insert and match state has its own distribution over the 20 amino acids, and cannot emit a δ. That is, only a delete state can emit a δ, and each delete state emits *only* δ.

If the emission probabilities for the match and insert states are uniform over the 20 amino acids, the model will produce random sequences that do not have much in common except possibly their lengths. At the other extreme, if each state emits one specific amino acid with probability one, and if further the transitions from m_i to m_{i+1} have probability one, then the model will always produce the same sequence. Somewhere in between these two extremes the parameters of the model can be set so that it produces sequences that are similar, thus producing what can be thought of as a "family" of sequences. Each choice of parameters produces a different family. This family can be rather "tight," meaning all sequences in it are very similar, or can be "loose," so that there is little similarity between the sequences produced. It is also possible that the similarity is high in some positions of the sequences produced and low in others. This will happen if some match states have distributions concentrated on a few amino acids while the others have distributions in which all amino acids are approximately equally likely. By contrast, the dynamic programming sequence alignment algorithms and BLAST allow one gap open penalty and use one substitution matrix uniformly across the entire length of the sequences compared. Allowing gap penalties and substitution probabilities to vary along the sequences reflects biological reality better. Alignments of related proteins generally have regions of higher conservation and regions of lower conservation. The regions of higher conservation are called functional domains, because their resistance to change indicates that they serve some critical function. Dynamic programming alignment and BLAST are essential for certain applications, such as pairwise alignments, or aligning a small number of sequences. But for modeling large families of sequences, or constructing alignments of many sequences, HMMs allow for efficiency, and at the same time exploit the larger data sets to increase flexibility.

In the HMM model of a protein family the transition (arrow) from a match state to an insert state corresponds to the gap open penalty, and the arrow from an insert state to itself corresponds to the gap extension penalty. Loosely speaking, the distribution over the amino acids for any state takes the place of a substitution matrix. The probabilities in the model can differ from position to position in the sequence, since each arrow has its own probability and each match and insert state has its own distribution. Thus the HMM model is sufficiently flexible to model the varying features of a protein along its length. While the model can be made even more flexible by adding further parameters, more data are needed to estimate these parameters effectively. The model described has proven in practice to provide a good compromise between flexibility and tractability. Such HMM models of are called *profile* HMMs.

All applications start with *training*, or estimating, the parameters of the model using a set of training sequences chosen from a protein family, such as the set of all globins in GenBank. This estimation procedure uses the Baum–Welch algorithm. The model is chosen to have length equal to the average length of a sequence in the training set, and all parameters are initialized by using uniform distributions (i.e., amino acids are given probability $\frac{1}{20}$, and transitions of the same type are given $\frac{1}{2}$ equal probabilities).[1]

12.3.2 Multiple Sequence Alignments

In this section we describe how to use the theory described above to compute multiple sequence alignments for a family of sequences. The sequences to be aligned are used as the training data, to train the parameters of the model. For each sequence the Viterbi algorithm is then used to determine a path most likely to have produced that sequence. These paths can then be used to construct an alignment. Amino acids are aligned if both are produced by the same match state in their paths. Indels are then inserted appropriately for insertions and deletions.

We illustrate this with an example. Consider the sequences CAEFDDH and CDAEFPDDH. Suppose the model has length 10 and their most likely paths through the model are

$$m_0 m_1 m_2 m_3 m_4 d_5 d_6 m_7 m_8 m_9 m_{10}$$

and

$$m_0 m_1 i_1 m_2 m_3 m_4 d_5 m_6 m_7 m_8 m_9 m_{10},$$

respectively. Then the alignment induced is found by aligning positions that were generated by the same match state:

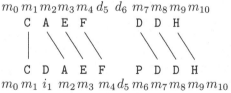

This leads to the alignment

$$\begin{array}{l} \text{C}-\text{AEF}-\text{DDH} \\ \text{C D AEF P DDH} \end{array} \, .$$

---

[1]What Krogh et al. (1994) do is somewhat more complicated. They allow the length of the model to change along with the parameters after each iteration of the re-estimation algorithm. They also must adjust the Baum–Welch algorithm from how we have described it, in order to handle the delete states. The reader interested in implementing these applications should refer to the literature for further details.

More generally, suppose we have the five sequences

$$CAEFTPAVH,$$
$$CKETTPADH,$$
$$CAETPDDH,$$
$$CAEFDDH,$$
$$CDAEFPDDH,$$

and the corresponding paths returned by the Viterbi algorithm are

$$m_0 m_1 m_2 m_3 m_4 m_5 m_6 m_7 m_8 m_9 m_{10},$$
$$m_0 m_1 m_2 m_3 m_4 m_5 m_6 m_7 m_8 m_9 m_{10},$$
$$m_0 m_1 m_2 m_3 d_4 m_5 m_6 m_7 m_8 m_9 m_{10},$$
$$m_0 m_1 m_2 m_3 m_4 d_5 d_6 m_7 m_8 m_9 m_{10},$$
$$m_0 m_1 i_1 m_2 m_3 m_4 d_5 m_6 m_7 m_8 m_9 m_{10}.$$

Then the induced alignment is

$$\text{C}-\text{AEF T P AVH}$$
$$\text{C}-\text{KET T P ADH}$$
$$\text{C}-\text{AE}- \text{T P DDH} .$$
$$\text{C}-\text{AEF}--\text{DDH}$$
$$\text{C D AEF}-\text{P DDH}$$

This technique can give ambiguous results in some cases. For example, if the model has length two and the sequences ABAC and ABBAC had

$$m_0 m_1 i_1 i_1 m_2 m_3 \text{ and } m_0 m_1 i_1 i_1 i_1 m_2 m_3$$

as paths, then the leading A's and trailing C's will be aligned, but it is not clear how to align the BA from the first sequence to the BBA from the second. In such cases Krogh et al. (1994) represent the ambiguous symbols with lowercase letters and do not attempt to give alignments of these regions.

This technique can be used to align many sequences with relatively little computing power. By contrast, dynamic programming algorithms cannot in practice align 50 or 100 long sequences. This is the value of a heuristic approach. Another advantage of the method is that it allows the sequences themselves to guide the alignment, rather than having a precomputed substitution matrix and gap penalties. Thus less bias should be introduced.

Krogh et al. (1994) tested this method on a family of 625 globin sequences. They used a published alignment of seven of these sequences constructed using knowledge of the three-dimensional structure of the sequences (Bashford et al. (1987)). (An alignment using a three-dimensional structure is considered reliable and so serves as a benchmark for testing multiple alignment algorithms.) They chose 400 of the 625 globins to train the model and then used the Viterbi algorithm to align all 625. These alignments were then compared to the induced alignment on the seven sequences from the Bashford alignment. The alignments agreed extremely well.

12.3.3 Pfam

Pfam is a web-based resource maintained by the Sanger Center (web URL (http://www.sanger.ac.uk/Pfam/). Pfam uses the basic theory described above to determine protein domains in a query sequence. A protein usually has one or more functional domains, namely portions of the protein that have essential function and thus have low tolerance for amino acid substitutions. Proteins in different families often share high homology in one or more domains. Entire protein families can be characterized by HMMs, as in the previous section, or one can characterize just functional domains. Pfam focuses on the latter.

Suppose that a new protein is obtained for which no information is available except the raw sequence. We wish to "annotate" this sequence. Annotation is the process of assigning to a sequence biologically relevant information, such as where the functional domains are, what their homology is to known domains, and what their function is. The typical starting point is a BLAST search. This will return all sequences in the chosen databases that have significant similarity to the query sequence. BLAST can return many such sequences. Though this is an important step in the annotation process, it is also desirable to have a database not of protein sequences themselves, but of protein domains. Pfam is not the first such database; however, previous domain databases do not use methods as flexible as HMMs and consequently tend not to model entire domains, but rather only the most highly conserved "motifs" that can be put in *ungapped* multiple sequence alignments. The use of HMMs allows for more effective characterization of full domains.

The domains in Pfam are determined based on expert knowledge, sequence similarity, and other protein family databases. Currently, Pfam contains 2,008 protein domains. For each domain a set of examples of this domain is selected. The sequences representing each domain are put into an alignment, and the alignments themselves are used to set the parameters; that is, Baum–Welch is not used. Recall that an alignment implies for each sequence in the alignment a path through the HMM, as described in the previous section. The proportion of times these paths take a given transition is used to estimate the transition probabilities, and likewise for the emission probabilities. These alignments are called "seed alignments" and are stored in the database. Given the HMMs for all of the domains, a query sequence is then run past each one using the forward algorithm. When a portion of the query sequence has probability of having been produced by an HMM above a certain cutoff, the domain corresponding to that HMM is reported. Furthermore, the sequence can be aligned to the seed alignment using the Viterbi algorithm as described above. For more details, see the Pfam web site.

12.3.4   Gene Finding

Genomic sequences with lengths on the order of many millions of bases are now being produced, and the sequences of entire chromosomes are becoming available. Such sequences consist of a collection of genes separated from each other by long stretches of nonfunctional sequence. It is of central importance to find where the genes are in the sequence. Therefore, computational methods that quickly identify a large proportion of the genes are very useful. The problem involves bringing together a large amount of diverse information, and there have been many approaches to doing this. Currently, a popular and successful gene finder for *human* DNA sequences is GENSCAN (Burge et al. (1997)), which is based on a generalization of hidden Markov models. We sketch below an algorithm similar in spirit to that in GENSCAN in order to illustrate the basic concept of an HMM human gene finder. To increase the accuracy of the procedure it is necessary to introduce many details that we do not describe here. The interested reader is encouraged to read Burge et al. (1997) and Burge (1997).

Semihidden Markov Models

Suppose that, in an HMM, p is the probability of the transition from any state to itself. The probability that the process stays in this state for n steps is $p^{n-1}(1-p)$, so that the length of time the process stays in that state follows a geometric distribution. For the gene model we construct, it is necessary to allow other distributions for this length. In a *semi-hidden Markov Model* (semiHMM, more logically called a hidden semi-Markov model) all transition probabilities from a state to itself are zero, and when the process visits a state it produces not just a single symbol from the alphabet but rather an entire sequence. The length of the sequence can follow any distribution, and the model generating the sequence of that length can be any distribution. The positions in the sequences emitted from a state need not be iid.

The model is formulated more precisely as follows. Each state S has associated with it a random variable L_S (L for "length") whose range is a subset of $0, 1, 2, \ldots$, and for each observable value ℓ of L_S there is a random variable $Y_{S,\ell}$ whose range consists of all sequences of length ℓ. When state S is visited a length ℓ is determined randomly from the distribution for L_S. Then the distribution for $Y_{S,\ell}$ is used to determine a sequence of length ℓ. Then a transition is taken to a new state and the process is repeated, generating another sequence. These sequences are concatenated to create the final output sequence of the semiHMM.

The algorithms involved in this model are an order of magnitude more complex than for a regular HMM, since given an observed output sequence not only do we not know the path of states that produced it, we also do not know the division points in the sequence indicating where a transition was made to a new state. The gene-finding application requires a generalization of the Viterbi algorithm. There is a natural generalization. However, since one is generally working with very long sequences, the natural generalization does not run in reasonable time. In practice, further assumptions must be made. Burge (1997) observed that if the lengths of the long intergenic regions can be taken as having geometric distributions, and if these lengths generate sequences in a relatively iid fashion, then the algorithm can be adjusted so that practical running times can be obtained. These assumptions are not unreasonable in our case, and so they should not greatly affect the accuracy of the predictions. We shall omit the technical details surrounding this issue. Our goal is to convey the main idea of how an HMM gene finder works.

A *parse* ϕ is a sequence of states $q_1, q_2, \ldots, q_r$ and a sequence of lengths $d_1, d_2, \ldots, d_r$. Given an observed sequence s from a semiHMM, the Viterbi algorithm finds an optimal parse ϕ_{opt} such that $\mathrm{Prob}(\phi_{\mathrm{opt}} \mid s) \geq \mathrm{Prob}(\phi \mid s)$ for all parses ϕ. In other words, ϕ_{opt} is a parse that is most likely to have given rise to the sequence s. As we will see, the optimal parse gives the gene predictions.

Gene Structure

We now outline the basic properties of human genes that are to be captured in the model. The statistical aspects arise because (1) characteristics shared by genes have similar but not identical properties and (2) signals that genes share can also exist randomly in the non-gene sequence. This issue has also been discussed in Section 5.3.

A gene consists mainly of a continuous sequence of the DNA that is copied, or "transcribed," into RNA, called "premessenger" RNA or pre-mRNA. This pre-mRNA consists of an alternating sequence of exons and introns. After transcription the introns are edited out of the pre-mRNA, and the final molecule, called "messenger RNA" or mRNA, is translated into protein. There can be some other editing and processing of an mRNA before translation. However, that will not be important for our purposes.

The region of the DNA before the start of the transcribed region is called the *upstream region*. This is where the *promoter* of the gene is, the region where certain specialized proteins bind and initiate transcription. There are different definitions of what constitutes the promoter region; often it is taken to be the 500 bases before the start of transcription. Here we shall be interested in only about 40 bases upstream from the start of transcription, since specific signals in the promoter region are extremely complex and are not well characterized. Our model, and the model used by Burge (1997), uses the so-called TATA box, which is a fairly common signal (approximately 70% of genes contain this signal), which is located 28–34 bases upstream from the start of transcription. We do not try to capture any other signal in the promoter region. For those genes without a TATA box, we rely on identifying the gene by the other signals in its transcribed region. The 5' *untranslated region* (5'UTR) follows the promoter. This is a stretch of DNA that does not get translated into protein. We call the first 8 bases of this region the *cap end* of the 5'UTR. Near the other end of the 5'UTR, just before the start codon in the first exon, is a signal that indicates the start of translation, called the translation initiation signal. We shall refer to the 18 bases just before the start codon as the *translation initiation end* (TIE) of the 5'UTR. This is followed either by a single exon or by a sequence of exons separated by introns. An intron may break a codon anywhere between its three nucleotides. Each intron has signals indicating its beginning and end. Modeling these signals well is crucial for correctly predicting the intron/exon structure. Following the final exon is the 3' *untranslated region* (3'UTR), which is another stretch of sequence that is transcribed but not translated. Near the end of the 3'UTR are one or more *Poly–A* signals signaling the end of transcription. A Poly–A signal is 6 bases long with the typical sequence AATAAA.

The Training Data

Each state of the model we construct is a model in its own right. It is necessary to train each state to produce sequence that models the corresponding part of an actual gene. To do this we start with a large set of training data consisting of long stretches of DNA where the gene structures have been completely characterized. Burge et al. (1997) compiled 2.5 million bases (Mb) of human DNA with 380 genes, consisting of 142 single-exon genes and a total of 1,492 exons and 1,254 introns. Many of these are complete genes consisting of both the upstream and downstream regions. In addition to this they included the coding region only (no introns) of 1,619 human genes.

The Model

We model a 5' to 3' oriented gene with a 13 state semiHMM as shown in Figure 12.2. The first row represents the intergenic region. The second row represents the promoter. The third row is the 5'UTR. The fourth row of five states represents the introns and exons. The final row is the 3'UTR and the Poly–A signal.

We first describe how each state is trained by giving the distributions of L_S and $Y_{S,\ell}$. We then discuss how the transition probabilities are set. The intergenic region between genes is labeled N in Figure 12.2. For completely uncharacterized sequence it is reasonable to assume that genes are randomly distributed. The number of genes in any given stretch of this uncharacterized sequence is therefore modeled by a Poisson distribution,

Intergenic
region

Promoter

5′UTR

Exons and
Introns

Posttranslational
region

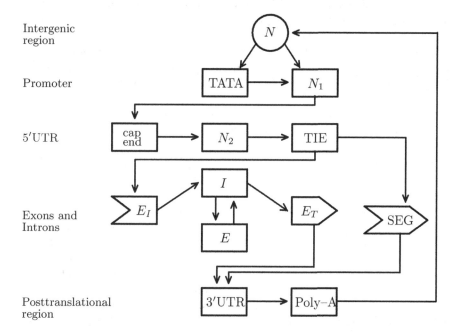

Figure 12.2.

and the distance between genes is modeled by an exponential distribution. Hence we model the length L_N with a geometric distribution (since it is the discrete analogue of the exponential) with mean equal to the size of the human genome (approximately 3 billion nucleotides) divided by the number of genes (approximately 60,000). Given an observed value ℓ of L_N, the sequence of length ℓ is generated by a fifth-order Markov model (as described in Section 11.3). The parameters of this model are set using the noncoding portions of the genes in the training set. This fifth-order model has $3 \cdot 4^5 = 3072$ parameters. Training so many parameters is possible only because there are millions of bases of DNA in the training set. There is known to be at least fifth-order Markov dependence in such noncoding DNA. A fifth-order Markov chain model is used in particular because it provides improved performance over lower-order models and because a higher-order model cannot be trained with the current training sets available. This model for producing sequence will be used for several of the states (those denoted by N_1 and N_2, as well as the entire 3′UTR), and will be referred to as the *intergenic null model*.

The TATA box is modeled with a 15-base weight matrix similar to that described in Section 5.3.2. The path can bypass the TATA box, which is necessary because only about 70% of genes have them.

The state following the TATA box produces a sequence following the null intergenic model. The length of the sequence L_{N_1} follows a uniform distribution from 28 to 34 bases.

The cap end state models the signal at the start of transcription. There has been shown to be information content in at least the first few bases of this sequence, and it is modeled with an 8-base weight matrix derived from the training data.

State N_2 is modeled with the intergenic null model with length given by a geometric distribution with mean 735 bases. This state is the main part of the 5'UTR.

The state labeled TIE contains the translation initiation signal. This state is modeled by a 18-base weight matrix derived from the training data.

The states on the next row are the exons and introns. There are two paths that can be taken through this row. The state on the right corresponds to single exon genes (SEG). It is necessary to give single exon genes their own state because the distribution of their lengths is quite different from the lengths of the exons of multiexon genes. The generation of codons is not, however, different from the multiexon genes. They both use a fifth-order Markov model, but in contrast to the intergenic region, this Markov model is nonhomogeneous. For each position in the codon there is a separate model. So in effect there are three times as many parameters as in the homogeneous intergenic model. This is partly why the training set was supplemented with the coding region of 1619 complete genes (with no introns), which gave approximately 2.3 million bases of coding sequence. The SEG generates a sequence starting with the start codon *atg* and ending with one of the three stop codons *taa, tag, tga*, chosen in accordance with their observed frequencies in the training set. The intervening codons are produced according to the fifth-order nonhomogeneous Markov model. The length of the single-exon gene is taken from an empirical distribution from the training set.

The situation with multiexon genes is more complicated. One complication arises from the fact that a single codon may be split between two exons. To handle this, the exon states in the model only produce a sequence that has length that is a multiple of three. The intron state produces one codon and then a further random number, either 0, 1, or 2, with probability $\frac{1}{3}$ of each. If this random number is 1, it places the first nucleotide at the beginning and the last two at the end; if 2, it places the first two nucleotides at the beginning and the last one at the end; and if it is 0, it does not include the codon at all.[2]

---

[2]Burge handles this in a different way, by creating three internal exon and three intron states, each corresponding to a different "phase." Due to the technical nature of the algorithms, that method may be more effective than the one described here. This method does, however, require that the model be complicated substantially in order to

Empirical distributions from the training set are used for the lengths of the initial, internal, and terminal exons. Intron length is modeled with a geometric distribution. The codons in the exons are modeled with the same fifth-order nonhomogeneous Markov model as the single-exon genes. The intron sequence is generated by the intergenic null model, except at the ends. At the beginning of an intron is a signal called the *donor splice signal*, and at the end is the *acceptor splice signal*. Each of these tells the machinery in the cell where to splice out the intron during the editing process. It is important to model these signals well in order to predict intron/exon structure correctly. The donor signal is taken to be the first six nucleotides of the intron, and the acceptor signal is the last 20. A weight matrix or first-order Markov approach is generally insufficient to capture these signals effectively. Ideally, we would like to have a complete joint probability distribution for these sequences. However, there is insufficient data to do this. For the donor signal we instead use the maximal dependence decomposition discussed in Section 5.3.4, and for the acceptor signal we use a second-order Markov model.[3] All of this is incorporated into the single intron state I.

The 3'UTR is modeled with the intergenic null model, with geometric length of mean approximately 450. The Poly–A signal has constant length 6 bases and is modeled with a weight matrix.

We now turn to the issue of assigning probabilities to the transitions. Most of the transitions have probability one. The exception is the transitions from N, from TIE, and from I. Since approximately 70% of genes have a TATA box, the transition probability from N to TATA is 0.7 and to N_1 is 0.3. The transition probability from the state labeled TIE to the state labeled SEG is taken from the proportion of single-exon genes. The transitions from the intron state I are also taken from the appropriate proportions observed in the training data.

With the model so defined, given an uncharacterized sequence of DNA, we apply the Viterbi algorithm to obtain an optimal parse. The parse gives a list of the states visited and the lengths of the sequences generated at those states. We thus get a decomposition of the original sequence into gene predictions, as well as predictions of complete gene structure for each predicted gene.

In practice, there are many more considerations in optimizing the performance. Perhaps one of the most important is that many of the probabilities that have been estimated depend on the *cg* content of the region of the DNA

---

avoid technically violating the semiHMM assumption. We made an effort here to give a model that satisfies the definition of a semiHMM, at the likely cost of some degree of accuracy.

[3] Burge models the donor splice signal also as dependent on the last three bases of the exon. Again, this violates the definition of semiHMM, so we have not included these dependencies in our model.

being searched. For example, regions with high cg content tend to contain a significantly higher density of genes. Burge's model takes this and several other factors into careful consideration, and the model continues to be refined as more and different types of data become available.

Problems

12.1 Define an HMM λ with the following parameters:

Three states, S_1, S_2, S_3, alphabet $A = \{1, 2, 3\}$,

$$P = \begin{bmatrix} 0 & 1/2 & 1/2 \\ 1 & 0 & 0 \\ 0 & 1 & 0 \end{bmatrix},$$

$$\pi = \begin{bmatrix} 1 \\ 0 \\ 0 \end{bmatrix},$$

$b_1(1) = \frac{1}{2}, b_1(2) = \frac{1}{2}, b_1(3) = 0,$
$b_2(1) = \frac{1}{2}, b_2(2) = 0, b_2(3) = \frac{1}{2},$
$b_3(1) = 0, b_3(2) = \frac{1}{2}, b_3(3) = \frac{1}{2}.$

What are all possible state sequences for the following observed sequences $\mathcal{O}$, and what is $p(\mathcal{O} \mid \lambda)$?

(a) $\mathcal{O} = 1, 2, 3.$
(b) $\mathcal{O} = 1, 3, 1.$

12.2 Given an HMM λ, suppose $\mathcal{O} = \mathcal{O}_1, \mathcal{O}_2, \ldots, \mathcal{O}_T$ is an observed sequence with hidden state sequence $q_1, q_2, \ldots, q_T$. For $t = 1, 2, \ldots, T$, let $\sigma_t = S_j$m where

$$j = \operatorname*{argmax}_i \ \mathrm{Prob}(\mathcal{O}_t \mid q_t = S_i).$$

In other words, σ_t is the state most likely to produce symbol $\mathcal{O}_t$. Construct an HMM λ with uniform initial distribution, three states, and an alphabet of size three, and give an observed sequence of length two, $\mathcal{O} = \mathcal{O}_1 \mathcal{O}_2$, for which

$$\mathrm{Prob}(\mathcal{O} \mid \lambda) > 0, \quad \text{but} \quad \mathrm{Prob}(q_1 = \sigma_1, q_2 = \sigma_2 \mid \lambda) = 0.$$

12.3 Given an observed sequence $\mathcal{O}$, we have given an efficient method for calculating

$$\underset{Q}{\mathrm{argmax}}\ \mathrm{Prob}(Q \mid \mathcal{O}).$$

One might ask why we are interested in this Q and not

$$\underset{Q}{\mathrm{argmax}}\ \mathrm{Prob}(\mathcal{O} \mid Q) \qquad (12.19)$$

instead. The latter state sequence Q, after all, is the one that if given has the highest probability of producing $\mathcal{O}$. To illustrate why this is the wrong Q to find, construct an HMM where the Q given by (12.19) is such that $\mathrm{Prob}(Q) = 0$, so could not possibly have produced $\mathcal{O}$.

12.4 Prove that the reestimation parameters (12.12)–(12.14) do indeed satisfy $\sum_i \bar{\pi}_i = 1$, $\sum_k \bar{P}_{jk} = 1$, and $\sum_a \bar{b}_i(a) = 1$.

12.5 Consider the five amino acid sequences

WRCCTGC, WCCGGCC, WCGCC, WCCCGCC,   WCCGC.

Suppose their respective paths through a protein model HMM of length 8 are

$$m_0 m_1 i_1 m_2 m_3 m_4 m_5 d_6 m_7 m_8,$$
$$m_0 m_1 m_2 m_3 m_4 m_5 m_6 m_7 m_8,$$
$$m_0 m_1 m_2 d_3 d_4 m_5 m_6 m_7 m_8,$$
$$m_0 m_1 m_2 m_3 m_4 m_5 m_6 m_7 m_8,$$
$$m_0 m_1 m_2 m_3 d_4 m_5 d_6 m_7 m_8.$$

Using the theory of Section 12.3.2, give the alignment of the sequences that these paths determine.

13

Gene Expression, Microarrays, and Multiple Testing

13.1 Introduction

13.1.1 Introduction to Microarrays

The major part of this book has dealt with sequence data; however, much of the important information about a gene is not evident directly from its sequence, for example exactly when, where, and how much it is expressed. For a typical cell at a given time, many proteins are not required, while others are required in varying abundances (see Section 3.5.2 and Appendix A). Consequently, in a given cell the cellular machinery is relatively inactive for some genes, at any given time, producing none or very few copies of the protein, while for other genes it is very active, producing many copies. Though we would like to measure the protein abundances directly, it turns out to be much easier to measure the relative mRNA levels in cells, and so in practice that is often what is done. It is important to keep in mind, however, that this is only an indirect measure of protein levels, since there are many other factors which determine how much and how fast the mRNA gets translated into protein.

Microarrays are a tool for measuring, in a given sample, the mRNA levels for thousands of genes simultaneously. A sample can be either a single cell, or a population of cells (e.g. a whole pancreas). In most cases samples consist of many cells because of the technical difficulties of performing single cell assays. When expression of a gene is evident from a sample consisting of a population of cells, then at least one cell in the sample must be expressing that gene. What can be concluded will then depend on the purity of the sample.

The analysis of microarray data is a relatively new field, dating back less than 10 years, and the theory at every level is still in the formative stages. Nevertheless, it is a rapidly growing field with published papers appearing almost daily utilizing and developing the theory. For a sense of the scope and development in the technology and application of this new research method the reader is referred to the review articles in *Nature Genetics* (2002). Our treatment is necessarily introductory, with a focus on the statistical aspects. For accounts of the broader scope of the field, see for example Schena (2003), Speed (2003), Knudsen (2002), Parmigiani et al. (2003), Baldi and Hatfield (2002), Draghici (2003), Grigorenko (2003), Kohane (2003), and Stekel (2003).

Due to the highly highly parallel and highly variable nature of microarray data, the statistical methods used to analyze these data are necessarily complex. To serve as an introduction, our focus is on three of the most basic questions which microarray data are used to address, and which involve statistical methods. These are:

(1) What genes are expressed in a given sample?

(2) Which genes are differentially expressed between different samples?

(3) How can one find different classes, or clusters, of genes which are expressed in a correlated fashion across a set of samples? How can one find different classes of samples based on their gene expression behavior?

Microarray data can be used to address many further questions, for example

(4) How can gene-gene interactions in cascades or networks of activity over time be discovered?

Besides being beyond the scope of this book, much of the theory for questions such as (4) has not settled down enough to be appropriate for an introductory textbook. Indeed work continues on all of the above questions, and the theory, as well as the technology itself, will continue to evolve.

Before addressing the questions listed above, it is important to understand the nature of the raw data obtained from a microarray experiment and how these raw data are pre-processed into a form suitable for statistical analysis. The data are technically complex, but it is necessary to go through these aspects in some detail in order to appreciate the fact that the data eventually analyzed have often been arrived at after several manipulations of the original data. We summarize various technical details associated with microarray data in the following sections.

13.1.2   Microarray Data

Most microarrays fall into one of two main categories. One is the "spotted array" and the other is the "probe-set" array (primarily produced by the Affymetrix company). Although spotted arrays and probe-set arrays differ from each other in various technical details, the two procedures share many common features and for the most part are designed to answer similar experimental questions.

Spotted Arrays

Spotted microarrays are either glass slides or nylon filters which are printed with thousands of "spots," where each spot contains millions of identical "probes" for a particular gene. A probe for a gene consists of a piece of single stranded DNA sequence which either has the same sequence as, or is complementary to, a segment of that gene's mature mRNA (of course the nucleotide "t" in DNA is replaced with "u" in mRNA). The length of the probes on spotted arrays range from around 70 nucleotides to the full length of mRNAs which can be thousands of nucleotides long. A typical array will have thousands of such spots representing thousands of genes.

Many genes share common features, so the probes must be chosen judiciously in order avoid ambiguously representing more than one gene. This can rarely if ever be accomplished perfectly, so some "cross-hybridization" is unavoidable.

There are two predominant approaches to selecting probes. The first is to choose them from libraries of cDNA clones. These are the libraries from which databases of ESTs are constructed (see Section 2.9.2). ESTs are small pieces of the sequence of the cDNA clones in the library, and by aligning them to known genes using BLAST we can obtain functional information about many of them. Currently we can get functional information in this way for about half of the cDNA clones in a typical library coming from a human tissue. A set of cDNA clones from various libraries is chosen, either based on the tissue source of the library, or of the sequence similarity to known genes, or any of many other criteria, depending on the intended use of the array. Usually uncharacterized clones are put on the array as well as the characterized ones, because the patterns of expression of uncharacterized clones can reveal novel genes that are important to the system being studied and can then be followed up for further investigation and characterization.

The second method for choosing the probes is to start with the genomic sequence, which is now becoming available for many species. First, putative genes are annotated on the sequence either manually, or computationally using an ab initio gene finding algorithm (see Section 12.3.4). Parts of each gene's coding sequence are then chosen to be probe sequences. They are chosen to be as unique to that gene as possible, to minimize cross-hybridization. Oligonucleotide probes are then synthesized representing the chosen sequences and spotted onto the array.

Once the probes for the spots are chosen, many arrays are printed as uniformly as possible. However, due to the printing process, there is some variation in spot size and probe concentration. This adds noise to the data, which can, to some extent, be corrected for, as will be discussed below.

Each array is then used to assay the gene expression for one sample. To do this, the mRNA is extracted from the sample and copied into DNA (so that one of the strands is complementary to the mRNA). One or both strands are labelled with a radioactive or fluorescent tag. The labelled DNA is then applied to the array in a denatured state (that is, separated into single strands). For the mRNAs which are *expressed* in the sample and which come from genes that are represented by a spot, a part of each corresponding labelled DNA will be complementary to the probes in that spot and so will hybridize to them. Probes in spots for *unexpressed* genes will at most pick up stray background signal or cross-hybridization.

A scanner is then used to generate a digital image so that the more labelled DNA that has hybridized to a spot, the brighter the spot will be on the image (see Figure 13.1 for an example of such an image from a glass slide microarray array with 1,936 spots and tagged with a green fluorescent dye). In order to avoid saturation effects the probe should be laid down far in excess of the numbers of labelled mRNA that are hybridized.

Statistical analysis requires numerical data, so the image is processed to "quantify" the spots into numerical intensities, where brighter spots are given higher intensities. Finding a suitable algorithm to accomplish this task is a difficult problem which involves some statistical issues, but is mainly a problem of an algorithmic nature, so we shall not discuss it here. However, when analyzing microarray data, one should have a good understanding of the algorithms used to quantify their data and the various issues involved because the quality of the results is very sensitive to this step (Schena (2003), Draghici (2003), Geschwind et al. (2002)).

Because of the way in which the arrays are printed, there is inevitable variation in the number of probe sequences delivered to the various spots, as well as variation in spot size. This can introduce significant noise into the data because a larger spot can tend to give a greater signal. To control for this, a reference sample is often labelled with a different dye (for example Cy3 ("green") can be used for the sample and Cy5 ("red") for the reference), and the two labelled extracts co-hybridized to the array simultaneously. The array is then scanned at two frequencies giving two digital images, a red and a green, which are separately quantified to give a red and a green signal for each spot. In the red/green ratio the variation due to spot size and concentration is greatly reduced. Suppose, for example, that a gene is expressed at the same level in both samples, and the two samples are assessed by a two-channel array, one sample per channel. An array with larger spots will tend to have a larger intensity in *both channels*, and there-

Figure 13.1. A scan of one channel of a two-channel glass slide microarray with 1936 spots

fore the ratio will be (roughly) equal to one regardless of spot size. Figures 13.2 and 13.3 illustrate the importance of this normalization. The graphs in Figure 13.2 are the individual channel intensities for one spot location from 34 two-channel microarrays. The species is *Plasmodium falciparum* and the green channel of the 34 arrays represent a partial time-course in the lifecycle of the organism. The red channel was used as a reference and is the same in all 34 arrays, it is a pool of the mRNA from all of the time points. The graph on the left is the red channel and the graph on the right is the green channel. Since the red channel of every array was hybridized with the same RNA, the graph on the left should be flat if measurements were perfect. For comparison, Figure 13.3 shows the graph of the ratios of the two channels shown in Figure 13.2. Even though there is over 4-fold change in intensity in each of the two channels, there is only a 1.6-fold change between the maximum and minimum ratios.

These so-called "two-channel" array experiments are among the most common types of arrays in use today. While introduction of the reference sample reduces one source of variation, it can force the investigator to design a more complex experiment. For example, if one is to compare sample types A to B, then in order to use the two-channel method one can introduce a third sample C and perform several A to C arrays and several B to C

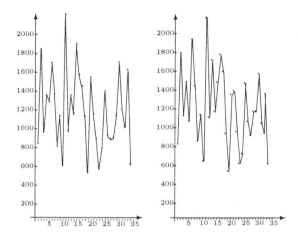

Figure 13.2. Graphs of the raw red (left) and green (right) channel intensities of Plasmodium falciparum gene PF11_0244 in 34 two-channel microarrays. The green channel consists of 28 different time-points in the Plasmodium falciparum asexual lifecycle (with replicate arrays for several time points), the red channel consists of a pool of all time points in the asexual lifecycle and is identical for every array.

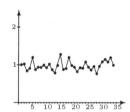

Figure 13.3. Graph of the ratios of the two channels of the same experiments as in Figure 13.2.

arrays, and compare the A/C and B/C ratios. This is known as a *reference design*. Alternatively one might perform several direct A to B comparisons using the two channels, one for A and the other for B. This is known as the *direct comparison design*. In the reference design we have reduced one type of variation while introducing another by adding a third sample type into the design. However, in the latter case we have restricted the types of statistical analyses we can apply, because we do not have reliable measurements from each of the two classes, so that we must find a method that works directly with ratios. Therefore, in the experimental design stage one must deal with several tradeoffs to try to perform the most efficient and effective experiment possible.

Probe-Set Arrays

The Affymetrix, or probe-set, approach is to use short probes, called "match" probes, on the order of 20 bases long, and to use a set of 10 to 20 different probes for each gene, each matching a different small segment of the mRNA. In addition one generally uses the same number of "mismatch" probes, which consist of the match probes with one base changed in each. A probe and its mismatch are called a "probe pair" (denoted PM and MM for "perfect match" and "missmatch"), and generally there are 10 to 20 probe pairs per gene. The mismatch probes allow for some estimation of the stray signal coming from cross hybridization and any other factors not due to the true signal. The use of mismatch probes also allows for a test of whether this probe-set represents an expressed gene (see Section 13.2.2).

The intensities from all of the match and mismatch probes are combined into one summary value measure for that gene. For each probe pair, the PM intensity contains the true signal as well as the stray signal. The MM probe is intended to measure the stray signal, and the difference of the PM and MM measurements (PM–MM) a measure of the "true" signal. It is, however, far from ideal, as MM will contain also some true signal, and might not represent accurately the true stray signal. On the other hand there are multiple probe pairs, generally 10–20, so one can hope that some of these effects will tend to average out. Therefore the summary value for an entire probe set should be some kind of average over the individual PM–MM values. The simple average can be negative, and is also not robust to outliers, so other more sophisticated forms of averaging have been suggested, which we now discuss. Desirable properties that the summary intensities should have is that of nonnegativity, that they have a linear relationship with actual mRNA concentration, and that they be robust to corrupted data. The most recent algorithm Affymetrix has suggested, the MAS 5.0 algorithm (see Affymetrix (2003)) adjusts the PM–MM values before taking the average. The algorithm retains probes where PM–MM is positive and substitutes the average of the positive PM–MM values for those PM–MM values which are negative. If all probe pairs are negative then an even more ad hoc approach is taken where the contribution of the MM probes is diminished. All averages are calculated using the Tukey biweight method, which is robust to outliers (Press (1992)).

Several benchmark data sets were generated to test how well the MAS 5.0 method satisfies the desired properties listed in the previous paragraph. In each of these data sets, a number of genes are spiked into an mRNA solution at varying concentrations. In this way the true abundances of the spiked in genes are known, and it can be checked how well the methods estimate the true values. From these data the MAS 5.0 method was shown to have reasonable properties. The reader is referred to the Affymetrix documentation for more information on this procedure (www.affymetrix.com).

When there are multiple arrays in a study, even if they are not replicates of the same condition, it is possible to construct a model, for each gene, with the abundance levels of the gene in the various arrays as parameters of the model. Estimates of those parameters then give abundance levels for each array. How well this works depends on how good the model is and how well we can estimate the parameters of that model.

The first model was proposed by Li and Wong (2001). It is based on the observation from spike-in benchmarks, such as those described above, that the intensities for each probe increase linearly with respect to the actual RNA abundance level in the sample, while the slope of line depends on the probe. The simplest such model is

$$PM_{ij} - MM_{ij} = \theta_i \phi_j + E_{ij}, \quad i = 1, \ldots, g, \quad j = 1, \ldots, I \quad (13.1)$$

where i indexes the array and j indexes the probes. (The gene in question is fixed.) The parameter θ_i is the abundance of the gene in the ith experiment. The ϕ_j's are the coefficients of the linear relationship between $PM_{ij} - MM_{ij}$ and θ_i for probe j. The E_{ij} terms are the random component of the model. A least-squares procedure is used to estimate the θ_i's and the ϕ_j's, as described in Section 8.4.3. The procedure assumes the distribution of the E_{ij} are independent of i and j. This assumption generally does not hold for microarray data; however, this model provides enough of an approximation to reality to gives meaningful results. Studies using the spike-in benchmark data sets show this method provides summary values that are less variable than the MAS 5.0 method, and which have good signal detection properties (Irizarry et al. (2003)).

It should be noted that the model (13.1) is overparameterized, and in order to solve it uniquely the equation

$$\sum_{j=1}^{I} \phi_j = I \quad (13.2)$$

is added to the system. Recall that a similar step was taken in Section 9.5. However, in that case we were testing a hypothesis, and adding this extra equation did not effect the validity of the procedure. We did not then attempt to estimate the actual values of the parameters. In the present case we are not testing a hypothesis; instead, we are interested in estimating the actual values of the θ_i's in the model. In this case the effect of adding equation (13.2) to the model is to put the estimated value of the probe set for one gene on a (possibly) different scale from the estimated value of the probe set for another. This does not cause problems for comparison of the estimated values for the same gene across arrays; however, it does cause a problem for the comparison of the estimated values between different genes on the same array. Therefore one must be careful when making such comparisons.

Many refinements to this procedure are available and the description above outlines the simplest of them. A description of more recent methods is provided by Irizarry et al. (2003).

In most of what follows we shall assume each gene is associated with a single intensity, as in the spotted array case, and will ignore the individual PM and MM intensities. In this way we shall take a perspective that does not treat Affymetrix data much differently from spotted array data.

13.1.3 Sources of Bias and Variation

Variation, Experimental versus Biological

There are very many sources of variation in microarray data. The first distinction to be made is between biological and experimental variation. Biological variation is the natural variation between different cells or different populations of cells. Experimental or technical variation is the variation deriving from the technical aspects of the procedure. There are many steps involved which are not reproducible exactly, even with the exact same mRNA, which lead to such technical error. These are due to errors such as:

(1) Array-specific effects: No two arrays are identical, so that there is a random array-to-array variation. Furthermore multiple printing pins are used to print each array, in order to speed up the printing process, often leading to pin specific variation in final observed intensities.

(2) Gene-specific effects: Hybridization conditions cannot be optimized simultaneously for many elements (genes/transcripts) at once.

(3) Dye-specific effects: Incorporation of fluorescent dye (used to detect hybridized material) varies. With two dyes, there might be differences in incorporation for the same transcript.

(4) Background noise and artifacts: There is always a low level background glow on any array, as well as dust, scratches, smears, etc.

(5) Preparation effects: operator, time/day of assay, weather, etc.

We want to capture the biological variation and minimize the experimental variation. For example, sometimes there is a significant day-to-day effect on the observed intensity levels for some genes, therefore, if possible, all hybridizations will be done on the same day. If this is not possible, then the statistical methods may have to account for this effect so that the extra data adds power to the methods in spite of the increased variation.

We also have to consider the nature of the inference we can draw from the data at hand. For example, if we were looking for a gene which is expressed differently between human cancer cells and normal cells, we would have to perform many experiments with a random sample of many different types of cancerous and normal cells, across many randomly chosen individuals, in order to draw general conclusions. Our inference relates only to that comparison for which we have an adequate random sample. If all of the cells come from one individual, then we can only draw conclusions about that individual.

Unfortunately, the form of replication necessary to draw general inferences is both expensive and very time consuming. This implies that often only as few as two or three replicates are available, with as many as ten or twenty being the exception rather than the rule.

There are also many non-systematic sources of variation that cannot be eliminated. Perhaps the most troublesome of these arises from the level of background noise which precludes the possibility of using microarrays to detect low levels of expression. Many important genes such as transcription factors, which induce the transcription of other genes, are expressed at low levels, and will generally be missed by microarrays. Furthermore, since two-channel data deals with ratios, the spots whose intensity levels are in the background level in both channels can introduce a large amount of variability into the distribution of these ratios.

Systematic Experimental Bias, and Calibration

Other types of variation cause systematic biases, and these can sometimes be corrected. For example with spotted two-channel Cy3/Cy5 microarrays, the green dye often has a tendency to be stronger than the red dye. Furthermore the magnitude of this effect varies from array to array. If we can measure this bias we can correct for it. A simple approach is to divide all ratios for each array by the mean or median of all ratios over the array. After this operation, the mean (or median) intensity over all arrays is one, thus putting all arrays on the same footing. This is an example of a global normalization where the same adjustment is performed for each spot. We now look at a method of local normalization.

There is a convenient method of displaying microarray data which has become fairly standard, and helps to visualize the spread between the two channels of a two-channel microarray, possibly revealing intensity dependent dye biases. If $G(g)$ is the Cy3 intensity for a gene g, and $R(g)$ is the Cy5 intensity for g, then we plot $M = \log_2(G(g)/R(g))$ on the vertical axis, against $A = (\log_2(G(g) + \log_2(R(g)))/2$ on the horizontal axis (see Figure 13.4). This is called an MvA plot in the literature. In this way the horizontal axis represents the intensity of the spot, as measured by the average of the logged values of the two channels, and the vertical axis represents the difference between the two logged expression values in the two channels. Having the logarithm on the vertical scale allows for symmetry in the graph about the $M = 0$ line.

If $M = 3$, for example, there is an eight-fold greater intensity in the green channel than in the red channel. If an experiment compares two

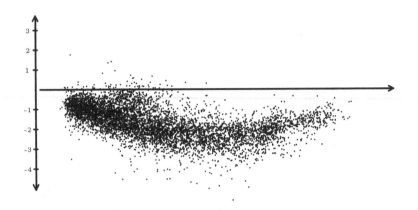

Figure 13.4. MvA plot

samples where most of the genes are expected to be expressed at roughly
the same levels, then the MvA plot should be roughly symmetric about the
horizontal line $M = 0$. If it is not, a loess curve (see Section 8.4.3) can be
fit to the plot and the intensities adjusted by subtracting the values on the
curve. This normalization, known as loess normalization, normalizes for the
intensity dependent dye bias present in the ratios of a two-channel array.
The procedure produces normalized *ratios* and does not provide normalized
values for the individual channels.

Typically a loess curve is fit to the graph, as described on page 288. This
can also be done print tip by print tip if one notices print tip dependent
biases, which often occur. Figure 13.5 shows an MvA graph, with a loess
curve fitted to it. Figure 13.6 shows the MvA plot after correcting. It is
important to note that what is changed in this correction are the M values.
It makes no sense to talk about the corrected A values, so the original A
values are used for the horizontal axis in Figure 13.6. Because of this it also
makes no sense to use the A and the normalized M values to solve back for
the two channels in the hope of obtaining a normalization for the individual
channels. If one must use individual channel data for some reason, for
example for trying to determine whether or not a gene is expressed in a
particular condition, then a further normalization would be required.

Another issue of normalization involves the spread of the M values across
the array, which may depend on the array itself and not on the biology.
The loess procedure described above does not normalize the spread of the
M values. Some packages are available to perform spread normalization
(e.g. Yang et al. (2001)) in the expectation that this will give better results
in the subsequent analysis. However, there is currently little theory, if any,
which justifies these normalizations or which investigates their effect on
subsequent analyses.

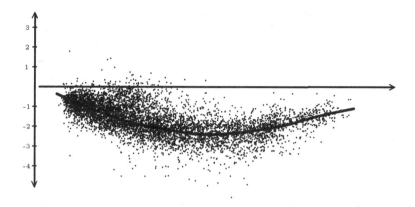

Figure 13.5. MvA plot with loess curve

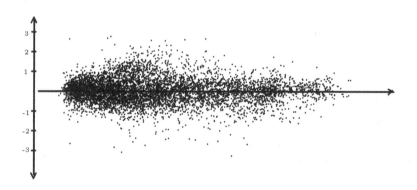

Figure 13.6. Normalized MvA plot of loess corrected data.

When one may not assume that most of the genes are unchanged between the two conditions, applying this method may normalize out true biological differences. For example, it is common to build arrays containing only genes known to be expressed in a particular tissue, for example a pancreas or a liver specific array. In such a case it is quite possible that a majority of the genes are differentially expressed and that there is not a balance between up and down regulation.

Unfortunately, when experiments are performed using such arrays, there are no good normalization methods which can correct for it, since there would be no way to tell the difference between a dye-effect and a biological effect. For example, if every gene on the array is x-fold more abundant in one condition and the dye effect causes another y-fold increase in the observed expression in that condition, then the MvA plot will be symmetric around the line $M = x + y$ and the loess correction will eliminate both the dye bias (y) and the biological effect (x). Arrays and experiments should be designed with this issue in mind, so that for example a number of control genes might be put on the array across a spectrum of intensities. Control genes are those that are expected not to change in the conditions being investigated. The loess curves can then be fit to just the control genes to perform the correction.

Another popular approach is to perform quantile normalization. In these methods the data are adjusted so that all arrays have the same quantiles. Again, data may not satisfy the assumption that the quantiles should be equal, as would occur for example if all genes on the array are upregulated.

Another popular normalization is to perform "dye-swap" experiments. In this case the solutions to be hybridized to the two channels are each divided into equal parts and two arrays are hybridized, with the dye assignments switched between the two arrays. The intensities from the two arrays are averaged, in the hope that the dye bias will be averaged out. This only works if the bias is not array specific. This method is not subject to the same assumptions as the previous ones, where differential expression should be symmetric or quantiles should be equal. However, the independence of the dye bias from the arrays is another strong assumption that generally does not hold, and in fact when it fails to hold, averaging dye-swaps can actually compound bias instead of reducing it.

Other Effects

In any microarray study there are certain factors of interest, such as the difference between normal and diseased tissue, or between various developmental stages. There can also be other effects that are undesirable, but unavoidable. For example, in the previous section we have discussed a dye bias and methods to normalize the data to remove it. There can be other effects such as a "day" effect where the observed intensities depend not only on the condition but on the day the experiment was performed. It is common for the experiments to be performed a few a day over a period of several days, or even much longer. ANOVA methods can be used to try to determine if there is a systematic effect on the intensities depending on the day, or on any other factor which might possibly have a systematic effect, as long the experiment is designed correctly. If two conditions are compared and all of the first condition experiments were performed on one day and all of the second condition on a second day, then the two effects (condition and day) are totally confounded and there can be no way to tease them apart. If however the conditions are randomly assigned to the two days, then an ANOVA analysis might reveal significant day effects. This would be done by the model (9.46), where "days" is taken as a random effect.

We have discussed in detail above the practical problems involved in microarray data generation, and some of the many assumptions and approximations that are used. We have done this in detail because the interpretation of microarray data via statistical methods, to which we now turn, must always be made with these assumptions and approximations in mind. For further comments of this type, see Miklos and Maleszka (2004).

13.2 The Statistical Analysis of Microarray Data: One Gene

13.2.1 Introduction

A microarray experiment in effect consists of thousands of single gene experiments run in parallel, so that in assessing which genes are differentially expressed between two samples one is conducting thousands of statistical tests in parallel. Because the analysis of the many gene case relies on an understanding of the single gene case, we defer consideration of the many-gene case to Section 13.3, and consider first the single gene case.

13.2.2 Determining Whether a Gene is Expressed

We start with the most basic question of whether a given gene is expressed in a given sample. We set the null hypothesis H_0 to be that the gene is not expressed, then with appropriate negative controls we can construct a method which accepts or rejects H_0 with appropriate error rate control. As discussed in Section 3.4.1, failure to reject H_0 does not allow any conclusions about the statement that the gene is not expressed; it simply means there was not enough evidence to conclude the gene is expressed. Negative controls can be gene specific, such as scrambling some proportion of the bases in the probe(s) for each gene, or they can be probes from a completely different species, not similar to any gene in the species being investigated. If the gene in question is actually expressed we should see a greater amount of hybridization to the perfectly matching probe than to the negative controls. With replicate measurements, a difference of the means of the two groups can be tested, as discussed below.

For Affymetrix arrays there are gene-specific negative controls provided by the mismatch probes. Since there are multiple match and mismatch probes for each gene, we can apply a statistical test to determine whether there is a higher overall expression level from the match as opposed to the mismatch probes. The Wilcoxon signed-rank test, described in Section 3.8.3, is used in the Affymetrix MAS 5.0 suite to test for a significant difference between the PM and MM intensities.

Furthermore, as discussed above, the failure to reject the null hypothesis does not imply that the null hypothesis is true. Nevertheless this method is used by Affymetrix to assign "Present" (P) and "Absent" (A) calls (as well as "Marginal" (M) calls when the P-value is between .04 and .06). Several caveats are in order when applying this method.

The first is that an A means something much less precise than a P. The method controls the Type I error rate only, so there is no control over the possibility of making a false A call. An A call should therefore be taken only as there being insufficient evidence that the gene is present.

The second caveat regards the way in which the null hypothesis must be formulated in order to apply the MAS 5.0 Wilcoxon test, as described in Section 3.8.4. The null hypothesis assumes complete equality of the distributions of the match and mismatch probe intensities. Because of this we might also, for example, call a gene present due to the fact that its PM and MM intensities have very different variances but do not have different means.

The third caveat is that since this method is applied to a single array, it does not take into account biological variability. Therefore care must be taken in the conclusions that are drawn. Several replicate arrays from the same condition can however be combined to give one P-value for the hypothesis that the gene is not expressed, as discussed in Section 3.12.

The fourth caveat is that there is no multiple testing correction done to the Wilcoxon P-values. Therefore, even if no genes are expressed, approximately 5% of them will be given P calls anyway.

13.2.3 Testing for Differential Expression

Possibly the most common aim of an investigator using microarrays is to find genes that are differentially expressed between two different samples, for example in normal versus disease cells. Consequently, the statistical theory for this problem has undergone considerable development, and accordingly we shall devote considerable attention to it.

Affymetrix provides a differential expression algorithm in the MAS 5.0 suite which is based on the Wilcoxon test and is similar in spirit to the method described in the previous section for making the "P" and "A" calls. As in the method in the previous section, this method utilizes only one array per condition. As such, the differential expression method suffers from the same caveats as the "is expressed" method. Another problem is that it might be difficult to determine how to normalize the two arrays to each other if one array is brighter overall than the other. In such cases, unless some normalization is made, there will be a tendency for all genes to be found to be differentially expressed. Another issue is how to handle one- versus two-sided tests. The Affymetrix solution to these problems is ad hoc and non-standard (see the Affymetrix technical guide Affymetrix (2003)), and as a result the significance measure returned is not a well-defined P-value. It is therefore not clear how to combine them to incorporate biological replication, making the conclusions that can be drawn quite narrow. Alternatively there are more standard approaches which use the summary values (see Section 13.1.2) and not the individual probe values themselves, and these approaches utilize replicate arrays per condition. This will be our approach to this problem from now on with regards to Affymetrix data.

Designing Experiments

Before considering the question of differential expression further, we must address experiment design issues. When comparing between two conditions, which we denote by condition A and condition B (e.g. cancerous versus normal cells), most of the statistical methods at our disposal assume a number of independent measurements from each condition. In other words, the data can be put into a matrix

$$
\begin{array}{c|cccc|cccc}
 & A_1 & A_2 & \cdots & A_m & B_1 & B_2 & \cdots & B_n \\
\hline
G_1 & a_{11} & a_{12} & \cdots & a_{1m} & b_{11} & b_{12} & \cdots & b_{1n} \\
G_2 & a_{21} & a_{22} & \cdots & a_{2m} & b_{21} & b_{22} & \cdots & b_{2n} \\
G_3 & a_{31} & a_{32} & \cdots & a_{3m} & b_{31} & b_{32} & \cdots & b_{3n} \\
\vdots & \vdots & \vdots & & \vdots & \vdots & \vdots & & \vdots \\
G_g & a_{g1} & ac_{g2} & \cdots & a_{gm} & b_{g1} & b_{g2} & \cdots & b_{gn}
\end{array}
\tag{13.3}
$$

where columns correspond to microarrays and rows correspond to genes. The columns labelled with the A_i's correspond to arrays from the first condition, and the columns labelled with the B_i's correspond to the arrays from the second condition. Designing an experiment to give data in such a form is natural when using one-channel arrays. With Affymetrix probe-sets arrays, data come in this format when the probe sets are converted into summary values, as described in Section 13.1.2. When two-channel arrays are used, some decisions must be made in order to put the data into this format. The individual channel data could be used, but the loss of the normalizing effect is generally too great. Alternatively, the reference design can be used: a third sample C is introduced as a common reference for one of the channels of every array. Several A versus C arrays and several B versus C arrays are generated, and the ratios A/C and B/C comprise the entries in the data matrix.

In the data matrix for a reference design, the information as to whether the ratio was formed by two high intensities, or from two low intensities, has been lost. Because of this, the low intensity background spots tend to introduce a lot of noise into the data matrix. Therefore it is not uncommon to perform a pre-processing of the data before analysis, for example by eliminating all spots with intensities in the background range for all arrays. Furthermore, because the distribution of the ratios of positive random quantities is often skewed to the right, it is common to perform a log transform of the ratios to remove the skewness of the distribution of the data before continuing analysis.

As an alternative to using a reference design, one could hybridize a collection of arrays with condition A in one channel and condition B in the other, yielding a set of ratios values directly comparing the two conditions, the direct comparison design. Because the channels should not be separated, data generated via the direct comparison design cannot generally be analyzed with the same kinds of statistical methods as with the reference design. However this design has the advantage of not introducing a third sample and therefore should lead to less variability in the data.

We start with the case where the data are in the form of (13.3).

Two-Class Data

The Choice of Test Statistic: One Gene

In this section we discuss aspects of the use permutation methods to test whether the expression levels in two conditions differ for any specific gene. Some permutation procedures appropriate for this were discussed in Section 3.8.1, and we first review some of the points made in that section.

The first point to recall is that, in contrast to the situation arising in parametric hypothesis testing procedures, there is no theoretical "best" test statistic to use in a permutation procedure. In practice, the t statistic (3.33), the t' statistic (3.37), the Mann–Whitney statistic of Section 3.8.2, or the ratio r of the average expression levels in the two groups are often used.

This observation leads to a second point discussed in Section 3.8.1, namely that the null hypothesis tested by a permutation procedure is that the complete probability distribution of any observation in the first condition is the same as that as the complete distribution for any observation in the second condition. Use of t, t', or r as the permutation statistic implies that a difference in the means of the observations in the two groups is of primary interest. However, since equality of these two means is not the exact hypothesis being tested in a permutation procedure, a significant result arising when one of these statistics is used does not necessarily imply that a significant difference in the means has been found.

Third, when the number of observations m in the first condition is equal to the number n in the second, the permutation distributions of t and t' are identical, so there is no benefit gained in using t' as the permutation test statistic. Indeed it can be argued that, since the null hypothesis tested in the permutation procedure is identity of distributions of the two groups of observations, use of the statistic t', which is designed to handle cases where the variances of observations in the two groups differ, is illogical.

Fourth, as shown in Section 3.8.1, for the case of a single gene the use of t as the permutation test statistic is equivalent to the use of the average of the observations in either condition. Use of this average rather than t involves significant savings in computation. When carrying out permutation procedures for many genes, however, use of t rather than an average can be preferred, as discussed below in Section 13.3.3.

Parametric and Non-Parametric Tests

Since for the moment we are focusing on differential expression tests for one gene, we consider the data in one row only of the data matrix (13.3). Following this we shall consider the single gene case of a direct comparison design where the data come as a set of ratios.

The first issue is how to formulate a null hypothesis about the gene's expression level in the two experimental conditions. Two possible null hypotheses are: "The means are the same in both groups" and "The distributions are the same in both groups." How we formulate the null hypothesis determines which methods we may apply to control the Type I error rate. There is a gain in using the latter formulation in that it will allow permutation methods to be applied, and there are often no apparent alternatives to permutation methods which do not make troublesome assumptions. For example, if we are willing to believe that the distribution of the gene's intensity in each of the two conditions is normal, with equal variance, then we can apply a standard t-test such as in Section 3.5.2 to test for equality of means. If the respective variances in test and control groups are possibly not equal, the test statistic t' defined in (3.37) could be used. However normality may rarely be assumed. Whether the assumption holds will depend on the particular gene and sample being studied. Since a population from which the samples are drawn (e.g. breast cancer cells) can be a mixture of more basic types of samples (e.g. different types of breast cancer), mixtures of distributions are possible.

Because the normality assumption might not be justified, the observed value of the t (or t') statistic should not be referred to t tables for significance. Instead, the statistic t is often used as the basis for a permutation test, as described in Section 3.8.1. That is, the value of this statistic is calculated for all $\binom{m+n}{m}$ permutations of the data, and the null hypothesis is rejected if the observed value of this statistic is a sufficiently extreme member of the various permutation values.

If it becomes necessary to test for the equality of means and not just complete equality of distributions, then one can use the bootstrap method, as described in Section 8.6. Applying this method requires more replicates per condition to achieve similar power.

The ANOVA Approach

In Section 9.5 the central ANOVA concept was discussed, that is the partition of a total amount of variation in a body of data into meaningful components, with an eventual assessment of the significance of the amount of variation in these various components. This concept has been used frequently in the analysis of microarray data. Normally ANOVA is used for the analysis of microarray data in the multi-gene context. However, it is convenient to introduce ANOVA briefly in the simpler one-gene context. Multi-gene analyses are discussed, in Section 13.3.7.

We here consider the one-gene analysis of Wolfinger et al. (2001), intended largely as a preliminary to a multi-gene analysis. For any gene g, Wolfinger et al. set up the model (in our notation)

$$X_{gik} = \mu + \tau_k + A_i + D_{ik} + E_{gik}. \tag{13.4}$$

This model is in the form of the mixed model (9.46). Here X_{gik} is the intensity for the gth gene, in the ith array, and in the kth condition. The random variable A_i refers to array i, the parameter τ_k to condition k ($k = 1, 2$), and the random variable D_{ik} to the interaction between array i and condition k. The notation reflects the fact that conditions are regarded as fixed effects and arrays as random effects. We are interested in testing the hypothesis $\tau_k = 0$, $k = 1, 2$, which can be accomplished via the methods of Section 9.5.

Direct Comparisons

We cannot apply the methods of the previous sections to compare two conditions with a set of two-channel arrays where one condition is in one channel and the other condition is in the other channel, unless we separate the channels. However, there is generally great variation introduced by spot size and other factors that are corrected for by taking ratios. Not using the correction afforded by taking ratios can dramatically affect the results and it is therefore recommended never to separate channels for a comparative analysis (see for example Yang and Thorne (2003)). It is better in this case to use instead a one-sample test, such as the one-sample t-test described in Section 3.5.1.

A permutation procedure can also be used in the one-sample case in the following way. Given a set of n ratios $a_i/b_i, i = 1, 2, \ldots, n$, it is possible to form $2^n - 1$ new sets of ratios by systematically replacing various ratios by their reciprocals. For example, if $n = 4$, two of the 15 new sets that can be formed in this way are $\{b_1/a_1, a_2/b_2, a_3/b_3, b_4/a_4\}$ and $\{a_1/b_1, a_2/b_2, b_3/a_3, b_4/a_4\}$. The test statistic of interest is then calculated for all 2^n possible sets, giving a permutation distribution of this statistic. From this the permutation P-value of the actual set of ratios can be obtained.

13.3   Differential Expression – Multiple Genes

13.3.1   Introduction

Microarrays are not used to determine whether a single gene is differentially expressed between two conditions, because they generate much noisier and less reliable data than other single gene methods. For example, "real time PCR" (RT PCR) will determine whether a particular gene is differentially expressed with a higher degree of accuracy. However RT PCR is usually too time consuming and expensive to perform on more than a small fraction of the genes that can be put on a microarray. RT PCR is suitable for answering the question: "Is this gene differentially expressed between the two conditions?" but is not suitable for answering the question: "*Which* genes are differentially expressed between the two conditions?"

Microarrays are more appropriate for the latter question, at least for those genes which are moderate or highly expressed. Microarrays are often combined with RT PCR, with microarrays being used as a first filter of the genes for a small set of putatively differentially expressed genes, which should contain a reasonably high proportion of truly differentially expressed genes, which can then be followed up with RT PCR for stronger verification. The statistical challenge is in defining a procedure for the microarray filtering phase which achieves acceptable control of false positive and false negative rates.

13.3.2   Ranked lists

A simple approach to help find the differentially expressed genes is just to produce a ranking of genes. For example, genes could be ranked in decreasing order of the absolute value of their corresponding t-statistics. This should tend to group the genes most likely to be differentially expressed at or near the top of the list. In this way investigators can work their way down the list from the top.

To the extent that the statistics corresponding to different genes have different null hypothesis distributions, a smaller value of the statistics for one gene may actually be more significant than a larger one for another, and therefore the ranking procedure might not produce a ranking by actual significance of the statistic. To overcome this one can rank the genes by the P-values of the statistics instead of the values of the statistics themselves. The P-values assigned by any single gene method do not have the same meaning in a multiple gene context; however, the P-values can still give a meaningful ranking.

In order to estimate P-values one can apply the permutation test described in Section 3.8.1 and then rank genes by their P-values as given by this test. If there are not sufficiently many replicates, the ranking by permutation P-values could give a less meaningful ranking than by the statistics themselves. For example, if there are only four replicates for each condition, there are only $\binom{8}{4} = 70$ different permutation P-values possible, as opposed to a continuum of possible P-values of the statistic. Therefore replacing the statistic by its permutation P-value increases granularity.

We illustrate this with an example. Suppose that there are 7,000 genes on the array of which ten are differentially expressed, and suppose that they are the top ten members of the ranked list of genes, as ranked by the absolute value of the statistic. If there are four replicates per condition, so that there are 70 permutations, then the smallest possible permutation P-value is $1/70 \cong .01428$, and roughly $7000/70 = 100$ genes will have this P-value just by chance. The ten truly differentially expressed genes will be lost in the noise in such a P-value ranking. Therefore, one would like to have at least as many permutations as there are genes. For an array of 7,000 genes this would require at least eight replicates per condition. Fortunately the number of permutations increases rapidly with the number of replicates; for example ten replicates per condition already gives 184,756 permutations.

If we are willing to make parametric assumptions about the data, then we can obtain parametric P-values, as in Section 3.5.2. Parametric assumptions are rarely valid for gene expression data; however for those genes whose differential expression is extreme, any reasonable method should find them.

13.3.3 The Choice of Statistic

In the multiple-testing case, the test statistic is replaced by a vector of statistics, one for each row of the data matrix (13.3). Examples are the t-statistic, or the ratio of the respective averages of the two groups. The issue of finding an optimal statistic is more difficult in the multiple gene case, because what might be optimal for one row (gene) may not be optimal for another.

If for each gene the expression level intensities are normally distributed and have equal variances in the two conditions, then the t-statistic puts all tests on an equal footing. The gene with the higher t-statistic, in absolute value, is the gene with the more significant data. If the normality and equal variance assumptions do not hold, this might not be the case, and indeed these assumptions cannot reasonably be assumed in gene expression data (Grant et al. (2002)). However, the t-statistic is still widely used, based on the belief that it is robust to the assumptions.

Attempts have been made to adjust the t-statistic in order to make its distribution less dependent on the gene. Storey and Tibshirani (2003) introduce a correction factor into the denominator of the t-statistic. The data are put into 100 bins based on the value of the usual t-statistic denominator. The correction factor is then chosen to minimize how much the dispersion of the adjusted t-statistic varies from bin to bin. This is a heuristic step for which there is no supporting theory. Empirical studies show, however, that multiple testing results can be dramatically effected by this parameter.

Another issue which arises in using the t-statistic is that the denominator, which is essentially the pooled sample variance of the data, itself has very large variance. This is particularly true when there are few replicates, as is usually the case with microarray data. In order to have better estimates, it has become popular to pool together genes with intensities at similar levels, and use the data from all such genes to derive one variance estimate to be used as the denominator of the t-statistic for all of them. This has become known as "borrowing strength across genes." As above however, there is little theory to support this.

13.3.4 Confidence Measures

The ranking procedures described above are sufficient for some studies, however, it is generally preferred, particularly when there is no a priori knowledge about how many genes are differentially expressed, to produce sets of genes, where there is some kind of confidence measure imposed on the sets to control the number of non-differentially expressed genes they contain.

A simple example illustrates the difficulty in interpreting the gene-by-gene P-values in parallel. Suppose that we have an array with 10,000 genes, 100 of which are differentially expressed, and we predict that all genes whose P-value is less than .05 are differentially expressed. In this case, approximately $(.05) \times 9,900 = 495$ false predictions will be made. Therefore, even if the procedure successfully predicts all 100 truly differentially expressed genes, they will be lost among a set of approximately 500 false predictions. Even though we started with a gene-by-gene P-value of .05, we end up with a set of predictions with about 83% false positives. If we lower the P-value cut-off to the more stringent value .01, then there will still be approximately 50% false positives in the set of predictions. If there are only a few genes which are differentially expressed, then even a P-value cut-off of .001 can give poor results. Since it is unknown in advance how many genes are truly differentially expressed, one does not know how low to set the cutoff to a achieve a desired confidence. This is one form of the multiple testing problem.

Since there is a large number of genes on the arrays and generally few replicates available, this problem must be considered in some depth. Several books and review articles are devoted to the general statistical problem of multiple testing, for example Westfall and Young (1993) and Hochberg and Tamhane (1987). Much of the theory in these publications is not relevant for microarray analysis, and some of the theory that is relevant is very recent, or is not included in traditional treatments. Therefore we tailor our treatment of multiple testing theory to the microarray context, although many of the principles discussed are more widely applicable.

Null Hypotheses in the Multiple Gene Context: Strong versus Weak Control

In the single gene case the null hypothesis specifies equal distributions (or perhaps means) of the expression levels in the two experimental conditions, as described in Section 3.5.2. We will refer to the null hypotheses for the individual genes as the gene-wise null hypotheses. One natural extension to multiple genes is to declare equality in both conditions for *every* gene. This is called the *complete* null hypothesis, denoted H^0. But generally it is not this hypothesis which we are interested in accepting or rejecting, because it does not in itself indicate which genes are differentially expressed, and can only show that some gene is differentially expressed.

In the multiple gene case we are interested in procedures that predict which genes are differentially expressed, and in proving that the procedure controls the number of false-positives in some way. There are two prevailing approaches to defining false positive rates in a differential expression prediction procedure, depending on whether or not we are willing to accept any false positives at all. The first is the *Family-Wise Error Rate* (FWER), and the second is the *False Discovery Rate* (FDR). Essentially the FWER is the *probability* of having even one (or more) false positives in the predicted set, while the FDR is the *expected* proportion of the predicted set which consists of false predictions. Both will be discussed below, but these are the kinds of false positive rates that should be kept in mind in the following discussion. For any procedure, if the FDR or FWER is known to be less than α, then the procedure is said to control the FDR or FWER error rate to level α.

The data matrix (13.3) gives rise to a g-dimensional vector V of test statistics, the elements of which have some unknown joint probability distribution F. Suppose V^0 is the subvector of V consisting of the components of V for which the null hypotheses are true, and suppose F' is a probability distribution of V which agrees with F on V^0. Because the distributions F and F' agree on the true null hypotheses, if the FWER or FDR of a procedure is controlled to level α assuming the distribution F', then the procedure also controls the error rate to level α assuming the distribution F, the true distribution. We will refer to any F' which agrees with F on V^0 as a *pivotal distribution* for V with respect to V^0.

Most procedures that control the FWER and FDR start with gene-by-gene P-values and then adjust them in some way, and reject those hypotheses for which the adjusted P-values are less than some cutoff. The cutoff is chosen so as to guarantee a desired bound on the error rate. The gene-by-gene P-values are calculated under some pivotal distribution F'. If a procedure can only be proven to control the error rate when all hypotheses are null, i.e. when $V = V^0$ (so that $F' = F$), then the procedure is said to give *weak control* of the error rate. If it is not necessary that $V = V^0$ for the procedure to control the error rate, then the procedure is said to give *strong control*.

The above discussion only concerns the false-positive aspect of the procedure. In order to be able to find the true-positives, the distribution F' must be chosen judiciously so that the observed values of the test statistic for the false hypotheses will tend to be extreme. We will refer to any F' which agrees with F on V', and for which the test statistic for the false hypotheses will tend to be extreme, as a *pivotal null distribution* for V. We use the notation F^0 for a pivotal null distribution, however, it should be kept in mind that F^0 is a distribution of the whole vector V, and not just the sub-vector V_0.

The permutation distribution of V obtained by permuting the columns of the data matrix in all possible ways approximately equals F on V^0, and will tend to produce extreme P-values for the truly differentially expressed genes. Therefore the permutation distribution provides a distribution in which to do the P-value calculations. A bootstrap procedure will also give a distribution with these properties. In the permutation and bootstrap procedures we use the data itself to obtain an approximate F^0, and in practice such an approximation is what is usually used in place of a pivotal null distribution. Therefore it is important to keep in mind, when using a bootstrap or permutation distribution in place of a true pivotal distribution, that the procedure only approximately controls the error rate to the desired level.

It is conceivable that one might also obtain a pivotal null distribution in a parametric manner. We now consider a very simplified and artificial example designed to illustrate this concept. Consider two genes, g_1 and g_2 in a cell, and suppose that the means μ_1 and μ_2 determine the distribution of the vector of intensities (X_1, X_2) as multivariate normal distribution (2.33)

$$f_{\mathbf{X}}(\mathbf{x}) = \frac{1}{(2\pi)^{n/2}|\Sigma|^{1/2}} e^{-\frac{1}{2}(\mathbf{x}-\mu)'\Sigma^{-1}(\mathbf{x}-\mu)}, \quad -\infty < x_i < +\infty, \quad (13.5)$$

with variance–covariance matrix

$$\Sigma = \begin{bmatrix} 1 & 0 \\ 0 & 1 \end{bmatrix}.$$

Suppose that we have two populations of cells P_1 and P_2, with genes g_1 and g_2 having intensities (X_1, X_2) in P_1 and (X_1', X_2') in P_2. Suppose (X_1, X_2) has mean (μ_1, μ_2), and (X_1', X_2') has mean (μ_1', μ_2'). The means μ_1, μ_2, μ_1', μ_2' are unknown to us. The variance of X_1, X_2, X_1', and X_2' is 1, regardless of the values of $\mu_1, \mu_2, \mu_1', \mu_2'$. Therefore the t statistic of either component, under the hypothesis that $\mu_1 = \mu_1'$ and $\mu_2 = \mu_2'$, follows the t distribution. In this case F^0 can be taken to be a product of two independent t distributions (T_1, T_2). If $\mu_1 = \mu_1'$, but $\mu_2 \neq \mu_2'$, the t statistic of the first component still follows the t distribution.

Suppose, however, that the variance–covariance matrix is

$$\Sigma = \begin{bmatrix} \mu_2 & 0 \\ 0 & \mu_1 \end{bmatrix}.$$

Now the form of the distribution of the statistic at either gene depends on the hypothesis at the other gene. In this case, if $\mu_1 = \mu_1'$ and $\mu_2 \neq \mu_2'$, then the t statistic at the first component no longer follows the t distribution as the variainces are not equal in the two groups, while if $\mu_2 = \mu_2'$ then the variances are equal and it does follow the t distribution. Therefore if g_1 is not differentially expressed and g_2 is differentially expressed, then F^0 is not pivotal, in that the F^0 obtained by setting $\mu_1 = \mu_1'$ and $\mu_2 = \mu_2'$ does not agree with the true distribution F on V^0. This problem exists even though the statistics are independent.

In the multiple testing literature such as Westfall and Young (1993) there is a general condition imposed upon the parametric complete null hypothesis known as "subset pivotality," which guarantees F^0 agrees with the true distribution F on V^0. This hypothesis is needed to apply most of the multiple testing methods in Westfall and Young, including the Bonferroni correction. In the literature these methods are often applied to microarray data with the subset pivotality issue glossed over. However, since most methods implicity or explicitly use permutation distributions, this generally has not effected the conclucions.

13.3.5   The Family-Wise Error Rate (FWER)

Suppose that we plan to carry out g statistical tests of g null hypotheses H_0^j, $j = 1, \ldots, g$, and that g_0 of the null hypotheses are true. The *family-wise error rate (FWER)* is the probability that at least one of the g_0 true null hypotheses will be rejected. The multiple testing problem was introduced in Section 3.11; however, the methods discussed were not shown to give strong control of the FWER.

Single-Step Methods

The simplest single-step method, known as the Bonferroni correction, is as follows. Gene-by-gene P-values are generated using some single-gene method, as described above, and we denote the P-value for gene i by p_i. The p_i are then "corrected" by dividing by g, and the corrected P-value $p_i g$ is denoted by $\tilde{p}_i$. The procedure rejects all hypotheses where $\tilde{p}_i < \alpha$. This method guarantees weak control of the FWER at level α as shown in Section 3.11.

If the gene-by-gene P-values are obtained assuming a pivotal null distribution (for example, the permutation distribution as in described in Section 3.8.1), then the method does give strong control. A proof is as follows: Suppose for notational convenience that the true null hypotheses (the non-differentially expressed genes) are hypotheses $i = 1, \ldots, g_0$, and that the false null hypotheses (the differentially expressed genes) are hypotheses $i = g_0 + 1, \ldots, g$. Of course the value of g_0 is unknown to us. Let F be the true distribution of the vector of statistics, and F^0 the pivotal null distribution of the vector under which the P-values are computed. Denote by H_0^i the hypothesis that gene i is not differentially expressed. Then at least one H_0^i, $i \leq g_0$, is rejected if

$$\min_{i \leq g_0} p_i g \leq \alpha.$$

We want to show that

$$\text{Prob}(\min_{i \leq g_0} p_i g \leq \alpha | F) < \alpha.$$

This probability is

$$\text{Prob}(\min_{i \leq g_0} p_i g \leq \alpha | F^0)$$

since F is equal to F^0 when $i \leq g_0$. Further,

$$\text{Prob}(\min_{i \leq g_0} p_i g \leq \alpha | F^0) < \text{Prob}(\min_i p_i g \leq \alpha | F^0),$$

which is less than α, by the theory in Section 3.11, because F^0 is exactly the distribution under which the P-values p_i were calculated. This completes the proof.

The Bonferroni correction described above is one of a family of corrections known as *"one-step"* methods, due to the fact that all of the P-values undergo the same equivalent correction, in this case multiplication by g. If g is very large, as is often the case for the analysis of microarray data, the Bonferroni correction will give a very conservative cutoff for significance.

In the case where all of the P-values can be assumed to be independent, a less conservative method can be applied. In this case

$$\text{Prob}(\min_i p_i \leq \alpha | F^0) = \text{Prob}(g(1-x)^{g-1} \leq \alpha | F^0),$$

by equation (2.88). In this one-step procedure, each of the P-values is compared to the Šidák significance point $K(g, \alpha)$, defined from equation (2.141), with g replacing n, as

$$K(g, \alpha) = 1 - \sqrt[g]{1 - \alpha}. \tag{13.6}$$

Those null hypotheses for cases in which the P-value is less than this significance point are rejected, while the remainder are not rejected. Strong control for this method can be established in a similar way to the Bonferroni correction.

A convenient way of formulating the above is in terms of *adjusted P-values*. In the Bonferroni method, for each i, define $\tilde{p}_i$ to be $p_i g$. In the case of independence and the use of the Šidák method, define $\tilde{p}_i = 1 - (1 - p_i)^g$.

Then in both cases the FWER is

$$\text{Prob}(\min_{i \le g_0} \tilde{p}_i < \alpha | F^0).$$

The formulation in terms of adjusted P-values is convenient, and we shall use them again below. However the terminology should not be taken to indicate that $\tilde{p}_i$ is an actual probability, in contrast to unadjusted P-values.

Step-Down Methods

In this section we describe the step-down methods of Westfall and Young (1993) for controlling the FWER at some chosen value α. These methods are less conservative than the one-step methods of the previous section. In these procedures different adjustments are made to the P-values of different genes. We start with the step-down version of the Bonferroni method.

First, the observed P-values are written in increasing order as $p_{(1)}$, $p_{(2)}, \ldots, p_{(g)}$, so that $p_{(j)} \le p_{(k)}$ if $j < k$. The various null hypotheses are then correspondingly ordered as $H_0^{(1)}$, $H_0^{(2)}, \ldots, H_0^{(g)}$. This ordering is fixed throughout the procedure, and the P-values are examined in turn, according to this ordering.

The initial step in the step-down procedure is to compare the smallest observed P-value $p_{(1)}$ with α/g, where α is the chosen FWER. If $p_{(1)} > \alpha/g$, then all null hypotheses are accepted and no further testing is done. If $p_{(1)} < \alpha/g$, the null hypothesis $H_0^{(1)}$ is rejected and the sequential procedure moves to step 2.

In step 2, if reached, $p_{(2)}$ is compared to $\alpha/(g-1)$. If $p_{(2)} > \alpha/(g-1)$, the null hypotheses $H_0^{(2)}$, $H_0^{(3)}, \ldots, H_0^{(g)}$ are all accepted and no further testing is done. If $p_{(2)} < \alpha/(g-1)$, the null hypothesis $H_0^{(2)}$ is rejected and the step-down procedure moves to step 3.

In general, if step j is reached, then if $p_{(j)} > \alpha/(g-j+1)$, then all further hypotheses are accepted and no further testing is done. If $p_{(j)} < \alpha/(g-j+1)$, hypothesis $H_0^{(j)}$ is rejected and the procedure moves to step $j+1$.

The rule at step 1 ensures that the FWER is controlled at the level α in the weak sense. The event that the smallest P-value is less than α/g is equivalent to the event that at least one P-value is less than α/g. If all null hypotheses are true, the probability of the latter event has been shown to be α.

To show that this method gives strong control of the FWER is more difficult. Assume as before that the true null hypotheses are hypotheses $i = 1, \ldots, g_0$, and that the false null hypotheses are hypotheses $i = g_0+1, \ldots, g$. Let F be the true distribution of the vector of test statistics, and F^0 the pivotal null distribution under which the P-values are calculated.

Suppose that $p_{i'} = \min_{i \le g_0} p_i$. Then $p_{i'} = p_{(j)}$, for some j necessarily less than g_0. Then in order for at least one H_0^i, $i \le g_0$ to be rejected, it must be that $p_{i'} \le \alpha/(g - j + 1)$ because the jth step is the first step where a true null hypothesis is considered in the procedure. Now

$$\text{Prob}(\min_{i \le g_0} p_i \le \alpha/(g - j + 1) \,|\, F)$$

is approximated by

$$\text{Prob}(\min_{i \le g_0} p_i \le \alpha/(g - j + 1) \,|\, F^0)$$

and this is less or equal to

$$\text{Prob}(\min_{i \le g_0} p_i \le \alpha/(g - g_0 + 1) \,|\, F^0),$$

which amounts to a single-step adjustment which has been shown to be less than α.

The condition (that $p_{(1)} > \alpha/g$) determining whether all null hypotheses are accepted is identical in both the one-step approach and the step-down approach. However, given that $H_0^{(1)}$ is rejected, the condition ($p_{(2)} < \alpha/(g - 1)$) that $H_0^{(2)}$ is rejected is less stringent in the step-down procedure than the corresponding condition ($p_{(2)} < \alpha/g$) in the one-step procedure. A similar remark holds for all further hypotheses. This observation justifies the claim for the increased power of the step-down procedure.

This is not to say that the step-down method has uniformly better properties than the one-step method. For example, when all null hypotheses are true, the probability that a specified null hypothesis is accepted under the one-step approach is larger than the corresponding probability under the step-down approach. This observation is in line with the fact that in statistics it is seldom the case that one procedure is better than another in all respects: Under some restrictions, or following some criteria, one procedure will be better, while under other restrictions, or following other criteria, the other will be better.

In spite of the increased power obtained by replacing a single-step Bonferroni correction by a step-down Bonferroni correction, the step-down procedure is generally still overly conservative. In the context of gene expression data, where g is so large, this is an issue. A less conservative alternative to the Bonferroni correction is needed.

We now consider an approach known as the minP method. We start with the single-step version of the procedure, and then generalize to the step-down version. For gene g_i, let T_i be the t-statistic and p_i the P-value of the observed value t_i under some pivotal null distribution F^0. A less conservative one-step adjustment than the Bonferroni is

$$\tilde{p}_j = \text{Prob}(\min_i P_i < p_j | F^0).$$

The procedure is to reject exactly those hypotheses H_0^i for which $\tilde{p}_i < \alpha$. It is evident directly from the definition of $\tilde{p}_i$ that this procedure controls the FWER in the weak sense. The verification that it also controls the FWER in the strong sense is similar to the proof on page 454 that the one-step Bonferroni method gives strong control.

To turn this into a step-down procedure, we rank the P-values as $p_{(1)}, \ldots, p_{(g)}$, where $p_{(1)}$ is the smallest P-value. In this case adjusted P-values are defined as follows:

$$\tilde{p}_{(1)} = \mathrm{Prob}(\min_{i \geq 1} P_i < p_{(1)} | F^0)$$

and

$$\tilde{p}_{(j)} = \max\{\tilde{p}_{(j-1)}, \mathrm{Prob}(\min_{i \geq j} P_i < p_{(j)} | F^0)\}, \quad \text{for } j \geq 2. \qquad (13.7)$$

The reason for taking successive maxima is to maintain the same order of adjusted and unadjusted P-values.

Rejection of those null hypotheses H_0^i for which $\tilde{p}_i < \alpha$ gives strong control of the FWER at level α. The proof of this claim is similar to that for the step-down Bonferroni method. For the algorithm describing how to actually calculate these adjusted P-values, see Ge et al. (2003).

There is some choice in how the P-values were adjusted above. The minimum in the probability term in (13.7) is calculated over a fixed set of indices $i \geq j$. One might do the calculation instead with the distributions of the true jth minimum. If the P-values are adjusted in this way, which might seem more natural, then strong control of the FWER actually no longer holds.

A further important caveat applying to all of the FWER methods above is that there are generally very few replicates available per condition, often as few as three or four. In this case it is unreasonable to use permutation P-values because of increased granularity, as discussed in Section 13.3.2. For example, if there are only three replicates per condition, there are only 20 permutations, and thus continuous t-statistic data will have been forced into 20 significance levels. If there are 100 differentially expressed genes out of a total of 10,000 on an array, there may be roughly 500 genes in the highest significance bracket, just by chance, obscuring the true 100 regardless of any procedure that works strictly with the P-values. Therefore in the case of few replicates, an alternative to the P-value approach is needed. In this case it is preferable to work directly with the t-statistics. They are ordered just as the P-values were, but now the most significant is the largest in absolute value. In this case we estimate

$$\mathrm{Prob}(\max_{i \leq j} |T_i| > |t_{(j)}| \,|\, F^0). \qquad (13.8)$$

This procedure, known as the maxT approach, will also give strong control of the FWER as well as the P-value approach, and preserves the granularity of the data. However, this method has the disadvantage that it loses some power to find those genes whose statistics have less variable distributions.

13.3.6   The False Discovery Rate (FDR)

Even when step-down methods are used, insisting on an FWER of the order of 5% can lead to quite conservative results. We might try raising the FWER if we are willing to accept *some* false positives, as is generally the case in practice. A problem with FWER methods, however, is that once they allow false positives, there is no control over how many they allow, because what is controlled is merely the probability that the number of false positives is greater than zero.

Because of this, the false discovery rate (FDR) approaches to multiple testing, introduced by Benjamini and Hochberg (1995), have gained popularity in the analysis of microarray data. This approach does not attempt to control the FWER; instead, it controls the proportion of false positives among the set of all genes predicted to be differentially expressed. If there are 10,000 genes on the array and 100 are differentially expressed, the researcher might tolerate a quite high FDR, since even an FDR of 50% would enrich the set of candidate differentially expressed genes from one true positive in 100 to 50 true positive in 100, giving a set of predictions appropriate for PCR follow-up verification.

Consider those genes for which the null hypothesis is rejected. Let V be the number of these genes for which the null hypothesis is true (and is thus falsely rejected) and S the number for which the null hypothesis is false (and is thus correctly rejected). Let $R = V + S$. The quantity Q is defined as

$$Q = \begin{cases} 0, & \text{if } V = R = 0 \\ V/R, & \text{if } R > 0. \end{cases} \tag{13.9}$$

Thus V, S, R, and Q are random variables, and even after the experiment is completed the value of Q is unknown to us. The original definition of the FDR is the expected value $E(Q)$ (Benjamini and Hochberg (1995)). In its simplest formulation, the aim of an FDR procedure is to control this expected value $E(Q)$ to be some prescribed value α (or less).

We note two properties of the FDR. First, if the complete null hypothesis is true, then $V = R$ so that $Q = 1$ whenever $R > 0$. The expected value $E(Q)$ in this case is Prob $(R > 0)$, which is just the FWER. Second, if some null hypotheses are not true, the FDR is less than the FWER.

It is important to keep in mind that an FDR is fundamentally different from a P-value, and is used for a very different purpose. A P-value is generally used to assess the significance of data. Published data claiming to be significant must have rigorously defined P-values achieving some acceptable significance level, generally .05 or .01. On the other hand, an FDR is generally used as a culling tool. As such, an FDR of .5 might be perfectly acceptable to the investigator, and it will not be problematic if the true FDR is only approximately .5. For example if the investigator chooses a method which is supposed to achieve an FDR of .1, and yet realizes an FDR of only .3, in practice that will generally be good enough for their purposes. Therefore, there is some latitude in the assumptions that can be made in FDR methods. That being said, it is preferable to find the best method possible to accurately control the FDR.

We start by describing the original procedure developed to control the FDR. We will then discuss some shortcomings of this procedure and of the definition of FDR itself, and then will investigate some alternatives.

Benjamini and Hochberg Step-Up Methods

The original method proposed to control the FDR was given in Benjamini and Hochberg (1995). This method is similar to FWER methods in that it starts with gene-by-gene P-values derived from g tests, under g individual null hypotheses, one for each gene. It is assumed that these P-values are independent. This independence assumption is one problem in applying this method to gene expression data.

By the definition of a P-value, for those tests where the null hypothesis is true, the individual P-values have uniform distribution. Let g_0 be the number of tests for which the null hypothesis is true and $g_1 = g - g_0$ be the number of tests for which the null hypothesis is false. Which hypotheses are true and which are false is unknown, and the value of g_0 is unknown as well. Let $H_1, H_2, \ldots, H_{g_0}$ be the true null hypotheses and $H_{g_0+1}, H_{g_0+2}, \ldots, H_g$ the false null hypotheses.

The g hypothesis tests will result in individual test P-values $P_1, P_2, \ldots, P_g$ respectively, which by assumption are independent. Each P_i is a random variable and we denote the observed value of P_i by p_i. Let $P_{(i)}$ be the ith smallest of the P_i (so that $P_{(1)} \leq P_{(2)} \leq \cdots \leq P_{(g)}$) and let $H_{(i)}$ be the hypothesis corresponding to $P_{(i)}$.

Let $q_i = \frac{i}{g}\alpha$, $i = 1, \ldots, g$, where α is the desired false discovery rate, and k be the maximum i such that $p_{(i)} \leq q_i$, where $p_{(i)}$ is the observed value of $P_{(i)}$. The Benjamini-Hochberg testing procedure is as follows. If there is no value i such that $p_{(i)} \leq q_i$ we accept all g null hypotheses. If $k \geq 1$ we reject the null hypotheses $H_{(1)}, \ldots, H_{(k)}$ and accept all others. Note that by the definition of k, there still may be $i' < k$ such that $p_{(i')} > q_{i'}$

With these preliminaries in place and Q defined as in (13.9), we have the following:

Theorem. $E(Q) = \frac{g_0}{g}\alpha.$

The expected value on the left is calculated under the true distribution of the data and not under any kind of complete null hypothesis, and therefore the theorem states that the FDR is controlled in the strong sense. Accordingly, the proof must show that $E(Q) \leq \alpha$ regardless of the value of g_0.

Proof of the theorem.

As above we assume that the hypotheses are ordered so that hypotheses $H_1, H_2, \ldots, H_{g_0}$ are true and $H_{g_0+1}, H_{g_0+2}, \ldots, H_g$ are false. Let $A_{v,s}$ be the event that exactly v true and s false null hypotheses are rejected. Then

$$E(Q) = \sum_{s=0}^{g_1} \sum_{v=1}^{g_0} \frac{v}{v+s} \operatorname{Prob}(A_{v,s}).$$

(If $v = 0$ then $Q = 0$, so that the inner sum starts from 1.)

For any i ($i = 1, 2, \ldots, g_0$) let $A_{v,s}^{(i)}$ be the event that H_i is rejected and that, in total, v true and s false null hypotheses are rejected. Then in the sum

$$\sum_{i=1}^{g_0} \operatorname{Prob}(A_{v,s}^{(i)}),$$

the probability of any particular combination of v rejected true null hypotheses occurs exactly v times, one for each choice of i, ($i = 1, 2, \ldots, g_0$). It follows that

$$\operatorname{Prob}(A_{v,s}) = \frac{1}{v} \sum_{i=1}^{g_0} \operatorname{Prob}(A_{v,s}^{(i)}). \tag{13.10}$$

From this,

$$E(Q) = \sum_{s=0}^{g_1} \sum_{v=1}^{g_0} \frac{v}{v+s} \frac{1}{v} \sum_{i=1}^{g_0} \operatorname{Prob}(A_{v,s}^{(i)})$$

$$= \sum_{i=1}^{g_0} \sum_{s=0}^{g_1} \sum_{v=1}^{g_0} \frac{1}{v+s} \operatorname{Prob}(A_{v,s}^{(i)}).$$

For any fixed $k = v + s$, this gives

$$E(Q) = \sum_{i=1}^{g_0} \sum_{k=1}^{g} \frac{1}{k} \operatorname{Prob}(A_k^{(i)}).$$

The event $A_k^{(i)}$ is the intersection of the events $B_{i,k}$ (that H_i is rejected) and $C_k^{(i)}$ (that of the remaining $g - 1$ hypotheses other than H_i, exactly $k - 1$ are rejected). Since the g tests are independent, the events $B_{i,k}$ and

$C_k^{(i)}$ are independent. Therefore

$$E(Q) = \sum_{i=1}^{g_0} \sum_{k=1}^{g} \frac{1}{k} \operatorname{Prob}(B_{i,k}) \cdot \operatorname{Prob}(C_k^{(i)})$$

$$= \sum_{i=1}^{g_0} \sum_{k=1}^{g} \frac{1}{k} \operatorname{Prob}(P_i \le q_k) \cdot \operatorname{Prob}(C_k^{(i)})$$

$$= \sum_{i=1}^{g_0} \sum_{k=1}^{g} \frac{1}{k} \cdot \frac{k\alpha}{g} \operatorname{Prob}(C_k^{(i)}),$$

the last equality holding because P_i is has a uniform distribution in $[0,1]$ for $i \le g_0$. Therefore

$$E(Q) = \sum_{i=1}^{g_0} \frac{\alpha}{g} \sum_{k=1}^{g} \operatorname{Prob}(C_k^{(i)})$$

or

$$E(Q) = \frac{g_0}{g} \alpha \sum_{k=1}^{g} \operatorname{Prob}(C_k^{(i)}).$$

But from the law of total probability, $\sum_{k=1}^{g} \operatorname{Prob}(C_k^{(i)}) = 1$. Therefore

$$E(Q) = \frac{g_0}{g} \alpha,$$

as was to be shown.   □

The above proof, like the original proof of Benjamini and Hochberg (1995), assumes that the g different tests are independent. In the microarray case this assumption is not reasonable. Benjamini and Yekutieli (2001) have extended the proof to the case of positive regression dependence between the tests. In the case of an arbitrary dependence structure, they show that the theorem holds (with FDR α) provided that q_i is redefined as

$$q_i = \frac{\alpha i}{g \sum_{j=1}^{g} \frac{1}{j}}.$$

Unfortunately there is not much evidence about what form of dependence there is between the expression levels of different genes. Therefore, making any assumptions about the form of dependence has unknown consequences.

From the proof above it can be seen that the FDR is actually controlled to the level $g_0 \alpha / g$, which is at most α.

The false discovery rate FDR defined above may be written as

$$E(V/R \mid R > 0) \operatorname{Prob}(R > 0).$$

Storey (2002) introduced the "positive false discovery rate," denoted pFDR, defined by

$$\text{pFDR} = E(V/R \mid R > 0).$$

This differs from the FDR by not including the term $\text{Prob}(R > 0)$, but by conditioning on the event $R > 0$ instead. Storey indicates various advantages of the pFDR over the FDR. If $\text{Prob}(R > 0)$ is not close to one, then these can be somewhat different. Given the latitude in acceptable FDR values, as discussed on page 459, and the fact that if $R > 0$ then generally $\text{Prob}(R > 0)$ is not small, this is not a grave issue, so we shall not investigate the pFDR further.

Permutation Methods

The Benjamini and Hochberg (1995) method described in Section 13.3.6 for controlling the FDR has several drawbacks. First, it depends on a knowledge of the various P-values associated with the various genes, and if a parametric test is used, these P-values are not usually known with high accuracy, since standard assumptions for parametric tests, for example t tests, seldom apply for microarray data. Second, the procedure assumes that the various tests are independent, and this is in effect never the case for microarray data. Next, the true FDR of the Benjamini and Hochberg (1995) procedure is not known, since the procedure controls the mean value of Q (defined in (13.9)) to be less than or equal to $g_0\alpha/g$, where α is the desired FDR and g_0 is the number of genes for which the null hypothesis is true. Since the value of the ratio g_0/g is not known, the Benjamini and Hochberg method controls the FDR at a value less than or equal to α, but does not estimate it. We now discuss how these problems might be addressed by using permutation methods, where we assume that the permutation is carried out using entire columns of the data matrix (13.3). Practical and theoretical aspects of the two-sample permutation procedure are discussed in Section 3.8.1, and we draw on these in discussing some permutation approaches to the problems listed above.

The permutation test for differential expression for one gene was discussed in Section 3.8.1 and its use for many genes was referred to briefly in Sections 13.3.2 and 13.3.3. The permutation P-value associated with the observed value of the test statistic is easily calculated from the permutation procedure. The problem of dependence between genes will be overcome by the permutation of entire columns of the data matrix. We now focus on the problem of the estimation of the FDR.

If it is assumed that $R > 0$, the FDR is defined as the mean of the ratio V/R. Both V and R are random variables and this ratio is also a random variable. Because the distribution of gene expression levels do not follow standard probability distributions, we cannot find the mean of V/R using parametric assumptions.

Since permutation of the columns of the data matrix (13.3) will tend to eliminate differential expression, the mean value of V can be estimated from the $P = \binom{m+n}{n}$ permutations of the data. Note that it is not enough just to take the mean of the permutation distribution of V, however with some adjustments an estimate of $E(V)$ can be obtained from this distrubution. On the other hand, R is simply the total number of genes declared to show significant differential expression, including both true and false positives, and thus the probability distribution of R cannot be estimated by the permutation procedure. Although bootstrap resampling can give an estimate of the distribution of R, neither permutation nor bootstrap methods can help us in estimating the distribution of V/R, and hence its mean. The values of V and R are correlated and the correlation between them is not known.

We therefore abandon the aim of estimating the mean of V/R and instead consider $E(V)/R$. The mean of V can be (roughly) estimated by permutations, as described above, and the value of R is known from our testing procedure. At the moment, this appears to be all that is possible in a permutation approach to estimating a quantity similar to the FDR.

We now describe a popular permutation method for estimating the FDR. This is the so-called statistical analysis of microarrays (SAM) approach, proposed by Tusher et al. (2001) and further developed by Storey and Tibshirani (2003). The procedure is similar to that described above but differs from it in several ways, discussed below.

The Tusher et al. (2001) SAM procedure starts with the data matrix (13.3). It then computes, for gene $i = 1, 2, \ldots, g$, the t-like statistic

$$d(i) = \frac{\bar{x}_i - \bar{y}_i}{s(i) + s_0}, \tag{13.11}$$

with $s(i)$ defined as in (3.34) and with a change in notation of s to $s(i)$, x_{1i} to x_i and x_{2i} to y_i. If $s_0 = 0$, $d(i)$ would be identical to a t statistic. The positive quantity s_0 is added to the t statistic denominator for reasons discussed below. The g genes are then ranked according to their respective $d(\cdot)$ values, and the notation is changed so that the gene with the largest $d(\cdot)$ is now called gene 1, the gene with the second largest $d(\cdot)$ value is now called gene 2, and so on.

The first step of the procedure is to assess which genes exhibit a significant difference between the groups compared. To do this, all $P = \binom{m+n}{n}$ permutations of the data are considered, with entire columns in the data matrix (13.3) being permuted together. (If m and n are large, a random sample of such permutations might be used.) The original data make up permutation 1. For each permutation $p = 1, 2, \ldots, P$ and for each gene $i = 1, 2, \ldots, g$ the values $d_p(i), p = 1, 2, \ldots, P$, are calculated following the prescription in (13.11). With this notation $d(i)$ would be written $d_1(i)$.

For any permutation p the genes are ranked as above, that is according to their respective $d_p(\cdot)$ values, with now the gene with the largest value of $d_p(\cdot)$ ranked first, the gene with the second-largest value of $d_p(\cdot)$ ranked second, and so on. The various $d_p(i)$ values now form a matrix. The entries in row i of this matrix, apart from the first entry, do not necessarily correspond to gene i, and often will not. Instead, they correspond, in permutation column $p = 2, 3, \ldots, P$ to whatever gene led to the ith largest $d_p(i)$ value in that column.

The average $d_E(i)$ of the entries in the ith row is then computed, and the difference $d(i) - d_E(i)$ is calculated for each gene.

A threshold value Δ is now chosen, and gene i is declared to be significantly differentially expressed if $|d(i) - d_E(i)| > \Delta$. Thus gene i is declared to be significantly differentially expressed if the value $d(i)$ corresponding to it differs sufficiently from the permutation estimate of the mean of the ith most differentially expressed gene. A small value of Δ leads to a large number of false positives and a large value of Δ to a small number of false positives. However the focus in the procedure is on the FDR and not on controlling the number of false positives.

The next aim is to estimate the FDR associated with this procedure. Suppose that the largest negative value of $d(i)$ among genes declared significantly differentially expressed is denoted by a and the smallest positive value of $d(i)$ among genes declared significantly differentially expressed is denoted by b. The SAM procedure then calculates, for permutation $p = 1, 2, \ldots, P$, the number of genes with a value of $d_p(\cdot)$ less than a or greater than b. The average over all permutations of the numbers so calculated is then found. The FDR is then estimated as the ratio of this average divided by the actual number R of genes declared significantly expressed. This procedure can be carried out for a range of values of Δ and a subjective choice made of a reasonable choice of Δ based on the array of FDR estimates generated by a variety of choices of Δ.

The procedure as described so far leads to an upward bias in the estimation of the FDR, since the null hypothesis is presumably not true for some genes, and yet the permutation procedure in effect makes all genes null. Storey and Tibshirani (2003) address this problem. Suppose that the proportion g_0/g of genes for which the null hypothesis is true is denoted by π_0. If π_0 were known, an improved estimate of the FDR is found by multiplying the estimate described above by π_0. However π_0 is not known, and Storey and Tibshirani estimate it as follows. Consider some small positive constant a (Storey and Tibshirani choose $a = 0.15$,) and let n_1 be the number of genes whose $d(\cdot)$ value lies in the range $(-a, +a)$. Those genes for which the null hypothesis is true will be more abundant in this range than genes for which the null hypothesis is not true. For every permutation of the data there will be some genes whose $d(\cdot)$ value lies in the range $(-a, +a)$. Define n_2 as the average (over all permutations) of the number of genes for which this is the case. Then it is reasonable to estimate π_0 by n_1/n_2.

The actual SAM approach to estimating the FDR has differences from that described above, and we now discuss some of these differences.

The SAM approach uses as test statistic the quantity $d(i)$ defined in (13.11), specifically in including the quantity s_0 in the denominator of what would otherwise be a t statistic. Since the values of the $d(i)$ are to be compared across all genes, it is considered desirable that the probability distribution of $d(i)$ should be independent of i. The value of s_0 is chosen so as minimize the estimated coefficient of variation of the $d(i)$'s. This procedure can at best be described as a reasonable heuristic. It cannot ensure that the variance of $d(i)$ is independent of gene expression.

The value of the above procedure for choosing s_0 is can be questioned in that what is of primary interest is the power of the method, i.e the number of differentially expressed genes it finds for a given FDR, and currently there is no theorem proving that this value of s_0 achieves maximum power, even approximately. In fact examples can be constructed for which the value of s_0 chosen is quite far from optimal, for example when the null genes have highly bimodal distributions.

One might therefore consider not using the s_0 parameter at all. However, it turns out that the method is very sensitive to this parameter, particularly when there are only a few replicates per condition. The reason for this is that s_0 dominates the denominator of $d(i)$ when $s(i)$ is relatively small compared to it. This has two effects. First, since there are generally many null genes, there are many null genes whose $s(i)$ is vanishingly small, by chance, and for these genes the t statistics blow up. The larger s_0 is, the more this effect will be mitigated. On the other hand, when s_0 is too large, the non-null genes with small $s(i)$ tend to get lost in the noise. Therefore, what value to set s_0 to depends on the nature of the differentially expressed genes as well as the non-differentially expressed genes. A moderate value of s_0 is usually optimal for finding the greatest number of differentially expressed genes.

There is no known formula that can be applied to the data matrix to determine the value of s_0 which maximizes the power. However, a power criterion to determine s_0 would be desirable.

The method of estimation of V in the SAM procedure differs from that described above, in which the expected value of V for any gene is estimated by data values solely for that gene. In the SAM approach this expected value is estimated pooling data from many other genes. This again is a heuristic with potentially unexpected consequences.

Relation Between the FWER Step-Down and FDR Step-Up Methods

The step-down procedure using Šidák significance points compares the jth smallest P-value $p_{(j)}$ with the jth Šidák significance point $K(g-j+1, \alpha)$. We reject null hypotheses $H_0^{(1)}, H_0^{(2)}, \ldots, H_0^{(m)}$ and accept all remaining null hypotheses if $p_{(j)} < K(g - j + 1, \alpha)$, $j = 1, 2, \ldots, m$ and $p_{(m+1)} > K(g - m, \alpha)$. This procedure yields an FWER of α.

In the FDR procedure we reject null hypotheses $H_0^{(1)}, H_0^{(2)}, \ldots, H_0^{(m)}$ and accept all remaining null hypotheses if $p_{(j)} < \frac{j\alpha}{g}$ and $p_{(m+1)} > \frac{(m+1)\alpha}{g}$. This procedure leads to an FDR of at most α.

To compare the two procedures we note that for large g, small j, $K(g - j + 1, \alpha) \cong \alpha/(g - j + 1)$. Thus to a close approximation the comparison of the properties of the two approaches reduces to a comparison of

$$\alpha/(g - j + 1) \text{ and } j\alpha/g.$$

For large g, small j, the second expression is about j times larger than the first. This implies that, with the same data in the two cases, the step-down procedure will reject fewer, perhaps far fewer, null hypotheses than will the FDR procedure. This is as we expect: the step-down approach is intended to be far more stringent than the FDR approach. The FDR approach trades a higher family-wide error rate for increased power.

13.3.7 The ANOVA Approach: Many Genes

In this section we outline some features of the ANOVA approach to the analysis of microarray data, focusing on the spotted array technique. The analysis of Kerr et al. (2000), may be taken as an example of the use of the ANOVA technique, so we refer to this analysis frequently below, using it to illustrate the advantages and disadvantages of the ANOVA approach.

For reasons discussed below, the logarithm of any expression level, rather than the expression level itself, is generally used in the ANOVA analysis. Thus to discuss ANOVA models of spotted microarray data we follow this practice and denote the logarithm of any expression level generically by X. Several ANOVA models, of a greater or lesser complexity, have been proposed for the analysis of microarray data. The model considered by Kerr et al. (2000) is

$$X_{ijkg} = \mu + A_i + \delta_j + \tau_k + \gamma_g + B_{ig} + \psi_{kg} + E_{ijkg}. \tag{13.12}$$

Here X_{ijkg} is the logarithm of a a gene expression level for array i, dye j, tissue type k, and gene g. On the right-hand side in (13.12), A_i is a random effect due to array i, δ_j is a fixed effect due to dye j, τ_k is a fixed effect due to tissue type k, γ_g is a fixed effect due to gene g while B_{ig}, and ψ_{kg} are (random) array×gene and (fixed) tissue×gene interactions respectively.[1] A slightly different model is discussed in Kerr and Churchill (2001).

Kerr et al. (2000) focus attention on the gene×tissue interaction, since if the F ratio testing for this interaction is significant, we have significant

---

[1]Our notation differs from that of Kerr et al. (2000) in that we denote fixed effects parameters with greek characters and random effects parameters with roman.

evidence that some genes are expressed differently in different tissues. Further enquiry is needed to determine which genes these are. One drawback of this approach is that interaction effects are difficult to determine in any ANOVA unless the interaction is quite strong, as discussed in Section 9.5.4.

To what extent does an microarray model of the form (13.12) satisfy ANOVA assumptions? The broad requirements for any ANOVA analysis were discussed in Section 9.5. Some of these were that the random variables involved each has a normal distribution, that the variances of all random variables be the same, that the various random variables involved be independent, and that the ANOVA linearity modeling assumption is acceptable.

It is generally accepted that microarray expression levels do not in practice have a normal distribution, and in the ANOVA approach the logarithm of expression levels are routinely used in the analysis. It is thus assumed in a formal ANOVA using F distribution tables that these logarithms have a normal distribution, at least to a sufficiently close approximation. On the other hand, non-parametric methods are possible which do not rely on the normal distribution assumption.

The logarithmic transformation does not however ensure that the variances of all expression levels considered will be equal. The case of unequal variances in ANOVA has been discussed extensively in the literature, the general conclusion being that when the variances are not extremely different from each other, significance levels found formally from F tables will differ only mildly from the true values.

When many genes are considered simultaneously the question of the independence assumption of gene action arises. Groups of genes sometimes act in a concerted fashion, and thus their expression levels will not be independent.

Next, the linearity assumption of the mean of the logarithm expression levels, as is assumed for example in the model (13.12), implies that a multiplicative model for the expression levels themselves is assumed, with the array, the dye, the tissue type, the gene and the various interactions all acting multiplicatively on expression level. This may or may not be a reasonable assumption.

Finally, various possible terms are not included on the right-hand side of (13.12). These include a number of interaction terms, as well as terms for primary effects; for example, no print-tip effects are included. Also the array $\times$ tissue interaction in the model (13.4) of Wolfinger et al. (2001) is not included in the model. The exclusion of various interaction terms is discussed below.

Clearly the use of an ANOVA model runs the risk of making implicit assumptions that might not be correct. Against this, ANOVA models are often rather robust against departures from assumption, as discussed in Section 9.5.3, and further, it is possible to test whether some of the assumptions made are reasonable.

An experimental design allowing for all possible combinations of dyes, tissue types, arrays, and genes might be impractically large. Kerr et al. (2000) discuss a design involving two arrays, two dyes (red and green), two tissue types (liver and muscle), and 1,286 genes. In array 1 liver tissue is labelled with red dye and muscle tissue with green dye, while in array 2 muscle tissue is labelled with red dye and liver with green. This *dye-swap* experiment is illustrated, for any one gene, in Table 1 of Kerr et al. (2000).

In this experiment only four of the possible eight combinations, namely {liver, red dye, array 1}, {liver, green dye, array 2}, {muscle, red dye, array 2}, and {muscle, green dye, array 1}, are carried out. Kerr et al. refer to this design as a (2×2) Latin square, but a 2×2 Latin square has no degrees of freedom for the estimation of error (Cochran and Cox (1957)), so we prefer to think of this design as being a fractional replication (see Section 9.5.6). If we denote array 1 by a, the red dye by b and liver by c, the only treatments carried out are abc, a, b, and c, and this is precisely the fractional replication design in Section 9.5.6. The defining contrast is ABC and so this design implies the aliasing of tissue with the array $\times$ dye interaction, of arrays with the tissue $\times$ dye interaction, and of dyes with the array $\times$ tissue interaction. These three interactions are all assumed to be zero and do not appear explicitly in the model (13.12). Thus it is assumed, for example, that all the variation due to the aliased effects of tissues and the array $\times$ dye interaction is due solely to differences between tissues. If then there had been a substantial array $\times$ dye interaction, this would lead to an inflated estimate of the tissues effect.

The ANOVA design for the experiment discussed by Kerr et al. (2000) involves 1,286 genes, with the design of the above table used for each gene. This implies a total of 5,144 observations and thus 5,143 degrees of freedom. It has a sum of squares for arrays, for dyes and for tissue types, each having one degree of freedom, a sum of squares for genes having 1,285 degrees of freedom, a sum of squares for the array $\times$ gene interaction having 1,285 degrees of freedom and a sum of squares for the tissue $\times$ gene interaction with 1,285 degrees of freedom. The remaining sum of squares, with 1,285 degrees of freedom, is treated as residual, or error. This implies that the dye $\times$ gene interaction sum of squares is part of the error sum of squares, and thus this interaction is assumed to be zero, while the assumption that the interactions referred to in the previous paragraph are all zero implies that all other sums of squares not listed above can be treated as treated as error. Kerr et al. (2000) focus on the F ratios for genes and tissue $\times$ gene interaction and find significant values for each. This implies that different genes are expressed significantly overall, and that some genes are expressed significantly more in some tissues than in others. They then conduct further tests to assess which genes are differentially expressed.

Although Kerr et al. (2000) do find a significant tissue $\times$ gene interaction effect, in general interactions in an ANOVA are harder to detect that main effects. If a large number of genes is considered when many of these genes are not expressed in the tissues being examined, the effects of those genes that are expressed in these tissues, and of those that are differentially expressed, might be swamped by the noise generated by unexpressed genes. Further, the assumption implicit in the ANOVA analysis described above, that the variance of the logarithm of the expression level is the same for all genes, cannot be expected to be true in general.

Many microarray ANOVA designs other than that of Kerr et al. (2000) are to be found in the literature, and it is not possible to survey them all here. General principles for microarray ANOVA designs are discussed in Churchill (2002), Speed (2003), and Glonek and Solomon (2002).

13.3.8   Comparing Two Groups by Discriminant Analysis

A natural approach to assessing the genes whose expression levels differ between two groups is that of discriminant analysis. We might first consider carrying out a T^2 test as described in (9.52), and if significance is obtained, indicating that the expression levels of some genes differ significantly between the two groups, attempt to find which genes do so differ by using the discriminant function (9.53). However, both these efforts will in general fail in the microarray context. The reason for this can be seen in two equivalent ways. First, in the microarray context the number of measurements made in the T^2 procedure is the number of genes analyzed, and this is usually several thousand. The values of n and m, the number of observations in each of the two groups considered, is far smaller than this, often being of order 10 or fewer. One of the degrees of freedom for the T^2 test then becomes negative, indicating that the test cannot be carried out. The second reason is associated with the first one, and is that the matrix S, whose inverse is needed in both the T^2 test and the discriminant function calculation, is singular. Thus no inverse matrix exists and the required mathematical procedures cannot be performed.

This problem is avoided by Dudoit et al. (2002), who compare expression levels of many genes between various groups. The number of observations for each gene is on the order of 80. Dudoit et al. first reduce the number of genes in the discrimination procedure by computing a between group to within group F statistic for every gene. They then consider only those genes that have the p highest absolute value of F. The choice of p depends on the data set considered, but must be well below 80 in order to overcome the degrees of freedom problem mentioned above. If p is chosen to be about 30, the matrix S in Section 9.6.4 is "ill-conditioned", so that the value of S^{-1} changes considerably with quite small changes in the data values. Dudoit et al. found that it is necessary to reduce the number p of genes considered to about 10 in order for the discriminant procedure to work well.

13.4 Principal Components and Microarrays

The concept of principal components was discussed in Section 8.5.3. In the microarray context, k (the number of observations) is the number g of genes considered on an array, and this is typically many thousands. Since one of the aims in an microarray analysis is to reduce the number of genes considered to those mainly relevant to the question at hand, a principal components analysis allows a significant reduction when the aim is to decide which genes have the largest variation in expression level.

One of the computational problems of the procedure outlined in Section 8.5.3 in the microarray context is that the matrix defined (8.53) and (8.54) is of size $g \times g$, and thus has g eigenvalues. However all but n of these are zero, and the nonzero eigenvalues, which are the only ones of interest, can be found by the technique of singular value decomposition (see Schott (1997) for a clear description of the technique). This technique also provides the eigenvectors associated with these eigenvalues.

Suppose that it is decided that the total variation in the data can be reasonably represented by some small number a of principal components $c_1, c_2, \ldots, c_a$ (see (8.59) for the case $a = 2$). These principal components have been called "eigengenes" (Speed (2003)). Suppose that the first principal component is $\sum a_i x_i$, where the a_i are the coefficients in that component and x_i is the expression level for gene i. Restricting attention to those genes for which $|a_i| > c$, for some chosen cut-off value c, allows us to focus on a small set of genes that might be used in a future microarray experiment. A similar procedure can be carried out for any of the a principal components considered.

One aim in a microarray analysis is to place the genes into clusters of genes with similar behavior across a set of arrays. Gene clusters can be found using a variety of methods, or principles, some of which we describe briefly in the section below.

One use for the principal components technique is as a preliminary step in the formation of clusters of genes. Formation of clusters of many thousands of genes can present visual problems, so that selecting a subset of genes on which to concentrate, for example those genes with a high coefficient weighting in the first principal component, can provide a substantial aid in the clustering process.

13.5 Clustering Methods

13.5.1 Hierarchical Clustering

Gene clusters can be found by a variety of measures. Some of these are based on a "distance" between any pairs of genes. One popular form of distance is based on correlation calculations. If the expression levels of two genes are taken in n individuals, an estimate of the correlation between these levels can obtained using the estimator (3.16). As discussed in Section 3.3.3, however, this is a biased estimator of the true correlation, the bias being of order n^{-1}. Since in microarray analysis the value of n is often quite small, this is a significant bias.

Given a estimate r of the correlation between expression levels of two genes, a distance between these genes can be defined in various ways. If the sign of r is deemed not to be important, two possible distance definitions are $1 - |r|$ and $\log|r|$. These are both 0 when $r = \pm 1$, indicating that genes that are judged to be perfectly correlated are taken as having no distance between them. When $r = 0$, the former distance is 1 and the latter infinite.

Genes can now be arranged in a hierarchical clustering as described by one or another distance-based method discussed in Chapter 15. The problem of the bias in the correlation estimate should, however, be kept in mind in such a process. Tree estimation with many species (here, cluster estimation with many genes) is sensitive to small changes in distance measures, so that any gene cluster formed using a correlation-based distance, especially with a small value of n, should be viewed with caution.

The hierarchical method is *unsupervised*, in that the number of clusters is not fixed in advance. The K-means approach described below is supervised, since the number of clusters is fixed extrinsically by the investigator.

13.5.2 Other Forms of Clustering

The clustering method just described has a hierarchical character to it. Other methods do not have such a character and can be thought of simply as partitioning methods, with no hierarchy implied between the various partitions. Two popular such methods are the *K-means* approach and the *self organizing maps* (SOM) approach. We now describe the first of these briefly.

In the K-means approach a fixed number K is decided upon, and each of the various genes is eventually to be allocated to exactly one of K groups. This method again relies on some distance measure between genes. A reasonable starting point is to make a common-sense allocation of each gene into one or other group. Then an average, or centered, value can be found for each group by a least squares procedure, and from this a "within-groups" sum of squares W, analogous to that in an ANOVA, can be calculated. This is found by summing the squares of the distances of each gene from its group average value, and then summing this sum over all groups.

Each gene is then potentially re-assigned to a new cluster, following the rule that the distance of every gene from every group average is calculated, and every gene is then assigned to the cluster which minimizes this distance. (Often a gene will be assigned to the cluster it is currently in.)

A new cluster average is then computed from a least squares procedure, using the genes now in that cluster, and the process above is repeated until no gene changes clusters.

The number of genes in each cluster is thus not fixed in advance. Further, the actual choice of group assignments of the various genes might depend on the initial allocation chosen for genes into groups, so that various possible initial assignments should be used.

Problems

13.1 Let X_1 and X_2 be independent random variables, each having the uniform distribution on $(0,1)$. Let $X_{(1)}$ and $X_{(2)}$ be the corresponding order statistics. Find an equation satisfied by A_1 and A_2, with $(A_1 \leq A_2)$, if we require that $\text{Prob}(X_{(1)} \geq A_1, X_{(2)} \geq A_2) = 1 - \alpha$, for some given constant α, $(0 < \alpha < 1)$. Show that the choice $A_1 = A_2 = K(2, \alpha)$ (defined in (2.141)) satisfies this equation.

13.2 *Continuation.* Suppose that an experiment consists of two distinct and independent sub-experiments, and that the P-values for each of these are denoted P_1 and P_2 respectively. Let $P_{(1)}$ and $P_{(2)}$ be the corresponding ordered P-values. Find an equation satisfied by two constants A_1 and A_2 $(A_1 \leq A_2)$, such that an family-wise Type I error of α is attained when both null hypotheses are accepted if and only if $P_{(1)} \geq A_1$ and $P_{(2)} \geq A_2$. If A_1 and A_2 are used respectively instead of $K(2, \alpha)$ and $K(1, \alpha)$ in a Westfall and Young step-down process, for what choices of A_1 and A_2 is the requirement of the strong control of the family-wise Type I error at α satisfied?

13.3 *Continuation.* Consider two distinct independent t-tests. Suppose the Westfall and Young step-down procedure is used to obtain an family-wise Type I error α. Find the probability that a specified one of the two null hypotheses is accepted, given that both null hypotheses are true. Show that this probability is less than the corresponding probability $K(2, \alpha)$ arising under the one-step procedure.

13.4 *Continuation.* Now consider g distinct independent t-tests. Suppose the Westfall and Young step-down procedure is used to obtain a family-wise Type I error α. Show that the probability that a specified one of the g null hypotheses is accepted, given that all null hypotheses are true, is

$$1 - \frac{\alpha}{g} - \frac{\alpha^2}{g} - \cdots - \frac{\alpha^g}{g}.$$

Compare this probability with the corresponding probability $K(g, \alpha)$ arising under the one-step procedure.

13.5 *Continuation.* It is not necessary to consider the Westfall and Young step-down procedure as being carried out in a sequential manner: which hypotheses are accepted and rejected is determined in one step by the location of the point $(p_{(1)}, p_{(2)}, \ldots, p_{(g)})$ in the $(P_{(1)}, P_{(2)}, \ldots, P_{(g)})$ space. For the case $g = 2$, and given a family-wise Type I error α, sketch the region R in the $(P_{(1)}, P_{(2)})$ plane having the property that both null hypotheses are rejected using this sequential procedure if $(p_{(1)}, p_{(2)})$ falls within R. Hence find the probability that both hypotheses are rejected when both null hypotheses are true. Check your answer using calculations deriving from the step-down procedure. What is the corresponding probability in the one-step procedure? Relate your answer to this last question to your answer to question 13.3.

13.6 *Continuation.* Compare the region R determined in Problem 13.5 with the region R' such that both null hypotheses are rejected using the one-step procedure if $(p_{(1)}, p_{(2)})$ falls within R'.

14
Evolutionary Models

14.1 Models of Nucleotide Substitution

The contemporary biological data from which so many inferences are made are the result of evolution, that is, of an indescribably complicated stochastic process. Very simplified models of this process are often used in the literature, in particular for the construction of phylogenetic trees, and aspects of these simplified models are discussed in this chapter. The emphasis is on introductory statistical and probabilistic aspects. A probabilistic approach has the merit of allowing the testing of various hypotheses concerning the evolutionary process. Hypothesis-testing questions in the evolutionary context are discussed in Section 15.9.

While there is clearly genetic variation from one individual to another in a population, there is comparatively little variation at the nucleotide level. For example, two randomly chosen humans typically have different nucleotides at only one site in about 500 to 1,000. To a sufficient level of approximation it is reasonable to assume, for the great majority of sites, that a single nucleotide predominates in the population. Indeed if this were not so, the concept of a paradigm "human genome" would be meaningless. Thus from now on, when we use the expression "the nucleotide at a given site in a population" we mean, more precisely, the predominant nucleotide at this site in the population.

Over a long time span the nucleotide at a given site might change. To analyze the stochastic properties of this change, we ignore the (perhaps comparatively brief) time period during which one nucleotide replaces another in the population, and imagine an effectively instantaneous change in frequency of a nucleotide from a value close to 0 to a value close to 1. We shall describe this event as the *substitution* of one nucleotide by another, meaning more precisely the substitution of the predominant nucleotide by another. The time unit chosen to evaluate the properties of this substitution process is arbitrary, but is often large, perhaps on the order of tens of thousands of generations.

We focus on DNA sequences and consider the predominant nucleotide at some specified site in some population. In Section 4.5 the states of a Markov chain were labeled according to the generic convention as states $E_1, E_2, E_3, \ldots, E_s$. In analyzing substitution processes for DNA sequences it is often convenient to change notation and to identify the states with the nucleotides a, g, c, and t, taken in this order. With this notation a Markov chain would, for example, be said to be in state g if, in the population considered, the predominant nucleotide at the site of interest is g. If unit time in the Markov chain is taken as, for example, 200,000 generations, a change from state a to state c in one time unit means the substitution of the nucleotide a by the nucleotide c after a period of 200,000 generations.

When this notational convention is used for a discrete-time Markov chain, symbols such as p_{ag} denote the probability of a change in the predominant nucleotide from a to g after one time unit. In the continuous-time models considered in Section 14.3, again with unit time taken at some agreed value, the probability of a change in the predominant nucleotide from a to g in a time period of length t is denoted by $P_{ag}(t)$.

The comparison of genetic data from two contemporary species often relies on a model of the evolutionary processes leading to these species. Various statistical procedures used in this comparison might require tracing up the tree of evolution from one species to a common ancestor and then down the tree to the other species. If the stochastic process assumed for tracing upwards is to be the same as that for tracing downwards, the stochastic process must be reversible. This is one reason why the concept of reversibility was introduced in Section 11.2.4. One aim of the analysis below is to check whether the various stochastic models that we introduce are indeed reversible.

14.2   Discrete-Time Models

14.2.1   The Jukes–Cantor Model

The simplest (and earliest) model of nucleotide substitution is the Jukes–Cantor model (Jukes and Cantor (1969)). The original version of this model was a continuous-time process. In this section we consider the parallel (and simpler) discrete-time version, and discuss the continuous-time process in Section 14.3.1.

The discrete-time Jukes–Cantor model is a Markov chain with four states a, g, c, and t. With the states labeled in this order, the transition matrix P for this model is given by

$$
P = \begin{bmatrix}
1 - 3\alpha & \alpha & \alpha & \alpha \\
\alpha & 1 - 3\alpha & \alpha & \alpha \\
\alpha & \alpha & 1 - 3\alpha & \alpha \\
\alpha & \alpha & \alpha & 1 - 3\alpha
\end{bmatrix}.
\tag{14.1}
$$

Here α is a parameter depending on the timescale chosen: If unit time were chosen as 100,000 generations, α would take a value smaller than it would be if unit time were chosen as 200,000 generations. Whatever time scale is chosen, it is clearly necessary that α be less than $\frac{1}{3}$.

The transition matrix P in (14.1) possesses several elements of symmetry, and in particular the model assumes that whatever the nucleotide in the population is at any time, the three other nucleotides are equally likely to substitute for it. Our first aim is to analyze the properties of this Markov chain by the spectral decomposition methods discussed in Appendix B.19.

The eigenvalue equation (B.45) shows that the matrix (14.1) has eigenvalues 1 (with multiplicity 1) and $1 - 4\alpha$ (with multiplicity 3). The left eigenvector corresponding to the eigenvalue 1 is $(.25, .25, .25, .25)$, and thus the stationary distribution is the discrete uniform distribution

$$
(\varphi_a, \varphi_g, \varphi_c, \varphi_t)' = (.25, .25, .25, .25)'.
\tag{14.2}
$$

This implies that after a long time has passed, the four nucleotides are essentially equally likely to predominate in the population, as might be expected from the symmetry of the model. The spectral expansion (B.49) of P^n is

$$
P^n = \begin{bmatrix}
.25 & .25 & .25 & .25 \\
.25 & .25 & .25 & .25 \\
.25 & .25 & .25 & .25 \\
.25 & .25 & .25 & .25
\end{bmatrix} + (1 - 4\alpha)^n \begin{bmatrix}
.75 & -.25 & -.25 & -.25 \\
-.25 & .75 & -.25 & -.25 \\
-.25 & -.25 & .75 & -.25 \\
-.25 & -.25 & -.25 & .75
\end{bmatrix}.
\tag{14.3}
$$

From this it follows that whatever the predominant nucleotide in the population is at time 0, the probability that this is also the predominant nucleotide at time n is

$$
.25 + .75(1 - 4\alpha)^n,
\tag{14.4}
$$

and the probability that some other specified nucleotide is the predominant nucleotide at time n is

$$
.25 - .25(1 - 4\alpha)^n.
\tag{14.5}
$$

These values confirm the stationary distribution given in equation (14.2), and also show that the rate of approach to the stationary distribution is determined by the numerical value of the eigenvalue $1 - 4\alpha$.

14.2.2   The Kimura Models

The highly symmetric assumptions implicit in the Jukes–Cantor model are not realistic. A *transition*, that is, the replacement of one purine by the other (for example of a by g) or of one pyrimidine by the other, is in practice more likely than a *transversion*, that is, the replacement of a purine by a pyrimidine or of a pyrimidine by a purine. Kimura (1980) proposed a (continuous-time) two-parameter model to allow for this. The transition matrix P for the discrete-time version of this model, with the same ordering of states as that used for the Jukes–Cantor model, is

$$\begin{bmatrix} 1-\alpha-2\beta & \alpha & \beta & \beta \\ \alpha & 1-\alpha-2\beta & \beta & \beta \\ \beta & \beta & 1-\alpha-2\beta & \alpha \\ \beta & \beta & \alpha & 1-\alpha-2\beta \end{bmatrix}. \tag{14.6}$$

Here α is the probability of a transition in one time unit, while β is the probability that a purine is substituted by a nominated pyrimidine in one time unit and is also the probability that a pyrimidine is substituted by a nominated purine in one time unit. It is required that $\alpha + 2\beta < 1$.

The eigenvalue equation (B.45) shows that the matrix (14.6) has eigenvalues 1 (multiplicity 1), $1-4\beta$ (multiplicity 1), and $1-2(\alpha+\beta)$ (multiplicity 2). The left eigenvector corresponding to the eigenvalue 1 is $(.25, .25, .25, .25)'$, so that in this model, as in the Jukes–Cantor model, the stationary distribution is again discrete uniform. The spectral expansion (B.49) of P^n is

$$P^n = \begin{bmatrix} .25 & .25 & .25 & .25 \\ .25 & .25 & .25 & .25 \\ .25 & .25 & .25 & .25 \\ .25 & .25 & .25 & .25 \end{bmatrix} + (1-4\beta)^n \begin{bmatrix} .25 & .25 & -.25 & -.25 \\ .25 & .25 & -.25 & -.25 \\ -.25 & -.25 & .25 & .25 \\ -.25 & -.25 & .25 & .25 \end{bmatrix}$$
$$+ (1-2(\alpha+\beta))^n \begin{bmatrix} .5 & -.5 & 0 & 0 \\ -.5 & .5 & 0 & 0 \\ 0 & 0 & .5 & -.5 \\ 0 & 0 & -.5 & .5 \end{bmatrix}. \tag{14.7}$$

This spectral expansion implies that whatever the predominant nucleotide at time 0 at any site, the probability that this is also the predominant nucleotide at time n is

$$.25 + .25(1-4\beta)^n + .5\left(1-2(\alpha+\beta)\right)^n. \tag{14.8}$$

If the initial nucleotide is a purine, the probability that at time n the predominant nucleotide is the other purine is

$$.25 + .25(1-4\beta)^n - .5\left(1-2(\alpha+\beta)\right)^n. \tag{14.9}$$

A parallel remark holds for pyrimidines. The probability that after n time units a purine has been substituted by a specific pyrimidine is

$$.25 - .25(1 - 4\beta)^n, \tag{14.10}$$

and the probability that it has been replaced by one or the other pyrimidine is

$$.5 - .5(1 - 4\beta)^n. \tag{14.11}$$

A parallel remark holds for the replacement of a pyrimidine by a purine. The rate of approach to the stationary distribution depends on the numerical values of α and β. If $\alpha > \beta$, as would normally be assumed, the largest nonunit eigenvalue is $1 - 4\beta$.

A generalization of the above original Kimura model is provided by the so-called Kimura 3ST model. Here the transition matrix is of the form

$$\begin{bmatrix} 1-\alpha-\beta-\gamma & \alpha & \beta & \gamma \\ \alpha & 1-\alpha-\beta-\gamma & \gamma & \beta \\ \beta & \gamma & 1-\alpha-\beta-\gamma & \alpha \\ \gamma & \beta & \alpha & 1-\alpha-\beta-\gamma \end{bmatrix}, \tag{14.12}$$

where α, β, and γ are all unknown parameters. This Markov chain also has a uniform stationary distribution. The detailed balance requirements (11.5) hold, so that the Markov chain is reversible. This also implies reversibility of the Jukes–Cantor Markov chain (14.1) and the Kimura two-parameter Markov chain (14.6), since these are special cases of this model.

14.2.3 Further Generalizations of the Kimura Models

Although the Kimura models are more realistic than the Jukes–Cantor model, they nevertheless possess symmetry assumptions that make them unrealistic. In particular, they have uniform stationary distributions, which is at variance with observation. Increasingly complex models have been proposed over the years to give a closer match between theory and observation. Each may be analyzed, in greater or lesser detail, by algebraic methods similar to those used for the Jukes–Cantor and the Kimura models. These have usually been proposed in the context of continuous-time models (discussed below), but the principles apply equally to discrete-time models, and we discuss them here in the discrete-time context.

An immediate generalization of the Kimura model (14.6) allows the purine to pyrimidine substitution probability to differ from the pyrimidine to purine substitution probability. This implies a transition matrix P of the form

$$\begin{bmatrix} 1-\alpha-2\gamma & \alpha & \gamma & \gamma \\ \alpha & 1-\alpha-2\gamma & \gamma & \gamma \\ \delta & \delta & 1-\alpha-2\delta & \alpha \\ \delta & \delta & \alpha & 1-\alpha-2\delta \end{bmatrix}. \tag{14.13}$$

The stationary distribution of the Markov chain implied by this model is

$$\left(\frac{\delta}{2(\delta + \gamma)}, \frac{\delta}{2(\delta + \gamma)}, \frac{\gamma}{2(\delta + \gamma)}, \frac{\gamma}{2(\delta + \gamma)} \right)'. \qquad (14.14)$$

This is not uniform if $\delta \neq \gamma$, although it does possess symmetry properties in that it allocates equal stationary probabilities to the two purines and to the two pyrimidines. The Markov chain defined by this transition matrix, like that defined by the Jukes–Cantor model and the Kimura model, is reversible.

A further generalization has been made by Blaisdell (1985), who allows different within-transition and within-transversion rates. The transition matrix P for this model is

$$\begin{bmatrix} 1 - \alpha - 2\gamma & \alpha & \gamma & \gamma \\ \beta & 1 - \beta - 2\gamma & \gamma & \gamma \\ \delta & \delta & 1 - \beta - 2\delta & \beta \\ \delta & \delta & \alpha & 1 - \alpha - 2\delta \end{bmatrix}. \qquad (14.15)$$

The stationary distribution of this Markov chain is found to be

$$\left(\frac{\delta(\beta+\gamma)}{\theta(\alpha+\beta+2\gamma)}, \frac{\delta(\alpha+\gamma)}{\theta(\alpha+\beta+2\gamma)}, \frac{\gamma(\alpha+\delta)}{\theta(\alpha+\beta+2\delta)}, \frac{\gamma(\beta+\delta)}{\theta(\alpha+\beta+2\delta)} \right)', \qquad (14.16)$$

where $\theta = \gamma + \delta$. This stationary distribution and the elements in the transition matrix (14.15) show that this model is not reversible. On the other hand, the elements in the stationary distribution can now all be different from each other, a property not enjoyed by any other model discussed above.

Another generalization has been made by Schadt et al. (1998), who introduced a continuous-time model whose discrete-time analogue has transition matrix P given by

$$\begin{bmatrix} 1 - \alpha - \gamma - \lambda & \alpha & \gamma & \lambda \\ \epsilon & 1 - \epsilon - \gamma - \lambda & \gamma & \lambda \\ \delta & \kappa & 1 - \beta - \delta - \kappa & \beta \\ \delta & \kappa & \sigma & 1 - \delta - \kappa - \sigma \end{bmatrix}.$$

This transition matrix is reversible if and only if

$$\beta\gamma = \lambda\sigma, \quad \alpha\delta = \epsilon\kappa. \qquad (14.17)$$

Further generalizations, steadily relaxing symmetry assumptions, have been made by Takahata and Kimura (1981) and Gojobori et al. (1982).

14.2.4   The Felsenstein Models

A different form of generalization of the Jukes–Cantor model was introduced by Felsenstein (1981), whose notation we adopt here. In the discrete-time version of this model, (often referred to as the F81 model, to distinguish it from the more general model discussed below), the probability of substitution of any nucleotide by another is proportional to the stationary probability of the substituting nucleotide. This implies a transition matrix P of the form

$$
\begin{bmatrix}
1 - u + u\varphi_a & u\varphi_g & u\varphi_c & u\varphi_t \\
u\varphi_a & 1 - u + u\varphi_g & u\varphi_c & u\varphi_t \\
u\varphi_a & u\varphi_g & 1 - u + u\varphi_c & u\varphi_t \\
u\varphi_a & u\varphi_g & u\varphi_c & 1 - u + u\varphi_t
\end{bmatrix},
\tag{14.18}
$$

where $(\varphi_a, \varphi_g, \varphi_c, \varphi_t)$ is the stationary distribution and u is a parameter of the model. (It will be shown in Section 14.2.7, and is easily checked directly, that the stationary distribution for the model defined by (14.18) is indeed $(\varphi_a, \varphi_g, \varphi_c, \varphi_t)$.)

A second Felsenstein model, (Felsenstein and Churchill (1996), see also Kishino and Hasegawa (1989)), often called the F84 model, is more general than that given by (14.18), and is the evolutionary model used in the PHYLIP package. This model has a transition matrix similar to that of (14.18), except that the upper-left 2×2 component of (14.18) is replaced by

$$
\begin{bmatrix}
1 - u + u\varphi_a - \frac{uK\varphi_g}{\varphi_a+\varphi_g} & u\varphi_g + \frac{uK\varphi_g}{\varphi_a+\varphi_g} \\
u\varphi_a + \frac{uK\varphi_a}{\varphi_a+\varphi_g} & 1 - u + u\varphi_g - \frac{uK\varphi_a}{\varphi_a+\varphi_g}
\end{bmatrix},
\tag{14.19}
$$

and the lower-right component 2×2 of (14.18) is replaced by

$$
\begin{bmatrix}
1 - u + u\varphi_c - \frac{uK\varphi_t}{\varphi_c+\varphi_t} & u\varphi_t + \frac{uK\varphi_t}{\varphi_c+\varphi_t} \\
u\varphi_c + \frac{uK\varphi_c}{\varphi_c+\varphi_t} & 1 - u + u\varphi_t - \frac{uK\varphi_c}{\varphi_c+\varphi_t}
\end{bmatrix},
\tag{14.20}
$$

The transition matrix defined by these amendments to the model (14.18), as with the model (14.18) itself, has stationary distribution $(\varphi_a, \varphi_g, \varphi_c, \varphi_t)$. From this it is easily shown that the model is reversible. The quantity K is positive and is a parameter of the F84 model not included in the F81 model: larger values of K increase transition substitution rates compared to those in the model (14.18).

Although the model (14.18) generalizes the Jukes–Cantor model, to which it reduces if $\varphi_a = \varphi_g = \varphi_c = \varphi_t = .25$, it does not generalize the Kimura two-parameter model. On the other hand the model defined by (14.18), (14.19), and (14.20) does generalize the Kimura two-parameter model (14.6), reducing to that model when the stationary distribution is uniform. (This requires the identifications of the parameters α and β in the Kimura model with $u(2K + 1)/4$ and $u/4$ respectively.) It also of course generalizes the model (14.18), to which it reduces when $K = 0$.

14.2.5   The HKY Model

A further model, introduced by Hasegawa et al. (1985), and called here the HKY model, assumes that the transition probability matrix P is of the form

$$
\begin{bmatrix}
1 - u\varphi_g - v\varphi_1 & u\varphi_g & v\varphi_c & v\varphi_t \\
u\varphi_a & 1 - u\varphi_a - v\varphi_1 & v\varphi_c & v\varphi_t \\
v\varphi_a & v\varphi_g & 1 - u\varphi_t - v\varphi_2 & u\varphi_t \\
v\varphi_a & v\varphi_g & u\varphi_c & 1 - u\varphi_c - v\varphi_2
\end{bmatrix}, \quad (14.21)
$$

where $\varphi_1 = \varphi_c + \varphi_t$, $\varphi_2 = \varphi_a + \varphi_g$. This model is an amalgam of the Kimura model (14.6) and the Felsenstein model, and includes these as particular cases.

The eigenvalues and eigenvectors of these matrices follow from those given by Hasegawa et al. (1985) for the analogous continuous-time process. The eigenvalues are

$$
\lambda_1 = 1, \ \lambda_2 = 1 - v, \ \lambda_3 = 1 - u\varphi_1 - v\varphi_2, \ \lambda_4 = 1 - v\varphi_1 - u\varphi_2. \quad (14.22)
$$

The corresponding left eigenvectors are

$$
\begin{aligned}
\boldsymbol{\ell}_1 &= (\varphi_a, \varphi_g, \varphi_c, \varphi_t) \\
\boldsymbol{\ell}_2 &= (\varphi_1\varphi_a, \varphi_1\varphi_g, -\varphi_2\varphi_c, -\varphi_2\varphi_t) \\
\boldsymbol{\ell}_3 &= (0, 0, 1, -1) \\
\boldsymbol{\ell}_4 &= (1, -1, 0, 0)
\end{aligned}
$$

and the corresponding right eigenvectors are

$$
\boldsymbol{r}_1 = (1, 1, 1, 1)'
$$

$$
\boldsymbol{r}_2 = \left(\frac{1}{\varphi_2}, \frac{1}{\varphi_2}, \frac{-1}{\varphi_1}, \frac{-1}{\varphi_1} \right)'
$$

$$
\boldsymbol{r}_3 = \left(0, 0, \frac{\varphi_t}{\varphi_1}, \frac{-\varphi_c}{\varphi_1} \right)'
$$

$$
\boldsymbol{r}_4 = \left(\frac{\varphi_g}{\varphi_2}, \frac{-\varphi_a}{\varphi_2}, 0, 0 \right)'.
$$

These have been normalized by the requirement (B.47), and this leads immediately to the spectral expansion (B.49) for the powers of the transition matrix (14.21). The left eigenvector $\boldsymbol{\ell}_1$ also shows that $(\varphi_a, \varphi_g, \varphi_c, \varphi_t)$ is the stationary distribution of the Markov chain for this model, as the notation anticipated. The HKY Markov chain is reversible: See Section 14.2.7. Thus the HKY model is the most flexible of those discussed for considering evolutionary processes. This matter is discussed further in Section 15.7.

14.2.6 Other Models

All of the above models have various simplifying assumptions built into them. At the other extreme, a model having no such properties might be more appropriate in practice; the numerical model (4.30), possessing no symmetry or other simplifying features, was introduced with this possibility in mind. In contrast to the algebraically defined models described above, the rate of approach to stationarity for almost every numerical model, and the stationary distribution itself, can only be found through a numerical spectral expansion such as in Section 11.2.3.

The simpler models discussed above are unrealistic in that, for example, they imply uniform nucleotide frequency distributions. More complex models, however, might not be reversible, and thus should not be used for the reconstruction of phylogenetic trees. The choice of one model over another is thus often a difficult matter. Tests of one model against another are discussed in Chapter 15.

14.2.7 The Reversibility Criterion

The discussion of Section 11.2.4 shows that the reversibility requirement is necessary for various inferential procedures, and equation (11.5) gives the reversibility requirement for any finite Markov chain. We now discuss the reversibility criterion in the context of evolutionary models, focusing on DNA substitutions and the 4×4 transition matrices used to describe these substitutions. We assume that all transition matrices considered refer to irreducible aperiodic Markov chains, so that any model discussed has a stationary distribution, denoted as above by $(\varphi_a, \varphi_g, \varphi_c, \varphi_t)$.

The general 4×4 transition matrix has twelve free parameters, namely three free transition probabilities in each of the four rows of the transition matrix. (The fourth transition probability in each row is determined by the remaining three.) However another parameterization, using twelve different free parameters, is more useful for our present purposes. This parameterization was given by Tavaré (1986), (see Swofford et al. (1996)), and under this parameterization the transition matrix is written in the form

$$\begin{bmatrix} 1 - uW & uA\varphi_g & uB\varphi_c & uC\varphi_t \\ uD\varphi_a & 1 - uX & uE\varphi_c & uF\varphi_t \\ uG\varphi_a & uH\varphi_g & 1 - uY & uI\varphi_t \\ uJ\varphi_a & uK\varphi_g & uL\varphi_c & 1 - uZ \end{bmatrix}. \tag{14.23}$$

Here $A, B, \ldots, L$ are the twelve free parameters, $(\varphi_a, \varphi_g, \varphi_c, \varphi_t)$ is the stationary distribution of the Markov chain, and

$$W = A\varphi_g + B\varphi_c + C\varphi_t, \ X = D\varphi_a + E\varphi_c + F\varphi_t,$$

$$Y = G\varphi_a + H\varphi_g + I\varphi_t, \ Z = J\varphi_a + K\varphi_g + L\varphi_c.$$

The necessary and sufficient condition for the Markov chain with transition matrix (14.23) to be reversible is that the equations

$$A = D, \quad B = G, \quad C = J, \quad E = H, \quad F = K, \quad I = L \qquad (14.24)$$

are all satisfied. When this condition holds, we call the model (14.23) the *general reversible process model*, following the terminology of Yang (1994) in the analogous continuous-time model. The general reversible model has six free parameters, which can be taken as A, B, C, E, F, and I, so one can think of paying for reversibility by losing the choice of six further parameters, or by losing six degrees of freedom.

It is easily checked that when the conditions (14.24) hold, $(\varphi_a, \varphi_g, \varphi_c, \varphi_t)$ is indeed the stationary distribution of the model (14.23) (see Problem 14.4).

The Felsenstein F81 model (14.18) is the particular case of (14.23) with $A = B = \cdots = L = 1$, the Felsenstein F84 model defined by (14.18), (14.19) and (14.20) is the particular case of (14.23) with $B = C = E = F = G = H = J = K = 1$, $A = D$, $I = L$, and the HKY model is a special case of (14.23) with $B = C = E = F = G = H = J = K$ and $A = D = I = L$. All three models are thus reversible, as claimed above. The choice of one of these models instead of the general reversible model may also be thought of as involving a loss of degrees of freedom. These degrees of freedom are relevant to the test of one these models against the general reversible model, a matter discussed in Section 15.9.4.

14.2.8   The Simple Symmetric Amino Acid Model

The simple symmetric matrix M_1 defined in in Section 6.5.4 is the "amino acid" 20×20 matrix generalization of the "nucleotide" Jukes–Cantor matrix (14.1). Like that model it is reversible. The n-step transition probabilities $\{m_{jk}^{(n)}\}$ given in equations (6.28) and (6.29) define the spectral expansion of this matrix, which is a direct analogue spectral expansion of the Jukes–Cantor matrix (14.1). As with the Jukes–Cantor model, the simple symmetric model is not realistic. We have analyzed it in some detail in Chapters 6 and 10 because it illuminates various properties of more realistic PAM models.

14.3   Continuous-Time Models

As justified in Section 11.7.1, it is assumed for all continuous-time evolutionary models considered that the transition probabilities are of the form described in (11.29) and (11.30).

14.3.1   The Continuous-Time Jukes–Cantor Model

The Jukes–Cantor and Kimura models were first proposed as continuous-time models, and in their continuous-time versions can be analyzed by the methods of Section 11.7. It is convenient to continue to label the states as a, g, c, and t, as in Section 14.1, and to introduce the further notation that i, j, and k are arbitrary members of the set $\{a, g, c, t\}$.

In the Jukes–Cantor model the instantaneous transition rates q_{ij} and q_i are defined by $q_{ij} = \alpha$ for all $i \neq j$, and $q_i = 3\alpha$. Replacing q_{kj} and q_j in equation (11.33) by α and 3α, respectively, we get

$$\frac{d}{dt} P_{ij}(t) = -3\alpha P_{ij}(t) + \alpha \sum_{k \neq j} P_{ik}(t). \tag{14.25}$$

Since $\sum_{k \neq j} P_{ik}(t) = 1 - P_{ij}(t)$, this equation simplifies to

$$\frac{d}{dt} P_{ij}(t) = \alpha - 4\alpha P_{ij}(t). \tag{14.26}$$

This is a linear differential equation and may be solved by standard methods. The solution involves a constant of integration, which is allocated by the boundary conditions $P_{ii}(0) = 1$, $P_{ij}(0) = 0$, $i \neq j$. Using these boundary conditions, the solution is found to be

$$P_{ii}(t) = .25 + .75e^{-4\alpha t}, \tag{14.27}$$

$$P_{ij}(t) = .25 - .25e^{-4\alpha t}, \quad j \neq i. \tag{14.28}$$

These equations show that as $t \to \infty$, both $P_{ii}(t)$ and $P_{ij}(t)$ approach 0.25. Thus in the stationary distribution all nucleotides have equal probability at any site, as is expected from the symmetry of the model. This stationary distribution can also be found by applying equations (11.34). The similarity of the expressions for $P_{ii}(t)$ and $P_{ij}(t)$ and the expressions given in equations (14.4) and (14.5) indicate why the present model is the continuous-time analogue of the discrete-time Jukes–Cantor model.

The fact that q_i is independent of i shows that if an event is defined as the substitution of one nucleotide for another, substitutions follow the laws of a homogeneous Poisson process as discussed in Section 4.1. The probability that j substitutions occur in time t is given by the Poisson probability distribution (4.10) with parameter 3α. It can be shown (see Problem 14.7) that if j transitions from one state to another have occurred in the continuous-time Markov process described by equation (14.26), the probability that the process has returned to its initial state after these j transitions is

$$\frac{1}{4} + \frac{3}{4}\left(-\frac{1}{3}\right)^j. \tag{14.29}$$

The probability that j transitions do occur in time t is given by (4.10) with the parameter λ replaced by 3α. Combining these results, the probability that at time t the original nucleotide is the predominant one is

$$\sum_{j=0}^{\infty} e^{-3\alpha t} \frac{(3\alpha t)^j}{j!} \left(\frac{1}{4} + \frac{3}{4} \left(-\frac{1}{3} \right)^j \right). \tag{14.30}$$

By equation (B.20), this reduces to the right-hand side in equation (14.27). Equation (14.28) is found similarly.

Suppose that two independent contemporary populations are available that descended from a common population t units of time ago, and that the evolutionary properties of these two populations since their common ancestor are described by the Jukes–Cantor model. We would like to use data from these two populations to estimate the evolutionary parameter α in this model, describing in a sense the rate at which one nucleotide replaces another in these populations. We would also like to estimate the time t, since this can be used in defining a distance between the two populations.

It is convenient to begin the analysis by considering the probability p that, at a given site, the predominant nucleotides in the two populations differ. Given two DNA sequences each consisting of N nucleotide sites, and with the assumption that the replacement processes at all sites are independent and have identical stochastic properties, the theory of Section 3.3.1 shows that $\hat{p}$, the observed proportion of sites at which the predominant nucleotides differ between the two populations considered, is an unbiased estimator of p. It is thus natural to estimate p by $\hat{p}$. Further, equation (2.77) shows that the variance of this estimate is

$$\text{variance of } \hat{p} = \frac{p(1-p)}{N}. \tag{14.31}$$

The simplicity of this calculation follows largely from the iid assumption made. This assumption is unlikely to hold in practice, and we later examine some of the consequences of its not holding.

We now use this estimate, and its variance, to discuss the estimation of t and of the parameter α in the Jukes–Cantor model. Poisson process theory shows that in this model, the mean number of substitutions, in t units of time, at any nucleotide site down the two lines of descent leading from the common founder to the two populations is $6\alpha t$. The mean number ν of substitutions down the two lines of descent at all N sites together is then $\nu = 6N\alpha t$.

The symmetry inherent in the Jukes–Cantor model implies that whatever the predominant nucleotide at time 0, the probability $I(t)$ that at time t the two descendent populations have the same predominant nucleotide is

$$I(t) = (P_{ii}(t))^2 + \sum_{j \neq i} (P_{ij}(t))^2,$$

where $P_{ii}(t)$ is given by (14.27) and $P_{ij}(t)$ is given by (14.28). These equations show that

$$I(t) = .25 + .75e^{-8\alpha t}. \qquad (14.32)$$

The probability p that the nucleotides are different in the two populations is thus

$$p = 1 - I(t) = .75(1 - e^{-8\alpha t}). \qquad (14.33)$$

This equation relating p to αt can be found in a second, more efficient, way. The reversibility of the Jukes–Cantor model implies that the properties of the stochastic process describing any line of descent is the same as that describing the process in reverse, that is, by considering the corresponding line of ascent. The elapsed time up the line of ascent from one of the two contemporary populations up to the founder population, and then down the line of descent from the founder population to the other contemporary population, is $2t$. Therefore, the probability that, at any nucleotide site, the same nucleotide occurs in both populations is given by equation (14.27) with t replaced by $2t$. This leads directly to the expression (14.32) and thus to (14.33).

The evolutionary parameters introduced above are all functions of the composite parameter αt. Standard practice in estimating αt is to invert the relation between p and αt in (14.33), yielding

$$\alpha t = -\frac{1}{8} \log \left(1 - \frac{4}{3} p \right),$$

and thus to estimate the composite αt by

$$\widehat{\alpha t} = -\frac{1}{8} \log \left(1 - \frac{4}{3} \hat{p} \right). \qquad (14.34)$$

This right-hand side is defined only if $\hat{p} < \frac{3}{4}$. When $\hat{p} \geq \frac{3}{4}$ the two sequences are more dissimilar than two random sequences will tend to be, and it is natural to define $t = +\infty$. Even when $\hat{p} < \frac{3}{4}$ the estimator (14.34) is a biased estimator of αt, and in fact, there is no unbiased estimator of αt. We consider this "within-model" bias further in Section 14.3.6. For the moment we follow the implications of the estimation procedure in 14.34).

Equation (14.34) shows that unless t can be estimated extrinsically, the aim of estimating the parameter α cannot be achieved. Similarly, unless α can be estimated extrinsically, the aim of estimating t cannot be achieved. All that is possible is estimation of the composite parameter αt. If on the other hand the parameter α can be regarded as being the same for all populations, at all times, and at all nucleotide sites, then proportional values for t can be found, and these can be used as surrogate distances in a phylogenetic tree estimation procedure as described in Section 15.5.

The estimate $\hat{\nu}$ of $\nu = 6N\alpha t$ is

$$\hat{\nu} = -\frac{3N}{4} \log\left(1 - \frac{4}{3}\hat{p}\right).$$ (14.35)

The issue of bias discussed above arises also for this estimator, but again we ignore it.

The estimation formula (14.35) has several interesting consequences. Perhaps the most interesting is that if $\hat{p}$ is small, the Taylor series approximation (B.25) shows that

$$\hat{\nu} \cong N\left(\hat{p} + \frac{2}{3}\hat{p}^2\right).$$

Thus $\hat{\nu}$ is marginally larger than $N\hat{p}$, the observed number of sites at which the predominant nucleotides in the two populations differ from each other. For example, if in 300 of $N = 3,000$ sites the predominant nucleotide differs between the two populations, then $\hat{p} = 0.1$ and $\hat{\nu} \cong 320$. Thus it is estimated that about 20 substitutions occurred but are not observed by counting the differences between the extant sequences. This is because more than one substitution can occur in the same position. The quantity ν accounts for *all* substitutions.

The statistical differentials variance formula (B.33) can be used to approximate the variances of the estimators (14.34), and (14.35). We illustrate this by considering the variance of $\hat{\nu}$. Equation (B.33) shows that this variance is approximately

$$\left(\frac{d\hat{\nu}}{d\hat{p}}\bigg|_{\hat{p}=p}\right)^2 (\text{variance of } \hat{p}),$$

which from (14.31) is

$$\left(\frac{d\hat{\nu}}{d\hat{p}}\bigg|_{\hat{p}=p}\right)^2 \frac{p(1-p)}{N}.$$

By equation (14.35), this equals

$$Np(1-p)e^{8\nu/3N}.$$ (14.36)

Since maximum likelihood estimators have optimality properties as discussed in detail in Chapter 8, it is appropriate to find the maximum likelihood estimate of the parameter ν. Suppose that of the N sites sampled there are $j = N\hat{p}$ sites for which different nucleotides appear in the two populations. Under the iid assumption the likelihood of the data in terms of ν is, from equations (14.32) and (14.33), proportional to

$$\left(\frac{3}{4}\left(1 - e^{-4\nu/3N}\right)\right)^j \left(\frac{1}{4} + \frac{3}{4}e^{-4\nu/3N}\right)^{N-j},$$ (14.37)

where $8\alpha t$ has been replaced by $4\nu/3N$. The derivative of the logarithm of this expression with respect to ν is

$$\frac{1}{N}e^{-4\nu/3N}\left(\frac{j}{p}-\frac{N-j}{1-p}\right),\qquad(14.38)$$

where

$$p=\frac{3}{4}\left(1-e^{-4\nu/3N}\right).\qquad(14.39)$$

It follows from this that the expression (14.35) for $\hat{\nu}$ in is the maximum likelihood estimate of ν if $\hat{p}$ is given by j/N.

This result also follows directly from the invariance property of maximum likelihood estimators, discussed in Section 8.3. The maximum likelihood estimator of p is $\hat{p}=j/N$, and the relation (14.39) between p and ν implies that the maximum likelihood estimate $\hat{\nu}$ is given by (14.35). In the following section we use this more direct approach when analyzing the continuous-time Kimura model.

A further differentiation of the log likelihood and the asymptotic variance formula (8.18) shows that the asymptotic variance of $\hat{\nu}$ is identical to the expression in (14.36).

14.3.2   The Continuous-Time Kimura Model

The analysis of the continuous-time Kimura model (14.6) is more complex than that for the Jukes–Cantor model, and the notation must be handled more carefully. The instantaneous parameters q_{ij} and q_i, introduced in equations (11.29) and (11.30), take the values

$$q_{ij}=\alpha \text{ for the } (i,j) \text{ pairs } (a,g),(g,a),(c,t), \text{ and } (t,c),$$
$$q_{ij}=\beta \text{ for the } (i,j) \text{ pairs } (a,c),(a,t),(g,c),(g,t),$$
$$(c,a),(c,g),(t,a), \text{ and } (t,g).$$

In all cases,

$$q_i=\alpha+2\beta.\qquad(14.40)$$

If these choices are made, the system of equations (11.33) yields a set of four simultaneous differential equations typified by

$$\frac{d}{dt}P_{aa}(t)=-(\alpha+2\beta)P_{aa}(t)+\alpha P_{ag}(t)+\beta\left(P_{ac}(t)+P_{at}(t)\right),\qquad(14.41)$$

where the notation now indicates specific nucleotide types. The boundary conditions for these equations are $P_{ii}(0)=1$ for all i, $P_{ij}(0)=0$ for all $j\neq i$. This system of four differential equations can be solved, and for any i the value of $P_{ii}(t)$ is found to be

$$P_{ii}(t)=.25+.25e^{-4\beta t}+.5e^{-2(\alpha+\beta)t}.\qquad(14.42)$$

The expression for $P_{ij}(t)$ depends on the choice of i and j. If the initial predominant nucleotide is a specified purine (respectively pyrimidine), then the probability that at time t the predominant nucleotide is the other purine (respectively pyrimidine) is

$$.25 + .25e^{-4\beta t} - .5e^{-2(\alpha+\beta)t}. \tag{14.43}$$

If the initial predominant nucleotide is a specified purine (respectively pyrimidine), then the probability that at time t the predominant nucleotide is a pyrimidine (respectively purine) is

$$.5 - .5e^{-4\beta t}. \tag{14.44}$$

The probabilities given by these expressions reduce to the corresponding formulae for the Jukes–Cantor model when $\alpha = \beta$. They are also the continuous time analogues of the discrete-time equations (14.8), (14.9), and (14.11). A further implication of these equations is that, as expected, in the stationary distribution for this model all nucleotides have equal probability.

The detailed balance requirements (11.35) hold for the Kimura model, so that the process is reversible. This implies that the Jukes–Cantor process is also reversible.

As in the Jukes–Cantor model, Poisson process theory is relevant to the evolutionary properties of the Kimura model. For the Kimura model q_i, defined in (14.40), is independent of i. Thus if an event is defined as a substitution from one nucleotide to another at any nucleotide site in the line of descent of any population, the probability that j events occur in time t is given by equation (4.10), with the generic parameter λ now given by $\alpha + 2\beta$.

The observed number of transition substitutions and the observed number of transversion substitutions can be used to estimate the parameters of the Kimura model using the reversibility property of the model and the "line of ascent and descent" argument described in Section 14.3.1. Here t is replaced by $2t$ in the right-hand sides of equations (14.42), (14.43), and (14.44). The resulting expressions then show that if

$$\gamma = 4(\alpha + \beta)t, \quad \phi = 8\beta t,$$

then the probability p_1 that at any site one purine (pyrimidine) occurs in one sequence and the other purine (pyrimidine) in the other sequence is

$$p_1 = .25 + .25e^{-\phi} - .5e^{-\gamma}.$$

The probability p_2 that at any site a purine (pyrimidine) occurs in one sequence and a pyrimidine (purine) in the other sequence is

$$p_2 = .5 - .5e^{-\phi}.$$

Suppose that in a DNA sequence consisting of N nucleotide sites there are n_1 sites where one purine (pyrimidine) occurs in the sequence of one population and the other purine (pyrimidine) in the sequence of the other population, and n_2 sites where a purine occurs in one sequence and a pyrimidine in the other. The observed proportions $\hat{p}_1 = n_1/N$ and $\hat{p}_2 = n_2/N$ are the maximum likelihood estimators of p_1 and p_2, respectively (see Problem 8.2). The invariance property of maximum likelihood estimators discussed in Section 8.3 implies that the maximum likelihood estimates $\hat{\gamma}$ and $\hat{\phi}$ satisfy the equations

$$\hat{p}_1 = .25 + .25e^{-\hat{\phi}} - .5e^{-\hat{\gamma}},$$

$$\hat{p}_2 = .5 - .5e^{-\hat{\phi}}.$$

These equations lead to the maximum likelihood estimates

$$\hat{\gamma} = -\log(1 - 2\hat{p}_1 - \hat{p}_2), \tag{14.45}$$

$$\hat{\phi} = -\log(1 - 2\hat{p}_2). \tag{14.46}$$

In the DNA sequence considered, the mean number ν of nucleotide substitutions through the course of evolution since the original founding population is

$$\nu = 2N(\alpha + 2\beta)t = N(.5\gamma + .25\phi).$$

From equations (14.45) and (14.46), the maximum likelihood estimator $\hat{\nu}$ of ν is

$$\hat{\nu} = -N\left(.5\log(1 - 2\hat{p}_1 - \hat{p}_2) + .25\log(1 - 2\hat{p}_2)\right). \tag{14.47}$$

This estimator is subject to the same qualifications as those made for the Jukes–Cantor estimator.

It is interesting to compare calculations derived from (14.47) with those from the parallel Jukes–Cantor model. If $N = 3,000$ and the data yield 210 transitional and 90 transversional changes, and thus 300 changes in total, then $\hat{p}_1 = 0.07$ and $\hat{p}_2 = 0.03$. The total number of changes from one predominant nucleotide to another at some time during the evolution of the two populations since their common founder population is then estimated from equation (14.47) to be about 326. This is somewhat larger than the corresponding estimate in the Jukes–Cantor model.

This implies that if a satisfactory estimate of α and β are available extrinsically, the estimate of t differs in the Kimura and the Jukes–Cantor models given the same data in each. Thus if the true model is the Kimura model and the Jukes–Cantor equation (14.34) is used to estimate t, then a "between-models" biased estimation procedure has been used. This indicates a problem with using an estimate of t as a surrogate for a between–population distance, as discussed above. In practice, the true evolutionary procedure was far more complex than that described by the Kimura model, so that even larger biases can be expected if a simple model such as the Jukes–Cantor model or the Kimura model is used for estimation of evolutionary parameters. This is discussed further in Section 14.3.6.

All the above estimators can also be found by writing down the likelihood and maximizing its logarithm with respect to γ and ϕ. In this case the likelihood is a multinomial, being proportional to

$$\left(.25 + .25e^{-\phi} + .5e^{-\gamma}\right)^{N-n_1-n_2} \left(.25 + .25e^{-\phi} - .5e^{-\gamma}\right)^{n_1} \left(.5 - .5e^{-\phi}\right)^{n_2}.$$
$$(14.48)$$

Maximization of this function with respect to ϕ and γ leads to estimators identical to those in (14.45) and (14.46). If the values of α and β are known, a similar procedure may be used to find the maximum likelihood estimate of t. A procedure very similar to this is discussed below in Section 14.3.5.

While the procedure of maximizing (14.48) is unnecessarily long so far as finding estimates is concerned, there is one advantage to it. Variance approximations for $\hat{\gamma}$ and $\hat{\phi}$ can be found from a second differentiation of the logarithm of the likelihood, using the information matrix (8.31), together with an approximation for the covariance of $\hat{\gamma}$ and $\hat{\phi}$ parallel to those for the variances of $\hat{\gamma}$ and $\hat{\phi}$. The parameter ν is a linear combination of γ and ϕ, and the variance of $\hat{\nu}$ can be found by applying equation (2.62).

14.3.3   The Continuous-Time Felsenstein Model

The discrete-time Felsenstein model was discussed in Section 14.2.4. That model is the discrete-time version of a continuous-time model originally proposed to address the problem of phylogenetic tree reconstruction, and so we now describe this continuous-time model.

The essential assumption of the discrete-time Felsenstein model (14.18) is that the probability of the substitution of one nucleotide by another is proportional to the stationary probability of the substituting nucleotide. This assumption is maintained in the continuous-time version of the model. If, for example, the predominant nucleotide at any given time t is a, the model assumes that for small h, the probability that the predominant nucleotide at time $t + h$ is X is $u\varphi_X h$, where X stands for any of the nucleotides g, c, and t, while the probability that the predominant nucleotide continues to be a at time $t + h$ is $1 - u(\varphi_g + \varphi_c + \varphi_t)h$, where in all the above expressions terms of order $o(h)$ are ignored. Parallel assumptions are made if the predominant nucleotide at time t is g, c, or t.

These assumptions define the instantaneous transition rates q_{jk} and q_j in equations (11.29) and (11.30), and this leads to specific forms for four simultaneous differential equations of the form (11.33). The solution of these equations is

$$P_{ii}(t) = e^{-ut} + (1 - e^{-ut})\varphi_i, \qquad (14.49)$$
$$P_{ij}(t) = (1 - e^{-ut})\varphi_j, \quad j \neq i. \qquad (14.50)$$

This solution implies that the continuous-time Felsenstein model satisfies the detailed balance equations (11.35), and is thus reversible.

14.3.4   The Continuous-Time HKY Model

The continuous-time analogue of the HKY model (14.21) is discussed in detail by Hasegawa et al. (1985). Its time-dependent solution is given by the spectral expansion corresponding to the discrete-time model (14.21), with λ_j^n replaced by $e^{(\lambda_j - 1)t}$. It shares all the desirable properties of its discrete-time analogue, and is used in phylogenetic studies, as discussed in Chapter 15.

14.3.5   Continuous-Time Amino Acid Model

The simple symmetric discrete-time model of Section 14.2.8 has a direct continuous-time analogue, found from the Kolmogorov equations (11.33) by putting $q_{jk} = \alpha$. We shall choose α so that the PAM requirement that the probability that the initially predominant amino acid after one time unit is still predominant is 0.99. With these choices the solutions of equations (11.33) are

$$P_{ii}(t) = .05 + .95e^{-20\alpha t}, \tag{14.51}$$

$$P_{ij}(t) = .05 - .05e^{-20\alpha t}, \quad j \neq i. \tag{14.52}$$

These equations are similar in form to the Jukes–Cantor equations (14.27) and (14.28). The requirement $P_{ii}(1) = 0.99$ yields $\alpha = 0.00053$.

These calculations allow maximum likelihood estimation of t, the time (in "PAM model" time units) back to an assumed common ancestor of two contemporary amino acid sequences, provided that the simple symmetric model can be assumed. Suppose, for example, that two contemporary sequences of length 10,000 are compared and that in this comparison there are 1,238 matches and 8,762 mismatches. Since the simple symmetric model is reversible and the time connecting the two sequences is $2t$, the invariance property of maximum likelihood estimators and equation (14.51) show that the maximum likelihood estimate $\hat{t}$ of t satisfies the equation

$$.1238 = .05 + .95e^{-40\alpha \hat{t}}. \tag{14.53}$$

The solution of this equation (with $\alpha = 0.00053$) is $\hat{t} = 120.5$. Of course, as discussed in Section 14.2.8 with respect to its discrete-time analogue, this simple model is quite unrealistic, so that a result such as this cannot be taken as applying to real evolutionary processes.

Further "amino acid" transition matrices may be found from the "amino acid" analogues of the Kimura, Felsenstein and HKY models. For example, the continuous-time amino acid analogue of the continuous-time F81 Felsenstein model (14.50) was discussed by Hasegawa and Fujiwara (1993), and has properties which are the direct "amino acid" extensions of those of the Felsenstein model.

Models based on amino acids can be far more complicated than those based on nucleotides. For example, Muse and Gaut (1994) consider a model with a 61×61 transition matrix whose states correspond to the non-terminating codons. In this model the probability of a substitution of codon i by codon j in time δt is of the form $\alpha \pi_n \delta t$ for a synonymous substitution (that is, if codons i and j correspond to the same the amino acid), is of the form $\beta \pi_n \delta t$ for a non-synonymous substitution for which codons i and j differ by a single nucleotide substitution, and is 0 if codons i and j differ by more than a single nucleotide substitution. Here π_n is the population frequency of the substituting nucleotide.

Further properties of amino acid models where reversibility is assumed are developed by Müller and Vingron (2000).

14.3.6 A Remark About Bias

Three different forms of bias arise in the discussion in Sections 14.3.1 and 14.3.2, and it is important to distinguish among them.

First, in Section 14.3.1 the estimator $\widehat{\alpha t}$ given by equation (14.34) is biased: its expected value is not the true value of αt. This bias is analogous to that arising for the maximum likelihood estimate of the parameter k described below equation (8.34). Although in both cases the correct probability distribution is assumed (the probabilities defined by the Jukes–Cantor model in Section 14.3.1, the gamma distribution in the case of estimating k), a bias in the estimation of a parameter still arises. We might call this a "within-model" bias.

We could hope to amend the estimator to lower or even remove the bias. In the case of the estimation of k as discussed below equation (8.34), the bias would be removed by using the method of moments estimator, or decreased by using the maximum likelihood estimation with a large sample size. In the case of equation (14.34), a bias reduction is possible by using a Taylor series approximation to the right-hand side. The second-order Taylor approximation (B.30) shows that the right-hand side in (14.34) may be approximated by

$$-\frac{1}{8} \log\left(1 - \frac{4}{3}p\right) + \frac{1}{6} \frac{\hat{p} - p}{1 - 4p/3} + \frac{8}{9} \frac{(\hat{p} - p)^2}{(1 - 4p/3)^2}. \qquad (14.54)$$

The mean of $\hat{p}$ is p and the variance of $\hat{p}$ is as given in equation (14.31). Thus the expected value of the right-hand side in (14.54) is

$$-\frac{1}{8} \log\left(1 - \frac{4}{3}p\right) + \frac{8}{9} \frac{p(1-p)}{N(1 - 4p/3)^2}. \qquad (14.55)$$

This implies that the estimator $\widehat{\alpha t}^*$ of αt, defined by

$$\widehat{\alpha t}^* = \widehat{\alpha t} - \frac{8}{9} \frac{\hat{p}(1 - \hat{p})}{N(1 - 4\hat{p}/3)^2}, \qquad (14.56)$$

removes some of the bias arising for the estimator $\widehat{\alpha t}$ defined in (14.34).

Second, the bias discussed in Section 14.3.2 arises for a different reason than that discussed above. In that section the true evolutionary model is the Kimura two-parameter model, but the parameter αt in that model is estimated by the procedure which assumes the Jukes–Cantor evolutionary model, that is through equation (14.34). Thus this bias arises through model misspecification, and the bias might be called a "between-models," or "systematic," bias. This form of bias is not removed by increasing the sample size. Nor is a bias reduction possible along the lines of that leading to (14.56) when the true evolutionary model is quite unknown.

In the case of the bias discussed in Section Section 14.3.2, if the value of α is known, the model misspecification described implies a bias in the estimation of the evolutionary time t. While the numerical value of the bias is not large, the two models involved, that is the Kimura two-parameter model and the Jukes–Cantor model differ from each other only through the value of one parameter. Thus the two models are "close." One can expect a far greater bias if the Jukes–Cantor model, or any other relatively simple evolutionary model, is used for estimation of evolutionary parameters when in fact a far more complicated model is appropriate.

Finally, in using any evolutionary model to estimate properties of a phylogenetic tree, independent evolutionary processes are often assumed at the various sites, with the same stochastic process properties assumed at all sites, at all times and in all species. These assumptions are unrealistic, and using them adds a further systematic bias in estimation procedures. This bias also is not removed by increasing the sample size or by methods similar to those leading to the revised estimator (14.56).

These various biases have obvious implications concerning the reliability of any phylogenetic tree reconstruction. This matter is discussed further in Chapter 15 in the context of such a reconstruction.

Problems

14.1 Prove equation (14.3) by induction on n. (For a discussion of proofs by induction, see Section B.18.)

14.2 Check that the stationary distribution of the Markov chain (14.15) is the vector given in (14.16).

14.3 Check that the Markov chains (14.12) and (14.13) are reversible by showing that they satisfy the detailed balance conditions (11.3), but that the chain defined by (14.15) does not satisfy these conditions and is thus not reversible.

14.4 Show that when the conditions (14.24) hold, the stationary distribution of the model (14.23) is $(\varphi_a, \varphi_g, \varphi_c, \varphi_t)$.

14.5 The Jukes–Cantor model (14.1) is a special case of (that is, is *nested within*) the Kimura model (14.6). It is also special case of (that is, is *nested within*) the Felsenstein model (14.18). Show that it is the only model that is nested within both the Kimura model (14.6) and the Felsenstein model (14.18), and that neither the Kimura model nor the Felsenstein model is nested within the other.

14.6 Show that the conditions (14.24) are necessary and sufficient for the transition matrix (14.23) to be reversible.

14.7 Derive the expression (14.29). (Hint: use the result of Problem 4.7.)

14.8 Check that the continuous-time Kimura model satisfies the detailed balance requirements (11.36).

14.9 Show that the sum in the expression (14.30) does reduce, as claimed, to the expression on the right-hand side of (14.27).

14.10 The aim of this question is to compare the numerical values of the two estimators $\hat{\nu}$ given respectively in equations (14.35) for the Jukes–Cantor model and (14.47) for the Kimura model. In this comparison we identify the values of $\hat{p}$ in the Jukes–Cantor model with $\hat{p}_1 + \hat{p}_2$ in the Kimura model.

  (i) Use the logarithmic approximation (B.24) to show that when $\hat{p}_1$ and $\hat{p}_2$ are both small, the two estimators are close, and differ by only a term of order $\hat{p}^2$.

  (ii) Use the more accurate approximation (B.25) to compare the values of the two estimators when (a) $\hat{p}_1 = \hat{p}_2$, (b) $\hat{p}_1 = 2\hat{p}_2$. Thus show that even when $\hat{p}_1 = \hat{p}_2$ the two estimates of ν differ.

14.11 Estimate the standard deviation of the estimate $\hat{t} = 120.5$ found by solving equation (14.53).

15
Phylogenetic Tree Estimation

15.1   Introduction

The construction, or more accurately the *estimation*, of phylogenetic trees is of interest in its own right in evolutionary studies. It is also useful in many other ways, for example in the prediction of gene function; aspects of this area of the emerging field of *phylogenomics* are discussed by Eisen (1998). In this chapter we give a brief introduction to this topic, on which there is a vast literature. Our focus tends to be on problems and difficulties in currently used procedures. Accounts of the whole field are provided by Hillis et al. (1996) and by Felsenstein (2004).

We deliberately use the expression "tree estimation" rather than the more commonly used "tree reconstruction," since the latter expression can be taken to imply an error-free process leading to the correct tree. In reality, tree estimation is prone to many potential forms of error, some of which we discuss here.

A phylogenetic tree is *binary* if each node except the *root* connects with either one or three branches, while the root, of which there can be at most one, connects with two branches. A tree without a root is called an *unrooted* tree.

The evolutionary relationships between a set of species is represented by a binary tree, and in this book we consider only binary trees. "Species" may refer either to organisms or to sequences such as protein or DNA. It is required that the set of "species" have a common ancestor, so that while one would not construct a tree relating a hemoglobin protein to a ribosomal protein, it would be natural to construct a tree relating the various hemoglobins.

The following are examples of rooted phylogenetic trees. The first relates five different species and the second relates five different hemoglobins. The unlabeled nodes represent common ancestors of the children nodes below them.

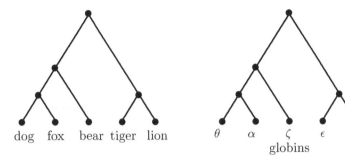

The lengths of the edges represent evolutionary time, so the longer an edge is, the longer is the evolutionary time separating the nodes at the ends of the edge. The node at the top, the root, represents a common ancestor to all species in the tree. In these examples the leaves represent extant species, so they are all at the same vertical height. This may not be so in a tree with extinct species, where the leaves are at possibly different levels, as in the following example:

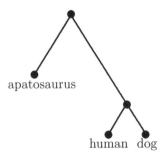

An unrooted tree shows the evolutionary relationships between the species at the leaves but does not indicate the direction in which evolutionary time flowed between the internal nodes. Each node connects either one or three edges in an unrooted tree. An example of an unrooted tree (with five leaves) is:

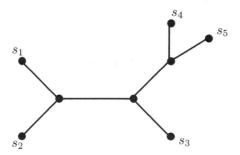

Given a set of species, there is some (unknown) phylogenetic tree connecting the members and giving their true evolutionary relationships. Our goal is to infer the tree as accurately as possible from information about the species. Before the advent of genetic information this was normally done using morphological information, such as the shape of the teeth and bones. The advent of genetic information in the form of DNA or protein sequences has fostered a more mathematical/algorithmic approach to tree estimation.

We discuss in turn three main types of tree-reconstruction methods, namely *distance* methods, *parsimony* methods, and *statistical* methods.

15.2   Distances

Several of the algorithmic procedures used in tree reconstruction are based on the concept of a *distance* between species. We thus introduce the concept of distance in this section and describe some properties of the distances that arise in phylogenetic reconstruction.

The most natural way to define the distance between species is in terms of years since their most recent common ancestor. If this distance were known, no other distance measure would be needed. In practice, this distance is seldom if ever known, and surrogate distances are used instead. The use of surrogates will affect the accuracy of tree reconstruction, as discussed in Sections 15.5 and 15.8. For the moment we assume that distances in years *are* known, and we call these *exact* distances.

Let S be a set of points. The standard requirements for a distance measure on S are that for all x and y in S, (i) $d(x, y) \geq 0$, (ii) $d(x, y) = 0$ if and only if $x = y$, (iii) $d(x, y) = d(y, x)$, and (iv) for all x, y, and z in S $d(x, y) \leq d(x, z) + d(z, y)$.

A distance measure $d(\cdot, \cdot)$ on a set of species is said to be *tree-derived*[1] if there is a tree with these species at the leaves such that the distance $d(x, y)$ between any pair of leaves x and y is the sum of the lengths of the edges joining them. This distance satisfies the three requirements above. However, not all distance measures on a set S are tree derived, since a tree-derived measure must satisfy certain additional requirements, as demonstrated by the unrooted tree shown:

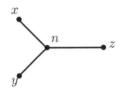

The distances in this tree must satisfy the *strict* inequality $d(x, z) < d(x, y) + d(y, z)$. For a tree-derived distance measure, this strict inequality must hold for all species x, y, and z in the tree.

For rooted trees joining extant sequences there are further requirements. In the rooted tree shown,

(15.1)

the distances between x, y and z must satisfy the equation $d(x, y) = d(x, z)$ and the additional requirements $d(y, z) < d(x, y)$, $d(y, z) < d(x, z)$. This requirement motivates a further abstract distance measure concept. Any distance measure on a set S that satisfies the requirement that of the three distances between any three members of the set, two are equal and are greater than the third, is said to be *ultrametric*.[2]

Thus a tree-derived distance measure on a (rooted) tree of extant species (with exact edge lengths) is ultrametric. The converse of this statement, that any ultrametric distance is tree derived, will be proved in Section 15.3. We shall show further how such a tree can be found, and that up to trivial changes this tree is unique. This allows us to go from a set of extant species to the unique *rooted* tree that relates them, as long as we know the exact distances between all pairs of species.

Not all tree-derived distance measures are ultrametric, as is the case when the tree represents only extant species and exact distances. However, if a distance measure is tree-derived it is still possible to recover a unique *unrooted* tree giving all distances between species. This is demonstrated

---

[1]The term *additive* is generally used. For our purposes "tree-derived" is more natural.

[2]Ultrametric distances arise in many contexts, not only biology, and the definition varies depending on the context. For phylogenetics, this is the correct definition.

in Section 15.4. In practice, distances used in tree reconstruction are not necessarily tree-derived, being surrogate distances that only estimate true distances. Properties of surrogate distances are discussed in Section 15.5.

15.3   Tree Reconstruction: The Ultrametric Case

In this section we prove that if an ultrametric distance measure between species is given, then there is a unique (up to trivial changes) rooted tree joining these species that gives these distances. (What "up to trivial changes" means will be made clear as we proceed.) This tree is found using a constructive method that differs from the well-known UPGMA (unweighted pair group method using arithmetic averages) algorithm discussed below.

We assume that a set of species $s_1, s_2, \ldots, s_n$ is given with an ultrametric distance measure $d(x, y)$ between all pairs of species x and y. The proof of the above claim is by induction. We first construct the correct tree relating two species and show it is unique. We then show that if we can reconstruct the tree correctly and uniquely for m species $s_1, s_2, \ldots, s_m$, for some $m \geq 2$, we can do this also for $m + 1$ species $s_1, s_2, \ldots, s_{m+1}$.

The first step is the construction of a tree relating s_1 and s_2. Both s_1 and s_2 must be equidistant from the root, so there is clearly a unique solution:

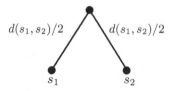

(In this and similar steps we ignore trivial changes such as interchanging s_1 and s_2.) Now suppose there are $m + 1$ species, $m \geq 2$. By the induction hypothesis, we can assume the tree can be uniquely reconstructed for $s_1, s_2, \ldots, s_m$. Consider the root of the tree and the two edges emanating from it. The species $s_1, s_2, \ldots, s_m$ divide into two sets, the set S_L consisting of the species down the left edge from the root and the set S_R consisting of the species down the right edge from the root. Let x be any element of S_L and y be any element of S_R.

(15.2)

Now consider species s_{m+1} and the three distances between x, y, and s_{m+1}. The ultrametric property implies that there are three possibilities, depending on which pair of these three is equidistant from the third. These possibilities are (1) $d(s_{m+1}, x) = d(s_{m+1}, y)$, (2) $d(s_{m+1}, y) = d(x, y)$, and (3) $d(s_{m+1}, x) = d(x, y)$.

If (1) holds, $d(x, y)$ is strictly less than $d(s_{m+1}, x) = d(s_{m+1}, y)$. In this case, we place s_{m+1} follows:

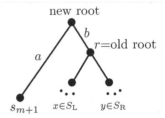

where $a = d(s_{m+1}, x)/2$ and $b = d(s_{m+1}, x) - d(x, y)/2$. Since $d(x, y) < d(s_{m+1}, x)$, b is positive, as required.

This gives a tree that gives the correct distances between s_{m+1}, x, and y. We must check that this placement gives the correct distance from s_{m+1} to any other species. For any species z in S_L or S_R, $d(z, \text{old root}) = d(x, y)/2$. Thus the distance from s_{m+1} to z in the new tree is $a + b + d(x, y)/2$, and it is required to show that this is equal to $d(s_{m+1}, z)$. Now, in the new tree

$$d(s_{m+1}, x) = a + b + \frac{d(x, y)}{2}, \tag{15.3}$$

so it is sufficient to show that $d(s_{m+1}, x) = d(s_{m+1}, z)$ for all z in S_L and S_R. Suppose $z \in S_L$. By the induction hypothesis, the tree does give the correct distance for x, y, and z, so $d(x, z) < d(x, y)$. Therefore, $d(x, z) < d(x, y) < d(s_{m+1}, x)$, the second inequality following from (15.3). Therefore, by the ultrametric property, $d(s_{m+1}, x) = d(s_{m+1}, z)$ as desired. The corresponding argument when z is in S_R is symmetric to this case.

In case (2), $d(s_{m+1}, y) = d(x, y)$. The ultrametric property then shows that $d(s_{m+1}, x) < d(x, y)$. Since x is in S_L and y is in S_R,

$$d(x, y) = d(x, z) \text{ for any } z \text{ in } S_R. \tag{15.4}$$

Therefore, $d(s_{m+1}, x) < d(x, z)$ for all z in S_R, and

$$d(s_{m+1}, z) = d(x, z) \text{ for all } z \text{ in } S_R. \tag{15.5}$$

Thus s_{m+1} is equidistant from every member of S_R. There are two possibilities: Either s_{m+1} is also equidistant from every member of S_L or it is not. If s_{m+1} is equidistant from every member of S_L, we add s_{m+1} to the tree, as shown:

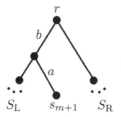

Here $a = d(s_{m+1}, x)/2$ and $b = d(s_{m+1}, y)/2 - a$. Since $d(s_{m+1}, x) < d(s_{m+1}, y)$ it follows that $b > 0$, as it must be.

We next check that this allocation gives the correct distances. For z in S_L we must show that $d(s_{m+1}, z) = 2a$, that is, that $d(s_{m+1}, z) = d(s_{m+1}, x)$. This requirement follows from the assumption that s_{m+1} is equidistant from every member of S_L. Now suppose that z is in S_R. Then the distance between s_{m+1} and z in the tree above is $a + b + \frac{1}{2}d(x, y)$. This simplifies to $\frac{1}{2}(d(s_{m+1}, y) + d(x, y))$. By (15.4) and (15.5), this is equal to $d(s_{m+1}, z)$, as required.

This leaves the case where s_{m+1} is equidistant from every member of S_R but *is not* equidistant from every member of S_L. If we remove the root and its two edges from the tree (15.2), we are left with two trees, one relating the species in S_L and one relating the species in S_R. We call these T_L and T_R, respectively. The tree T_L contains $m - 1$ or fewer leaves, so we know by induction that s_{m+1} can be uniquely added to this tree and all the distances between s_{m+1} and the rest of the leaves in T_L are correct. Call this new tree T'. We know further that since s_{m+1} is not equidistant to every leaf in T_L, T' has the same root as T_L. Therefore, we can put back together the whole tree relating all of $s_1, s_2, \ldots, s_{m+1}$ as

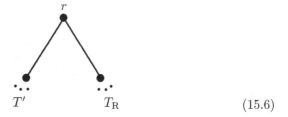

$$(15.6)$$

We now show that s_{m+1} has the correct distances to the rest of the species $s_1, \ldots, s_m$. Since we know that T' has been correctly reconstructed, this is true for all species in S_L. Suppose z is in S_R. Now x is in S_L, so the tree constructs the distance from s_{m+1} to z to be $d(x, z)$. We must therefore show that $d(x, z) = d(s_{m+1}, z)$. We know that $d(x, s_{m+1}) < d(x, y)$ (this follows from assumption 2), and we know that $d(x, y) = d(x, z)$. Therefore, $d(x, s_{m+1}) < d(x, z)$. The ultrametric property therefore gives us $d(x, z) = d(s_{m+1}, z)$, as desired.

The proof for case (3) is identical in form to this proof.

□

This proof is constructive in that it provides an algorithm for reconstructing the tree. One adds the species to the tree one by one in no particular order, with the inductive step showing explicitly where each next species should be added.

We now sketch the UPGMA algorithm of Sokal and Michener (1958), which is used frequently for tree reconstruction. In the case of ultrametric distances, this reconstructs the same tree as that derived by the methods described above. This algorithm can be used for any set of distances, ultrametric or not; however, if the distance is not ultrametric, it cannot reconstruct a tree that gives back the distance. It is often used anyway, as an approximation.

Two species are *neighbors* if the path between them contains only one node. Thus in the tree below, x and y are neighbors while x and z are not.

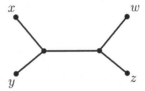

For a tree whose derived distance satisfies the ultrametric property, pairs of species that are closest together with respect to the distance metric are neighbors.

The UPGMA algorithm uses the concept of groups of species. We start with the set of groups where each group consists of a single species. Let $d(i,j)$ be the distance between species i and j. We define the distance between groups G_u and G_v, with respective sizes n_1 and n_2, by

$$d^*(G_u, G_v) = \frac{1}{n_1 n_2} \sum_{\substack{x \in G_u \\ y \in G_v}} d(x,y).$$

Suppose at the first step that the two groups with minimal distance between them are groups $G_r = \{x\}$ and $G_s = \{y\}$. These are now joined with a two-leaf rooted tree, with root "species" r_1, as shown in the diagram.

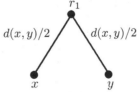

Groups G_r and G_s are now replaced by the single group $G_{rs} = \{x, y\}$. The algorithm continues with this new group, with the distance from G_{rs} to any remaining group being defined as above.

The group G_{rs} is now added to the previous set of groups and G_r and G_s removed. The process is repeated with this new set of groups. In general, at each step the pair of groups having smallest distance between them is replaced by a new group consisting of the union of these two groups.

This process is repeated, building the tree in a bottom-up way until it is entirely reconstructed.

This procedure gives the tree topology. One then can solve for the correct lengths for all of the edges in the tree. For example, in the first step species x and y are joined at the node r_1, the distance from any other species z to r_1 is

$$\frac{1}{2}(d(x, z) + d(y, z) - d(x, y)). \tag{15.7}$$

If the distance is ultrametric, this procedure builds the (unique) correct tree. Unfortunately, when distances are not ultrametric, as in practice will almost always be the case, this tree reconstruction method can give quite misleading results. The tree that the algorithm returns is necessarily a tree relating extant species, and therefore will give an ultrametric distance, so if the original distance is not ultrametric an incorrect tree must be returned. One might think that this tree will at least have the correct topology, if not the correct edge lengths; however as is shown in the diagram below this is not always the case. In this tree, x and y are the two closest leaves in the tree, but they are not neighbors.

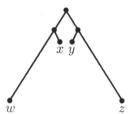

This problem leads to the tree construction method described in the following section.

15.4   Tree Reconstruction: the Neighbor-Joining Approach

In this section we show that given any tree-derived distance measure on a set of species, a tree can be constructed giving this distance measure. This method was introduced by Saitou and Nei (1987).

The proof of this claim uses the quantity $\delta(x, y)$, defined by

$$\delta(x, y) = (N - 4)d(x, y) - \sum_{z \neq x, y} (d(x, z) + d(y, z)), \tag{15.8}$$

where N is the number of species. The function $\delta(x, y)$ is not a distance measure because it may take negative values. However, the following is true.

Theorem. Suppose S is a set of species and d is a tree-derived distance on S obtained from an unrooted tree. If x and y are such that $\delta(x, y)$ is minimum, then x and y are neighbors.

Proof. We follow the proof of Studier and Keppler (1988). Consider four species as in the following tree:

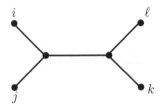

We have

$$\delta(i, j) = -d(i, k) - d(i, \ell) - d(j, k) - d(j, \ell)$$

and

$$\delta(i, k) = -d(i, j) - d(i, \ell) - d(k, j) - d(k, \ell),$$

so that

$$\delta(i, k) - \delta(i, j) = -d(i, j) - d(k, \ell) + d(i, k) + d(j, \ell).$$

The above diagram shows that the positive terms on the right-hand side in this equation include the length of the internal edge twice, whereas the negative terms do not include it. Therefore, $\delta(i, k) - \delta(i, j) > 0$, that is, $\delta(i, k) > \delta(i, j)$. Every leaf in an (unrooted) four-leaf tree has a neighbor. Therefore, any value of δ arising for a pair of non-neighbors x and y is greater than that for neighbors x and z.

Suppose now that $N > 4$. In this case not all species have neighbors in the sense defined above. Suppose i and j are such that $\delta(i, j)$ is minimum, but i and j are not neighbors. If i has a neighbor k, then

$$\delta(i,k) - \delta(i,j)$$
$$= (N-4)d(i,k) - (N-4)d(i,j)$$
$$- \sum_{z \neq i,k} d(i,z) - \sum_{z \neq i,k} d(k,z) + \sum_{z \neq i,j} d(i,z) + \sum_{z \neq i,j} d(j,z)$$
$$= (N-4)d(i,k) - (N-4)d(i,j)$$
$$- \left(d(i,j) + \sum_{z \neq i,j,k} d(i,z) \right) - \left(d(k,j) + \sum_{z \neq i,j,k} d(k,z) \right)$$
$$+ \left(d(i,k) + \sum_{z \neq i,j,k} d(i,z) \right) + \left(d(j,k) + \sum_{z \neq i,j,k} d(j,z) \right)$$
$$= (N-3)d(i,k) - (N-3)d(i,j) - \sum_{z \neq i,j,k} d(k,z) + \sum_{z \neq i,j,k} d(j,z)$$
$$= \sum_{z \neq i,j,k} (d(i,k) + d(j,z) - d(i,j) - d(k,z)). \tag{15.9}$$

The following tree shows that each quantity in the summand is negative.

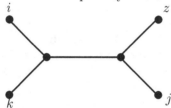

Therefore, $\delta(i,k) < \delta(i,j)$, contradicting the minimality assumption on $\delta(i,j)$. It follows that neither i nor j can have a neighbor if $\delta(i,j)$ is minimal, and the tree must be as shown:

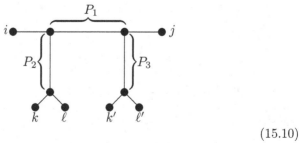

$$\tag{15.10}$$

In this tree P_1, P_2, and P_3 are paths, not edges, but all other line segments represent edges. The path to any leaf other than those shown joins this diagram along one of the paths P_1, P_2, or P_3. The key observation is that we can assume the existence of both k and ℓ down path P_2, because if there were only one leaf down path P_2, it would have to be a neighbor of i. A similar observation holds for k' and ℓ', because neither is a neighbor of j.

We focus for the moment on i, j, k, and ℓ.

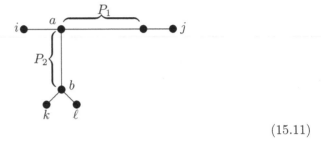

$$(15.11)$$

A calculation similar to the one that led to (15.9) gives

$$\delta(k, \ell) - \delta(i, j)$$
$$= \sum_{z \neq i,j,k,\ell} \Big([d(i, z) + d(j, z) - d(i, j)] - [d(k, z) + d(\ell, z) - d(k, \ell)] \Big).$$

$$(15.12)$$

The proof of this claim is left as an exercise (Problem 15.3).

By our assumption that $\delta(i, j)$ is minimal, the left-hand side of (15.12) is non-negative, so that the sum on the right-hand side is nonnegative. Any leaf z different from i, j, k, and ℓ joins the diagram (15.11) at path P_1 or at path P_2. Let $p(z)$ be the point where the path from z meets this diagram above. This leads to the two possibilities shown below.

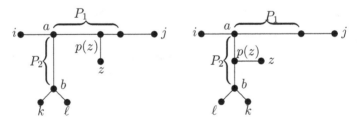

We calculate the summand in (15.12) in both cases. In the left-hand case this summand is equal to $-2d(a, b) - 2d(a, p(z))$, and in the right-hand case it is equal to $2d(a, p(z)) - 2d(b, p(z))$. Given one leaf of each type, say leaf z_1 of the type in the left-hand diagram and leaf z_2 of the type in the right-hand diagram, addition of the two summands gives

$$-2d(a, b) - 2d(a, p(z_1)) + 2d(a, p(z_2)) - 2d(b, p(z_2)),$$

and since $-d(a, b) + d(a, p(z_2)) < 0$, this is negative. It follows that there must be more leaves of type 2 than there are of type 1. Tree (15.10) then shows that there are strictly more leaves joining P_2 than P_3.

However, exactly the same argument can be applied to show that more leaves must join P_3 than P_2. This contradiction shows that the assumption that i and j are not neighbors is impossible, so that the theorem is proved.$\Box$

The above proof shows how an unrooted tree can be reconstructed when the distances $d(i, j)$ are tree-derived. Let x and y be such that $\delta(x, y)$ is the minimum of all $\delta(i, j)$ values, where $\delta(i, j)$ is defined in (15.8). We know from the above that x and y are neighbors. Let r_1 be the node on the path that joins them. Since $d(i, j)$ is assumed to be a tree-derived metric, the distance $d(r_1, z)$ for any leaf $z \neq x, y$ is calculated as defined in (15.7). Using any such leaf, $d(x, r_1)$ and $d(y, r_1)$ are determined by

$$d(x, r_1) = \frac{d(x, z) - d(y, z) + d(x, y)}{2},$$

and

$$d(y, r_1) = \frac{d(y, z) - d(x, z) + d(x, y)}{2}.$$

We next replace x and y in our original set of species with r_1, and repeat the process. Eventually the entire tree is reconstructed.

The "neighbor-joining" algorithm of tree reconstruction for distances that are possibly not tree-derived is based on the above procedure. For each distance $d(i, j)$, we calculate the quantity $\delta(i, j)$, and the species x, y for which this quantity is minimized are first paired. Distances from the node r_1 joining species x and y are then calculated using (15.7). Species x and y are then replaced by a new species r_1 and the procedure repeated starting from a new set of values $\delta(i, j)$ including this node. Eventually, a tree is reconstructed. If the distances between species are not tree-derived, and are given for example as inferred distances as discussed in Section 15.5, there is no guarantee that the tree developed is the correct tree.

15.5 Inferred Distances

Unfortunately, exact distances between species are seldom if ever known, and inferred distances must be used instead. We now discuss some properties of these distances.

If $d(x, y)$ is the exact distance between extant species x and y and $d'(x, y)$ is an inferred distance between x and y, and if there is a constant C such that $d(x, y) = C d'(x, y)$ for all x and y, then since d is tree-derived, d' will be also. The relationship between trees and tree-derived distance measures discussed in Section 15.4 shows that a tree reconstructed from d' has the same topology, that is, the same branching structure and same labeling of the leaves, as that found using d.

We now discuss inferred distances used in practice using DNA information derived from each species considered. The molecular clock need not run at the same speed along different lineages. This can happen for several reasons, including differing generation lengths in different species, different degrees of importance of the biological sequence to the particular species, and so on. Thus a distance measure reflecting the molecular clock might lead to the following tree joining mouse, elephant, and elephant shrew:

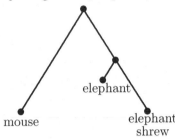

In this example the molecular clock ran faster in the mouse and shrew lineages than in the elephant lineage, so that proportions were not kept with the actual exact distances. If an inferred distance measure perfectly captures this difference in rates, it will be tree-like, and the neighbor-joining algorithm can be used to reconstruct the tree that gave these inferred distances. This tree will have the same branching structure as the true tree, often its most interesting feature.

In practice, such reconstruction is seldom possible. An inferred distance between two extant species is sometimes found by using aligned DNA sequences taken from each, using an inferred distance measure proportional to

$$-\log\left(1 - \frac{4}{3}p\right),$$
(15.13)

where p is the proportion of nucleotides where the two sequences differ. This estimate derives from equations such as (14.34), and thus implicitly assumes a Jukes–Cantor model of evolution. It is thus subject to the comments made below equation (14.47), so that if some more complicated model is appropriate, these estimators might have undesirable properties.

A more complicated inferred distance may be derived from the Kimura two-parameter model of Section 14.3.2, using the analogue of (15.13) for that model. This is done in the example of Section 15.8. If, as is almost certain to be the case in practice, a model more complex than the Kimura two-parameter model is appropriate, then inferred distances using this model might also have undesirable properties.

There are further problems concerning inferred distance estimates. First, proteins and protein-coding DNA sequences have subregions that are highly conserved and other regions that change more readily. After a long time the latter regions will diverge significantly between sequences, while the conserved regions remain more or less the same. Second, the rates at which these sequences evolve also depend on the generation times of the species involved. Finally, measures such as (15.13) are only as good as the alignment on which they are based. "Nonfunctional" sequences, when distantly related, are difficult to align correctly.

As a result of these problems, inferred distances calculated from genetic data of extant species will seldom if ever give a distance proportional to the true distance, and need not even satisfy the three fundamental properties of distance measures given in Section 15.2.

Despite these problems it is still possible to apply the UPGMA and neighbor-joining algorithms. Because inferred distances do not necessarily satisfy the three basic properties of distances, negative distances can arise in the inferred trees. Further, the UPGMA and neighbor-joining trees need not agree, as is shown in the example of Section 15.8.

15.6   Tree Reconstruction: Parsimony

Under the parsimony approach a total "cost" is assigned to each tree, and the optimal tree is defined as that with the smallest total cost. The focus is on finding an optimal *topology*, or shape, and not on edge lengths. The parsimony approach to tree construction is based on the assumption that the tree (or trees) with least cost should be used to estimate the phylogeny of the species involved.

We describe the cost calculation for the case of DNA sequences when unit cost is made for each nucleotide substitution. A central step in the procedure is to allocate sequences to the internal nodes in the tree. For any such set of sequence allocations the total cost of the tree is the sum of the costs of the various edges, where the cost of an edge joining two nodes, or a node and a leaf, is the number of substitutions needed to move from the sequence at one to the sequence at the other.

In principle the optimal tree is found in two steps. First, all possible tree topologies must be listed, including the allocation of the species from which the data are found to the leaves. Second, for any such choice the labeling of the internal nodes of the tree that minimizes the cost of the tree must be found. This second step is accomplished by *Fitch's algorithm* (Fitch (1971)). This algorithm provides the optimal set of internal nodes for a fixed topology. Once these are found the minimum "cost" of a tree with a given fixed topology can be found. The final tree is that which, over all choices of topology, has overall minimum cost.

In practice, the most difficult problem associated with this procedure concerns the listing of all possible topologies of the tree. For trees joining a small number of species the optimal tree can be found by complete enumeration of all possible tree topologies. For three species there is only one form for the topology, namely that shown in (15.1), with three choices for the labeling of the species x, y and z. Figure (6.3) on page 246 shows the five optimal allocations of internal nodes for the three aligned sequences AA, AB, and BB, found by complete enumeration. With four species there are two different forms of topology for rooted trees, as shown. There are three essentially different labelings for the left-hand tree (three choices for the species chosen as the neighbor of the left-most species) and twelve for the right-hand tree (four choices for the rightmost species, three remaining choices for the next to rightmost species), giving fifteen essentially different labeled trees in all. Once again complete enumeration is straightforward.

Although complete enumeration methods are possible when the number of species is small, the number of possible trees increases extremely rapidly as the number of species increases, and complete enumeration is not possible with more than a small number of species. When there are s species at the leaves of a rooted tree there are $1\times\ 3\times\ 5\times\ \cdots\ \times(2s-3)$ $=(2s-3)!/2^{s-2}(s-2)!$ essentially different tree topologies. For 20 species this is about 8×10^{21} topologies. Despite this, some researchers use parsimony methods to reconstruct trees with as many as 500 or more species, and this requires the use of heuristic methods (see Durbin et al. (1998)).

The parsimony method does not construct arm lengths. Nor does it use any explicit evolutionary model. Proponents of the method see the latter as an advantage: to the extent that any evolutionary model used in tree reconstruction might be misleading, a method not dependent on any model has some merit. Others see the lack of an evolutionary model, and the lack of constructed arm lengths, as disadvantages to the method.

15.7   Tree Estimation: Maximum Likelihood

In this section we discuss the maximum likelihood estimation of the phylogenetic tree leading to a number of contemporary species from some common ancestor species, given genetic data from the contemporary species and some continuous-time evolutionary model such as one of those discussed in Section 14.3. Maximum likelihood phylogenetic tree estimation was introduced by Edwards and Cavalli-Sforza (1964), but much of the theory was introduced by, or follows from, Felsenstein (1981).

Suppose first that the topology of the tree is given and that the aim is to find the maximum likelihood estimate of the various arm lengths in the tree, given some continuous-time evolutionary model. This is done by writing down the likelihood of the data in terms of these lengths as parameters, and then maximizing this likelihood with respect to these lengths. This maximization process is, in practice, usually extremely difficult, even when many simplifying assumptions are made. One such simplifying assumption often made is the iid assumption, that the substitution processes at different sites within any species are described by the same stochastic model and that these processes are independent from one site to another. The assumption of independent evolution at different sites allows an analysis of the evolution at the various sites separately, with an overall likelihood obtained by multiplying individual site likelihoods, or in practice an overall log likelihood obtained by summing log likelihoods. It is also frequently assumed that time homogeneity of the common stochastic process applies, and that different species evolve independently. These assumptions can hardly be expected to approximate reality closely, and much recent research attempts to remove them or at least to assess the biases involved when they are incorrectly made. We refer to some of these analyses in Section 15.9.4. On the other hand, in view of the comments about modeling in Section 4.10, in particular the fact that one can expect a simple initial model to be refined by further work, these assumptions form a reasonable starting point for the stochastic approach to the estimation of phylogenetic trees.

We now discuss an analysis in which the iid assumption is made. Initially, no specific assumption is made below about the evolutionary model chosen.

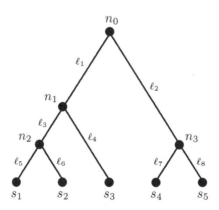

Figure 15.1.

As a specific example, suppose that we have data from five species $s_1, \ldots, s_5$ and that the topology of the phylogenetic tree connecting them is as given in Figure 15.1. Then at any particular site the five respective nucleotides in these five species are known, and these comprise all the data available for this site. Denote these five nucleotides by A, B, C, D, E. We do not know the nucleotides at the internal nodes n_0, n_1, n_2, and n_3, nor do we know the lengths $\ell_1, \ldots, \ell_8$ of the various arms in the tree. The aim of the process is to estimate $\ell_1, \ldots, \ell_8$, which are lengths of time.

Suppose that the nucleotide at the root n_0 is W and is given, and that the nucleotides at the nodes n_1, n_2, and n_3 are X, Y, and Z, respectively. Then there is a certain likelihood for the arm lengths $\ell_1, \ldots, \ell_8$, deriving from whichever nucleotide evolutionary model is chosen. The stationary probability of the nucleotide W is denoted by φ_W and the probability, under the model assumed, of a substitution of nucleotide A by nucleotide B after time ℓ is denoted by $P_{AB}(\ell)$. Then the joint probability that W is indeed the nucleotide at the root of the tree, that the nucleotides X, Y, and Z occur at the nodes as indicated, and that the arm lengths take the values $\ell_1, \ldots, \ell_8$ is

$$\varphi_W P_{WX}(\ell_1) P_{WZ}(\ell_2) P_{XY}(\ell_3) P_{XC}(\ell_4) P_{YA}(\ell_5) P_{YB}(\ell_6) P_{ZD}(\ell_7) P_{ZE}(\ell_8).$$
$$(15.14)$$

Expressions of this form are now computed for all 64 possible combinations of nucleotides at the internal nodes. The sums of the resulting expressions give the likelihood of $\ell_1, \ldots, \ell_8$ conditional on the assumption that the nucleotide at the root of the tree is W. This calculation is then made for all four possible nucleotides at the root of the tree, and the sum of these four expressions is, for the site in question, the likelihood of $\ell_1, \ldots, \ell_8$.

This procedure is now repeated over all the nucleotide sites in the data, and the overall likelihood is computed from the product of the various site likelihoods. This overall likelihood can now, at least in principle, be maximized.

This procedure can now be done, again in principle, for all possible topologies, and then the maximum likelihood tree is that which, taken over all possible topologies, has the maximum likelihood.

These comments gloss over difficult computational problems. There are, however, some computational simplifications possible. In the first part of the process, the various summations implied by the above operations can be reorganized in a way that saves considerable computational effort, as pointed out by Felsenstein (1981). A second computational saving was also noted by Felsenstein. If the stochastic process of nucleotide substitution assumed is reversible, it is impossible to estimate the location of the root of the tree. This "pulley principle" implies that in the reversible case the estimation procedure relates to unrooted trees. Since with s observed species there are $(2s - 3)!/2^{s-2}(s - 2)!$ rooted trees, as noted in Section 15.6, and the smaller number $(2s - 5)!/2^{s-3}(s - 3)!$ of unrooted trees (Edwards and Cavalli-Sforza (1964), Felsenstein (1978a)), one might think that the use of unrooted trees would imply some useful economy in computation. However, even for comparatively small s the number of unrooted trees is still prohibitively large for a complete search, and thus heuristic methods are necessary for the second part of the process.

Felsenstein (1981) suggests as one heuristic method the strategy of building up an unrooted tree starting with two species and successively adding species. If at any stage there are $k - 1$ species in the tree, there are $2k - 5$ segments to which species k can be attached. Each of these is tried in turn, the maximum of the likelihood for each possible attachment is calculated, and the topology with the highest likelihood accepted. Before the next species is tried, local rearrangements of the current tree can be carried out to see whether any rearrangements increase the likelihood. If so, the rearrangement maximizing the likelihood is accepted.

This process does not necessarily lead to the tree with the maximum likelihood, and the tree constructed by this process will depend on the initial two species chosen and the order in which species are added. Nevertheless, with reasonably self-consistent data the same tree will tend to arise whatever order of species is chosen (Felsenstein (1981)). Different orderings can be tried, and if more than one tree is obtained from the different orderings, that with the highest likelihood can be chosen. Heuristic methods such as this overcome some of the computationally difficult problems of tree estimation.

In some cases the maximum-likelihood tree can be shown to be identical to the tree found by parsimony methods. This is interesting because the first attempt at phylogenetic tree reconstruction by maximum-likelihood methods (Edwards and Cavalli-Sforza (1964)) proved to be beyond the computing power of the time, and Edwards and Cavalli-Sforza resorted to an algorithmic parsimony approach to the problem. The links between maximum likelihood estimates and parsimony estimates are examined in detail by Tuffley and Steel (1997). Identity of the two will occur only under the simplest, and probably unrealistic, evolutionary models. Further aspects of the relation between the maximum likelihood and parsimony tree reconstruction methods are given by Lewis (1998).

When the evolutionary process possesses simplifying properties, for example iid substitution processes at the various sites, time homogeneous substitution processes, and identical processes along the various arms of the phylogenetic tree, one would expect that all methods discussed above would lead to similar trees, all of which should be close to the true tree. This would be especially so if the stochastic substitution process assumed in the data analysis were the same as that generating the actual substitutions. Kuhner and Felsenstein (1994) generate data at the tips of various trees under all the simplifying assumptions listed above. The substitution process at any site used in deriving the simulated data was chosen as the Kimura two-parameter model discussed in Section 14.3.2, and this was in effect the model assumed in the data analysis. Further, the transition/transversion ratio assumed in the analysis was set equal to that in the simulation. A further feature of the analysis was that the mean number of substitutions at any one site in the entire phylogenetic tree was about 0.02 in the "low" substitution and about 0.9 in the "high" substitution model. However, even at the high rate 0.9, the probability of two or more substitutions at a given site in the entire tree is, from the Poisson approximation (4.11), about $1 - e^{-.9}(1 + .9) = .23$, so that on average two or more substitutions will arise in less than one quarter of cases. This implies that the tree reconstruction process is simpler than those arising when many substitutions can be expected to occur, for which there will be an increased level of scrambling of data at the tips of the trees.

The simulation results of Kuhner and Felsenstein (1994) confirm that under these many simplifying assumptions, all standard tree estimation methods lead to similar results and to trees either identical to, or very close to, the true tree, with accurate estimates of arm lengths and other features. Kuhner and Felsenstein also carry out simulations when substitution rates are allowed to vary between branches of the tree and when substitution rates are allowed to vary between different sites. In this case inaccuracies and biases arise in the various estimation procedures considered. This indicates the danger of tree reconstruction methods using real data when the simplifying iid and constancy of rates assumptions are made, since in practice these assumptions are unlikely to hold. This is illustrated by the example of Section 15.8.

For real data sets a different picture can emerge. Yang (1994), (1996a), (1996b), (1997a), (1997b) analyzed a variety of data sets from various primate species. Although it is not easy to summarize the results of these analyses briefly, some broad conclusions do emerge. First, simple models such as the Jukes–Cantor, Kimura, and and the simple Felsenstein model (14.18) can result in severe estimation errors of branch lengths. The HKY model must perform better than these models since they are all special cases of this model, and the added complexity of the HKY model does appear to add significant flexibility to the modeling process. However the HKY model itself sometimes performs significantly worse than the general reversible process model (see Section 14.2.7), of which it is itself a particular case. Thus increasing the complexity of the model often seems to be necessary and to be worth the loss of degrees of freedom involved. On the other hand the completely general evolutionary model (14.23) does not seem usually to perform significantly better than the general reversible process model. Thus given the desirability of the reversibility criterion, the latter model appears often to provide a reasonable compromise between simplicity and analyzability on the one hand and complexity and reality on the other.

Further aspects of Yang's analysis refer to the question or substitution rate heterogeneity, discussed in Sections 15.9.3 and 15.9.4.

15.8 Example

The various tree construction methods outlined above use different approaches to construction and different optimality criteria. Those methods based on distances suffer from the fact that the inferred distances normally used need not be proportional to the true distances, and in fact may not even be tree-like. Further, all methods make either explicit or implicit assumptions about the evolutionary process. These are explicit in the maximum likelihood approach but implicit in other approaches. For example, the logic of the parsimony approach relies implicitly on an assumption of equal evolutionary rates in all edges of the tree and at all times. These assumptions clearly are not correct for the real historical evolutionary process, and the explicit assumptions used in many maximum likelihood approaches are also not likely to reflect biological reality closely. Given this, the "between-model" biases that can arise as described in Section 14.3.6 become relevant.

It is therefore not surprising that the four methods described, namely ultrametric reconstruction, neighbor joining, parsimony and maximum likelihood, will often produce different trees when real (as opposed to simulation) biological data are used, even when the same data are used in the four methods. This is illustrated in the example in this section, which investigates the evolutionary relationships between 14 species of mammals. We will see that even though all four methods disagree to some extent, there are divisions that they all do agree on. This leads to the concept of a *consensus tree*, which tries to capture this information. We do not give the details of consensus trees, but we illustrate the basic idea in this example. The DNA used in the tree construction is from the interphotoreceptor retinoid binding protein (Stanhope et al. (1996)). Sequences for the 14 species were taken from Genbank, aligned using CLUSTAL W (Thompson et al. (1994)), and a 532 nucleotide ungapped subalignment extracted by eye and used as input to the various tree algorithms. The species are

Marsupial Mole	Whale	Insectivore
Wombat	Dolphin	Human
Rodent	Pig	Sea Cow
Elephant Shrew	Horse	Hyrax
Elephant	Bat	

and the alignment is

```
gctccagcaaatgatcaagtaccaggtattggagggcaatgtgggttacctaagagtggactacatccctggccag
gctccagcaaatgatcaagtaccaggtactggagggtaatgtgggttacctgagagtggactacatccctggccag
gctacagaggaatattcaccatgaggttctggagggcaacttgggttacctatgggtggacgatctcttgggccag
gctggagagaagcatgagctacaggattctggatggtaatgtgggctacttgcagatagacaacatcccaggccag
gctgcagacaagcatgagctacaaggttctggagggcaacgtgggctacctgcgggtagacaacatcccaggccag
gctgcagaacggcctccgccatgaggttctggaaggcaatgtgggctacctgcgggtggacgacatcccaggccag
gctgcagaacggcttccgccatgaggttctggaaggcaatgtgggctacctgcgggtggacgacatcccgggccag
gctgcacaatagtctccgccatgaggttctggaaggcaatgtgggctacctgcgggtggacgacatcccaggccag
```

```
gctgcaggagggcatccgctatgacattctggagggcgacgtgggctacttgcgagtggacaacatcccgggccag
gctgcaaaaggccatccactacaatgttctggagggcaacgtgggctactttcgggtggacgacatcccgagccag
gctgcagagggccatccgctaccaggttctggcggccaatgtgggctacctggggagggataaacctccccggtcag
gctgcaaagggggcctccgccatgaggttctggagggtaatgtgggctacctgcgggtggacagcgtcccgggccag
gctgcagaccagcatgagctacaaggttctggatggcaatgtgggctacctgcgggtagacaacatccctggccag
actgcagacaagcatgagctacaaggttctggagggcaacgtgggttacctgcgggtagacaacattccgggtcaa
```

```
gaggtagtagaaaaagtcggggagttcctggtgaatgacatctggaagaagctcatggggacatcctctctagtgc
gaggtggtagagaaagtcggggagttcctggtgaatgatgtctggaagaagctcatggggaccctcttctctggtgt
gaggtactgagtaagctcggggggattcctggtggcccacatgtggggggcagctcatgaatacctctggcttggtgc
gaggtactgagccgactaggggccttcctggtggcccatgtctggagacagctcatgggcacctctgctttggtgt
gaggtgctgaaccagctgggggccttcctggtgactcacgtctggaagcagcttatgggctcctctgccttagtgc
gaggtgatgagcaagctgaggagcttcctggtggccaacgtctggaggaagctcatgggcacctctgccttggtgc
gaggtgatgagcaagctgaggagcttcctggcggccaacgtctggaggaagctcatgggcacctctgccttggtgc
gaggtgatgaacaagctgggggagcttcctggtagtcaacgtctgggaaaagctaatgggcacctctgccttggtgc
gaggtggtgagcaagctggggggcttcctggtggacaatgtctggaggaagctcatgggcacctctgccttggtgc
gaggtggtgagcaatcttggggggcttcctcgtggacaatttctggaggaagctcctgggcacctctgccttggtgc
gaggtggtgaccatactgggggctctcctggtggccaatgtctgggggaagctcatagccacctctcccttggtgc
gaggtgctgagcatgatggggggagttcctggtggcccacgtgtgggggaatctcatgggcacctccgccttagtgc
gaggtgctgagccgtctgggggcttcctggtgactcacatctggaagcagctcatgggctcctctgccttagtcc
gatgtgctgaaccagctggggggcttcctggtgactcatgtgtggaagcagctcatgggctcctctgccttagtgc
```

```
tagatctccagcacagcacagggggtgaagtttcgggaatcccctttgtcatttcctatctacatcagggggatat
tggatctccagcacagcacgggggaggcgaagtttcaggaatcccgtttgtcatttcctacctacaccagggggataa
tagatctccggcactgtactgggggggcatgtttctggtattccctatgtcatctcctacttgcaccccgggaacac
tggacctgcgcggcagtgcacaggaggccatgtttccagcatcccttaccttatttcctacctgcacccagcgggcac
tggacctgcgacactgcacagggggccatgtctccagcatcccttacctcatttcctacctgcacccggggcggcac
tggacctccgccattgcactggggggcacatttctggcatccccctatgtcatctcctacctgcacccggggaacac
tggacctccgccactgcactggcggccacatttccggcatccccctatgtcatctcctacctgcacccagggaacac
tagacctccgggcactgcaccaggggccacgtttctggcatccccctatgtcatctcctacctgcacccagggaacac
tggacctccgggcactgcactgggggccacgtttccggcatccccctatatcatctcctacctgcacccaggaaacac
tagacctcccacactgcactgggggggcacgtttctgggatctcctatgtcatctcctacttgcaccgagggaacac
tggacctccgacactgcactggggggccatgtctctgggatccccctacgtcatctcctacctgtacccaggaaacac
tggatctccgggcactgcacaggaggccaggtctctggcattccctcacatcatctcctacctgcacccagggaacac
tggacctgcggcactgtatgggtggccatgtctccagcatcccttacatcatctcctacctacaccccggaggagc
tggacctaaggcactgcacgggggggccatgtctccagtatcccttacctcatctcctacctgcatccagggagcac
```

```
cctgctccatgtagacacagtttatgaccggccatcaaacactaccacagagatctggacccagcctcaggtgctg
tctgctgcatgtagacacagtttatgaccggccatcaaacaccaccacagagatctggaccctgccccaggtgttg
aatcatgcatgtgaacaccatctatgatcggccctctaataccaccacagagatctggaccttggccaaggtcctg
ggtcctgcacgttgacaccatttacaaccgtccctctaacacaaccactgagctctggactttgcctcaggtgctt
cgtgctgcacgtggacaccatttacaaccgcccctccaatacgactacggagctctggaccttgccccaggtgctg
agtcctgcacgtggataccatctatgatcgcccctctaatacgaccactgagatctggaccctgcccgaagtccta
agtcctgcatgtggataccatctacgatcgcccctctaatacgaccactgagatctggaccctccccgaagtccta
ggtcctgcacgtggacaccatctatgaccgtccctccaatacgaccactgagatctggaccctgcccgaagtcctg
ggtcctgcacgtggacaccatctacgaccgcccctccaatacgaccactgagatctggaccctgcccgaggtcctg
cgtcctgaatgtggacccactctatgaccccccctccaacacgaccacagagatctggaccctgccccaggtcctg
ggtcctgcatatggacaccatctatgaccgcccctccaatatcaccactgagctctggaccctgcccagctccag
catcctgcacgtggacactatctacaaccgcccctccaacaccaccacgggagatctggaccttgccccaggtcctg
agtgctgcatgtggacaccatttacaaccgcccctccaatacgactactggggtctggaccttgccccaagtgctg
tgtgctgcacgtggacaccatttacaaccgcccctccaatacaactactgagctctggaccttgccccaggtgctg
```

```
ggtgagaggtatggaggggagaaggacatggtggttctcaccagccatcatactgtaggggtagctgaggatatcg
```

```
ggtgagaggtacggtggggagaaggacgtggtggtcctcaccagccatcacacggtcggggtagcagaggatattg
ggggagaggtacagtgctgacaaggatgtggtggtcctcaccagtggccacactggaggagtgggtgaggacattg
ggggagagatacagtgctgagaaggatgtggtggtcctcaccagtggtcaaacccggggtgtggctgaggacattg
ggggagaggtatagcgccgacaaggatgtggtggtcctcaccagtggccacaccaggggcgtggccgaggacatcg
ggagagaactacggtgccgataaggatgtggtggtcctcaccagtggtcgcaccgggggtgtggctgaggacatcg
ggagacaactacggtgccgataaggatgtggtggtcctcaccagtggtcgcacgggggggtgtggctgaggacatct
ggagacaggtacagtgcggataaggacgtggtggtcctcaccagcagccacacaggggcgtggctgaggacatcg
ggagagaggtacagtgccgacagggatgtggtggtcctcaccagtggccacaccgggggcgtggccgaggacattg
ggagagaggtacagtgctgacaaggatgttgtggtcctcaccagtggccacactggaggagtggctgaggacattg
ggagagcggtacggtgcagacaaggatgtggtggtcctcatcagcgaccacactgggggtgtggctgaggacatta
ggagaaaggtacggtgccgacaaggatgtggtggtcctcaccagcagccagaccaggggcgtggccgaggacatcg
ggagaaaggtacagtcccaacaaggatgtggtggtcctcaccagtggccacaccaggggcgtggccgaagacatcg
ggggagagatacagtgctgacaaggatgtggtggtcctcaccatgggccacaccaggggtgtggccgaggacatcg
```

```
cctatattctcaagaagatgcgccgggccattgtggtgggagagcagactctgggaggggccctagatctccggaa
cctacatcctcaagaagatgcgccgggccattgtggtgggagagcagactctgggaggggccctagatctccggaa
cctatatcctcaaacagatgcgcagggccatcatggtgggtgagcagactgaaggtggtgccctggacctccagaa
tctacatcctcaagcagatgggcagggccatagtggtgggtgaacgtactgggggggtctccctggacctccagaa
tctacatcctcaagcagatgggcagggccatcgtggtgggcgagcggactgagggtggtgccctggacctccagaa
cttatatcctcaaacaaatgcgcagggccattgtggtgggcggaggactgtggggggggccttggacctccagaa
cttatatcctcaaacagatggacagggccatcgtggtggacgaacggactgtggggggggccttggacctccagaa
cctacatcctcaaacagatgcgcagggccattgtggtcggcgagcgaactgtggggggtgccctggacctccagaa
cttacatcctcaaacagatgcgcaggaccatcgtggtgggtgagcggaccgtgggaggtgccctggacctccagaa
cttacatcctcaaacagatgcgcagggccattgtggtgggtgagcagactgtggggggtgccctggacctccagaa
cttacatcctcaaacagatgcgccgggctattgtggtgggcgagcagactgtggggggctgctctggacctccagaa
cgcacatccttaagcagatgcgcagggccatcgtggtgggcgagcggactggggagggggccctggacctccggaa
ttcacatccttaagcagatgggcagggccatagtggtgggcgagaagacggaggcaggtgccctgcacctccagaa
tctacatcctcaagcagatgggcagggccattgtggtaggcgagcggaccgagggtggtgccctggacctccagaa
```

```
gctgcgcatcggtcagtcagactttttcatcactgtgcccgtgtcacgctccctgagccccccttggtggggggagt
gcttcgtattggtcagtcagactttttcatcactgtgcccgtgtcccgttctctgagcccccctcagtgggggggagc
actgaggataggccagtccaacttcttcctcacagtgcctctggcgatgtctctggggccgatgggtggaggtggc
gctaaggatagccaactctgacttcttcctcactctacctgtgtccaggtccttggggcctctgggtggaggcacc
gataggccactctgacttcttcctcactctgcctgtgtctaggtccttaggcccctgggcgggggaagccagaca
gataggccagtctgacttcttttctcaccgtgcccgtgtccaggtccctggggcccctgggcaagggcagtcagact
gataggccagtctgagttcttttctcacagtgcccgtgtccaggtccctggggcccctgggcaagggcagccagact
gataggccagtccgacttctttctcaccgtgcctgtgtccaggtccctggggcccctgggtgagggcagccagaca
gataggccagtccgacttcttcctcaccgtgcccgtgtccaggtccctgggtctgcgcgaggtcctcatgcataac
gataggccagtctgacttcttcctcactgtgcctgtgtcctaggtccctgggctctggggtgggggcaggcagaca
gataggccagtctgacttcttcatcactctgcctgtgtctccaggtctctggggactctgggcggggggcagccagaca
gataggcgagtctgacttcttcttcacggtgcccgtgtccaggtccctgggggccccttggtggaggcagccagacg
gataggtcactctgatttctttctcactctgcctgtgtccaggtccttggggcctttgggcaggggaagccagaca
aataggtcactcagacttcttttttcactctgcctgtgtccaggtcactgggcccccttaggcaggggaagccagaca
```

Tree reconstruction using the parsimony, maximum likelihood, UPGMA, and neighbor-joining methods was carried out using PHYLIP (Phylogeny Inference Package, Felsenstein (1980–2000)). We give below the output as it is given from the package. The distances used in the UPGMA and neighbor-joining methods are those given by the distance matrix in PHYLIP, and are based on the maximum likelihood estimates of the divergence times between any two species under the continuous time version of the second Felsenstein model of Section 14.2.4. These distances are:

```
mars.mole   .00 .11 .41 .44 .39 .37 .40 .37 .41 .36 .40 .37 .42 .39
wombat      .11 .00 .39 .40 .36 .33 .35 .33 .35 .32 .33 .33 .38 .34
rodent      .41 .39 .00 .33 .30 .24 .25 .22 .28 .23 .25 .23 .32 .31
elph.shrew  .44 .40 .33 .00 .20 .26 .26 .25 .28 .28 .28 .26 .20 .21
elephant    .39 .36 .30 .20 .00 .22 .23 .22 .25 .25 .24 .21 .11 .12
whale       .37 .33  24 .26 .22 .00 .03 .10 .16 .17 .18 .17 .22 .24
dolphin     .40 .35 .25 .26 .23 .03 .00 .11 .16 .19 .18 .17 .22 .25
pig         .37 .33 .22 .25 .22 .10 .11 .00 .17 .18 .19 .17 .24 .24
horse       .41 .35 .28 .28 .25 .16 .16 .17 .00 .17 .21 .20 .25 .26
bat         .36 .30 .23 .28 .22 .17 .19 .18 .18 .00 .15 .20 .27 .27
insectivor  .40 .33 .25 .28 .26 .18 .18 .19  21 .15 .00 .19 .26 .26
human       .37 .33 .28 .26 .21 .17 .17 .17 .23 .20 .19 .00 .22 .23
sea cow     .42  38 .32 .20 .11 .22 .22 .24 .25 .27 .26 .22 .00 .14
hyrax       .39 .34 .31 .21 .12 .24 .25 .24 .26 .27 .26 .24 .14 .00
```

Because these distances are not tree-derived, the trees found from the
UPGMA and neighbor-joining methods need not be identical (and, as
shown in Figure 15.2, are not in this case), and will only approximate the
real tree. These two trees are shown in Figure 15.2, and the trees derived
from the parsimony and maximum likelihood trees are shown in Figure
15.3.

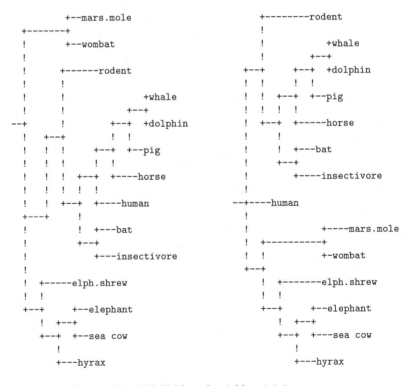

Figure 15.2. UPGMA and neighbor-joining trees

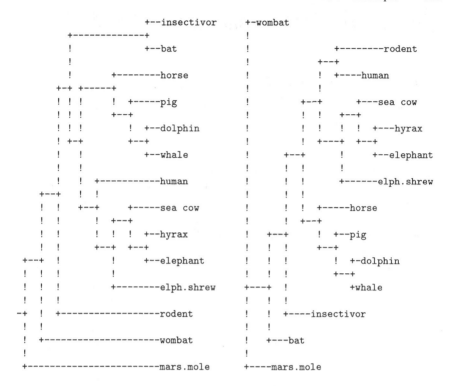

Figure 15.3. Parsimony and Maximum Likelihood trees

It is interesting to note that some relationships are preserved in all four trees, while others vary from tree to tree. The human species has the most ambiguous placement. The UPGMA tree groups human with horse, pig, dolphin, and whale; the neighbor-joining tree places human further away from those species; the parsimony tree places humans with the group containing elephant, sea cow, hyrax, and elephant shrew; and the maximum likelihood tree places humans with rodents. These wide differences highlight the effects of the different modeling assumptions, the different optimality criteria, and the different broad approaches to tree construction. The problems associated with modeling and hypothesis testing in tree construction are thus significant, and are discussed further in Section 15.9.

15.9   Modeling, Estimation, and Hypothesis Testing

15.9.1   Estimation and Hypothesis Testing

Suppose that a phylogenetic tree has been estimated by one of the methods discussed in Section 15.8 or by some other reasonable method. How much confidence can we place in the various features that this tree possesses? We illustrate approaches to this question by discussing the concept of monophyletic groups and the estimation of the entire tree topology.

Given any internal node in a phylogenetic tree, there is a set of leaves, that is, contemporary species, that descend from that node. We call any such a set of leaves a *monophyletic group*. A tree reconstructed from a set of species can be thought of as a set of predictions of monophyletic groups. Suppose that the investigator is a priori interested in the possible existence of some monophyletic group, and this monophyletic group does arise in a given tree estimation process. How much confidence can be placed in the claim that this represents a true monophyletic group? As a related question, how much confidence can be placed in the estimate of the entire topology of the tree?

Any tree estimation method implicitly or explicitly assumes certain properties of the evolutionary process leading to the data at the leaves of the tree from which the evolutionary reconstruction is made. If these assumptions are incorrect, a misleading assessment of monophyletic groups and the overall tree topology can arise. If several different estimation procedures all rely on similar but nevertheless incorrect implicit assumptions about the evolutionary process, or if resampling estimates are based on the same, but incorrect, evolutionary assumptions, consistency of a monophyletic group or of the tree topology from one estimation procedure to another does not add confidence to the claim that this represents a true group or topology. Thus the fact that {Elephant, Elephant Shrew, Hyrax, Sea Cow} forms a monophyletic group in all four of the tree estimation examples in Section 15.8 does not mean that we can assign any particular measure of confidence to the statement that Elephant, Elephant Shrew, Hyrax, and Sea Cow really do form a monophyletic group. This is the first and most important point to make when assessing the accuracy of evolutionary tree estimation.

Even if an estimation procedure correctly assumes the properties of the evolutionary process leading to the data at the leaves of the tree, it is not certain that correct monophyletic groups or the correct tree will be estimated from these data. This follows from the stochastic nature of the evolutionary process and of the sampling procedure.

Despite these problems, it is often asked how much confidence one can place in some feature of an estimated phylogenetic tree. Bootstrap methods are (incorrectly) sometimes used for this purpose, and we now discuss their properties in the phylogenetic estimation context.

15.9.2 Bootstrapping and Phylogenies

There is some confusion in the literature about what a bootstrap procedure can do, and what the properties of the procedure are. For example, bootstrap methods are sometimes said to lead to confidence intervals, or at least bootstrap confidence intervals, for some monophyletic group. However this use of terminology is misleading and is avoided here: a confidence interval properly defined is a range of real numbers, whereas in the monophyletic group context what is sought in the bootstrap procedure is a measure of confidence in the claim that an estimated monophyletic group is a real one.

In the bootstrap procedure of Section 8.6.4, a collection of n "observations" is drawn from n data values $(x_1, x_2, \ldots, x_n)$ with replacement, and the procedure is then repeated a large number R of times. This leads for example to confidence intervals using (8.66).

By analogy, the bootstrap procedure in the phylogenetic context proceeds as follows. Suppose that a phylogenetic tree has been estimated by some given method, for example maximum parsimony, using as data the nucleotides at n aligned nucleotide sites in each of m species at the leaves of the tree. We can think of the m-tuple of m nucleotides at site i in these species as a data vector $\mathbf{x}_i$, and the entire collection of the data at n aligned sites as a set of vector values $(\mathbf{x}_1, \mathbf{x}_2, \ldots, \mathbf{x}_n)$. Following the bootstrap paradigm, n vectors are drawn from $(\mathbf{x}_1, \mathbf{x}_2, \ldots, \mathbf{x}_n)$ with replacement from the original data. Some vectors, that is some sites, might not be represented at all in this sample, some might be represented once, some twice, and so on. The phylogenetic tree is then estimated using this bootstrap sample as "data," employing the same estimation procedure as for the original data. The process is then repeated a large number R of times, and a phylogenetic tree is estimated from each bootstrap sample by the same method as that used for the original data. What can these R trees tell us?

Before answering this question a preliminary comment is in order. The general bootstrap procedure of Chapter 8.6 assumes that the data $(x_1, x_2, \ldots, x_n)$ are the observed values of iid random variables. Thus in applying the bootstrap procedure as described above one assumes implicitly that the evolutionary processes at the various sites are independent and have identical stochastic properties. These are strong assumptions that are likely to be violated in practice. Newton (1996) emphasizes the importance of the independence assumption, and makes the reasonable point that sampling from separated sites in the DNA sequence available, rather than from adjoining sites, might be used to try to achieve the independence requirement.

Suppose nevertheless that the bootstrap procedure is carried out. Clearly, any feature that does not appear in most of the R trees is not interesting. What can be said about those features that do occur in all, or almost all, of the R trees? To discuss this we consider the concepts of the *preci-*

sion, repeatability, and *accuracy* of phylogenetic bootstrap procedures, as introduced by Hillis and Bull (1993).

The concept of precision relates to the bootstrapping procedure itself, and measures the extent to which an inference derived from R bootstrap samples can be expected to agree with the corresponding inference derived from an infinite collection of bootstrap samples of the same data. This is a question internal to the bootstrap procedure and has no direct relation to inferences concerning the unknown phylogeny itself, and we do not consider it further here.

The next concept is that of repeatability. Repeatability is defined as the probability that some feature of interest (for example some monophyletic group) which was obtained from an actual sample taken from some set of n sites would also arise using data hypothetically obtained from another set of n sites. The initial aim of the bootstrapping procedure as introduced by Felsenstein (1985) was to estimate repeatability using bootstrap data from the actual sample. Although the data at the two sets of n sites arise from the same phylogenetic tree, the properties of the stochastic processes at the two sets of sites need not in principle be the same: parameters such as mutation rates can differ, for example, between the two sets of sites. Clearly one can only expect a bootstrap procedure of the actual data to estimate repeatability if the properties of the stochastic processes leading to the actual data and the hypothetical data are the same. The calculations described below assume that this assumption, which we call the iid assumption case, holds.

Suppose for example that one is interested in some group of species, and the feature of interest is whether this group is monophyletic. An alternative definition in the iid assumption case is that repeatability is the probability that, given that this group is estimated as being monophyletic when the data at hand are used, it will be estimated again as being monophyletic using data from a re-run of the stochastic evolutionary process that led to the original data, with the same phylogenetic tree and the same tree estimation algorithm used in the re-run as applied for the original data. This definition allows simple simulation estimation of repeatability, and thus a comparison of this estimate with its bootstrap estimate. We discuss this further below.

We now turn to the most important question, namely that of accuracy. Suppose again that one is interested in some group of species, and the question of interest is again whether this group is monophyletic. The accuracy of a tree estimation procedure is the probability that, if in fact this group truly is monophyletic, that the group will appear as monophyletic in the estimated tree.

There are two aspects concerning the question of accuracy. First, as discussed in Sections 14.3.6 and 15.8, any estimation procedure makes implicit or explicit evolutionary assumptions that might differ from those applying for the true historical evolutionary tree. If there is a substantial difference between the true evolutionary process and that assumed either implicity or explicitly in the estimation process, one can expect incorrect tree estimation to arise. Any error in the estimated tree will tend to be shared with bootstrap tree estimates, so that close agreement between an estimated tree and a large proportion of bootstrap estimates does not automatically imply confidence that the estimated tree is accurate. Thus even if a high proportion of bootstrap tree estimates lead to the same monophyletic group as that obtained from the original data, one has no increased confidence that the monophyletic group is a real one if there is a systematic bias in the tree estimation procedure. This obvious, but important, point is often overlooked in the literature.

Second, several authors have asked, when no such systematic bias arises, what the bootstrap procedure can tell us about the repeatability and accuracy of a phylogenetic tree estimate. Hillis and Bull (1993), for example, claim on the basis of many simulations that even when no systematic biases arise, bootstrap estimates when taken as estimates of accuracy are biased and often conservative. This matter, and the question of the extent to which a bias correction is possible, is discussed at length by Felsenstein and Kishino (1993) and by many other authors.

The simple evolutionary model of Newton (1996) allows explicit calculations bearing on the above questions. This model has the same general nature as the continuous time Jukes–Cantor evolutionary model, but with one of only two, rather than four, possible "nucleotides", called "0" and "1," possible at any site. As the the Jukes–Cantor model, extremely simple properties are assumed: specifically, the parameter q_{ij} of equation (11.29) is taken as a fixed constant α, being the same at all sites, all times, and in all branches of the phylogenetic tree. Further, independent evolution is assumed at different sites. As a further simplification, only three species, A, B, and C, were considered. This model is then the simplest of all possible evolutionary models.

With three species, only three (binary rooted) topologies are possible, denoted ((AB)C), ((AC)B), and ((BC)A). The correct topology is the first of these, that is where A and B form a monophyletic group. In this topology the time backwards from the present to the branching of species A and B is t_0 and the time from the present back to the initial branching of the group (AB) with C is 1.

The data at the various sites observed fall into four groups. For any site in the first group the types observed in all three species are the same (that is, all are "0" or all are "1"). For any site in the second group the same type ("1" or "0") arises in species A and B and the other type ("0" or "1") arises in species C. For any site in the third (resp. fourth) group, species C shares the same type with species A (resp. B) that is different from the type in species B (resp. A). Sites in the first group are *uninformative*, and do not support any specific topology. Sites in the second group support the correct topology and sites in groups 3 and 4 support one or other incorrect topology. Sites in groups 2, 3 and 4 are *informative* for topology estimation.

The respective probabilities that any given site fall into these groups are denoted by p_1, p_2, p_3 and p_4. By symmetry, $p_3 = p_4$. Any sample of n sites will yield a data vector (y_1, y_2, y_3, y_4) of the numbers of sites falling into these four respective types. Apart from an unimportant combinatorial constant, the probability of this data vector is, from equation (2.30),

$$p_1^{y_1} p_2^{y_2} p_3^{y_3 + y_4}. \tag{15.15}$$

Under the (incorrect) respective topologies (AC)B and (BC)A, this probability becomes, respectively,

$$p_1^{y_1} p_2^{y_3} p_3^{y_2 + y_4}, \ p_1^{y_1} p_2^{y_4} p_3^{y_2 + y_3}. \tag{15.16}$$

Newton chooses as the topology estimate the maximum likelihood estimator, that is the topology that maximizes the likelihood of the data. Thus the correct topology will be inferred if the expression in (15.15) exceeds both those in (15.16). Algebraic rearrangement shows that this occurs if

$$\left(\frac{p_2}{p_3} \right)^{y_2 - y_3} > 1, \ \left(\frac{p_2}{p_3} \right)^{y_2 - y_4} > 1. \tag{15.17}$$

In all practical cases the probability p_2 that the data support the correct topology exceeds the probability p_3 that the data support some specific incorrect topology. In this case (15.17) becomes

$$y_2 > y_3, \ y_2 > y_4. \tag{15.18}$$

In this simple model, this is also the criterion for choosing the correct topology under a parsimony approach.

The results of Problem 11.10 may be used (see Problem 15.6) to show that

$$p_1 = \frac{1 + 2e^{-4\alpha} + e^{-4\alpha t_0}}{4}, \ p_2 = \frac{1 - 2e^{-4\alpha} + e^{-4\alpha t_0}}{4}, \ p_3 = p_4 = \frac{1 - e^{-4\alpha t_0}}{4}. \tag{15.19}$$

Our aim is to find the probability P of accuracy, that is the probability that the inequalities (15.18) are satisfied, given the various multinomial probabilities in (15.19).

Newton (1996) conducted simulations with various values of t_0, n and of α (his λ is identical to our 2α) to estimate P. However, his criterion for a correct tree estimation was not (15.18) but

$$y_2 \geq y_3, \ y_2 > y_4. \tag{15.20}$$

Using this criterion, he found by simulation that when $\alpha = 1/2, t_0 = 2/3, n = 100$, the estimate $\hat{P}$ of P is about 0.74. (Use of (15.18) yields an estimate of about 0.71. (This implies that the probability that two trees have the same likelihood is about 6%.) Using his criterion, the average bootstrap estimate $\hat{P}_B$ was 0.63, suggesting a downward bias in the bootstrap estimation for this case of about 14%. This arises despite the extraordinarily simple model considered and fact that the bootstrap iid assumption was made in the simulations. At least for this simple case, this bias confirms the claim that bootstrap estimators of accuracy are conservative.

We now consider the relation between P and repeatability. Repeatability is the probability that either the correct tree is found on two independent re-runs of the evolutionary process or that the same incorrect topologies are found in two such re-runs. Assuming that an unambiguous tree is estimated on each re-run, this probability is

$$P^2 + 2((1 - P)/2)^2 = (3P^2 - 2P + 1)/2. \tag{15.21}$$

This is less than P when $P > 1/3$, the relevant case for the tree under consideration, and when $P = 0.74$ takes the value 0.58. The bootstrap estimate $\hat{P}_B$ is about 9% greater than this.

When the number n of sites examined is increased, both the accuracy and the repeatability approach 1, as does the bootstrap estimate of repeatability. Given the extreme simplicity of the model, and the fact that all assumptions needed for the bootstrap are assumed to hold, this is to be expected.

In a model as simple as this it is possible to make progress theoretically, and to do this we consider an analysis drawing on that of Zharkikh and Li (1992a).

From the inequality (15.18) the probability P that an estimated tree is accurate is the probability of the event $Y_2 > \max(Y_3, Y_4)$. This event can be written as the event $Y^* > 0$, where

$$Y^* = Y_2 - \left(\frac{Y_3 + Y_4}{2} + \frac{|Y_3 - Y_4|}{2} \right). \tag{15.22}$$

From equation (2.66), the mean of Y^* is the sum of the means of the three terms on the right-hand side of (15.22). The means of Y_2 and $(Y_3 + Y_4)/2$ are found immediately: that of Y_2 is np_2 and, since $p_3 = p_4$, that of $(Y_3 + Y_4)/2$ is np_3. The mean of $|Y_3 - Y_4|$, however, is not found so immediately. $Y_3 - Y_4$ has mean 0 and, from (2.31) and (2.62), has variance $2np_3(1 - p_3)$ $+2np_3^2 = 2np_3$, the second term in the sum in this expression arising from the covariance between Y_3 and Y_4. If a random variable X is $N(0, \sigma^2)$, equations (1.55) and (1.64) show that

$$E(|X|) = \int_{-\infty}^{+\infty} |x| \frac{1}{\sqrt{2\pi}\sigma} e^{-\frac{x^2}{2\sigma^2}} dx. \tag{15.23}$$

The symmetry of the integrand around $x = 0$ shows that this integral is

$$2 \int_0^{+\infty} x \frac{1}{\sqrt{2\pi}\sigma} e^{-\frac{x^2}{2\sigma^2}} dx,$$

and the change of variable $y = x^2/2\sigma^2$ shows that this is

$$\sigma\sqrt{\frac{2}{\pi}} \int_0^{+\infty} e^{-y} dy = \sigma\sqrt{\frac{2}{\pi}}. \tag{15.24}$$

Thus the mean of $|Y_3 - Y_4|/2$ is $\sqrt{np_3/\pi}$, and from this the mean of Y^* is

$$np_2 - np_3 - \sqrt{np_3/\pi}. \tag{15.25}$$

The bootstrap procedure mimics that in (15.22). If we denote the bootstrap numbers corresponding to Y_2, Y_3, and Y_4 by Y_{B2}, Y_{B3}, and Y_{B4}, we define the bootstrap analogue Y_B^* of Y^* by the equation

$$Y_B^* = Y_B - \left(\frac{Y_{B3} + Y_{B4}}{2} + \frac{|Y_{B3} - Y_{B4}|}{2} \right). \tag{15.26}$$

The probability that the bootstrap sample supports the correct topology is the probability that $Y_B^* > 0$. There is however one important difference between this calculation and the parallel calculation with Y^*. Since $p_3 = p_4$, the mean of $Y_3 - Y_4$ is 0. However it is not necessarily the case, and indeed usually will not be, that $y_3 = y_4$, the mean of $Y_{B3} - Y_{B4}$, is 0. This implies an increase in the mean of $Y_{B3} - Y_{B4}$ compared to that of $Y_3 - Y_4$, so that the mean of Y_B^* is less than that of Y^*. This implies that the probability that $Y_B^* > 0$ is less than the probability that $Y^* > 0$. This implies in turn that the bootstrap probability of P is smaller than the true value, in line with the above simulation results.

In summary, even in the extremely simple evolutionary model discussed above, standard estimation procedures can give biased estimators both of accuracy and repeatability when the number of sites sampled is not large. More important than this is the fact that the real historical biological process is infinitely more complicated than the simple model just discussed, or indeed with any evolutionary model used in practice. Thus potentially large problems of systematic bias, as discussed in Section 14.3.6, must be expected. These biases will be shared by bootstrap tree estimates. Further, when the iid assumptions necessary for the bootstrap do not hold, as will likely be the case in practice, further problems with bootstrap inferences about the true phylogenetic tree can be expected. Clearly bootstrap methods, if employed at all with real biological data, should be used with the utmost caution and with a full understanding of these problems.

15.9.3 Assumptions and Problems

The fact that in the example of Section 15.8 the four tree construction methods lead to four quite different trees should be a cause for concern. The differences arise because of the different approaches that the four methods adopt. Further, the methods based on distance measures assumed the Kimura two-parameter stochastic evolutionary model. This is an oversimplified model, which cannot be expected usually to describe reality well.

As discussed in Chapter 14, distances estimated under one model when another more complex model is appropriate are biased. Of course, it is difficult in practice to know what a reasonable model might be, so that it is important to assess the problems that arise if a tree is estimated under one model when some other model is appropriate. In this section we discuss aspects of modeling problems in the phylogenetic context and in the following section discuss hypothesis testing involving models.

The concept of a consistent estimator of a parameter was introduced in Section 8.2. Steel (1994) demonstrated that if the frequencies of the various nucleotides (for DNA sequences) or amino acids (for protein sequences) can be taken as known, then under an iid assumption of the substitutions at different sites, a consistent estimator of a phylogenetic tree can be found. However, Steel et al. (1994) show that if the frequencies of the various nucleotides (for DNA sequences) or amino acids (for protein sequences) are unknown, then with site-to-site variation in substitution rates it can be impossible in principle to estimate a phylogenetic tree consistently, even with the most extensive data.

Similarly, Chang (1996b) demonstrated that when the correct evolutionary model is used, and under further mild restrictions, a consistent estimate of an evolutionary tree topology, together with the various branch lengths, can be achieved. These models assume identical substitution processes at all sites. However, when different substitution processes apply at different sites, for example with site-to-site variation in substitution rates, inconsistent tree estimation can occur (Chang (1996a)).

Yang (1993) approached the problem of site-to-site variation in substitution rates in the Felsenstein model by allowing the rate parameter u to be a site-to-site random variable following a gamma distribution with mean 1 and variance $1/\beta$, for which, from Section 1.10.5,

$$f_U(u) = \frac{\beta^\beta u^{\beta-1} e^{-\beta u}}{\Gamma(\beta)}, \quad u \geq 0. \tag{15.27}$$

Yang's analysis assumes the Felsenstein model of substitution, and a similar analysis can be made for other models. Yang also discusses properties of a $-2\log \lambda$ testing procedure to assess whether a site-to-site variation model provides a significantly better fit than the "fixed u" model. This procedure is discussed further in Section 15.9.4.

Lake (1987) also addressed the problem of site-to-site variation in substitution rates and developed a model allowing this variation for which consistent estimation is possible. However, this model is specialized and may not apply widely. In general, the problem of site-to-site variation in substitution rates appears to be difficult to overcome. Questions concerning the reconstruction of phylogenetic trees by algorithmic methods in the presence of site-to-site variation in substitution rates are discussed by Steel et al. (1994).

The analysis described in Section 15.7, and other analyses of phylogenetic tree estimation in the literature, assume that even if rates differ from site to site, at least the evolutionary processes at the various sites are independent. However, population genetics theory shows that very close sites evolve in a dependent manner. This matter has been addressed by various authors, in the context of both maximum likelihood estimation and algorithmic estimation. Dixon and Hillis (1993) discuss the use of weights to accommodate non-independence in a parsimony analysis. Others have examined substitution processes in regions rather than at nucleotide sites when various correlated and other independent substitution processes structures are assumed (Muse (1995), Schöniger and von Haeseler (1994), (1995)). Since correlated substitutions are known to be the case, more research on this matter is necessary.

A further interesting fact is that there can be more than one maximum likelihood tree, and that the topologies of the various trees having maximum likelihood can be different. This was demonstrated by Steel (1994), and disproved a claim (Fukami and Tateno (1989)) that a unique maximum likelihood tree exists. It can be argued that while Steel's result is true in principle, it is very unlikely in practice that multiple maximum likelihood trees will arise, especially with extensive data sets. The simulations of Rogers and Swofford (1999) suggest that the multiple maximum phenomenon will rarely occur in practice. Nevertheless, the very possibility of multiple maxima should cause concern, especially when the various optimality properties of maximum likelihood estimators are invoked in phylogenetic tree estimation.

Most of the models described in this section are simple ones, and it is agreed that they cannot describe the real historical evolutionary process with any accuracy. It can however be argued that simpler rather than more complex models are desirable when investigating phylogenetic trees. There are three arguments in favor of this viewpoint (Sullivan and Swofford, 2001). First, even the most complex model likely to be used will still be far from the correct model describing the actual historical evolutionary process, so that not much might be gained in terms of reality with the more complicated model. Second, a substantial increase in the number of parameters in a model makes parameter estimation quite difficult, so that unreliable parameter estimates, leading to unreliable tree estimates, will arise when a model with many parameters is used. Finally, if the tree estimation procedure is comparative robust, so that the estimate is not strongly biased when an incorrect model is used, then a simpler model might be preferred to a more complex one. While these arguments are worth serious consideration, we have seen that a model specification that differs only in small details from the correct model can lead to biased tree estimation. The difficult question of a "best" approach to tree estimation is still unresolved.

15.9.4   Phylogenetic Models and Hypothesis Testing

In this section we discuss aspects of hypothesis testing in the context of phylogenetic tree construction. The derivation of tests of hypotheses in this area, and the analysis of properties of these tests, is an active area of current research and we refer to aspects of this later. Here we consider comparatively simple aspects of the hypothesis-testing question.

The testing vehicle that we discuss is the $-2 \log \lambda$ test statistic discussed in Section 9.4. It is tempting to use this statistic in cases where various parameters of the phylogenetic tree of interest have been estimated by maximum likelihood, since this test statistic is a function of the ratio of two maximum likelihoods. However, inappropriate use of this statistic can cause problems, as we discuss below.

Suppose that a set of aligned DNA sequences from a collection of species is given, and we wish to test the (alternative) hypothesis that these data are significantly better explained by the Kimura model of Section 14.3.2 than by the Jukes–Cantor model of Section 14.3.1. The null hypothesis is that the Kimura model does not explain the data better than the Jukes–Cantor model. In statistical terms, this is a test of the null hypothesis that in the Kimura model, $\alpha = \beta$.

It is important to note two aspects of this test. First, the two hypotheses are nested, since the Jukes–Cantor model is a special case of the Kimura model. Thus the "nesting" requirement of the $-2 \log \lambda$ testing method is met. Second, if the actual topology of the phylogenetic tree, which can loosely be thought of as an unknown parameter, is also estimated as part of the maximizing procedure, then the requirement that the parameters of interest be real numbers is not met. Because of this, the asymptotic null hypothesis distribution of $-2 \log \lambda$ might not be chi-square. Thus for the moment we assume that the tree topology is given, and that the estimation procedure relates only to the various arm lengths.

Even this restriction does not guarantee that the asymptotic null hypothesis chi-square distribution of $-2 \log \lambda$ can be assumed. Part of the estimation procedure relates to the unknown DNA sequences at the internal nodes of the tree, and if we take these as unknown parameters, the question of estimating discrete parameters arises. It is therefore important to assess what the asymptotic null hypothesis distribution of $-2 \log \lambda$ is. It appears that this is very difficult to do theoretically, and that it is necessary to resort to computer simulation. Here we present the results of Whelan and Goldman (1999), who follow the simulation approach.

Whelan and Goldman take the topology of the phylogenetic tree as given, and carry out a random procedure of DNA substitution at the various sites in a DNA sequence, assuming that the substitution process is described by the Jukes–Cantor model. They then find the maximum of the likelihood of the derived data in the species at the leaves of the tree under both the Jukes–Cantor model and the Kimura model, and from this they compute

$-2 \log \lambda$. This procedure is then replicated a large number of times, and from these replicates an accurate assessment of the null hypothesis distribution of $-2 \log \lambda$ can be found. It is found that this distribution is very close to that of chi-square with one degree of freedom.

It is therefore reasonable to assume in this case, where the topology of the tree is given, that the $-2 \log \lambda$ statistic has an approximate chi-square distribution with one degree of freedom under the null hypothesis. Suppose that for a given set of data we do not reject the null hypothesis when carrying out this test. It is important to be aware of what this implies. It does not necessarily imply that the Jukes–Cantor model provides a satisfactory explanation of the data. What it does imply is that the Kimura model, with one further free parameter, does not give a significantly better fit to the data than the Jukes–Cantor model. Thus the test is a comparative one rather than an absolute one, and we have not shown in any absolute sense that the Jukes–Cantor model provides an adequate fit to the data. It is important to make this point, since claims of absolute fit rather than relative fit are sometimes made by carrying out the $-2 \log \lambda$ procedure in similar cases.

Whelan and Goldman (1999) carry out a variety of similar assessments, and we outline three of their results here. First, if the stationary probabilities φ_a, φ_g, φ_c, and φ_t in the Felsenstein model of Section 14.3.3 are all equal to .25, then that model becomes the Jukes–Cantor model, so that the Jukes–Cantor model is nested within the Felsenstein model. Since $\varphi_a + \varphi_g + \varphi_c + \varphi_t = 1$, there are three more free parameters under the Felsenstein model than under the Jukes–Cantor model. It then becomes reasonable to hope that, if the null hypothesis is the Jukes–Cantor model and the alternative hypothesis is the Felsenstein model (with arbitrary stationary probability values), and where the topology of the tree is given, the null hypothesis distribution of $-2 \log \lambda$ is approximately chi-square with three degrees of freedom. The simulations of Whelan and Goldman (1999) support this conclusion.

Second, suppose that the null hypothesis is the Jukes–Cantor model and the alternative hypothesis is the Felsenstein model with stationary probability values equal to the observed values in the data. Then neither model is nested within the other and we have no theoretical support for the claim that the null hypothesis distribution of $-2 \log \lambda$ is chi-square. Whelan and Goldman (1999) show that the null hypothesis distribution of $-2 \log \lambda$ is not close to a chi-square with three degrees of freedom, and that in fact negative values of $-2 \log \lambda$ can arise, an impossibility for a random variable truly having a chi-square distribution. This matter is referred to again in Problem 15.7.

Third, in Section 15.9.3 we discussed models that allow unequal substitution rates at various sites. Since the assumption of equal rates leads to simpler models, it is important to find a test for it. Such tests are important because, for example, an unjustified assumption of equal substitution rates can lead to biases in substitution matrix estimation (Kelly and Churchill (1996)).

Whelan and Goldman (1999) and Goldman and Whelan (2000) discuss this matter. They consider the test of a null hypothesis Jukes–Cantor model, with substitution rate 1 at all sites, against an alternative hypothesis Jukes–Cantor model for which the substitution rate at any site is a random variable having the gamma distribution (15.27) with mean 1 and variance $1/\beta$. They also consider the parallel test for the Felsenstein model.

Although in both models the relevant null and alternative hypotheses differ by one free parameter, and the null hypothesis is nested within the alternative, the simulations show that the null hypothesis distribution of $-2\log\lambda$ is not chi-square with one degree of freedom. This occurs because the null hypothesis value of β is $+\infty$, corresponding to a variance of zero. The value $+\infty$ is not covered by the asymptotic $-2\log\lambda$ theory, since that theory requires β to be a real number. The $-2\log\lambda$ testing procedure has been misused in the literature to assess whether significant site-to-site variation in substitution rates occurs within the Felsenstein model.

Self and Liang (1987) suggest that in this case the null hypothesis distribution of $-2\log\lambda$ is approximately that of a random variable taking the value 0 with probability $\frac{1}{2}$ and having a chi-square distribution with one degree of freedom also with probability $\frac{1}{2}$, and their simulations support this suggestion. On an associated point, Ota et al. (1999) show that boundary maximum likelihood estimates such as those arising in Example 3 of Section 9.4 of parameters cannot have asymptotic normal distributions, and the theory of Chapter 8 then implies that the asymptotic $-2\log\lambda$ testing theory cannot hold. A simple example of such a boundary estimate, and its clearly non-normal asymptotic distribution, is provided by Example 3 of Section 9.4.

The null hypothesis chi-square mixture of Self and Liang (1987) is not always applicable. Ota et al. (2000) show that when there are two or more boundary parameter estimates, the null hypothesis distribution of $-2\log\lambda$ cannot be expressed as a linear combination of any number of chi-square distributions. This remark is relevant because models with two or more boundary maximum likelihood estimates are beginning to appear (Huelsenbeck and Nielsen (1999)).

These observations indicate the dangers in unthinking application of tests using $-2\log\lambda$. A further and more serious problem arises, as discussed above, when the maximum likelihood procedure involves estimating the topology of the phylogenetic tree as well as its branch lengths, a case not covered by the $-2\log\lambda$ theory. Here the null hypothesis distribution of $-2\log\lambda$ is far from chi-square, essentially because the tree topology is not a real number able to take values in some interval. This matter is discussed further by Goldman (1993), who notes that several authors have made an inappropriate chi-square assumption in the literature when the tree topology is estimated as part of the testing procedure.

The chi-square distribution of $-2\log\lambda$ is an asymptotic one, relating to an indefinitely increasing sample size. With a sufficiently large sample size one might expect that the true tree topology will be inferred, so that the problem just mentioned would not arise. However, the sample size necessary for this situation to occur is probably immense, and far larger than any realistic sample size arising in practice.

Goldman et al. (2000) provide a general and comprehensive discussion of the use of likelihood ratio tests of topologies in phylogenetics. The main point that they make is that the frequently used Kishino–Hasegawa test (Kishino and Hasegawa (1989) designed to decide between competing phylogenetic topology hypotheses is valid only only certain assumptions, in particular that the topologies being compared are specified a priori. This test has often been misused, in particular in cases where one of the topologies being compared is the maximum likelihood topology and is thus not specified a priori), and Goldman et al. show that severe biases can arise in the test in such cases.

The bootstrap testing procedure is often thought of as one that overcomes problems of an unknown null hypothesis distribution. However Andrews (2000) has shown that when a maximum likelihood estimate is a boundary value, problems can arise even with the bootstrap operation.

Much of the discussion above concerning the problems of hypothesis testing in the phylogenetic tree context has used as an example the test of the (null hypothesis) Jukes–Cantor model against the (alternative hypothesis) Kimura model, within which it is nested. These problems also arise in other tests. Yang (1994) discusses tests of the Felsenstein and the HKY models against the general reversible model, within which they are nested. He also considers the test of the general reversible process model itself against a completely arbitrary model. The number of degrees of freedom used in these tests are derived from the number of degrees of freedom by which the respective models differ, as discussed in Section 14.2.7. The extent to which a more restrictive model can be accepted depends on the data analyzed, so that general conclusions are not easily obtained. Yang does however claim that, broadly speaking, the general reversible process model can be recommended and that the completely arbitrary model is not recommended.

It is clear that the hypothesis tests considered above barely scratch the surface of what is possible and necessary. A list of possible tests, and the outcomes of these tests, is given by Huelsenbeck and Rannala (1997). Among the many results that they present, Huelsenbeck and Rannala confirm that the distribution of $-2\log\lambda$ is far from chi-square when the tree topology is estimated as part of the testing procedure. They also give the results of several tests where $-2\log\lambda$ is correctly applied to real data sets, to assess what modeling assumptions may reasonably be made in practice. As one important example, they show that substitution rates vary significantly from one lineage to another, thus rejecting the "molecular clock hypothesis" favored by many population geneticists that the substitution rate per generation is constant in all branches of the phylogenetic tree. This hypothesis is further discussed by Tourasse and Li (1999). They show that the Kimura model fits observed data significantly better than the Jukes–Cantor model, so that transition substitution rates differ significantly from transversion substitution rates. Models allowing different substitution rates between sites fit observed data significantly better than models that assume the same rate at all sites. Models that allow different substitution rates in different regions of the genome fit observed data significantly better than models that assume the same rate in all regions, a result confirmed by the recent chromosome 22 data (Dunham et al. (1999)). Liò and Goldman (1998) discuss further aspects of model testing, and the use of hidden Markov models, in analyzing models with different substitution rates at different sites.

Testing for monophyly provides an important case of hypothesis testing involving tree structure. Here the null hypothesis to be tested is that some group of species is monophyletic. The maximum likelihood tree can be constructed under the (null) hypothesis of monophyly of a certain set of species and can also be constructed without this restriction, and the two likelihoods compared. Huelsenbeck et al. (1996) and Huelsenbeck and Crandall (1997) point out that the null hypothesis distribution of $-2\log\lambda$ is far from chi-square when the null hypothesis claim of monophyly is correct. This again arises because tree topology is not a real number taking values in some interval. Indeed, it is not even clear how many degrees of freedom $-2\log\lambda$ would have if it did have a null hypothesis chi-square distribution. If $-2\log\lambda$ is to be used in a test of monophyly, its null hypothesis distribution must be found empirically, using simulation methods.

An excellent description of these problems is given by Huelsenbeck and Crandall (1997), who also provide a summary of maximum likelihood estimation procedures used in phylogenetic analysis, and their properties. A further description of statistical problems in phylogenetic analysis is provided by Holmes (1999), who gives an amusing table listing "translations" of expressions frequently used in biological articles into the corresponding standard statistical terminology.

Problems

15.1 Prove that a tree-derived distance satisfies the four properties of a distance given on page 499.

15.2 Prove that for a tree whose derived distance satisfies the ultrametric property, pairs of species with the smallest distance between them are neighbors.

15.3 Prove equation (15.12).

15.4 For five species a, b, c, d, and e, with distance given by

	a	b	c	d	e
a	0	2	8	8	8
b		0	8	8	8
c			0	4	4
d				0	2
e					0

reconstruct the tree using both algorithms in Section 15.3. For the first algorithm, use the order a, b, c, d, e. Show the partial tree for each step, together with edge lengths.

15.5 For five species a, b, c, d, and e, with distance given by

	a	b	c	d	e
a	0	9	8	7	8
b		0	3	6	7
c			0	5	6
d				0	3
e					0

reconstruct the tree using the neighbor-joining algorithm of Section 15.4. Show the partial tree in each case, together with edge lengths. Now reconstruct the tree, together with edge lengths, using the UPGMA algorithm of Section 15.3. Compare your answers for the two reconstructions.

15.6 Use the result of Problem 11.10 to derive the expressions (15.19).

15.7 In what circumstances will negative values of $-2\log\lambda$ arise in testing the Jukes–Cantor model against the Felsenstein model in which the stationary probabilities are set to the observed values?

Appendix A
Basic Notions in Biology

We outline here the basic notions from biology that are needed in the book. Deoxyribonucleic acid (DNA) is the basic information macromolecule of life. It consists of a polymer of nucleotides, in which each nucleotide is composed of a standard deoxyribose sugar and phosphate group unit, connected to a nitrogenous base of one of four types: adenine, guanine, cytosine, or thymine (abbreviated here a, g, c, and t, respectively). Because of similarities in the chemical structure of their nitrogenous bases, adenine and guanine are classified as purines, while cytosine and thymine are classified as pyrimidines. Adjacent nucleotides in a single strand of DNA are connected by a chemical bond between the sugar of one and the phosphate group of the next. The classic double-helix structure of DNA is formed when two strands of DNA form hydrogen bonds between their nitrogenous bases, resulting in the familiar "ladder" structure. Under normal conditions, these hydrogen bonds form only between particular pairs of nucleotides (referred to as base pairs): Adenine pairs only with thymine, and guanine pairs only with cytosine. Two strands of DNA are complementary if the sequence of bases on each is such that they pair properly along the entire length of both strands (see Figure A.1). The sequence in which the different bases occur in a particular strand of DNA represents the genetic information encoded on that strand. By virtue of the specificity of nucleotide pairing, each of the two strands of any DNA molecule contains all of the information present in the other. There is also a chemical polarity to polynucleotide chains such that the information contained in the a, g, c, and t bases is synthesized and decoded in only one direction.

```
...  ----------------------------------------  ...
     a  c  c  g  t  a  t  a  a  c  g  a  t  c  c  t  c  t  g  a
     :  :  :  :  :  :  :  :  :  :  :  :  :  :  :  :  :  :  :  :
     t  g  g  c  a  t  a  t  t  g  c  t  a  g  g  a  g  a  c  t
...  ----------------------------------------  ...
```

Figure A.1. Example of a portion of DNA sequence. The dashed lines represent the sugar-phosphate backbone, and the letters represent the nitrogenous bases. Dotted lines connecting base pairs denote hydrogen bonds. Note the specificity of base-pairing (a to t, c to g).

In the cell, DNA is organized into chromosomes, each of which is a continuous length of double stranded DNA that can be hundreds of millions base pairs long. Most human cells contain 23 pairs of chromosomes, one member of each pair paternally inherited and the other maternally inherited. The two chromosomes in a pair are virtually identical, with the exception of the sex chromosome, for which there are two types, X and Y. Nearly every cell in the body contains identical copies of the full set of 23 pairs of chromosomes. An organism's total set of DNA is referred to as its genome; the human genome contains more than three billion base pairs.

A human chromosome consists mostly of non-protein-coding DNA, the function of which is only just becoming understood (see for example Gibbs (2003)). Interspersed in the DNA are protein-coding genes. These genes constitute only approximately 2% of the human genome; however, they are the classic focus of attention of geneticists. Genes themselves are often organized into exons, which are the sequences that will eventually be used by the cell, alternating with introns, which will be excised and discarded. The human genome is currently thought to contain approximately 30,000–40,000 genes (International Human Genome Sequencing Consortium (2001)). The information in these genes will go on to be encoded in RNA (ribonucleic acid), and in many cases ultimately in proteins.

The first step in this process is transcription, the creation of an RNA molecule using the DNA sequence of a gene as a template. Transcription is initiated at non-coding sequences called promoters, located immediately preceding the gene. Like DNA, RNA is made up of a series of nucleotides, but with several important differences: RNA is single-stranded, contains the sugar ribose, and substitutes the nitrogenous base uracil for thymine. After post-transcriptional modification, which includes the removal of introns, the RNA will go on to various fates within the cell. Of particular interest is mRNA (messenger RNA), which will be translated into protein.

A protein is comprised of a sequence of amino acids. There are twenty amino acids which commonly appear in proteins. Each of these amino acids is represented by one or more sequences of three RNA nucleotides known as a codon; for example, the RNA sequence aag encodes the amino acid lysine. The combination of four possible nucleotides in groups of three results in 4^3 or 64 codons, meaning that most amino acids are coded for by more than one codon. An organelle known as the ribosome performs the translation of mRNA into protein. The ribosome pairs each codon in the RNA sequence with the appropriate amino acid, and then adds the amino acid onto the growing protein. The process of translation is mediated by two special types of codon: start codons signal the location on the RNA molecular where translation should begin, while stop codons signal the location where translation should terminate. Once the sequence of amino acids that make up a particular protein is assembled, the protein dissociates from the ribosome and folds into a specific three-dimensional form. The function of a protein ultimately depends on both its three-dimensional structure and its amino acid sequence. Proteins go on to perform a variety of functions in the cell, covering all aspects of cellular functions from metabolism to growth to division.

Currently, functions have been assigned to only a small proportion of the genes in even the best understood of model organisms. In order to assign function to the remaining genes, it is helpful to examine the expression patterns of these genes in various tissues. Microarray technology developed over the past several years now allows the measurement of mRNA levels for tens of thousands of genes simultaneously. This provides an efficient and convenient way to determine the expression patterns of genes in many different types of tissues, but at the same time provides new challenges in information cataloging and statistical analysis.

Appendix B
Mathematical Formulae and Results

B.1 Numbers and Intervals

It is a nontrivial matter to define the real numbers carefully, and for our purposes this is not necessary. We will instead take the real numbers as a starting point. Intuitively, they represent all the numbers that correspond to lengths, together with their negatives and zero. The real numbers can be put in one-to-one correspondence with the points on a line.

The *integers* are all the numbers $\ldots, -3, -2, -1, 0, 1, 2, 3, \ldots$. The *positive* integers, sometimes called the *natural numbers*, are the numbers 1, 2, 3, $\ldots$, whereas the *non-negative* integers are the numbers $0, 1, 2, 3, \ldots$. The *rational* numbers are those that can be written as ratios of integers, i.e., in the form $\frac{a}{b}$, where a and b are integers. The *irrational* numbers are the real numbers that cannot be so written. It can be shown, for example, that $\sqrt{2}$, π, and e are all irrational.

We use four kinds of intervals in this book. The first kind consists of *open* intervals. These intervals are denoted by round brackets:

$$(a, b) = \text{ set of all real numbers } x \text{ such that } a < x < b.$$

When we write such an interval, we always assume $a < b$; otherwise, the interval would be empty. We also allow a to be $-\infty$ and/or b to be ∞. The second kind consists of *closed* intervals. These intervals are denoted by square brackets:

$$[a, b] = \text{ set of all real numbers } x \text{ such that } a \leq x \leq b.$$

In this case we do allow $a = b$, but we do not allow a or b to be infinite. The final kind consists of *half-open* intervals, of the form

$$(a, b] = \text{set of all real numbers } x \text{ such that } a < x \leq b$$

and

$$[a, b) = \text{set of all real numbers } x \text{ such that } a \leq x < b.$$

For the first kind a can be $-\infty$ and in the second b can be ∞. We require $a < b$ in all such intervals.

B.2 Sets and Set Notation

The sets we need to consider in this book are subsets of n-dimensional space, for $n = 1, 2, \ldots$. By n-dimensional space we mean the set of n-tuples $(x_1, x_2, \ldots, x_n)$ where $x_1, x_2, \ldots, x_n$ are real numbers. Other sets we consider are subsets of these sets. For example, S might be the set of all n-tuples $(x_1, x_2, \ldots, x_n)$ where x_n is positive. If $n = 2$, then this is the half-plane above the horizontal axis. We refer to the tuples in S as *members* or *elements* of S. If s is a member of S, we write $s \in S$, and if s is not a member of S, we write $s \notin S$. If S_1 and S_2 are two sets in the same space, their *union* is the set of elements that are in S_1 or S_2 (or both), written $S_1 \cup S_2$. Their *intersection* is the set of elements that are in both S_1 and S_2, written $S_1 \cap S_2$.

Suppose S_1 is a subset of m-dimensional space and S_2 is a subset of n-dimensional space. Then the Cartesian product of S_1 and S_2 is the subset of $(m + n)$-dimensional space consisting of $(m + n)$-tuples whose first m components form an element of S_1 and whose last n components form an element of S_2. The Cartesian product of S_1 and S_2 is written $S_1 \times S_2$. If $S_1 = [a, b]$ and $S_2 = [c, d]$, then $S_1 \times S_2$ is a rectangle.

B.3 Factorials

Let n be a positive integer. Then we define $n!$ (pronounced "*n factorial*") to be

$$n! \stackrel{\text{def}}{=} n(n-1)(n-2) \cdots 3 \cdot 2 \cdot 1.$$

It will be apparent later that it is convenient to define $0! = 1$. Note that $n! = n(n-1)!$, and thus $\frac{n!}{(n-1)!} = n$.

B.4   Binomial Coefficients

Let r be *any real number* and let k be a positive integer. We define the *binomial coefficient* $\binom{r}{k}$ by

$$\binom{r}{k} \stackrel{\text{def}}{=} \frac{r(r-1)(r-2)\cdots(r-k+1)}{k!},$$

and for $k = 0$, put $\binom{r}{k} = 1$. The reason why it is called a "coefficient" will be explained in Appendix B.6.

When n is a positive integer and $k \leq n$, $\binom{n}{k} = \frac{n!}{(n-k)!k!}$ takes on a combinatorial meaning, which we shall discuss shortly.

B.5   The Binomial Theorem

In this section we describe the basic version of the binomial theorem, and in a later section we shall revisit it in more generality. The basic version states that for any real numbers a and b, and any positive integer n,

$$(a+b)^n = \sum_{k=0}^{n} \binom{n}{k} a^k b^{n-k}. \tag{B.1}$$

In particular,

$$(x+1)^n = \sum_{k=0}^{n} \binom{n}{k} x^k. \tag{B.2}$$

We provide a proof of (B.1) later, but first we need to discuss some *combinatorial* issues.

B.6   Permutations and Combinations

Consider n distinguishable balls in an urn, numbered 1 to n. Let $k \leq n$ be a positive integer. We are interested in the following two "combinatorial" quantities:

(1) In how many orders can we choose k of the balls *without replacement*? Equivalently, how many k-tuples of distinct integers $(n_1, n_2, \ldots, n_k)$ are there, where $1 \leq n_i \leq n$?

(2) In how many ways can we choose k of the balls from the urn, without replacement?

The key distinction between these two counting problems is that the first considers the order in which the objects were chosen, whereas the second does not. The first quantity is called the number of *permutations* of n

objects taken k at a time, and we denote it by $_nP_k$. The second quantity is called the number of *combinations* of n objects taken k at a time and is often referred to as "n choose k." We do not introduce a notation for "n choose k," since we will soon show that it is equal to $\binom{n}{k}$.

We can easily list all the possible permutations of 3 things taken 2 at a time as

$$(1,2),\ (1,3),\ (2,3),\ (2,1),\ (3,1),\ (3,2),$$

so that $_3P_2 = 6$. On the other hand, there are 3 combinations of 2 things,

$$\{1,2\},\quad \{1,3\},\quad \{2,3\}.$$

In general, there can be n choices for the first element in the permutation, and for each of those there are $n-1$ choices for the second element, etc., down to $n-k+1$ choices for the kth element in the permutation. Therefore,

$$_nP_k = n(n-1)(n-2)\cdots(n-k+1) = \frac{n!}{(n-k)!}.$$

It should be clear that for each permutation of n objects taken k at a time, there will be $k!$ that involve the same k objects, each arranged in a different order. Therefore, it follows that the number of combinations of n things taken k at a time (i.e., the answer to question 2 above) is equal to

$$\frac{_nP_k}{k!} = \binom{n}{k}.$$

Thus, it is natural that $\binom{n}{1} = n$ and $\binom{n}{n} = 1$. The motivation behind the definition of $\binom{n}{0} = 1$ should now be clear.

We can now give a combinatorial proof of the binomial theorem. Consider the product

$$(a+b)^n = \overbrace{(a+b)(a+b)\cdots(a+b)}^{n\ \text{times}}.$$

The right-hand side can be expanded by repeated application of the distributive property. We end up with one term of the form $a^k b^{n-k}$ for every way we can choose a's from k of the terms in the product. Therefore, the coefficient of $a^k b^{n-k}$ is $\binom{n}{k}$. This explains why $\binom{n}{k}$ is called a binomial *coefficient*.

B.7 Limits

We cannot give a rigorous introduction to limits, since the subject has many subtleties and complexities that would take us unreasonably far afield. There are two kinds of limits we consider in this book: *discrete* and *continuous* limits. Continuous limits arise for functions defined on the real numbers. Discrete limits are for functions defined on the natural numbers. The latter kind of functions are usually called *sequences* and are usually denoted using subscripts, typically with a notation of the form $a_1, a_2, a_3, \ldots, a_n, \ldots$, instead of the usual functional notation $f(1), f(2), f(3), \ldots, f(n), \ldots$. When we want to address the entire function (sequence) at once, we write $\{a_n\}_{n=m}^{\infty}$, where m is the first index in the sequence, or as $\{a_n\}$ when the domain of n is clear. In the discrete case we are mainly interested in the limit as n goes to infinity:

$$\lim_{n \to \infty} a_n.$$

An important discrete limit, proven in most basic calculus books by an application of L'Hospital's rule, is that for any fixed t,

$$\lim_{n \to \infty} \left(1 + \frac{t}{n} \right)^n = e^t. \tag{B.3}$$

Certain uniform approximations can be derived from this. For example, for all $|t| \leq 1$, the approximation

$$\left(1 + \frac{t}{n} \right)^n \approx e^t$$

is accurate to within 1% for all $n \geq 50$, to within .1% for all $n \geq 500$, and to within .01% for all $n \geq 5{,}000$.

For functions of a real variable $f(x)$, we consider limits as x approaches any real value (in the domain of the function), or $\pm\infty$. A limit of importance for us is the limit of the function $x \log x$ as $x \to 0$ from the right, usually written

$$\lim_{x \to 0^+} x \log x.$$

It can be shown that this limit is equal to 0. Thus when an expression of the form $\sum_x x \log x$ arises in some application, where 0 is a possible value for x, we put that the term in the sum corresponding to $x = 0$ equal to 0.

B.8   Asymptotics

Computational issues often require an analysis of the asymptotic behavior of functions. To be more precise, let $f(t)$ and $g(t)$ be functions that are defined on the non-negative real numbers and takes values in the positive real numbers. We are interested in methods for describing how $f(t)$ and $g(t)$ behave with respect to each other as t approaches some limit (possibly infinity). To this end, five basic relations between such functions are defined.

(1a) $f = O(g)$ at ∞,   if there are constants $C, K \geq 0$ such that
                                     $\frac{f(t)}{g(t)} \leq C$ for all $t > K$.

(1b) $f = O(g)$ at a, $a < \infty$ if there are constants $C, h \geq 0$ such that
                                     $\frac{f(t)}{g(t)} \leq C$ for all t within h of a.

(2) $f = o(g)$ at a,   if $\lim_{t \to a} \frac{f(t)}{g(t)} = 0$.

(3) $f \sim g$ at a,   if $\lim_{t \to a} \frac{f(t)}{g(t)} = 1$.

(4) $f \, \Omega \, g$ at a,   if $f = O(g)$ and $g = O(f)$ at a.

(5) $f \approx g$ at a,   if $\lim_{t \to a}(f(t) - g(t)) = 0$.

Conditions 1a and 1b are read "f is big 'oh' of g as t approaches a," condition 2 is read "f is little 'oh' of g as t approaches a," condition 3 is read 'f is asymptotic to g as t approaches a," and condition 4 is read "f is 'omega' of g as t approaches a."

The intuition behind 1a is that if it holds, then for sufficiently large values of t, $f(t)$ is no bigger than some fixed constant times $g(t)$. The intuition behind 1b is similar, replacing "sufficiently large" with "sufficiently close."

Even though the function $f(t) = 2t$ becomes arbitrarily larger than $g(t) = t$, as t gets larger, $f = O(g)$ at infinity. When $a = \infty$, and the functions involved are polynomials, the relevant terms are the largest powers of t that occur in each polynomial. If the largest power in g is at least as large as the largest power in f, then $f = O(g)$ at infinity, and if the largest power in f is at least as large as the largest power in g, then $f = O(g)$ at zero. On the other hand, if f is a polynomial, and g is an exponential function with base greater than 1, such as e^t, then it is never the case that $g = O(f)$ at infinity. These statements are usually verified by repeated applications of L'Hospital's rule.

If $f = o(g)$ at a, then f is in some sense "eventually" (i.e., for t sufficiently close to a) smaller than g. For example, $t^2 + t = o(t^3)$ at infinity. However, even though $t^2 - t$ is smaller than t^2 by more than a constant factor, $t^2 - t$ is *not* $o(t^2)$ at infinity (in fact $t^2 - t \sim t^2$ at infinity).

If $f \sim g$, then f and g grow at roughly the same rate. This does not, however, mean that the $f(t) \approx g(t)$ as $t \to \infty$. For example if $f(t) = t^2$ and $g(t) = t^2 + t$, then $f \sim g$; however, $g(t) - f(t) = t \to \infty$. One must therefore be careful to use the exact definitions when using these asymptotic concepts.

For the purposes of bioinformatics, the most important asymptotic classes of functions are the ones that are asymptotic to either polynomials (or more generally, to any power of t), logarithms, or exponentials. We say that a function asymptotic to some polynomial as $t \to \infty$ has *polynomial growth*, one asymptotic to a logarithm as $t \to \infty$ has *logarithmic growth*, and one asymptotic to an exponential as $t \to \infty$ has *exponential growth*. Obviously, other types of growth exist, such as "doubly exponential," "factorial," etc. One should note that a function with polynomial growth is not necessarily a polynomial, for example $f(t) = t + \sqrt{t}$ is not a polynomial, but $f(t) \sim t$ at infinity, and $g(t) = t$ is a polynomial.

An order of magnitude relation used often in this book is of the form $f(h) = \lambda h + o(h)$ for h small. This means that if we write $f(h) = \lambda h + g(h)$, then $g(h)/h \to 0$ as $h \to 0^+$. Terms of order $o(h)$ are always written as $+o(h)$ in such cases, even though they may well be negative.

B.9   Stirling's Approximation

The concepts of the previous section were framed in terms of continuous limits, but they can all be translated into the discrete case by substituting sequences for the functions of real variables. Our next two results are discrete asymptotic equations for $n!$ and $\binom{n}{k}$. The first is Stirling's approximation:

$$n! \sim \sqrt{2\pi}\, n^{n+1/2} e^{-n}. \tag{B.4}$$

From this it follows that

$$\binom{n}{k} \sim \frac{1}{\sqrt{2\pi}} \frac{\sqrt{n}}{\sqrt{k(n-k)}} \left(\frac{k}{n}\right)^{-k} \left(1 - \frac{k}{n}\right)^{-(n-k)}. \tag{B.5}$$

B.10   Entropy as Information

Suppose that some number between 1 and 64 (inclusive) is chosen at random by person A, and person B is required to find this number. By asking questions using a strategy in which the number of possibilities is halved with the answer to each question (the first being, perhaps, "is the number 32 or less?"), it is clear that six questions are sufficient to determine unambiguously which number was chosen. The precise form of the question is not fixed: For example, the first question might be, "is the number odd or even?" but the form of the questions, in which the number of possible cases is halved after each question, is clear. We can say that six bits of information are sufficient to determine the number chosen if this form of question is used. Further, no other system of questions can consistently outperform this one, in the sense that the mean number of questions asked by any other system must be at least 6.

It is clear that when a person must determine one of 2^k numbers, k questions, or k bits of information, are needed. A more complicated calculation arises in other cases. Suppose, for example, that one of the 20 amino acids is chosen at random. If the amino acids are numbered $1, 2, \ldots, 20$ in some agreed order, the first two questions, and their answers, might be, "is the number of the amino acid 10 or less?" (yes), "is the number of the amino acid 5 or less?" (no). At this point the next question might be "is the number 6, 7, or 8?" If the answer is "no," only one further question is needed to determine whether the number is 9 or 10. If the answer is "yes," the next question might be, "is it either 6 or 7?". If the answer is "no," no further questions are needed to determine the number. If the answer is "yes," one further question is needed. Thus often four but in some cases five questions are needed. (It is an interesting exercise to prove that the mean number of questions that need to be asked is 4.4.) In general, if a number is chosen from a large number N of numbers, the mean number of questions needed to determine which one was chosen is approximately $\log_2 N$. This is confirmed by the fact that $\log_2 20 = 4.32$. The number of questions is exactly $\log_2 N$ when N is of the form 2^k, for some positive integer k.

If the random variable Y discussed in the entropy formula (1.117) takes one of N possible values, each with equal probability, and 2 is used as the base of the logarithms in the formula, then the entropy defined is $\log_2 N$. Thus in this case the entropy measures the number of bits of information needed to find any given value of this random variable. The more general formula (1.117) allows for an extension of this argument to the case where Y does not have a uniform distribution.

B.11 Infinite Series

We present the definition of what a series is and give a survey of the few most relevant results for us.

Let $\{a_k\}_{k=1}^{\infty}$ be a sequence of real numbers. Using the a_k's we define a new sequence

$$s_n = a_1 + a_2 + \cdots + a_n,$$

and s_n is called the nth *partial sum* of the sequence $\{a_k\}$. A shorthand notation is $s_n = \sum_{k=1}^{n} a_k$. We define the *infinite series* $\sum_{k=1}^{\infty} a_k$ by

$$\sum_{k=1}^{\infty} a_k \overset{\text{def}}{=} \lim_{n \to \infty} s_n,$$

as long as the limit on the right-hand side exists. Thus the symbol $\sum_{k=1}^{\infty} a_k$ is the limit of the partial sums s_n as $n \to \infty$. If the limit does not exist, we say that the series diverges. There are many ways in which a series can diverge, but if in particular the limit on the right-hand side tends to infinity, we say that the series diverges to infinity. As an example of a series that diverges, but not to infinity, consider $\sum_{k=1}^{\infty} (-1)^k$; for this series $s_n = 0$ if n is even and $s_n = -1$ if n is odd.

A basic fact about infinite series is that

$$\sum_{k=m}^{\infty} a_k \text{ converges} \Rightarrow \lim_{k \to \infty} a_k = 0.$$

The essence of the proof is that $a_k = s_k - s_{k-1}$, but $\lim s_k = \lim s_{k-1}$, so the limit of a_k as $k \to \infty$ is zero.

The converse is not, however, true. The classic example is the harmonic series

$$\sum_{k=1}^{\infty} \frac{1}{k}.$$

This series diverges, albeit very slowly. If s_n is the sum of the first n terms in this series, then it can be shown that $s_n \sim \log n$, so s_n "diverges like" $\log n$. In fact, more can be shown. The limit

$$\lim_{n \to \infty} \left(\sum_{k=1}^{n} \frac{1}{k} - \log n \right) \tag{B.6}$$

exists, and is approximately equal to $.5772156649\ldots$. This is the famous *Euler's constant*, denoted by γ, that appeared in Chapter 2 as the mean of the extreme value distribution (2.128). Because of the limiting relationship (B.6), it is common practice, for large n, to use the approximation

$$\sum_{k=1}^{n} \frac{1}{k} \approx \log n + \gamma, \tag{B.7}$$

since the right-hand side is easier to calculate than the left-hand side. More important, use of this approximation often simplifies and clarifies the implication of some mathematical formulae involving $\sum_{k=1}^{n} 1/k$ when n is large (see, for example, Problem 2.24).

Although the harmonic series diverges, the series

$$\sum_{k=1}^{\infty} \frac{1}{k^p} \tag{B.8}$$

converges when $p > 1$. As might be suprising, when p is an even integer, the sum of the series involves π: for example, when $p = 2$,

$$\sum_{k=1}^{\infty} \frac{1}{k^2} = \frac{\pi^2}{6}. \tag{B.9}$$

This is used in Chapter 2.

Another important kind of series used often in this book is the *geometric* series. These are any series of the form

$$\sum_{k=m}^{\infty} r^k, \tag{B.10}$$

where $-1 < r < 1$. In contrast to the situation for the series in the expression (B.9), there is a simple formula for the sum of a geometric series, namely

$$\sum_{k=m}^{\infty} r^k = \frac{r^m}{1-r}.$$ (B.11)

In particular,

$$\sum_{k=0}^{\infty} r^k = \frac{1}{1-r}.$$ (B.12)

The most basic example is

$$1 + \frac{1}{2} + \frac{1}{4} + \frac{1}{8} + \frac{1}{16} + \cdots = \sum_{k=0}^{\infty} \frac{1}{2^k} = \frac{1}{1-\frac{1}{2}} = 2.$$

B.12   Taylor Series

Until now we have been discussing series of fixed terms. It is also necessary to consider series of variable terms. For example, we can consider the left side of equation (B.12) as a function of r, where r ranges over the (open) interval from -1 to 1, and thus consider equation (B.12) as an equation of functions. To emphasize this point we rewrite (B.12), replacing r with x, as

$$\sum_{k=0}^{\infty} x^k = \frac{1}{1-x}, \quad \text{for } -1 < x < 1.$$ (B.13)

This is an example of a *Taylor* series, and in this section we state without proof various relevant results from the theory of Taylor series. There is a class of functions, called *analytic functions*, that have various important properties. In particular, they are infinitely differentiable. That is, the first, second, third, etc. derivatives exist at every point in their domain. We denote the kth derivative of $f(x)$ at the point a by $f^{(k)}(a)$.

Definition. A function $f(x)$ defined on an open interval containing zero is analytic at zero if $f^{(k)}(0)$ exists for all k, and for all x in some open interval containing zero (this interval is possibly smaller than the domain of $f(x)$),

$$f(x) = \sum_{k=0}^{\infty} \frac{f^{(k)}(0)}{k!} x^k.$$ (B.14)

In the expression the zeroth derivative of the function $f(x)$ is taken as the function itself. Equation (B.14) is called the *Taylor series expansion of $f(x)$ at zero*.

The motivation for the expression (B.14) is as follows. The linear polynomial $p_1(x) = f(0) + f'(0)x$ is the tangent line to the graph of $f(x)$ at the point $(0, f(0))$. This linear function has the same value and same derivative at zero as $f(x)$. This is the best linear approximation to $f(x)$ at zero (see figure B.1). The polynomial $p_2(x) = f(0) + f'(0)x + \frac{f'(0)}{2}x^2$ has the

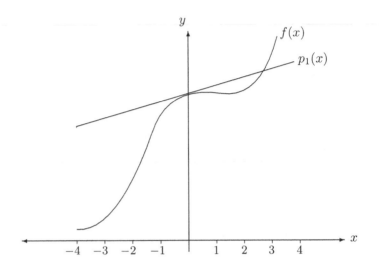

Figure B.1. The linear approximation $p_1(x) = f(0) + f'(0)x$ to a function $f(x)$.

same value, derivative, and second derivative at zero as does $f(x)$. Just as $p_1(x)$ is the best linear approximation to $f(x)$ near zero, $p_2(x)$ is the best quadratic approximation to $f(x)$ near zero (i.e., the best approximation by a parabola). In general,

$$p_n(x) = \sum_{k=0}^{n} \frac{f^{(k)}(0)}{k!} x^k \qquad (B.15)$$

has the same value and derivatives as f at zero, up to the nth derivative, and is the best approximation to f, near zero, by an nth degree polynomial. That is, the nth-order Taylor polynomial approximation to the function $f(x)$ is

$$f(x) \approx \sum_{k=0}^{n} \frac{f^{(k)}(0)}{k!} x^k. \qquad (B.16)$$

If $f^{(k)}(0)$ exists for all k, then if one takes the limit $n \to \infty$ one might hope that (B.14) always holds, but this is not the case; not all functions are analytic. However, the cases where this procedure fails do not arise for the functions considered in this book, so whenever we need to rewrite a function in terms of its Taylor series expansion (B.14), we shall feel free to do so.

Another useful fact is that if $f(x) = \sum_{k=0}^{\infty} a_k x^k$ converges in an open interval containing zero, then $f'(x) = \sum_{k=1}^{\infty} k a_k x^{k-1}$ in the same open interval. In other words, we can differentiate the series term by term. This fact implies that if a series $\sum a_k x^k$ converges for all x in some open interval containing zero, then it is an analytic function on that interval and is its own Taylor series expansion at zero.

We conclude by giving some examples useful for bioinformatics, and we also give the generalization of (B.14) to other points in the domain besides zero. The first example, already given in (B.13), is the geometric sum formula, which we can now see in a new light: The right-hand side is the Taylor expansion of $\frac{1}{1-x}$. Differentiating term by term gives

$$f'(x) = \sum_{k=1}^{\infty} \frac{f^{(k)}(0)}{k!} k x^{k-1} = \sum_{k=1}^{\infty} \frac{f^{(k)}(0)}{(k-1)!} x^{k-1}. \tag{B.17}$$

Using this, we can easily get a new Taylor series from the old one. For example, from (B.13) we get

$$\frac{1}{(1-x)^2} = \sum_{k=1}^{\infty} k x^{k-1}, \text{ for } -1 < x < 1, \tag{B.18}$$

and

$$\frac{2}{(1-x)^3} = \sum_{k=2}^{\infty} k(k-1) x^{k-2}, \text{ for } -1 < x < 1. \tag{B.19}$$

Another basic Taylor series is the one for the exponential function e^x, namely

$$e^x = 1 + x + \frac{x^2}{2} + \frac{x^3}{3!} + \cdots = \sum_{k=0}^{\infty} \frac{x^k}{k!}, \tag{B.20}$$

for all real numbers x. If $|x|$ is small, this implies that

$$e^x \cong 1 + x, \tag{B.21}$$

or, more accurately,

$$e^x \cong 1 + x + \frac{1}{2} x^2. \tag{B.22}$$

Another important Taylor series is that for the (natural) logarithm. If $-1 < x \le 1$, this is

$$\log(1+x) = x - \frac{x^2}{2} + \frac{x^3}{3} - \frac{x^4}{4} + \cdots = \sum_{k=1}^{\infty} (-1)^{k+1} \frac{x^k}{k}. \tag{B.23}$$

If $|x|$ is small, this implies that

$$\log(1+x) = x + o(x) \text{ at zero}, \tag{B.24}$$

or more accurately,

$$\log(1+x) = x - \frac{x^2}{2} + o(x^2) \text{ at zero.} \tag{B.25}$$

As a final example of a Taylor series expansion, we revisit the binomial theorem. From the fact that $(1+x)^\alpha$ is analytic at zero, we get the general version of the binomial theorem:

Theorem (General Binomial Theorem). For $-1 < x < 1$ and α any real number,

$$(1+x)^\alpha = \sum_{k=0}^{\infty} \binom{\alpha}{k} x^k. \tag{B.26}$$

A function can be analytic at points other than zero, as the following more general definition shows.

Definition. Let $f(x)$ be a function defined on an open interval containing a. Then f is analytic at a if (1) $f^{(k)}(a)$ exists for all k and (2) for all x in some open interval containing a (this interval being possibly smaller than the domain of f), we have

$$f(x) = \sum_{k=0}^{\infty} \frac{f^{(k)}(a)}{k!}(x-a)^k. \tag{B.27}$$

The nth-order Taylor approximation following from this, generalizing equation (B.16), is

$$f(x) \approx \sum_{k=0}^{n} \frac{f^{(k)}(a)}{k!}(x-a)^k, \text{ for } x \text{ close to } a. \tag{B.28}$$

As two particularly important cases, the first-order, or linear, approximation is

$$f(x) \cong f(a) + (x-a)f'(a), \text{ for } x \text{ close to } a, \tag{B.29}$$

and the second-order approximation is

$$f(x) \cong f(a) + (x-a)f'(a) + \frac{(x-a)^2}{2}f''(a), \text{ for } x \text{ close to } a. \tag{B.30}$$

We have thus far defined the concept of a function's being "analytic" at a point. We call a function $f(x)$ analytic in an interval if it is analytic at every point in the interval. If $f(x)$ is analytic at every point in its domain we say that $f(x)$ is an *analytic function*.

B.13   Uniqueness of Taylor Series

One of the reasons why Taylor series are so useful is the fact that they satisfy certain uniqueness properties. In particular, if $f(x) = \sum_{n=0}^{\infty} a_n x^n$ and $g(x) = \sum_{n=0}^{\infty} b_n x^n$ and $f(x) = g(x)$ for all x in some open interval where both f and g converge, then $a_k = b_k$ for all k. This is because if $f(x) = g(x)$ for all x in an open interval, then this interval must contain zero, and then $f^{(k)}(0) = g^{(k)}(0)$ for all k; thus

$$0 = f^{(k)}(0) - g^{(k)}(0) = k!(a_k - b_k),$$

which implies $a_k = b_k$ for all k.

This is a convenient fact. From it we see that a random variable with positive integer values is characterized by its pgf in the following sense. Let Y_1 and Y_2 be discrete random variables taking positive integer values and suppose that $p_1(t)$ and $p_2(t)$ are their respective pgfs. If $p_1(t) = p_2(t)$ on an open interval containing 1, then the probability distributions of Y_1 and Y_2 are identical. This useful observation is used, often implicitly, many times in this book. In the next section we shall see that this is true even if we drop the assumption that the random variable takes only positive integer values.

B.14   Laurent Series

A Laurent series is a generalization of Taylor series, being of the form

$$\sum_{k=-\infty}^{\infty} a_k (x - b)^k.$$

For us the interesting case is $b = 0$ so we restrict attention to Laurent series of the form

$$\sum_{k=-\infty}^{\infty} a_k x^k.$$

If a random variable can take negative as well as positive integer values, its pgf is a Laurent series of this form, with t replacing x and with $\sum a_k = 1$. Our interest in this series is for values of t close to 1. When the series $\sum_{k=-\infty}^{\infty} a_k t^k$ converges in an open interval containing $t = 1$, as it does in all cases we consider in this book, then similar properties hold for these series as held for Taylor series. In particular, a probability distribution is still characterized by its pgf.

B.15   Numerical Solutions of Equations

Many equations arising in practice cannot be solved exactly, and the best that is possible is to find arbitrarily accurate numerical solutions. For example, the calculation of λ from equation (10.3) will normally require these methods. Here we outline Newton's method for the approximate numerical solution of some equation.

Suppose that we wish to find the solution of the equation $f(x) = 0$. Let a be a first estimate of this solution. The equation (B.29) shows that an approximation x_1 to the required solution is given by

$$x_1 \cong a - \frac{f(a)}{f'(a)}.$$

We now take x_1 as a new estimate of the solution and find an improved estimate x_2 from the equation

$$x_2 \cong x_1 - \frac{f(x_1)}{f'(x_1)}.$$

This procedure can be repeated until an estimate of the solution having any desired degree of accuracy is found. In some cases the process can lead to a diverging rather than a converging sequence of values, so some care must be exercised in using it.

B.16   Statistical Differentials

Suppose that X_1 is a random variable with mean μ and small variance (so that X_1 is unlikely to differ much from μ). Equation (B.29) then shows that if f is analytic,

$$f(X_1) - f(\mu) \cong (X_1 - \mu)f'(\mu). \qquad \text{(B.31)}$$

Let $X_2 = f(X_1)$. Then taking expectations throughout in the approximation (B.31),

$$\text{mean of } X_2 \cong f(\mu). \qquad \text{(B.32)}$$

Squaring in equation (B.31) and then taking expectations leads to the first-order approximation

$$\text{variance of } X_2 \cong (f'(\mu))^2 \text{ variance of } X_1. \qquad \text{(B.33)}$$

The approximations made in deriving this formula imply that it must be used with some caution.

One use of (B.33) is to derive a random variable X_2 whose variance is approximately independent of its mean. The variance of X_2 is approximately independent of the mean of X_2 if the transformation $f(\cdot)$ is chosen so that the right-hand side in (B.33) is independent of μ. This will arise if

$$X_2 = f(X_1) = \text{const} \int^{X_1} \frac{1}{\sigma(X_1)} d\mu, \qquad (\text{B.34})$$

where $\sigma(X_1)$ is the standard deviation of X_1, assumed to depend on μ. For example, if X_1 has a Poisson distribution, for which the mean and variance ar equal, the required transformation is $X_2 = \text{const} \int^{X_1} \frac{1}{\sqrt{\mu}} d\mu = \text{const} \sqrt{X_1}$. When the standard deviation of X_1 is proportional to its mean, use of (B.34) shows that $X_2 = \log X_1$ has a variance approximately independent of its mean. (See Problem 1.38 for an example of a case where the logarithmic transformation leads to a random variable whose variance is completely independent of the mean.)

B.17 The Gamma Function

The *gamma function* is an analytic function defined for positive u by the integral

$$\Gamma(u) = \int_0^\infty e^{-t} t^{u-1} dt. \qquad (\text{B.35})$$

The gamma function extends the factorial function to all positive numbers, as can be seen from the facts that

$$\Gamma(u+1) = u\Gamma(u), \text{ for } all \text{ positive real numbers } u, \qquad (\text{B.36})$$

and

$$\Gamma(1) = 1, \qquad (\text{B.37})$$

which imply

$$\Gamma(n) = (n-1)! \quad \text{when } n \text{ is a positive integer.} \qquad (\text{B.38})$$

When u is positive but not an integer, equation (B.36) can be used to find $\Gamma(u)$, provided that values of the gamma function for values of u in (say) the interval $(1, 2)$ are available. These values are listed in Table 6.1 of Abramowitz and Stegun (1972). A useful value of the gamma function is

$$\Gamma\left(\frac{1}{2}\right) = \sqrt{\pi}. \qquad (\text{B.39})$$

We need two further facts about the derivative of the gamma function. First,

$$\left(\frac{d\,\Gamma(u)}{du}\right)_{u=1} = -\gamma, \qquad (\text{B.40})$$

where γ is Euler's constant defined in (B.6). Second,

$$\left(\frac{d^2 \log \Gamma(u)}{d\,u^2}\right)_{u=1} = \frac{\pi^2}{6}. \tag{B.41}$$

The fact that $\Gamma(n) = (n-1)!$ when n is a positive integer allows a rapid evaluation of the integral

$$\int_0^\infty \lambda x^n\, e^{-\lambda x} dx, \tag{B.42}$$

which arises in connection with finding the moments of the exponential distribution (1.66). The change of variable $u = \lambda x$ leads to the calculation

$$\int_0^\infty \lambda^{-n} u^n\, e^{-u} du = n! \lambda^{-n}, \tag{B.43}$$

for the nth moment of this distribution.

Fractional moments are also sometimes of interest: Thus if X is a random variable having the exponential distribution (1.66),

$$E(X^{1/2}) = \int_0^\infty \lambda^{-1/2} u^{1/2} e^{-u} du$$
$$= \frac{\Gamma(3/2)}{\sqrt{\lambda}} = \frac{\sqrt{\pi}}{2\sqrt{\lambda}} \tag{B.44}$$

from (B.36) and (B.39).

B.18   Proof by Induction

One of the most useful methods of proving some proposition $P(n)$ about a positive integer n is a proof by induction. In a proof of this type we (i) first prove that $P(n)$ is true for the particular case $n = 1$, and then (ii) show that the truth of $P(n)$ for the case $n = m$ implies its truth for the case $n = m + 1$. Using (ii) for the case $m = 1$, (i) and (ii) together imply the truth of $P(n)$ for the case $n = 2$. Then using this result and (ii) for the case $m = 2$ implies the truth of $P(n)$ for the case $n = 3$, etc. Thus (i) and (ii) together imply that $P(n)$ is true for all positive integers.

A classic simple example of such a proof is as follows. Define $S(n)$ by $S(n) = 1 + 2 + 3 + \cdots + n$ and then define the proposition $P(n)$ as $S(n) = n(n+1)/2$. This proposition is true for the case $n = 1$, since $1 = 1(1+1)/2$. Suppose that $P(n)$ is true for $n = m$, so that $1 + 2 + \cdots + m = m(m+1)/2$. Then $1 + 2 + \cdots + m + (m+1) = (1 + 2 + \cdots + m) + (m+1) = m(m+1)/2 + (m+1) = (m+1)(m+2)/2$. This shows that $P(n)$ is also true for the case $n = m + 1$.

B.19   Linear Algebra and Matrices

In this section we quote without proof several results from the theory of linear algebra that are relevant to material in the text. It is assumed that the reader is familiar with the basics of linear algebra, including the concepts of matrices and their eigenvalues and eigenvectors, of vectors, of matrix and vector products, and of the transposition operation (denoted here by "'"). The vector v is assumed to be a column vector, so that the vector v' is taken as a row vector.

Let P be an $s \times s$ matrix. This matrix has s *eigenvalues*, possibly not all distinct, denoted by $\lambda_1, \lambda_2, \ldots, \lambda_s$, found as the s solutions to the determinantal equation (in λ)

$$|P - \lambda I| = 0. \tag{B.45}$$

It is possible that there are solutions of equation (B.45) having multiplicity exceeding 1: For our purposes we assume that all eigenvalues of P are distinct.

Corresponding to the eigenvalue λ_j is a *right eigenvector* r_j and a *left eigenvector* ℓ_j', both defined only up to a multiplicative constant, for which

$$P r_j = \lambda_j r_j, \quad \ell_j' P = \ell_j' \lambda_j. \tag{B.46}$$

We now assume that these eigenvectors are normalized so that

$$\ell_j' r_j = 1 \text{ for all values of } j, \quad j = 1, 2, \ldots, s. \tag{B.47}$$

Although this normalization does not uniquely determine ℓ_j' and r_j, it is sufficient for our purposes. With this normalization, P can be expressed in terms of its eigenvalues and eigenvectors in a *spectral expansion* as

$$P = \sum_{j=1}^{s} \lambda_j r_j \ell_j'. \tag{B.48}$$

If λ_j is an eigenvalue of P with corresponding right eigenvector r_j and left eigenvector ℓ_j', then for any positive integer n, $\lambda_j{}^n$ is an eigenvalue of P^n, with right and left eigenvectors r_j and ℓ_j', respectively. This implies that the spectral expansion of P^n is

$$P^n = \sum_{j=1}^{s} \lambda_j{}^n r_j \ell_j'. \tag{B.49}$$

This result has relevance to us when P is the transition matrix of a Markov chain, since the spectral expansion of P^n gives useful information about n-step transition probabilities of this chain.

The theory is more complex when some eigenvalues of P have multiplicity greater than 1, but in all cases of interest to us, a spectral expansion of the form (B.49) exists and is used often in Markov chain theory in the text.

Appendix C

Computational Aspects of the Binomial and Generalized Geometric Distribution Functions

The expression for the generalized geometric cumulative distribution function $F_Y(y)$ given in equation (6.3) can be written as

$$\frac{(1-p)^{k+1}}{k!}\left(\frac{k!}{0!} + \frac{(k+1)!p}{1!} + \frac{(k+2)!p^2}{2!} + \cdots + \frac{y!p^{y-k}}{(y-k)!}\right). \tag{C.1}$$

In Section 6.3 we consider the calculation of $\mathrm{Prob}(Y \leq y-1)$, which we denote here by $Q(y)$, for the values $y = k+1, k+2, \ldots$. It is thus of interest to find an efficient way of carrying out this calculation for any given values of p and k and a collection of values of y. The expression (C.1) shows that

$$Q(y+1) - Q(y) = (1-p)^{k+1}\binom{y}{k}p^{y-k}. \tag{C.2}$$

Replacing y by $y + 1$ in (C.2) we get

$$Q(y+2) - Q(y+1) = (1-p)^{k+1}\binom{y+1}{k}p^{y+1-k}, \tag{C.3}$$

and from (C.2) and (C.3) we obtain

$$Q(y+2) - Q(y+1) = \frac{y+1}{y-k+1}p\left(Q(y+1) - Q(y)\right). \tag{C.4}$$

This leads to the recurrence relation (Wilf (2003))

$$Q(y+2) = \frac{(y+1)p + y - k + 1}{y-k+1}Q(y+1) - \frac{y+1}{y-k+1}pQ(y) \tag{C.5}$$

This can be solved rapidly for any chosen collection of values of y, using the initial conditions

$$Q(k+1) = (1-p)^{k+1}, \; Q(k+2) = (1-p)^{k+1}\left(1 + (k+1)p\right).$$

It can also be shown that

$$Q(y) = \sum_{j=k+1}^{y} \binom{y}{j}(1-p)^j p^{y-j}. \tag{C.6}$$

This is a sum of *binomial* probabilities. The discussion in Section 1.3.6 relating the cumulative distribution functions of the binomial and the generalized geometric show why this is so. The event that $y-1$ or fewer trials occur before the trial at which failure $k+1$ arises is exactly the same as the event that there are $k+1$ or more failures in the first y trials. $Q(y)$ is defined as the probability of the former event, and so it is also the probability of the latter event. This is what equation (C.6) states. This observation allows us to use the recurrence relation (C.5) to calculate the distribution function of the binomial distribution (1.8) for specific values of y and p.

Arratia et al. (1986) in effect show that for the case where y is large and p is small, a close approximation for $Q(y)$ is

$$Q(y) = \text{Prob}(Y \le y - 1) \cong 1 - \frac{y^k p^{y-k}(1-p)^k}{k!}. \tag{C.7}$$

This is exact for $k = 0$ and when $k = 1$ differs from the exact value by an amount p^y, which, when p is small and y is large, is very small. For larger values of k the approximation (C.7) is also very accurate, although the accuracy decreases slightly as k increases. The approximation (C.7) is useful in many applications, and often saves us from making an exact but tedious numerical calculation.

Appendix D
BLAST: Sums of Normalized Scores

Our aim in this appendix is to use the joint density function (10.39) for the case $r = 2$, namely

$$f(s_1, s_2) = \exp\left(-(s_1 + s_2) - e^{-s_2}\right), \qquad (D.1)$$

to derive the "$r = 2$" case of equation (10.40), namely

$$f(t) = \frac{e^{-t}}{2} \int_0^{+\infty} \exp(-e^{(y-t)/2}) \, dy, \qquad (D.2)$$

for the density function $T_2 = S_1 + S_2$, the sum of the highest two normalized scores in a BLAST calculation.

To do this we introduce the dummy variable Y, defined by $Y = S_1 - S_2$. The domain of the two random variables S_1 and S_2 in the (S_1, S_2) plane is $S_1 \geq S_2$, and this transforms into the domain $Y \geq 0$ in the (T_2, Y) plane. The Jacobian J of the transformation from (S_1, S_2) to (T_2, Y) is 2, and we can write $s_1 + s_2 = t$ and $s_2 = (t - y)/2$. Equations (2.145) and (D.1) then show that the joint density function of T_2 and Y is

$$f(t, y) = \frac{1}{2}\exp(-t - e^{(y-t)/2}). \qquad (D.3)$$

The density function of T_2 is found by integrating out y over the range $y \geq 0$ in this joint density function, and this leads immediately to equation (D.2).

References

Abramowitz, M. and I. Stegun (1972). *Handbook of Mathematical Functions*. Dover, New York.

Affymetrix Statistical Algorithms Reference Guide (2003). **http://www.affymetrix.com/support/technical/manuals.affy**

Altschul S.F. (1991). Amino acid substitution matrices from an information theoretic perspective. *J. Mol. Biol.* **219**, 555–565.

Altschul S.F. (1993). A protein alignment scoring system sensitive at all evolutionary distances. *J. Mol. Evol.* **36**, 290–300.

Altschul, S.F. and W. Gish (1996). Local alignment statistics. *Methods in Enzymology* **266**, 460–480.

Altschul, S.F., T.L. Madden, A.A. Schaffer, J. Zhang, Z. Zhang, W.Q. Miller, and D.J. Lipman (1997). Gapped BLAST and PSI-BLAST: a new generation of protein database search programs. *Nucleic Acids Research* **25**, 3389–3402.

Andrews, D.K.W. (2000). Inconsistency of the bootstrap when the parameter is on the boundary of the parameter space. *Econometrica* **68**(2), 399–406.

Arratia, R., L. Gordon, and M.S. Waterman (1986). An extreme value theory for sequence matching. *Ann. Statist.* **14**, 971–983.

Arratia, R., L. Goldstein, and L. Gordon (1989). Two moments suffice for Poisson approximations: the Chen-Stein method. *Ann. Prob.* **17**, 9–25.

Arratia, R., E.S. Lander, S. Tavaré, and M.S. Waterman (1991). Genomic mapping by anchoring random clones: a mathematical analysis. *Genomics* **11**, 806–827.

Babu, G.J and A.R. Padmanabhan (2002). Re-sampling methods for the non-parametric Behrens–Fisher problem. *Indian J. Statist* **64**, 677–691.

Bacro, J.H., J.J. Daudin, S. Mercier and S. Robin (2002). Back to the local score in the logarithmic case: a direct and simple proof. *Ann. Inst. Stat. Math.* **54**, 748–757.

Bailey, T.L. and C. Elkan (1998). Fitting a mixture model by expectation maximization to discover motifs in biopolymers. *Proceedings of the Second International Conference on Intelligent Systems for Molecular Biology*, AAAI Press, Menlo Park, CA.

Bailey, T.L. and M. Gribskov (1998). PROSITE: a dictionary of sites and patterns in proteins. *J. Comp. Biol.* **5**, 211–221.

Baldi, P. and G.W. Hatfield (2002). *DNA Microarrays and Gene Expression*. Cambridge University Press, Cambridge, UK.

Benjamini, Y. and Y. Hochberg (1995). Controlling the false discovery rate: a practical and powerful approach to multiple testing. *Proc. R. Statist. Soc. Series B*, **57**, 289–300.

Benjamini, Y. and D. Yekultieli (2001). The control of the false discovery rate in multiple testing under dependency. *Annals of Statistics* **29**, 1165–1188.

Benner, S.A., M.A. Cohen, and G.H. Gonnet (1994). Amino acid substitution during functionally constrained divergent evolution of protein sequences. *Protein Engineering* **7**, 1323–1332.

Berk, R.H. and A. Cohen (1979). Asymptotically optimal methods of combining tests. *J. Amer. Stat. Assoc.* **74**, 812–814.

Bernaola-Galván, P., I. Grosse, P. Carpena, J.L. Oliver, R. Román-Roldán, and H.E. Stanley (2000). Finding borders between coding and non-coding

DNA regions by an entropic segmentation method *Phys. Rev. Lett.* **85**, 1342–1345.

Biaudet, V., M. El Karoui, and A. Gruss (1998). Codon usage can explain GT-rich islands surrounding Chi sites on the *Escherichia coli* genome. *Mol. Microbiol.* **29**, 666–699.

Bishop, M.J. and A.E. Friday (1985). Evolutionary trees from nucleic and protein sequences. *J. Roy. Soc. London B.* **226**, 271–302.

Blaisdell, B.E. (1985). A method of estimating from two aligned present-day DNA sequences their ancestral composition and subsequent rates of substitution, possibly different in the two lineages, corrected for multiple and parallel substitutions at the same site. *J. Mol. Evol.* **22**, 69–81.

Blom, G. (1982). On the mean number of random digits until a given sequence occurs. *J. Appl. Prob.* **19**, 136–143.

Blom, G. and D. Thorburn (1982). How many random digits are required until given sequences are obtained? *J. Appl. Prob.* **19**, 518–531.

Breen, S., M.S. Waterman, and N. Zhang (1985). Renewal theory for several patterns. *J. Appl. Prob.* **22**, 228–234.

Burge, C. (1997). Identification of complete gene structures in human genomic DNA. Ph.D. thesis, Stanford University, Stanford, CA.

Burge, C. and S. Karlin (1997). Prediction of complete gene structures in human genomic DNA. *J. Mol. Biol.* **268**, 78–94.

Burge, C. (1998). Modeling Dependencies in Pre-mRNA Splicing Signals, Section II.8 in *Computational Methods in Molecular Biology*, S. Salzberg, D. Searls, and S. Kasif, eds., Elsevier Science.

Bussemaker, H.J., H. Li, and E.D. Siggia (2000). Building a dictionary for genomes: identification of presumptive regulatory rate by statistical analysis *Proc. Nat. Acad. Sci.* **97**, 10096–10100.

Chan, H.P. (2003). Upper bounds and importance sampling of *p*-values for DNA and protein sequence alignments. *Bernoulli* **9**, 183–199.

Chang, J.T. (1996a). Inconsistency of evolutionary tree topology reconstruction methods when substitution rates vary across characters. *Math. Biosci.* **134**, 189–215.

Chang, J.T. (1996b). Reconstruction of Markov chain models on evolutionary trees: identifiability and consistency. *Math. Biosci.* **137**, 51–73.

Chedin, F., P. Noirot, V. Biaudet, and S. Ehrlich (1998). A five-nucleotide sequence protects DNA from exonucleolytic degradation by AddAB, the RecBCD analogue of *Bacillus subtilis*. *Mol. Microbiol.* **31**, 1369–1377.

Chen, L.H.Y (1975). Poisson approximation for dependent trials. *Ann. Prob.* **3**, 534–545.

Chernick, M. (1999). *Bootstrap Methods: A Practitioner's Guide.* Wiley, New York.

Cheung, V.G., J.P. Gregg, K.J. Gogolin-Ewens, J. Bandong, C.A. Stanley, L. Baker, M.J. Higgins, N.J. Nowak, T.B. Shows, W.J. Ewens, S.F. Nelson and R.S. Spielman (1998). Linkage-disequilibrium mapping without genotyping. *Nature Genetics* **18**, 225–230.

Churchill. G.A. (2002). Fundamentals of experimental design for cDNA microarrays. *Nature Genetics* **32**, 490–495.

Cleveland, W.S. (1979). Robust locally weighted regression and smoothing scatterplots. *J. Amer. Stat. Assoc* **74**, 829–836.

Cleveland, W.S. and S.J. Devlin (1988). Locally weighted regression: an approach to regression analysis by local fitting. *J. Amer. Stat. Assoc.* **83**, 596–610.

Cochran, W.G. and and G.M. Cox (1957). *Experimental Designs.* Wiley, New York.

Cowan, R. (1991). Expected frequencies of DNA using Whittle's formula. *J. Appl. Prob.* **28**, 886–892.

Daudin, J.J. and S. Mercier (2000). Distribution exacte du score local d'une suite de variables indépendent et identiquement distribuées. *C. R. Acad. Sci. Paris* **329**, 815–820.

Daudin, J.J., M.P. Etienne, and P. Vallois (2001). Asymptotic behaviour of the local score of independent and identically distributed random sequences. *Pub. Inst. Elie Cartan* **28**, 1–25.

Daudin, J.J., M.P. Etienne, and P. Vallois (2003). Asymptotic behaviour of the local score of independent and identically distributed random sequences. *Stoch. Proc. and their Applns.* **107**, 1–28.

Davis, C.S. (2003). *Statistical Methods for the Analysis of Repeated Measurements.* Springer, New York.

Davison, A.C. and D.V. Hinkley (1997). *Bootstrap Methods and their Applications.* Cambridge University Press, Cambridge, UK.

Dayhoff, M.O., R.M. Schwartz, and B.C. Orcutt (1978). A model of evolutionary change in proteins, pp. 3435–358 in *Atlas of Protein Sequence and Structure,* **5**, Supplement 3.

Dixon, M.T. and D.M. Hillis (1993). Ribosomal RNA secondary structure: compensatory mutations and implications for phylogenetic analysis. *Mol. Biol. Evo.* **10**, 256–267.

Dembo, A., S. Karlin, and O. Zeitouni (1994a). Critical phenomena for sequence matching with scoring. *Ann. Prob.* **22**, 1993–2021.

Dembo, A., S. Karlin, and O. Zeitouni (1994b). Limit distribution of maximal non-aligned two-sequence segmental score. *Ann. Prob.* **22**, 2022–2039.

Draghici S. (2003). *Data Analysis Tools For DNA Microarrays.* Chapman and Hall, New York.

Dudoit, S., Y.H. Yang, M.J. Callow, and T.P. Speed (2000). Statistical methods for identifying differentially expressed genes in replicated cDNA microarray experiments. Stanford University technical report 578, Department of Biochemistry. www.stat.berkeley.edu/users/terry/Zarray/Html/matt.html.

Dudoit, S., J. Fridlyand, and T.P. Speed (2002). Comparison of discrimination methods for the classification of tumors using gene expression data. *J. Amer. Statist. Soc.* **97**, 77–87.

Durbin, R., S. Eddy, A. Krogh, and G. Mitchison (1998). *Biological Sequence Analysis.* Cambridge University Press, Cambridge, UK.

Edwards, A.W.F. (1992). *Likelihood.* Johns Hopkins Press, Baltimore, MD.

Edwards, A.W.F. and L.L. Cavalli-Sforza (1964). Reconstruction of phylogenetic trees. *Phenetic and Phylogenetic Classification,* V.H. Heywood and J. McNeil, eds., Systematics Association Publication 6, 67–76.

Efron, B. (1982). *The Jackknife, the Bootstrap and Other Resampling Plans.* SIAM Applied Mathematics Publication. Philadelphia, PA.

Efron, B. and R.J. Tibshirani (1993). *An Introduction to the Bootstrap.* Chapman and Hall, New York.

Eisen, J.A. (1998). Phylogenomics: improving functional predictions for uncharacterized genes by evolutionary analysis. *Genome Research* **8**, 163–167.

Ewens, W.J., R.C. Griffiths, S.N. Ethier, S.A. Wilcox, and J.A. Marshall Graves (1992). Statistical analysis of *in situ* hybridization data: derivation of the z_{max} test. *Genomics* **12**, 675–682.

Feller, W. (1968). *An Introduction to Probability Theory and its Applications.* Volume I, 3rd edition, Wiley, New York.

Felsenstein, J. (1978a). The number of evolutionary trees. *Systematic Biology* **27**, 27–33.

Felsenstein, J. (1978b). Cases in which parsimony or compatibility methods will be positively misleading. *Syst. Zool.* **27**, 401–410.

Felsenstein, J. (1981). Evolutionary trees from DNA sequences: A maximum likelihood approach. *J. Mol. Evol.* **17**, 368–376.

Felsenstein, J. (1983). Statistical inference of phylogenies. *J. Roy. Statist. Soc. A.* **146**, 246–272.

Felsenstein, J. (1985). Confidence limits on phylogenies: an approach using the bootstrap. *Evolution* **39**, 783–791.

Felsenstein, J. (1980–2000). Phylogeny Inference Package. Web resource. (http://evolution.genetics.washington.edu/phylip.html).

Felsenstein, J. (2004). *Inferring Phylogenies.* Sinauer Associates, Sunderland, MA.

Felsenstein, J., and G.A. Churchill (1996). A hidden Markov model approach to variation among sites in rate of evolution. *Mol. Biol. Evol.* **13**, 93–104.

Felsenstein, J. and H. Kishino (1993). Is there something wrong with the bootstrap on phylogenies? A reply to Hillis and Bull. *Syst. Biol.* **2**, 193–200.

Fisher, R.A. (1950). *Statistical Methods for Research Workers*. Oliver and Boyd, London.

Fitch, W.M. (1971). Toward defining the course of evolution: minimum change for a specified tree topology. *Systematic Zoology* **20**, 406–416.

Fukami, K. and Y. Tateno (1989). On the uniqueness of the maximum likelihood method for estimating molecular trees: uniqueness of the likelihood point. *J. Mol. Evol.* **28**, 460–464.

Ge, Y., S. Dudoit, and T.P. Speed (2003). Resampling-based multiple testing for microarray data hypothesis. *Test* **12**(1), 1-44.

Geschwind, D.H. and J.P. Gregg (2002). *Microarrays for the Neurosciences*. MIT Press, Boston, MA.

Gibbs, W.W. (2003). The unseen genome: gems among the junk. *Scientific American* **289**, Number 5, 46–53.

Glonek, G.F.V. and P.J. Solomon (2002). **http://www.maths.adelaide.edu.au/MAG**

Gojobori, T., K. Ishii, and M. Nei (1982). Estimation of the average number of nucleotide substitutions when the rate of substitution varies with nucleotide. *Mol. Biol. Evol.* **13**, 93–104.

Goldman, N. (1993). Statistical tests of models of DNA substitution. *J. Mol. Evol.* **36**, 182–198.

Goldman, N. (2002). Personal communication.

Goldman, N. and S. Whelan (2000). Statistical tests of gamma-distributed rate heterogeneity in models of sequence evolution in phylogenetics. *Mol. Biol. Evol.* **17**, 975–978.

Goldman, N., J.P. Anderson, and A.G. Rodrigo (2000). Likelihood-based tests of topologies in phylogenetics. *Syst. Biol.* **49**, 652–670.

Golub, T.R., D.K. Slonim, P. Tamayo, C. Huard, M. Gaasenbeek, J.P. Mesirov, H. Coller, M.L. Loh, J.R. Downing, M.A. Caligiuri, C.D. Bloomfield, and E.S. Lander (1999). Molecular classification of cancer: class discovery and class prediction by gene expression monitoring. *Science* **286(5439)**, 531-537.

Gotoh, O. (1982). An improved algorithm for matching biological sequences. *J. Mol. Biol.* **162**, 705–708.

Grant, G.R., V.G. Cheung, E. Manduchi, and W.J. Ewens (1999). Significance testing for discrete identity-by-descent mapping. *Ann. Hum. Genet.* **63**, 441-454.

Grant, G.R., E. Manduchi, and C.J. Stoeckert Jr. (2002). Using nonparametric methods in the context of multiple testing to identify differentially expressed genes. *Methods of Microarray Data Analysis*. Kluwer Academic Publishers, Boston.

Grigorenko, E.V. (ed.) (2003). *DNA – Technologies and Experimental Strategies*. Chapman & Hall.

Hall, P. (1992). *The Bootstrap and Edgeworth Expansion*. Springer–Verlag, New York.

Hasegawa, M., H. Kishino, and T. Yano (1985). Dating of the human–ape splitting by a molecular clock of mitochondrial DNA. *J. Mol. Evol.* **22**, 160–174.

Hasegawa, M. and M. Fujiwara (1993). Relative efficiencies of the maximum likelihood, maximum parsimony, and neighbor-joining methods for estimating protein phylogeny. *Mol. Phylogen. Evol.* **2**, 1–5.

Henikoff, S. and J.G. Henikoff (1992). Amino acid substitution matrices from protein blocks. *Proc. Nat. Acad. Sci.* **89**, 10915–10919.

Hertz, G.Z. and G.D. Stormo (1999). Idenfifying DNA and protein patterns with statistically significant alignments of multiple sequences. *Bioinformatics* **15**, 563–577.

Hillis, D.M. and J.-J. Bull (1993). An empirical test of bootstrapping as a method for assessing confidence in phylogenetic analysis. *Syst. Biol.* **2**, 182–192.

Hillis, D.M., C. Moritz, and B.K. Mable, (eds). (1996). *Molecular Systematics*. Sinauer Associates, Sunderland, MA.

Holmes, S.P. (1999). Phylogenies: an overview. pp. 81–118 in *Statistics in Genetics*, M.E. Halloran and S. Geisser, eds., Springer–Verlag, New York.

Huelsenbeck, J.P. and K. Crandall (1997). Phylogeny estimation and hypothesis testing using maximum likelihood. *Ann. Rev. Ecol. Syst.* **28**,

437–466.

Huelsenbeck, J.P., D.M. Hillis, and R. Nielsen (1996). A likelihood ratio test of monophyly. *Syst. Biol.* **45**, 546–558.

Huelsenbeck, J.P. and B. Rannala (1997). Phylogenetic methods come of age: testing hypotheses in an evolutionary context. *Science* **276**, 227–232.

Huelsenbeck, J.P. and R. Nielsen (1999). Variation in the pattern of nucleotide substitution across sites. *J. Mol. Evol.* **48**, 86–93.

International Human Genome Sequencing Consortium (2001). Initial sequencing and analysis of the human genome. *Nature* **409**, 860–921.

Irizarry, R.A., B.M. Bolstad, F. Collin, L.M. Cope, B. Hobbs, and T. Speed (2003). Summaries of Affymetrix GeneChip probe level data. *Nucleic Acids Research* **31**, e15.

Jensen, S.T. and J.S. Liu (2004). BioOptimizer: a Bayesian scoring function approach to motif discovery. To appear in *Bioinformatics*.

Jensen, S.T., X.S. Liu, Q. Zhou, and J.S. Liu (2004). Computational discovery of gene regulatory binding motifs: a Bayesian perspective. To appear in *Statistical Science*.

Jonassen, I., J.F. Collins, and D.G. Higgins (1995). Finding flexible patterns in unaligned protein sequences. *Protein Science* **4**, 1587–1595.

Jukes, T.H. and C.R. Cantor (1969). Evolution of protein molecules. pp. 21–132 in *Mammalian Protein Metabolism*, H.N. Munro (ed.), Academic Press, New York.

Karlin, S. (1994). Statistical studies of biomolecular sequences: score-based methods. *Phil. Trans. R. Soc. Lond. B* **344**, 391–402.

Karlin S. and S.F. Altschul (1990). Methods for assessing the statistical significance of molecular sequence features by using general scoring schemes. *Proc. Nat. Acad. Sci.* **87**, 3364–2268.

Karlin S. and S.F. Altschul (1993). Applications and statistics for multiple high-scoring segments in molecular sequences. *Proc. Nat. Acad. Sci.* **90**, 5873–5877.

Karlin, S. and V. Brendel (1992). Chance and statistical significance in protein and DNA sequence analysis. *Science* **257**, 39–49.

Karlin, S. and V. Brendel (1996). New directions in education: computational tools for the molecular biologist and biological sources for the mathematician. pp. 408–427 in *Education in a Research University*, K.J. Arrow et al. eds., Stanford University Press, Stanford, California.

Karlin, S., C. Burge, and A.M. Campbell (1992). Statistical analyses of counts and distributions of restriction sites in DNA sequences. *Nucl. Acids Res.* **20**, 1363–1370.

Karlin S. and A. Dembo (1992). Limit distributions of maximal segmental score among Markov-dependent partial sums. *Adv. Appl. Prob.* **24**, 113–140.

Karlin, S., F. Ghandour, F. Ost, S. Tavaré, and L.J. Korn (1983). New approaches for computer analysis of nucleic acid sequences. *Proc. Nat. Acad. Sci.* **80**, 5660–5664.

Karlin, S. and C. Macken (1991a). Assessment of inhomogeneities in an *E. coli* physical map. *Nucleic Acids Research* **19**, 4241–4246.

Karlin, S. and C. Macken (1991b). Some statistical problems in the assessment of inhomogeneities of DNA sequence data. *J. Amer. Stat. Assoc.* **86**, 27–35.

Karlin, S. and H. Taylor (1975). *A First Course in Stochastic Processes*. Academic Press, New York.

Karlin, S. and H. Taylor (1981). *A Second Course in Stochastic Processes*. Academic Press, New York.

Kelly, C. and G. Churchill (1996). Biases in amino acid replacement matrices and alignment scores due to rate heterogeneity. *J. Comp. Biol.* **3**, 307–318.

Kerr, M.K., M. Martin, and G.A. Churchill (2000). Analysis of variance for gene expression microarray data. *J. Comp. Biol.* **7**, 819–837.

Kerr, M.K. and G.A. Churchill (2001). Statistical design and analysis of gene expression microarray data. *Genet. Research* **77**, 123–128.

Kimura, M. (1980). A simple method for estimating evolutionary rate in a finite population due to mutational production of neutral and nearly neutral base substitution through comparative studies of nucleotide sequences. *J. Molec. Biol.* **16**, 111–120.

Kishino, H. and M. Hasegawa (1989). Evaluation of the maximum likelihood estimate of the evolutionary tree topologies from DNA sequence data, and the branching order in *Hominoidea*. *J. Molec. Evol.* **29**, 170–179.

Knudsen, S. (2002). *A Biologist's guide to Analysis of DNA Microarray Data.* John Wiley & Sons, New York.

Kohane, I.S., A.T. Kho, and A.J. Butte (2003). *Microarrays for an integrative genomics.* MIT Press, Boston, MA.

Krogh, A., M. Brown, I.S. Mian, K. Sölander, and D. Haussler (1994). Hidden Markov Models in Computational Biology: applications to protein modeling. *J. Molec. Biol.* **235**, 1501–1531.

Kuhner, M. and J. Felsenstein (1996). Simulation comparison of phylogeny algorithms under equal and unequal evolutionary rates. *J. Biol. Evol.* **11**, 459–468.

Kullback, S. (1978). *Information Theory and Statistics.* Dover, New York.

Lake, J.A. (1987). Reconstructing evolutionary trees from DNA and protein sequences: evolutionary parsimony. *Mol. Biol. Evol.* **4**, 167–191.

Lander, E.S. and M.S. Waterman (1988). Genomic mapping by fingerprinting random clones: a mathematical analysis. *Genomics* **2**, 231–239.

Lawrence, C.E, S.F. Altschul, M.S. Boguski, J.S. Liu, A.F. Neuwald, and J.C. Wootton (1993). Detecting subtle sequence signals: a Gibbs sampling strategy for multiple alignment. *Science* **262**, 208–214.

Lehmann, E.L. (1986). *Testing Statistical Hypotheses.* Wiley, New York.

Lehmann, E.L. (1991). *Theory of Point Estimation.* Wadsworth, Pacific Grove, CA.

Leung, M.Y., G.M. Marsh, and T.P. Speed (1996). Over and underrepresentation of short DNA words in Herpes virus genomes. *J. Comp. Biol.* **3**, 345–360.

Lewis, P.O. (1998). Maximum likelihood as an alternative to parsimony for inferring phylogeny using nucleotide sequence data. pp. 132–163 in *Molecular Systematics of Plants II*, P. Soltis, D. Soltis, J. Doyle eds., Kluwer,

Dordrecht.

Li, C. and W.H. Wong (2001). Model-based analysis of oligonucleotide arrays: expression index computation and outlier detection. *Proc. Nat. Acad. Sci.* **98**, 31–36.

Liò, P. and N. Goldman (1998). Models of molecular evolution and phylogeny. *Genome Research* **8**, 1233–1244.

Littell, R.C. and J.L. Folks (1973). Asymptotic optimality of Fisher's method of combining independent tests. II. *J. Amer. Stat. Assoc.* **68**, 193–194.

Liu, L, D.M. Hawkins, S. Ghosh, and S.S. Young (2003). Robust singular value decomposition analysis of microarray data. *Proc. Nat. Acad. Sci.* **100**, 13167–13172.

Liu, X.S., D.L. Brutlag, and J.S. Liu (2001). BioProspector: discovering conserved DNA motifs in upstream regulatory regions of co-expressed genes. *Pacific Symposium on Biocomputing* **6**, 127–138.

Liu, J.S, D.L. Brutlag, and X. Liu (2001). An algorithm for finding protein-DNA interaction sites with applications to chromatin immunoprecipitation microarray experiments. *Nature Biotechnology* **20**, 835–839.

Liu, J.S, A.N. Neuwald, and C.E. Lawrence (2001). Bayesian methods for multiple local sequence alignment and Gibbs sampling strategies. *J. Amer. Statist. Assoc.* **90**, 1156–1170.

Manduchi E., G.R. Grant, S.E. McKenzie, G.C. Overton, S. Surrey, and C.J. Stoeckert (2000). Generation of patterns from gene expression data by assigning confidence to differentially expressed genes. *Bioinformatics* **16**, 685–698. See also http://www.cbil.upenn.edu/PaGE

Manly, B.F.J. (1997). *Randomization, Bootstrap and Monte Carlo Methods in Biology.* Chapman and Hall, London.

Markowski C.A. and E.P. Markowski (1990). Conditions for the effectiveness of a preliminary test of variance. *Amer. Stat.* **44**, 322–326.

Mercier, S. and J.J. Daudin (2001). Exact distribution of the local score of one iid random sequence. *J. Comp. Biol* **8**, 373–380.

Miklos, G.L.G. and R. Maleszka (2004). Microarray reality checks in the context of complex disease. *Nature Biotechnology* **22**, 615–621.

Modarres, R., J.L. Gastwirth, and W.J. Ewens (2004). A cautionary note on the use of permutation and nonparametric tests in the analysis of environmental data. (To appear in *Environmetrics*).

Mott, R. and R. Tribe (1999). Approximate statistics of gapped alignments. *J. Comp. Biol.* **6**, 91–112.

Mudholker, G.S. and E.O. George (1979). The logit method for combining probabilities. pp. 345–366 in *Symposium on Optimizing Methods in Statistics*, (J. Rustagi, ed.) Academic Press, New York.

Müller, T. and M. Vingron (2000). Modeling amino acid replacement. *Journal of Computational Biology*. **7**, 761–776.

Muse, S.V. (1995). Evolutionary analysis when nucleotides do not evolve independently. pp. 115–124 in *Current Topics on Molecular Evolution*, M. Nei and N. Takahata eds., Pennsylvania State University Press, College Park, PA.

Muse, S.V. and B.S. Gaut (1994). A likelihood approach for comparing synonymous and nonsynonymous substitution rates, with application to the chloroplast genome. *Mol. Biol. Evol.* **11**, 715–724.

National Bureau of Standards (1953). *Probability Tables for the Analysis of Extreme-Value Data*. Applied Mathematics Series, Washington.

Nature Genetics (2002), Volume 32. *The chipping forecast*.

Neuwald, A.F. and P. Green (1994). Detecting patterns in protein sequences. *J. Mol. Biol.* **239**, 698–712.

Newton, M. (1996). Bootstrapping phylogenies: large deviations and dispersion effects. *Biometrika* **83**, 315–328.

Nicodème, P. (2001). Fast approximate motif statistics. *J. Comp. Biol.* **8**, 235–248.

Norris, J.R. (1997). *Markov Chains*. Cambridge University Press, Cambridge, UK.

Ota, R., P. Waddell, and H. Kishino (1999). Statistical distribution for testing the resolved tree against star tree. pp. 15–20 in *Proceedings of the*

574   References

Annual Joint Conference of the Japanese Biometrics and Applied Statistics Societies. Sinfonica, Minato-ku, Japan.

Ota, R., P. Waddell, M. Hasagawa, H. Shimodaira, and H. Kishino (2000). Appropriate likelihood ratio tests and marginal distributions for evolutionary tree models with constraints on parameters. *Mol. Biol. Evol.* **17**, 783–803.

Parmigiani, G. E.S Garrett, R.A. Irizzarry, and S.L. Zeger (eds.) (2003). *The Analysis of Gene Expression Array Data*. Springer Verlag, New York.

Pearson, W.R. (1998). Empirical statistical estimates for sequence similarity searches. *J. Mol. Biol.* **276**, 71–84.

Press, H.W., S.A. Teukolsky, W.T. Vetterling, and B.P. Flannery (1992). pp. 275–286 in *Numerical recipes in C* (2nd edition), Cambridge University Press, Cambridge, UK.

Press, H.W., B.P. Flannery, S.A. Teukolsky, and W.T. Vetterling (1992). *Numerical Recipes in FORTRAN: The Art of Scientific Computing, 2nd ed.* Cambridge University Press, Cambridge, UK.

Quackenbush, J. (2002). Microarray data normalization and trransformation. *Nature Genetics* **32**, 496–501.

Rabiner, L.R. (1989). A tutorial on hidden Markov models and selected applications in speech recognition. *Proc. IEEE.* **77**, 257–286.

Régnier. M. (2000). A unified approach to word occurrence properties. *Discrete Applied Mathematics* **104**, 259–280.

Reinert, G., S. Schbath, and M.S. Waterman (2000). Probabilistic and statistical properties of words: an overview. *J. Comp. Biol.* **7**, 1–46.

Rigoutsos, I. and A. Floratos (1998). Motif discovery without alignment or enumeration. In *Proc. RECOMB98*, S. Istrail, P. Pevzner, and M.S. Waterman eds., 211–227.

Robin, S. (2002). A compound Poisson model for word occurrences in DNA sequences. *J. Roy. Statist. Soc. C* **51**, 1–15.

Robin, S. and J.J. Daudin (1999). Exact distribution of word occurrences in a random sequence of letters. *J. Appl. Prob.* **36**, 179–193.

Robin, S. and S. Schbath (2001). Numerical comparison of several approximations of the word count distribution in random sequences. *J. Comp. Biol.* **8**, 349–359.

Robinson, A.B. and L.R. Robinson (1991). Distribution of glutamine and asparagine residues and their near neighbors in peptides and proteins. *Proc. Nat. Acad. Sci.* **88**, 8880–8884.

Rogers, J.S. and D.L. Swofford (1999). Multiple local maxima for likelihoods of phylogenetic trees: a simulation study. *Mol. Biol. Evol.* **16**, 1079–1085.

Rosenthal, R. (1978). Combining results of independent studies. *Psychol. Bull.* **85**, 185–193.

Saitou, N. and M. Nei (1987). The neighbor-joining method: a new method for reconstructing phylogenetic trees. *Mol. Biol. Evol.* **4**, 406–425.

SAS Institute SAS/STAT (1999). SAS/STAT User's Guide, Version 8, SAS Institute Inc., Cary, NC.

Schadt, E.E., J.S. Sinsheimer, and K. Lange (1998). Computational advances in maximum likelihood methods for molecular phylogeny. *Genome Research* **8**, 222–233.

Schäffer, A.A., L. Aravind, T.L. Madden, S. Shavirin, J.L. Spouge, Y.I. Wolf, E.V. Koonin, and S.F. Altschul (2001). Improving the accuracy of PSI-BLAST protein database searches with composition-based statistics and other refinements. *Nucleic Acids Research* **29**, 2994–3005.

Schbath, S. (1997). Coverage processes in physical mapping by anchoring random clones. *J. Comp. Biol.* **4**, 61–82.

Schbath, S., N. Bossard, and S. Tavaré (2000). The effect of nonhomogeneous clone length distribution on the progress of an STS mapping project. *J. Comp. Biol.* **7**, 47–57.

Schena, M. (2003). *Microarray Analysis*. John Wiley & Sons, New York.

Schenker, N. (1985). Qualms about bootstrap confidence intervals. *J. Amer. Stat. Assoc.* **80**, 360–361.

Scholz, F.W. (1982). Combining independent *P*-values. In *A Festschrift for Erich L. Lehmann*, P. Bickel, K. A. Doksum and J.L. Hodges, eds.)

Wadsworth, Belmont, CA.

Schöniger, M. and A. von Haeseler (1994). A stochastic model for the evolution of autocorrelated DNA sequences. *Mol. Phylogeny Evol.* **3**, 240–247.

Schöniger, M. and A. von Haeseler (1995). A stochastic model for the evolution of autocorrelated DNA sequences. *Syst. Biol.* **44**, 533–547.

Schott, J.R. (1997) *Matrix Analysis for Statistics.* Wiley, New York.

Self, S.G. and K.-L. Liang (1987). Asymptotic properties of maximum likelihood estimators and likelihood ratio tests under non-standard conditions. *J. Amer. Stat. Assoc.* **82**, 605–610.

Servadio, A., A. Poletti and F. Taroni, (eds.) (2001). *Brain Research Bulletin*, **56**, Issue 3–4.

Shaffer, J.S. (1995). Multiple hypothesis testing. *Annu. Rev. Psych.* **46**, 561–584.

Siegmund, D. and B. Yakir (2000). Approximate p-values for local sequence alignments. *Ann. Stat.* **28**, 657-680.

Smith, T.F. and M.S. Waterman (1981). The identification of common molecular sub-sequences. *J. Mol. Biol.* **147**, 195–197.

Smythe, R.T. (2004). On runs in independent sequences. To appear in *J. Iranian Stat. Soc.*

Sokal, R.R. and C.D. Michener (1958). A statistical method for evaluating systematic relationships. *University of Kansas Scientific Bulletin* **28**, 1409–1438.

Sokal, R.R. and F.J. Rohlf (1995). *Biometry.* Freeman, New York.

Speed, T.P. (ed.) (2003). *Statistical Analysis of Gene Expression Microarray Data.* Chapman and Hall/CRC Press, London.

Sprent, P. (1998). *Data Driven Statistical Methods.* Chapman and Hall, London.

Stanhope, M.J., M.R. Smith, V.G. Waddell, C.A. Porter, M.S. Shivji, and M. Goodman (1996). Mammalian Evolution and the Interphotoreceptor Retinoid Binding Protein (IRBP) Gene: Convincing Evidence for Several

Superordinal Clades. *J. Mol. Evol.* **43**, 83–92.

Steel, M.A. (1994). The maximum likelihood point for a phylogenetic tree is not unique. *Syst. Biol.* **43**, 560–564.

Steel, M.A., L.A. Székely, and M.D. Hendy (1994). Reconstructing trees when sequence sites evolve at variable rates. *J. Comp. Biol.* **1**, 153–163.

Stekel, D. (2003) *Microarray Bioinformatics.* Cambridge University Press, Cambridge, UK.

Storey, J.D. (2002). A direct approach to false discovery rates. *J. Roy. Statist. Soc.* **84**, 479–498.

Storey, J.D. and D. Siegmund (2001). Approximate *P*-values for local sequence alignments: numerical studies. *J. Comp. Biol.* **8**, 549–556.

Storey, J.D. and R. Tibshirani (2003). SAM thresholding and false discovery rates for discovering differential gene expression in DNA microarrays. pp. 272–290 in *The Analysis of Gene Expression Data*, ed. by G. Parmigiani, E.S. Garrett, R.A. Irizarry, and S. Zeger. Springer, New York.

Studier, J.A. and K. Keppler (1988). A note on the neighbor-joining algorithm of Saitou and Nei. *Mol. Biol. Evol.* **5**, 729–731.

Sullivan, J. and D.L. Swofford (2001). Should we use model-based methods for phylogenetic inference when we know that assumptions about among-site rate variation and nucleotide substitution pattern are violated? *Syst. Biol.* {bf 50, 723–729.

Swofford, D.L., G.J. Olsen, P.J. Waddell, and D.M. Hillis (1996). Phylogenetic inference. pp. 407–514 in *Molecular Systematics*, ed. by D.M. Hillis, C. Moritz, and B.K. Mable. Sinauer, Sunderland, MA.

Takahata, N. and M. Kimura (1981). A model of evolutionary base substitutions and its application with special reference to rapid change of pseudogenes. *Genetics* **98**, 641–657.

Tanushev, M.S. and R. Arratia (1997). Central limit theorem for renewal theory for several patterns. *J. Comp. Biol.* **4**, 35–44.

Tavaré, S. (1986). Some probabilistic and statistical problems in the analysis of DNA sequences. *Lectures on Mathematics in the Life Sciences* **17**, 57–86.

Tavaré, S. and B.W. Giddings (1989). Some statistical aspects of the primary structure of nucleotide sequences. pp. 117–132 in *Mathematical Methods for DNA Sequences*, ed. by M.S. Waterman. CRC Press, Boca Raton, Florida.

Thompson, J.D., D.G. Higgins, and T.J. Gibson (1994). CLUSTAL W: improving the sensitivity of progressive multiple sequence alignment through sequence weighting, position-specific gap penalties and weight matrix choice. *Nucleic Acids Res.* **22**, 4673–4680.

Tourasse, N.J. and W.-H. Li (1999). Performance of the relative-rate test under nonstationary models of nucleotide substitution. *Mol. Biol. Evol.* **16**, 1068–1078.

Trends Guide to Bioinformatics (1998). M. Patterson and M. Handel eds., Elsevier Science, New York.

Tuffley, C. and M. Steel (1997). Links between maximum likelihood and maximum parsimony under a simple model of site substitution. *Bull. Math. Biol.* **59**, 581–607.

Tusher, V.G., R. Tibshirani and G. Chu (2001). Significance analysis of microarrays applied to the ionizing radiation response. *Proc. Nat. Acad. Sci.* **98**, 5116–5121.

Velikanov, M.V. and R. Krapal (1999). Polymerase chain reaction: a Markov process approach. *J. Theoret. Biol.* **201**, 239–249.

von Ende, C.N. (2001). Repeated-measures analysis. pp. 134–157 in *Design and Analysis of Ecological Experiments*, S.M. Scheiner and J. Gurevitch, eds., Oxford University Press, Oxford, UK.

Venter, J.C. et al. (2001). The sequence of the human genome. *Science* **291**, 1304–1351.

Wasserman, L. (2004). *All of Statistics*. Springer, New York.

Waterman, M.S. (1995). *Introduction to Computational Biology*. Chapman and Hall, New York.

Westfall, P.H. and S.S. Young (1993). *Resampling-based Multiple Testing: Examples and Methods for P-value Adjustment*. Wiley, New York.

Whelan, S. and N. Goldman (1999). Distribution of statistics used for the comparison of models of sequence evolution in phylogenetics. *Mol. Biol.*

Evol. **16**, 1292–1299.

Wilf, H. (2003). Personal communication.

Wilks, S.S. (1962). *Mathematical Statistics*. Wiley, New York.

Wolfinger, R.D., G. Gibson, E.D. Wolfinger, L. Bennett, H. Hamedeh, P. Bushel, C. Afshari, and R.S. Paules (2001). Assessing gene significance from cDNA microarray expression data via mixed models. *J. Comp. Biol.* **8**, 625–637.

Yang, Y.H., S. Dudoit, P. Liu and T.P. Speed (2001). Normalization for cDNA Microarray Data. SPIE BiOS, San Jose.

Yang, Y.H. and N. Thorne (2003). Normalization for two-color cDNA microarray data. pp. 403–418 in *Science and Statistics. A Festschrift for Terry Speed*, D. Goldstein ed. IMS Lecture Notes Monograph Series, Vol. 40.

Yang, Z. (1993). Maximum-likelihood estimation of phylogeny from DNA sequences when substitution rates differ over sites. *Mol. Biol. Evol.* **10**, 1396–1401.

Yang, Z. (1994). Estimating the pattern of nucleotide substitution. *J. Mol. Evol.* **39**, 105–111.

Yang, Z. (1996a). Among-site rate variation and its impact on phylogenetic analysis. *Trends Ecol. Evol.* **11**, 367–372.

Yang, Z. (1996b). Maximum-likelihood models for combined analyses of multiple sequence data. *J. Mol. Evol.* **42**, 587–596.

Yang, Z. (1997a). *Phylogenetic analysis by maximum likelihood (PAML)*, *ver. 3*. Department of Biology, University College London.

Yang, Z. (1997b). PAML: A program package for phylogenetic analysis by maximum likelihood. *Comm. Appl. Biosci.* **13**, 555–556.

Zharkikh, A. and W.-H. Li (1992a). Statistical properties of bootstrap estimation of phylogenetic variability from nucleotide sequences. I. Four taxa with a molecular clock. *Mol. Biol. Evol.* **9**, 1119–1147.

Zharkikh, A. and W.-H. Li (1992b). Statistical properties of bootstrap estimation of phylogenetic variability from nucleotide sequences. II. Four taxa without a molecular clock. *J. Mol. Evol.* **35**, 356–365.

Zhu, J., J.S. Liu, and C.E. Lawrence (1998). Bayesian adaptive sequence alignment algorithms. *Bioinformatics* **14**, 25–39.

Author Index

Subject Index

594   Subject Index